T0191808

Environmental Contamination Remediation and Management

Series Editors

Erin R. Bennett, Great Lakes Institute for Environmental Research, University of Windsor, Salem, MA, USA

Iraklis Panagiotakis, Environmental Engineer & Scientist, ENYDRON – Environmental Protection Services, Athens, Greece

There are many global environmental issues that are directly related to varying levels of contamination from both inorganic and organic contaminants. These affect the quality of drinking water, food, soil, aquatic ecosystems, urban systems, agricultural systems and natural habitats. This has led to the development of assessment methods and remediation strategies to identify, reduce, remove or contain contaminant loadings from these systems using various natural or engineered technologies. In most cases, these strategies utilize interdisciplinary approaches that rely on chemistry, ecology, toxicology, hydrology, modeling and engineering.

This book series provides an outlet to summarize environmental contamination related topics that provide a path forward in understanding the current state and mitigation, both regionally and globally.

Topic areas may include, but are not limited to, Environmental Fate and Effects, Environmental Effects Monitoring, Water Re-use, Waste Management, Food Safety, Ecological Restoration, Remediation of Contaminated Sites, Analytical Methodology, and Climate Change.

More information about this series at http://www.springer.com/series/15836

Nurudeen A. Oladoja · Emmanuel I. Unuabonah
Editors

Progress and Prospects in the Management of Oxyanion Polluted Aqua Systems

 Springer

Editors
Nurudeen A. Oladoja
Department of Chemical Sciences
Adekunle Ajasin University
Akungba, Nigeria

Emmanuel I. Unuabonah
Chemical Sciences
Redeemer's University
Ede, Nigeria

ISSN 2522-5847　　　　　　　ISSN 2522-5855　(electronic)
Environmental Contamination Remediation and Management
ISBN 978-3-030-70759-0　　　　ISBN 978-3-030-70757-6　(eBook)
https://doi.org/10.1007/978-3-030-70757-6

This Springer imprint is published by the registered company Springer Nature Switzerland AG
The registered company address is: Gewerbestrasse 11, 6330 Cham, Switzerland

This book is dedicated to the memory of late Prof. Augustine E. Ofomaja, Department of Chemistry, Vaal University of Technology, South Africa, a studious and fastidious water research scientist

Preface

Polluted water or wastewater is usually a horde of chemical species that span all the genre of chemical strata. Among the possible assortment of pollutants found in polluted water is a class known as the oxyanions (or oxoanions). These are anions of oxidized chemical elements, which have natural, geogenic and anthropogenic origins. They are large class of chemical species with diverse, rich and interesting chemistry. Consequently, their fate, dynamics, interactions and impact in aqua systems are highly complicated, which necessitated an innovative approach to their management.

This book is a compendium of research efforts and findings on the source, occurrence, hydrochemistry and the impact of oxyanions in aqua systems. Several emergent innovative strategies proffered to ameliorate the effects of these hazardous chemical species in water, and those that have been deployed to transform their roles from being hazardous to being beneficial, were expounded in this book. The innovative management strategies for oxyanions in aqua systems were presented from the perspectives of laboratory and field experiences.

Since the book is aimed at providing insights into the spate of progress and pitfalls on the innovative management strategies of oxyanions, it opened with a chapter on the role of oxyanions, as a friend or foe, followed by another chapter that highlighted the ecological impacts of these oxyanions. In order to understand the chemical states of oxyanions in groundwater, a rich potable water resource for millions in several developing countries in the world, a chapter was dedicated to their occurrence, dynamics and fate in groundwater. The progress in the use of emerging management strategies (e.g. microbial fuel cells, photocatalysis, wetlands) was chronicled in other chapters. Some chapters were also dedicated to the occurrence and management of the often neglected and uncommon oxyanions (e.g. selenium, antimoniate, borates, carbonates, molybdate, plumbate, halogen derivatives). In order to present a broader perspective on the management of oxyanions, a chapter was dedicated to the challenges of the global laws and economic policies in the abatement of oxyanion pollution in aqua systems.

This book on the *Progress and Prospects in the Management of Oxyanion Polluted Aqua Systems* is presented from a broad and all-encompassing perspective, rather than from a streamlined overview, which makes this book relevant to professionals from varying backgrounds and interests.

Akungba, Nigeria   Nurudeen A. Oladoja
Ede, Nigeria   Emmanuel I. Unuabonah

Contents

Contributors

Bashir Adelodun Department of Agricultural and Biosystems Engineering, University of Ilorin, Ilorin, Nigeria;
Department of Agricultural Civil Engineering, Kyungpook National University, Daegu, South Korea

James R. Adewumi Department of Civil and Environmental Engineering, Federal University of Technology, Akure, Nigeria

Rana Ahmed Institute of Chemistry, University of Potsdam, Potsdam, Germany

Fidelis O. Ajibade Department of Civil and Environmental Engineering, Federal University of Technology, Akure, Nigeria;
Research Centre for Eco-Environmental Sciences, Chinese Academy of Sciences, Beijing, PR China;
University of Chinese Academy of Sciences, Beijing, PR China

Temitope F. Ajibade Department of Civil and Environmental Engineering, Federal University of Technology, Akure, Nigeria;
University of Chinese Academy of Sciences, Beijing, PR China;
Institute of Urban Environment, Chinese Academy of Sciences, Xiamen, PR China

Zeeshan Ajmal MoA Key Laboratory for Clean Production and Utilization of Renewable Energy, MoST National Center for International Research of BioEnergy Science and Technology, College of Engineering, China Agricultural University, Beijing, People's Republic of China

Abdallah Albourine Laboratory of Materials and Environment, Faculty of Sciences, Ibn Zohr University, City Dakhla, Agadir, Morocco

Moses O. Alfred African Centre of Excellence for Water and Environmental Research (ACEWATER), Redeemer's University, PMB 230, Ede, Osun State, Nigeria;
Department of Chemical Sciences, Redeemer's University, PMB 230, Ede, Osun State, Nigeria

Jafar Ali Department of Biochemistry and Molecular Biology, University of Sialkot, Punjab, Pakistan;
Key Laboratory of Environmental Nanotechnology and Health Effects, Research Center for Eco-Environmental Sciences, Chinese Academy of Sciences, Beijing, China;
University of Chinese Academy of Sciences, Beijing, People's Republic of China

Eric T. Anthony Department of Chemistry, University of Fort Hare, Alice, South Africa

Pervez Anwar Department of Biochemistry and Molecular Biology, University of Sialkot, Punjab, Pakistan

Muhammad Arif MoA Key Laboratory for Clean Production and Utilization of Renewable Energy, MoST National Center for International Research of BioEnergy Science and Technology, College of Engineering, China Agricultural University, Beijing, People's Republic of China

Amitabh Banerji Institute of Chemistry, University of Potsdam, Potsdam, Germany

Ajibola A. Bayode Department of Chemical Sciences, Faculty of Natural Sciences, Redeemer's University, PMB 230, Ede, Osun State, Nigeria;
African Center of Excellence for Water and Environmental Research (ACEWATER), Redeemer's University, PMB 230, Ede, Osun State, Nigeria;
Laboratório de Química Analítica Ambiental e Ecotoxicologia (LaQuAAE), Departamento de Química e Física Molecular, Instituto de Química de Sao Carlos, Universidade de Sao Paulo, Sao Carlos, Brazil

Claudia L. Bianchi Department of Chemistry, University of Milan, Milan, Italy

Bianca M. Bresolin Department of Separation Science, Lappeenranta University of Technology, Mikkeli, Finland

Y. I. Bulu Department of Plant Science and Biotechnology, Adekunle Ajasin University, Akungba-Akoko, Nigeria

Rasaki S. Dauda African Centre of Excellence for Water and Environmental Research (ACEWATER), PMB 230, Ede, Nigeria;
Department of Economics, Redeemer's University, PMB 230, Ede, Nigeria

Ridha Djellabi Department of Chemistry, University of Milan, Milan, Italy;
Department of Chemistry, Università Degli Studi Di Milano, Milan, Italy

Renjie Dong MoA Key Laboratory for Clean Production and Utilization of Renewable Energy, MoST National Center for International Research of BioEnergy Science and Technology, College of Engineering, China Agricultural University, Beijing, People's Republic of China

Awoke Guadie Research Centre for Eco-Environmental Sciences, Chinese Academy of Sciences, Beijing, PR China

Brigitte Helmreich Technical University of Munich, Chair of Urban Water Systems Engineering, Garching, Germany

Peter Hesemann ICGM, Univ Montpellier - CNRS, Montpellier, France

Ahmad Hosseini-Bandegharaei Faculty of Health, Sabzevar University of Medical Sciences, Sabzevar, Iran

Abdelghani Hsini Laboratory of Materials and Environment, Faculty of Sciences, Ibn Zohr University, City Dakhla, Agadir, Morocco

Muhammad Kashif Irshad Department of Environmental Science and Engineering, Government College University, Faisalabad, Pakistan

Aroosa Khan Department of Microbiology, Quaid-I-Azam University, Islamabad, Pakistan

Khursheid Ahmed Khan Ghazi University City Campus, Dera Ghazi Khan, Pakistan

Michael Klink Faculty of Applied and Computer Science, Vaal University of Technology, Vanderbijlpark, South Africa

Daniel T. Koko African Centre of Excellence for Water and Environmental Research (ACEWATER), Redeemer's University, PMB 230, Ede, Osun State, Nigeria;
Department of Chemical Sciences, Redeemer's University, PMB 230, Ede, Osun State, Nigeria

Mohamed Laabd Laboratory of Materials and Environment, Faculty of Sciences, Ibn Zohr University, City Dakhla, Agadir, Morocco

Kayode H. Lasisi Department of Civil and Environmental Engineering, Federal University of Technology, Akure, Nigeria;
University of Chinese Academy of Sciences, Beijing, PR China;
Institute of Urban Environment, Chinese Academy of Sciences, Xiamen, PR China

Isiaka A. Lawal Faculty of Applied and Computer Science, Vaal University of Technology, Vanderbijlpark, South Africa

Philippe Moisy CEA, DES/ISEC/DMRC, Univ Montpellier, Marcoule, Bagnols sur Cèze, France

Artur J. Motheo São Carlos Institute of Chemistry, University of São Paulo, São Carlos, Brazil

Yassine Naciri Laboratory of Materials and Environment, Faculty of Sciences, Ibn Zohr University, City Dakhla, Agadir, Morocco

Muhammad Nauman State Key Laboratory of Chemical Resource Engineering, Beijing University of Chemical Technology, Beijing, People's Republic of China

Nathaniel A. Nwogwu Department of Agricultural and Bioresources Engineering, Federal University of Technology, Owerri, Nigeria

Aemere Ogunlaja Department of Biological Sciences, Redeemer's University, PMB 230, Ede, Osun State, Nigeria;
African Centre of Excellence for Water and Environmental Research, Redeemer's University, PMB 230, Ede, Osun State, Nigeria

Nurudeen A. Oladoja Hydrochemistry Research Laboratory, Department of Chemical Sciences, Adekunle Ajasin University, Akungba Akoko, Nigeria

Ngozi C. Ole African Centre of Excellence for Water and Environmental Research (ACEWATER), PMB 230, Ede, Nigeria;
Faculty of Law, Redeemer's University, PMB 230, Ede, Nigeria

Damilare Olorunnisola Department of Chemical Sciences, Faculty of Natural Sciences, Redeemer's University, PMB 230, Ede, Osun State, Nigeria;
African Center of Excellence for Water and Environmental Research (ACEWATER), Redeemer's University, PMB 230, Ede, Osun State, Nigeria

Gang Pan Key Laboratory of Environmental Nanotechnology and Health Effects, Research Center for Eco-Environmental Sciences, Chinese Academy of Sciences, Beijing, China;
Centre of Integrated Water-Energy-Food Studies, School of Animal, Rural and Environmental Sciences, Nottingham Trent University, Brackenhurst Campus, Southwell, UK

Moses Gbenga Peleyeju Faculty of Applied and Computer Science, Vaal University of Technology, Vanderbijlpark, South Africa

Grace S. Peter Department of Biological Sciences, Redeemer's University, PMB 230, Ede, Osun State, Nigeria;
African Centre of Excellence for Water and Environmental Research, Redeemer's University, PMB 230, Ede, Osun State, Nigeria

Abdul Qadeer National Engineering Laboratory for Lake Pollution Control and Ecological Restoration, Chinese Research Academy of Environmental Science, Beijing, China

T. D. Saliu Hydrochemistry Research Laboratory, Department of Chemical Sciences, Adekunle Ajasin University, Akungba-Akoko, Nigeria

Florence A. Sowo Department of Microbiology, University of Ibadan, Ibadan, Oyo State, Nigeria

Andreas Taubert Institute of Chemistry, University of Potsdam, Potsdam, Germany

Chidinma G. Ugwuja Department of Chemical Sciences, Faculty of Natural Sciences, Redeemer's University, PMB 230, Ede, Osun State, Nigeria;

African Center of Excellence for Water and Environmental Research (ACEWATER), Redeemer's University, PMB 230, Ede, Osun State, Nigeria

Adamu Y. Ugya Department of Environmental Management, Kaduna State University, Kaduna State, Nigeria;
College of New Energy and Environment, Jilin University, Changchun, PR China

Emmanuel I. Unuabonah African Center of Excellence for Water and Environmental Research (ACEWATER), Redeemer's University, PMB 230, Ede, Osun State, Nigeria;
Department of Chemical Sciences, Faculty of Natural Sciences, Redeemer's University, PMB 230, Ede, Osun State, Nigeria

Aijie Wang Research Centre for Eco-Environmental Sciences, Chinese Academy of Sciences, Beijing, PR China

Hong C. Wang Research Centre for Eco-Environmental Sciences, Chinese Academy of Sciences, Beijing, PR China

Lei Wang Key Laboratory of Environmental Nanotechnology and Health Effects, Research Center for Eco-Environmental Sciences, Chinese Academy of Sciences, Beijing, China

Hassan Waseem Department of Biochemistry and Molecular Biology, University of Sialkot, Punjab, Pakistan

Chapter 1
Oxyanions in Aqua Systems—Friends or Foes?

Nurudeen A. Oladoja and Brigitte Helmreich

Abstract Oxyanions or oxoanions are pervasive and important constituents of all aqua systems. In the present discourse, a systematic effort is made to present the multifunctional roles (either as beneficial chemical species (i.e., as friends) or deleterious chemical species (i.e., as foes)) of common oxyanions in aqua systems and to decipher the contributory factor(s) that dictate the role assumed in each case. In order to achieve the aim of the discourse, the hydrochemistry of oxyanions is discussed from a general viewpoint before the features, mode of occurrence and the health implications of the oxyanions are specifically discussed. A critical review of reported literatures, that were compiled from laboratory and clinical trials, on the impacts of common aqua phase oxyanions and the variables that defined these impacts, show that three variables are the determinants of the choice(s) adopted. The variables identified include aqua phase concentration of the oxyanion, aqua phase pH value, which determines the available oxyanion specie in the aqua system, and the valency or oxidation state of the oxyanion that determines the general chemistry in the aqua matrix.

1.1 Introduction

The aqueous phase inorganic species are diverse, and the group denoted as oxyanions (also known as oxoanion) is a prominent and an important class. Oxyanions are anions of oxidized elements in aqua systems (e.g., in surface waters, groundwater, and drinking water) which have natural, geogenic (e.g., bicarbonate $[HCO_3]^-$, orthosilicate $[SiO_4]^{4-}$ and arsenate $[AsO_4]^{3-}$), and anthropogenic origins

N. A. Oladoja (✉)
Hydrochemistry Research Laboratory, Department of Chemical Sciences, Adekunle Ajasin University, Akungba Akoko, Nigeria
e-mail: nurudeen.oladoja@aaua.edu.ng

B. Helmreich
Technical University of Munich, Chair of Urban Water Systems Engineering, Am Coulombwall 3, 85748 Garching, Germany

© Springer Nature Switzerland AG 2021
N. A. Oladoja and E. I. Unuabonah (eds.), *Progress and Prospects in the Management of Oxyanion Polluted Aqua Systems*, Environmental Contamination Remediation and Management, https://doi.org/10.1007/978-3-030-70757-6_1

(e.g., bromate $[BrO_3]^-$, chlorate $[ClO_3]^-$, perchlorate $[ClO_4]^-$, nitrate $[NO_3]^-$). One cannot suggests that an oxyanion, that has a natural source is not harmful, just as one cannot say that an anthropogenic oxyanion must automatically be harmful. Examples of the anthropogenic formation of oxyanions are the process of disinfection of drinking water and the elimination of organic trace substances in wastewater by oxidation processes. In order to ensure safe wastewater discharge, well-assessed treatment systems currently exist for the classical issues in wastewater treatment plants (WWTPs), such as removal of biodegradable organic substances and nutrients (phosphorus and nitrogen) or transformation to innocuous forms. However, polar and semi-polar micropollutants are not, or only incompletely removed by these classical technologies. Although, oxidation processes (e.g., ozonation, advanced oxidation processes) are considered as promising technologies to remove such organic micropollutants [1], but they could also result in the oxidation of inorganic ions like chlorine and bromine, which form undesirable oxyanions (i.e., chlorate and bromate).

Irrespective of the source and history, aqueous systems are replete with a plethora of chemical species (i.e., both organic and inorganic species) that defines its characteristics and ultimately its value. Technically, it has been proven that the presence of chemical species, oxyanions inclusive, in aqua matrix could be likened to a coin with two different sides. It has both the good side (i.e. as friend) or the bad side (i.e. as foe). The determinants of the side of the coin that is displayed at a particular time need to be properly understood and critically appraised to attain the desired safe and eco-friendly aqua system status. In this discourse, a critical appraisal of the effects of the presence and absence of oxyanions in aqua systems is performed to provide water professionals with an in-depth knowledge on the influence of these important chemical constituents in aqua matrices. The insight gained from the information provided herein shall serve as a premise for developing strategies for the effective management of the derivable benefits and shortcomings from the presence and absence of these chemical species in aqueous system. In order to achieve the aim of this discourse, the hydrochemistry of oxyanions shall be discussed from a general perspective and the features, mode of occurrence, economic and health implications of different oxyanions that are commonly found in aqueous system shall be elucidated. The factors that influence their role, either as friend or as foe in aqua systems shall be systematically explicated.

1.2   The Hydrochemistry of Oxyanions

In chemical terminologies, the ionic entity with the general formula $M_xO_y^{z-}$ (M is the central chemical element, O is the oxygen atom, $y = 1, 2, 3, 4$ or 6, and z the valence of the ions, respectively) is known as oxyanion. Oxyanions correspond to their oxyacid with the formula $H_zM_xO_y$. A plethora of oxyanions abound in aqua matrices and the amount of oxygen, y, (i.e., the oxo group) bonded to the central element depends on the two properties of the central elements, viz. the size (i.e., the coordination number); and the oxidation state (i.e., the valency).

In an aqueous system, some positively charged elements exist as hydroxides (e.g., boric acid ($B(OH)_3$) or oxides (e.g., nitrogen oxide (NO), boron trioxide (B_2O_3)) and not as the expected cationic species. Some other elements act as oxoacids (e.g., carbonic acid (H_2CO_3), nitric acid (HNO_3), phosphoric acid (H_3PO_4), sulfuric acid (H_2SO_4), chloric acid ($HClO_3$), perchloric acid ($HClO_4$) etc.), which ionize to form oxyanions. Oxyanions are found either in the monomeric or the polymeric form. The two foremost factors (i.e., the oxidation state and the position of the central element) determine the formation of either forms of the oxyanions. The oxidation state of the central element and its position in the periodic table dictate the monomeric ($x = 1$) oxyanions $[MO_y]^{z-}$. Oxyanions in the 1st row of the periodic table have a maximum coordination number of four (although no oxyanion exists in this coordination number in the 1st row). Examples are carbonate $[CO_3]^{2-}$, nitrate $[NO_3]^-$ and nitrite $[NO_2]^-$. Oxyanions in the 2nd row are all linked with four oxygen atoms, e.g., sulfate $[SO_4]^{2-}$, phosphate $[PO_4]^{3-}$ and perchlorate $[ClO_4]^-$. In the 3rd row, a coordination number of six is possible but because they would carry a too high electrical charge, these elements form a tetrahedral oxyanion (e.g., arsenate $[AsO_4]^{3-}$, molybdate $[MoO_4]^{2-}$) [2]. In an aqueous system, highly charged oxyanions can undergo condensation reactions to produce the polymeric oxyanions (e.g., the formation of triphosphate $[P_3O_{10}]^{5-}$ or dichromate $[Cr_2O_7]^{2-}$). The polyoxyanions contain multiple oxyanion monomers that are joined by sharing corners or edges [3].

The hydration or the hydrolysis reaction of oxyanions occurs via electrostatic attraction between the lone pair of electrons on the oxygen atoms of the oxyanion and the partially positive charge that resides on the hydrogen atoms of the water molecule. The process of oxyanion hydrolysis is exothermic, and the magnitude of the energy generated depend on the charge density and mass of the oxyanion. The magnitude of the energy of hydration is progressively enhanced as the value of the ionic charge is increased but this is conversely reduced with increasing size of the anion. On the strength of a robust interaction between the anionic specie and the hydrogen of the water molecule, the water molecule dissociates, the hydrogen ion produced reacts with the anionic specie and the hydroxide ion produced also turns the solution basic thus [3]:

$$[MO_y]^{z-} + H_2O \rightleftharpoons [MO_{(y-1)}OH]^{(z-1)-} + OH^-$$

The three variables that affect the basicity of oxyanions include ionic charge, number of the bonded oxo group and value of the electronegativity of the central element. Increasing the charge on anionic specie enhances the hydrolysis propensity and the formation of alkaline solution. Some elements display variable oxidation state; thus, the tendency to form oxyanions with different number of oxygen atoms that are bonded to the central element is very high in such elements. Consequently, different oxyanions of an element with different number of oxygen atom differ substantially in their basicity. The basicity reduces with increasing number of oxygen atom bonded to the central element. The basicity of an oxyanion increases with the reduction in the value of the electronegativity of the atom.

1.3 Features, Mode of Occurrence, and Health Implications

1.3.1 Antimony Oxyanions

Paucity of information exist on the speciation of antimony (Sb) in aqua system, but the available information cum the thermodynamic predictions indicate that the pentavalent oxyanion antimonite $[Sb(OH)_6]^-$ is favored as the predominant specie [4]. Antimonates can be considered to be derivatives of the antimonic acid (H_3SbO_4), or combinations of metal oxides and antimony pentoxide (Sb_2O_5). Antimony itself is used in the production of semiconductor devices in the electronic industry and production of antimony alloy. Antimony alloys have found applications in batteries, as type metal (in printing presses), bullets and cable sheathing. Different antimony compounds serve as raw materials in the production of flame-retardants, paints, enamels, glass, and ceramics [5].

Antimony (Sb) enter the aquatic environment mainly through anthropogenic route and the majority are in the form of antimony trioxide (Sb_2O_3), which has an application mainly as flame retardant but also in the glass industry. Annually, it has been estimated that about 6400 tonnes of antimony are transported into the oceans [2]. The entrants of Sb_2O_3 into the environmental matrix occurs through the burning of coal or one of the components of fly ash during the smelting of antimony-containing ores [6]. Chemically, the behavior of antimony is very similar to its neighbour (arsenic) in the periodic table [7]. It has been posited that antimony could be a natural co-contaminant with arsenic in some drinking water [8]. Soluble forms of antimony are highly mobile in aqueous system, but the insoluble forms get attached to extractable iron and aluminium that is found in suspended and sedimented particulates [9]. There is no limit value for antimonite, only for antimony for drinking water. The US EPA recommended a maximum contaminant level of 6 $\mu g/L$ for Sb in drinking water [10]. The current recommended WHO guideline for Sb in drinking water is 20 $\mu g/L$ [11]. The European Union also set maximum admissible concentrations of 5 $\mu g/L$ for Sb in drinking water [12].

An overview of the concentrations of antimony in aqua system shows that the possibility of its occurrence in significantly higher concentrations in natural waters is very low, except in regions where there is contamination of acid mine drainage [13]. Concentrations between 0.1 and 0.2 $\mu g/L$ has been reported in groundwater and surface water [5], while the concentration in the marine system is approximately 0.15 $\mu g/L$ [14]. For groundwater, it is reported [15] that in oxic groundwaters (at recharge zones of the aquifers), antimony(V) occurs as the oxyanion $[Sb(OH)_6]^-$, whereas in sulfidic groundwaters, the thioantimonite species $[HSb_2S_4]^-$, and to a lesser extent, $[Sb_2S_4]^{2-}$, are predicted to be important forms of dissolved Sb. In contrast to the effluent from glass or metal processing industries, domestic wastewater is practically devoid of antimony [16]. Despite the fact that antimony is considered a possible replacement for lead in solders, no evidence of any significant contribution to drinking water concentrations from this source has been reported [13]. The concentrations in drinking water are less than 5 µg/L [17, 18], and the daily oral ingestion of antimony ranges between 10 and 70 µg [13].

Thus far, the toxicity of antimony has been ascribed to two salient features that include the solubility in water and the oxidation state of the antimony species [19]. In general, Sb(III) is more toxic than Sb(V) and the inorganic forms are more toxic than the organic forms [20]. An irritation on the gastrointestinal mucosa that triggered nausea has been linked with the oral ingestion of soluble antimony salts [9]. Other effects that have been documented include abdominal cramps, diarrhea, and cardiac toxicity [21]. Repeated oral exposure to therapeutic doses of Sb(III) has been linked with optic nerve destruction, uveitides, and retinal bleeding. Specific symptoms of antimony toxicity also include headache, coughing, anorexia, troubled sleep, and vertigo [20]. On the basis of findings from different experimental studies, some authors concluded that the ingestion of meglumine antimonite (Sb(V)) could not produce mutagenic or carcinogenic effects in human beings [15].

The most common route with which antimony gets into drinking water is through the leaching from metal plumbings and fittings, and the toxicity greatly depends on the form of antimony in drinking water [15].

1.3.2 Arsenic Oxyanions

Oxyanions of arsenic (As) are found in the environmental media in oxidation states, As(V) and As(III) as arsenate $[AsO_4]^{3-}$ and arsenite $[AsO_3]^{3-}$, respectively. Both the organic and the inorganic forms exist, but the foremost inorganic species include the arsenate $[AsO_4]^{3-}$ and arsenite $[AsO_3]^{3-}$ while the main organic species are monomethylarsenic acid and dimethylarsenic acid [22]. Albeit, the inorganic forms are more toxic while As(III) is more toxic and mobile than As(V). Under oxidic conditions, in an aqua system, inorganic arsenic acid (H_3AsO_4) prevails at extremely low pH (<2); and within a pH range of 2–11, it ionizes to $[H_2AsO_4]^-$ and to $[HAsO_4]^{2-}$. Under mildly reduced conditions and low pH, inorganic arsenious acid (H_3AsO_3) is formed, but at pH values greater than 12, it becomes ionized to $[H_2AsO_3]^-$ and $[HAsO_3]^{2-}$ [23].

In many toxicity test reports, As(III) is significantly more toxic than As(V) but some other reports have discountenanced this fact [24] and posited that both inorganic forms are equally very toxic. In addition, the common assumption that the inorganic species are more toxic than the organic form should be considered on the premise that environmental conditions can promote the degradation of the organic species to the more toxic inorganic forms.

The ubiquity and toxicity of arsenic and its compounds have made it a pollutant of note among the priority pollutants. In more than 70 countries, over 150 million people are estimated to be at the risk of consuming arsenic contaminated water [25]. The origin of arsenic in aqua system has been traced to the products of geological processes, biological activity, volcanic emissions, and anthropogenic activities (e.g., arsenic compound as ingredients in the formulations of pesticides, herbicides, and crop desiccants) [26]. Long-term exposure to arsenic in drinking water has been reported to cause dermal challenges, circulatory problems, degenerative diseases, and cancer [25]. Consequent upon the findings from the different studies, the acceptable value of arsenic in drinking water is, according to the US EPA, the WHO, and the European Union, 10 μg/L [10–12].

1.3.3 Boron Oxyanions

Boron (B)-containing oxyanions are referred to as borate, which stands for a large number of boron–oxygen compounds (i.e. as borates or boric acid), and it is commonly used in this form as chemical, for metal welding, in ceramics production, and in the pharmaceutical industry. Naturally, boron occurs as the borate minerals and borosilicates. Borates are used as compounding ingredient in ceramic technology to increase bulk, reduce energy consumption and to reduce cost. Among the different borate ions, the simplest is orthoborate $[BO_3]^{3-}$. Borates form salts with metallic elements, and many of the metallic borate salts readily get hydrated to include structural hydroxide groups. Thus, they are often considered as hydroxoborates. Structurally, they are made up of either trigonal planar BO_3 or tetrahedral BO_4 structural units, linked by shared oxygen atoms and may be cyclic or linear in structure [27].

The different forms of borates that are found in aqua system are pH dependent. At pH values below 7 (i.e., acidic and near neutral pH), it is found as boric acid (H_3BO_3 or $B(OH)_3$). Boric acid does not dissociate in aqueous solution, but act as a Lewis acid by accepting the electron pair of an hydroxyl ion produced by autoprotolysis of water. Thus, $B(OH)_3$ is acidic, forming the tetrahydroxyborate complex $[B(OH)_4]^-$ and releasing the corresponding proton left by water autoprotolysis, viz [28]:

$$B(OH)_3 + 2H_2O \rightleftharpoons [B(OH)_4]^- + [H_3O]^+$$

At neutral pH 7, boric acid undergoes condensation reactions to form polymeric oxyanions, e.g. the formation of tetraborate $[B_4O_5(OH)_4]^{2-}$

$$2\,B(OH)_3 + 2[B(OH)_4]^- \rightleftharpoons [B_4O_5(OH)_4]^{2-} + 5H_2O$$

The tetraborate anion that is found in the mineral borax is an octahydrate, $Na_2[B_4O_5(OH)_4]\cdot8H_2O$ (borax). It is made up of two tetrahedral and two trigonal boron atoms that are symmetrically assembled in cyclic structure. The two tetrahedral boron atoms are linked by a common oxygen atom, and each also bears a negative net charge brought by the hydroxyl groups laterally attached to them.

Other well-known borate oxyanions are diborate $[B_2O_5]^{4-}$ and triborate $[B_3O_7]^{5-}$.

The entrants of boron into environmental media have been traced mainly to industrial wastewater discharge [29, 30], especially from the ceramic industry. Due to the high solubility of boron in aqueous system and toxicity to plants and animals above certain concentration, it is regarded as highly polluting. Elevated concentration of boron (≥ 2 mg/L) may be toxic to plants, except where the tolerance of the plant in question to the toxicity of boron is very high [31]. Naturally, boron may be present in groundwater and the magnitude in the matrix is a function of the geological features of the system. The presence of boron in surface water has been reported. It is found generally in natural water as boric acid, $B(OH)_3$ and/or borate, $[B(OH)_4]^-$ [32]. Normally, boron exists in minimal concentration in soil and irrigation waters, but it accumulates in soils irrigated with boron containing wastewater. Boron concentration <1 mg/L in irrigation water is required for sensitive crops.

The complex of boron with heavy metals (e.g., Pb, Cd, Cu, Ni) is considered more toxic than the heavy metal precursors. Lower concentration of boron in soil serves as nutrient to some plants, but higher amounts adversely affect the growth of many agricultural products [33]. Owing to the mild bactericidal and fungicidal properties of boric acid, it is used as a disinfectant and food preservative. The herbicidal properties have also been exploited for weed control on railways, in timber yards and on other industrial sites. Borax is used widely in welding and brazing of metals, and at present, it has also found applications in hand-cleansing, high-energy fuels, cutting fluids, and catalysts [34].

Although there are limited toxicological data of boron to human health, recommended values for drinking water for boron are 1000 µg/L for the European Union [12] and 2400 µg/L for the WHO [11]. There is no limit value at US EPA.

1.3.4 Bromine Oxyanions

The oxyanions of bromine (Br) are hypobromite $[BrO]^-$, bromite $[BrO_2]^-$, bromate $[BrO_3]^-$ and perbromate $[BrO_4]^-$. Bromate $[BrO_3]^-$ is not a common constituent of natural waters, and its common salts are potassium $K[BrO_3]$ and sodium bromate $Na[BrO_3]$. These two bromate salts are strong oxidizing agents that are used mainly in permanent wave neutralizing solutions and in the dying of textiles using sulfur dyes [35]. Potassium bromate is a chemical reagent that is equally used as an oxidizer and dough conditioner during baking [36, 37]. It is applied in treating barley during the production of beer and improving the quality of fish paste products in Japan [38]. Bromine oxyanions are formed in the presence of bromide Br^- in natural waters as by-product during disinfection for drinking water or oxidation of wastewater treatment effluents [39, 40] using ozone. In many natural waters, low levels of bromide (<20 mg/L) are found. It is a precursor for bromate formation. For bromide levels in the range of 50–100 mg/L, excessive bromate formation may already become a problem [40]. Bromate formation also occurs under favorable conditions, when concentrated hypochlorite solutions are used to disinfect water containing high bromide concentrations [41]. Among other factors, the conversion of bromide to bromate during ozonation may be influenced by the presence of natural organic matter, pH value, and temperature. It has been reported that during this process (ozonation), the rate of formation of bromate ions may increase with temperature [42, 43] and increased alkalinity [44]. The presence of bromide in the precursor (i.e., chlorine and sodium hydroxide) to the synthesis of sodium hypochlorite and the high pH of the concentrated solution were also adduced to be the initiators of the bromate formation in this regard. The formation of bromate also occurs in electrolytically generated hypochlorous acid solutions, when bromide is present in the brine [45].

Bromate concentrations that ranged between 2 and 16 μg/L have been reported in the finished water from European water utilities [41]. Health Canada [46] reported, in different studies, an average level of 1.71 μg/L, with a range of 0.55–4.42 μg/L and an average concentration of 3.17 μg/L, with a range of 0.73–8.0 μg/L in drinking water from public utilities. The annual mean bromate concentration reported in the USA in finished surface water was 2.9 μg/L, with a range of 0.2–25 μg/L. An overview of the exposure of humans to bromate is unlikely to be significant for most people. In the case of ozonation, as the procedure for disinfection of drinking water, intake of bromate might range from 120 to 180 μg/day [47]. Owing to the strong oxidizing power of bromate, it reacts with organic matter in the system to form a bromide ion. It does not volatilize and its only slightly adsorbed onto soil or sediment in the aqua system [47].

A large number of human poisoning from bromate occur via the ingestion of home permanent wave solutions. When bromate is ingested, it is absorbed from the gastrointestinal tract [48]. The reversible toxic effects of bromate salts that have been documented include nausea, vomiting, abdominal pain, anuria and diarrhea, central nervous system depression, haemolytic anemia, and pulmonary edema. The irreversible effects that manifest in the ingestion of elevated concentration of bromide (185–385 mg of bromate per kg of body weight) include renal failure and deafness [49].

An acceptable level of bromate in drinking water is, according to the US EPA, the WHO and the European Union, 10 μg/L [10, 11].

1.3.5   Carbon Oxyanions

Carbonates and bicarbonates are carbon (C) oxyanions and the salts of carbonic acid (H_2CO_3) that contain the carbonate ion $[CO_3]^{2-}$ or bicarbonate ion $[HCO_3]^-$. These oxyanions made up of carbon (C) atom are surrounded by three oxygen atoms, in a trigonal planar arrangement. Carbonate is the conjugate base of the bicarbonate, which in turn is the conjugate base of carbonic acid. In aqueous system, carbonate, bicarbonate, carbon dioxide, and carbonic acid exist together in a dynamic equilibrium depending on the pH value [50]. In strong basic conditions, the carbonate ion predominates, while in weakly basic conditions, the bicarbonate ion is the dominant specie. In more acid conditions, aqueous carbon dioxide, $[CO_2]_{(aq)}$, is the main form, which is in water in equilibrium with carbonic acid, but the equilibrium lies strongly toward the predominance of carbon dioxide [50]. Carbonates are important constituents of natural minerals and rocks. The most common materials are calcite and limestone ($CaCO_3$), dolomite ($[CaMg(CO_3)]_2$), and siderite ($FeCO_3$) [51].

A very important issue in water chemistry has been attributed to the presence of carbonate and bicarbonate as components in water hardness. The type of water hardness caused by the presence of dissolved bicarbonate and carbonate is the so-called temporary hardness (carbonate hardness). Carbonate minerals also yield calcium and magnesium cations (Ca^{2+}, Mg^{2+}). The sum of calcium and magnesium ions is referred as total hardness [50]. Hardness reduces the effectiveness of soaps and detergents and contributes to scale formation in pipes and boilers, but low hardness contributes to the corrosiveness of water. Albeit, hardness is not considered a health hazard but hard water requires softening, by lime precipitation or ion exchange, to broaden the applications. According to the WHO "there does not appear to be any convincing evidence that water hardness causes adverse health effects in humans" [52]. Bicarbonate is a vital component of the pH buffering system of the human body [53]. The presence of hydroxyl and bicarbonate ions has been attributed to be the main cause of alkalinity in water. Alkalinity in water has a buffer effect, stabilizes, and prevents fluctuations in the water pH value. Inadequate alkalinity in water system (<80 mg/L) enhances the corrosive tendencies of the water system. High alkalinity is also required in wastewater treatment because anaerobic digestion requires sufficient alkalinity to ensure that the pH does not drop below 6.2, to ensure the optimal performance of the methane bacteria. For the digestion process to operate successfully, the alkalinity must range between 1000 and 5000 mg/L as calcium carbonate. Alkalinity in wastewater is also important when chemical treatment is used, in biological nutrient removal, and whenever ammonia is removed by air stripping.

In order to prevent the situation where water corrodes pipe or deposits calcium carbonate, which produces an encrusted film, treated water is stabilized through a process known as recarbonation (i.e., the adjustment of the ionic condition of a water) so that it will neither corrode pipes nor deposit calcium carbonate [54]. During or after the lime-soda ash softening process, carbon dioxide is introduced into the water to recarbonate it. Lime softening of hard water supersaturates it with calcium carbonate and the treated water may have a pH value that is greater than 10. In order to correct this, pressurized carbon dioxide is bubbled into the water to lower the pH and remove the dissolved calcium carbonate [54]. Water recarbonation also helps to remove the bitter taste that is synonymous with treated water with high pH.

1.3.6 Chlorine Oxyanions

The oxyanions of chlorine (Cl) are hypochlorite $[ClO]^-$, chlorite $[ClO_2]^-$, chlorate $[ClO_3]^-$ and perchlorate $[ClO_4]^-$. Perchlorates, as the oxyanions with the highest oxidation number (+7) of chlorine, are colourless, odourless, negatively charged molecular specie that are derived from a combination of four oxygen atom and central chlorine atom. Generally, these inorganic salts are bonded to a positively charged group such as ammonium or an alkali or alkaline earth metal. The perchlorate anion is a closed shell atom that is well protected by the four oxygen atoms surrounding it, which makes the reaction of perchlorate to be slow.

Lithium perchlorate, which decomposes exothermically to produce oxygen, is used in oxygen candles on spacecraft, submarines, and in other situations where a reliable backup oxygen supply is needed. The perchlorate salts that are largely produced for commercial purposes include magnesium perchlorate, potassium perchlorate, ammonium perchlorate, sodium perchlorate, and lithium perchlorate. The dominant use of perchlorate is as oxidizer in propellants for rockets, fireworks and highway flares [55]. It is used to control static electricity in food packaging, e.g., when sprayed onto containers it stops statically charged food from clinging to plastic surfaces [56]. In smaller applications, perchlorates are used extensively within the pyrotechnic industries, certain munitions and for the manufacture of matches.

Albeit, perchlorate is highly soluble in aqueous medium, but it also exists as a solid in the absence of water. With the exception of potassium perchlorate, which has low solubility in water (1.5 g in 100 ml of water at 25 °C), perchlorates dissolve and dissociate in water to give the perchlorate anion and the associated cation from the perchlorate salt. Owing to the fact that perchlorate salts are readily soluble in both aqueous and non-aqueous solutions, when these salts get solvated, especially ammonium perchlorate, they are capable of undergoing redox reactions and subsequent release of gaseous contaminants.

Perchlorates are relatively stable and mobile in the environment [57]. A large proportion of naturally occurring perchlorates have been reported in some locations in the continent of USA (e.g., west Texas and northern Chile). Results from surveys of groundwater, ice, and relatively unperturbed deserts have been used to estimate a 100,000–3,000,000 tonnes "global inventory" of natural perchlorate presently on Earth [58]. Consequent upon the ionic nature of perchlorates, they do not volatilize; thus, they are persistent in the environment. Microorganisms found in soils and water have been reported to reduce perchlorates to other substances but if perchlorates are air borne, they ultimately settle out of the air, especially in rainfall.

Perchlorates in drinking water is of concern because of uncertainties about toxicity and health effects at low levels [59]. It is highly soluble, exceedingly mobile and persists in the environment. It impacts on ecosystems, an indirect exposure pathway for humans due to accumulation in vegetables [59]. Detected perchlorates in environmental samples originate from disinfectants, bleaching agents, herbicides, and mostly from rocket propellants. Low levels of perchlorate have been detected in both drinking water and groundwater in 26 states in the USA in concentrations of 2–5 μg/L [57]. The direct ecological effect of perchlorate is not well known, and its impact can be influenced by several factors (e.g., rainfall and irrigation, dilution, natural attenuation, soil adsorption, and bioavailability) [60].

In humans, the main target organ for perchlorate toxicity is the thyroid gland [61]. Perchlorate has been shown to partially inhibit the thyroid's uptake of iodine, the building block for the synthesis of the hormone. It is anticipated (not yet demonstrated in humans) that persistent and high dosage exposure to perchlorate could lead to reduction in the production of thyroid hormones (i.e., hypothyroidism). Premised on the ability of perchlorates to lower thyroid hormone levels, it has been prescribed as a drug to manage overactive thyroid glands (i.e., hyperthyroidism). The observed effects in a some of the treated patients include skin rashes, nausea, and vomiting. The report of a study of adults in Nevada [61] found that the number of cases of thyroid disease in a group of people who drank water contaminated with perchlorates was no different from the number of cases found in a group of people who drank water without perchlorate. This is an indication that levels of perchlorate in the water were not the cause of the thyroid disease, and a search of the literature confirmed no evidence of perchlorate inducing thyroid disease. In another study, it has been shown that the effects of perchlorate in people depends on gender, the length of exposure, and the magnitude of the ingested perchlorates.

While perchlorates are used as a chemical substance itself and be found in waters because of its high-water solubility, the presence of chlorate $[ClO_3]^-$ and chlorite $[ClO_2]^-$, in water, where chlorine has an oxidation number of (+5) and (+3), respectively, can be traced to different routes. The common route is the disinfection by-products of chlorine dioxide (ClO_2). Chlorine dioxide is commonly used as disinfectant for drinking water [62] and for odour/taste control in water. In addition to the use as disinfectant in water industry, chlorine dioxide is a common industrial bleaching agent (e.g., cellulose, paper pulp, flour, oils and leather). The taste and odor threshold for chlorine dioxide in drinking water is 400 μg/L [63]. Under alkaline condition, the stability of chlorine dioxide is very low in aqueous system. Thus, it rapidly decomposes into chlorite, chlorate, and chloride ions in treated water, and chlorite is the predominant species.

To achieve low concentrations of chlorate and chlorite in drinking water, a maximum Contaminant Level Goal for 800 μg/L ClO_2 was prescribed [10]. The public health goal in the US EPA for chlorite is also 800 μg/L [10]. Other oxidants like ozone or OH-radicals can also result in the formation chlorate, e.g., during drinking water disinfection or treatment of trace organic chemicals [47]. In addition, chlorate and chlorite are also generated when the solution of sodium hypochlorite slowly decomposes in water. In recent studies, relatively high concentrations of natural chlorate deposits have been reported in arid and hyper-arid regions globally [63]. The evaluation of the chlorate level in rainfall samples showed that the amount detected was similar to that of the perchlorate level in the same sample. Thus, it was posited that the two oxyanions may share a common natural formation mechanism and could be a part of the chlorine biogeochemistry cycle [64]. Chemical compounds containing the two anions chlorite and chlorate are also used in an array of industrial operations. Sodium chlorite ($NaClO_2$) is used in on-site generation of chlorine dioxide, bleaching agent, and in the manufacture of waxes, shellacs, and varnishes. Sodium chlorate ($NaClO_3$) is used in the preparation of chlorine dioxide; raw materials in dyes, matches and explosives manufacture, leather production, and in herbicides and defoliants [65, 66].

It has been concluded that chlorite is not classifiable with respect to its carcinogenicity to humans [67]. The primary and most consistent finding that arises from exposure to chlorite and chlorate is oxidative stress that results in changes in the red blood cells [68, 69]. It has been reported that a long-term study is currently in progress that should provide more information on the effects of chronic exposure to these oxyanions of chorine [70].

1.3.7 Chromium Oxyanions

Chromium (Cr) is known to be a common groundwater contaminant at hazardous waste sites, where there is subsurface contamination from industrial operations (e.g., metal plating, leather tanning, and pigment production). Being a transition metal, it exists in varying oxidation state and the most common forms are Cr(0), Cr(III), and Cr(VI). The oxyanions of chromium in the oxidation state +6 are chromate $[CrO_4]^{2-}$, dichromate $[Cr_2O_7]^{2-}$, trichromate $[Cr_3O_{10}]^{2-}$ and tetrachromate $[Cr_4O_{13}]^{2-}$ which are all based on tetrahedral chromium. They are moderately strong oxidizing agents and can be interconvertible in an aqueous solution [7]. The chromate and dichromate ions are reduced during oxidation process it to oxidation state +3. In acid solution Cr^{3+} ion is produced. Relative to Cr(III), the mobility and toxicity of Cr(VI) is very high. It penetrates the cell wall easily and poison the cell itself, and it has been fingered as a culprit in various cancer diseases [71, 72]. The EU Council Directive and the WHO recommend a maximum concentration of 50 µg/L for total chromium in drinking waters [11, 12], while the EPA public health goal of 100 µg/L was recommended for total chromium [10].

The toxicity of Cr(VI) emanates from the oxidizing ability and the tendency to form free radicals during the process of reduction of Cr(VI) to Cr(III) that occurs inside the cell [73]. Above the permissible level, at short-term exposure it induces skin and stomach irritation and sometimes ulceration, while at long-term exposure it causes dermatitis, damage to liver, kidney circulation, nerve tissue damage, and death in large doses [74, 75]. In contrast, Cr(III) is less toxic and is nearly insoluble at neutral pH [76]. It has been listed as an essential element and micronutrient required in maintaining good health and the normal metabolism of glucose, cholesterol, and fat in human bodies [77]. Consequent upon the high toxicity of Cr(VI), the total removal from aqua stream or reduction to Cr(III) are regarded as the fundamental management strategies of Cr(VI) in aqua system.

1.3.8 Molybdenum Oxyanions

Molybdenum oxide MoO_3 in the oxidation number +6 is soluble in strong alkaline water, forming the oxyanion molybdate $[MoO_4]^{2-}$, which is the simplest of a series of molybdates. Molybdates are weaker oxidants than chromates [2]. A very large range of molybdate oxyanions exists, whose structural features could either be discrete or polymeric extended structures. These structurally complex oxyanions (Poly-molybdates) such as heptamolybdate $[Mo_7O_{24}]^{6-}$ and octamolybdate $[Mo_8O_{26}]^{4-}$

are formed by condensation of molybdate at lower pH values. Polymolybdates can incorporate other ions, forming polyoxometalates [78, 79].

The formation of the phosphorus-containing heteropolymolybdate $P[Mo_{12}O_{40}]^{3-}$ is used for the spectroscopic detection of phosphorus, as the complex is dark-blue [80]. The industrial applications of Mo ranges from the manufacture of special steels, electrical contacts, spark plugs, X-ray tubes, filaments, screens and grids for radio valves, and in the production of tungsten, glass-to-metal seals, non-ferrous alloys and pigments. Molybdenum disulfide MoS_2 (oxidation number +4), which is naturally found as molybdenite, has unique properties as a lubricant additive. MoS_2 is sparingly soluble in water and get easily oxidized to give more soluble molybdate oxyanions, which are stable in water in the absence of a reducing agent [81].

Molybdenum compounds have also found applications in agriculture either for the direct treatment of seeds or in fertilizer formulation to prevent molybdenum deficiency [82]. Consequent upon the fact that molybdenum generally occurs at very low concentrations in drinking water; it is therefore not considered necessary to set a formal guideline value [69]. At molybdenum concentration of ≥ 10 mg/L, ammonium molybdate imparts a slightly astringent taste to water [81].

The investigation of surface water samples from some major river basins in the USA revealed the presence of Mo in 32.7% of the surface water samples at concentrations that range between 2 and 1500 µg/l (mean value = 60 µg/l) [83]. A survey of the presence of Mo in groundwater samples in the USA showed concentration values that ranged between values that are undetectable and 270 µg/L [84]. When a survey of finished water supplies in the USA was conducted, the magnitude of Mo detected ranged from undetectable value and 68 µg/L (median value = 1.4 µg/L) [85]. It has been posited that the concentrations of Mo in drinking water are not usually >10 µg/L [86], except regions that are close to Mo mining sites where the Mo concentration in finished water was reported to be as high as 200 µg/L.

The rate of gastrointestinal absorption of Mo is influenced by its chemical form and the animal species [87]. It has been reported that Mo^{+6} is readily absorbed while Mo^{+4} is not readily absorbed into the physiological system [88]. In humans, 30–70% of dietary intake is absorbed from the gastrointestinal tract [89]. Following gastrointestinal absorption, Mo rapidly appears in the blood and most organs. Highest concentrations are found in the liver, kidneys and bones [88], and no apparent bioaccumulation in human tissues has been reported. Mo is an essential trace element in animals and humans, and safe and adequate intake dosage have been prescribed (e.g., 0.015–0.04 mg/day for infants, 0.025–0.15 mg/day for children aged 1–10 and 0.075–0.25 mg/day for all individuals above the age of 10) for the different strata of the populace [90]. Abnormal distribution of urinary metabolites, neurological disorders, dislocated ocular lenses and failure to thrive has been found in infant with inborn deficiency of the molybdoenzymes sulphite oxidase and xanthine dehydrogenase [91]. Dietary Mo deficiency, which results in impaired function of the two molybdoenzymes has been ascribed to the development of tachycardia, tachypnoea, severe headaches, night blindness, nausea, vomiting, central scotomas, generalized edema, lethargy, disorientation and coma in a Crohn disease patient receiving total parenteral nutrition [92].

1.3.9 Nitrogen Oxyanions

There are two naturally occurring oxyanions of nitrogen (N) that are part of the nitrogen cycle: Nitrite $[NO_2]^-$ with nitrogen in the oxidation state of $+3$ and nitrate $[NO_3]^-$ with nitrogen in the oxidation state of $+5$. Among the two species, the nitrate ion is the stable form especially in surface water while the nitrite ion is relatively unstable. Albeit, nitrate is inert but the valency/oxidation state can be reduced by microbial actions (denitrification) to nitrogen (N_2) [93]. In the case of nitrite, chemical and biological actions can either transform it to various compounds or oxidize it to nitrate [93, 94]. In domestic wastewater, nitrate and nitrite are generated when the constituent proteins and urea in the wastewater are decomposed to ammonia, which will be oxidized by microbial action (nitrification) to the oxyanions under aerobic conditions [93]. Nitrification is a two-step process: (1) the oxidation of $[NH_4]^+$ to $[NO_2]^-$ by *Nitrosomonas* bacteria, and (2) the oxidation of $[NO_2]^-$ to $[NO_3]^-$ by *Nitrobacter* bacteria [50, 93].

The entrants of nitrates and nitrites into the drinking water matrix (i.e., groundwater and surface water) occur via diverse routes that spanned from agricultural practices, industrial and municipal effluents [95]. Nitrate is one of the major components of inorganic fertilizers, and pure potassium nitrate is an important raw material in glass making, as an oxidizer and in the production of explosives. An additional source of nitrate from fertilizer is nitrification of ammonium containing fertilizer during percolation into groundwater in the soil layer. $[NO_3]^-$ is not retained in the soil and is readily infiltrated into groundwater [50].

Nitrate is secreted in saliva and then converted to nitrite by oral microflora. In the food industry, sodium nitrite is used as a food preservative and in some cases, nitrate is added to food as a reservoir for nitrite. In mammals, nitrate and nitrite are formed endogenously. Nitrite is formed in drinking water distribution pipes by *Nitrosomonas* bacteria during stagnation of nitrate-laden but oxygen depleted water in galvanized steel pipes or in an uncontrolled use of chloramine as a disinfectant [96]. The concentrations of nitrite in drinking water are usually less than 0.1 mg/L, but chloramination can enhance the concentration because of the possibility of formation of nitrite within the distribution system. The nitrification process, initiated by the chloramination in the distribution systems, can increase nitrite levels, usually by 0.2–1.5 mg/L of nitrite, but potentially by more than 3 mg/L of nitrite [97].

In industrial estates, the concentrations of nitrate in rainwater of up to 5 mg/L have been reported [98], but the concentrations were lower in the rural areas. In the pristine state, the nitrate concentration in surface water is usually low (0–18 mg/L) but as a result of pollution (e.g., agricultural runoff, refuse dump runoff or contamination with human or animal wastes) the concentration becomes higher. In natural groundwater system, under aerobic conditions, nitrate concentration is usually low. Concentration levels that range between 4 and 9 mg/L for nitrate and values that do not exceed 0.3 mg/L for nitrite has been reported in the USA [99]. In recent years, continual increase of nitrate levels in groundwater resource has been observed, and this has been ascribed to the increasing use of artificial fertilizers, the disposal of wastes (especially from animal husbandry), and changes in land use. At the global level, it has been reported that in agricultural areas, individual hand-dug wells are contributors to nitrate contamination problems and nitrate levels in water from such hand-dug wells are often greater than 50 mg/L [96].

Nitrate poisoning is the cause of the undesirable blue baby syndrome that sometimes lead to death in infants. The human toxicity of nitrate is attributed to its reduction to nitrite and the toxicity of nitrite stems from the oxidation of normal haemoglobin (Hb) to methaemoglobin (metHb) [96]. The oxidized product (i.e., metHb) is unable to transport oxygen to the tissues and clinically manifest when metHb concentration is at 10% of normal Hb concentrations or higher. This clinical state is referred to as methemoglobinemia, and it manifests as cyanosis and asphyxia at low and high concentrations, respectively. In adults, cases of methemoglobinemia have been reported when high doses of nitrate (4–50 g of nitrate, equivalent to 67–833 mg of nitrate per kilogram of body weight) were consumed [100]. Human toxicity have been reported when food containing nitrite was consumed, and the oral lethal dose for humans was estimated at values that ranged between 33 and 250 mg of nitrite per kilogram of body weight [96]. The toxic dosage that gives rise to methemoglobinemia ranges between 0.4 and 200 mg/kg of body weight [100].

In human stomach, the reaction of nitrite with nitrosatable compounds to form N-nitroso compounds (a chemical compound that has been tested to be carcinogenic in animals), has been confirmed. It is assumed that the N-nitroso compounds that has been found to be carcinogenic in some animals are also carcinogenic in humans but the data from epidemiological studies are at best, only suggestive [96]. In Australia, consumption of water with high nitrate concentrations has been attributed (claims not confirmed) to the cause of congenital malformations. However, no other study have been able to establish a link between innate abnormalities and nitrate ingestion [101]. Studies that related the occurrence of cardiovascular diseases to the consumption of nitrate-rich water gave inconsistent results [102]. Consequent upon the fact that nitrate inhibit iodine uptake, probable link between nitrate ingestion and the influence on thyroid was studied and both experimental and epidemiological studies indicated antithyroid effect of nitrate in humans [96]. Using humans as a case study, it has been reported that nitrite (0.5 mg of sodium nitrite per kilogram of body weight per day, for 9 days) reduced the production of adrenal steroids, as reflected by the decreased concentration of 17-hydroxysteroid and 17-ketosteroids in urine [103]. Based on epidemiological evidences for methemoglobinemia in infants, the guideline value for nitrate has been fixed at 50 mg/L, and based on human data on infants, the guidelines for nitrite was fixed at 3 mg/L.

In addition to the health implications of nitrate in drinking water, it is also one of the priority pollutants that contributes to eutrophication. Nitrogen-saturated terrestrial ecosystems contribute both inorganic and organic nitrogen to freshwater, coastal, and marine eutrophication, where nitrogen is typically a *limiting nutrient*. It is also noteworthy that the success of biological wastewater treatment systems hinges on the presence of nitrogenous compounds, including nitrite and nitrate.

The limit values for nitrate and nitrite in drinking water are different worldwide. The US EPA recommend a maximum contaminant level of 1 mg/L for $[NO_2]^-$ and

10 mg/L for $[NO_3]^-$ in drinking water [10]. The current recommended WHO guideline is 3 mg/L for $[NO_2]^-$ and 50 mg/L for $[NO_3]^-$ [11], while the European Union set its maximum admissible concentrations of 0.5 mg/L for $[NO_2]^-$ and 50 mg/L for $[NO_3]^-$ in drinking waters [12].

1.3.10 Phosphorus Oxyanions

Phosphorus (P) has different oxyanions derived from orthophosphoric acid (H_3PO_4) and from phosphorous acid (H_3PO_3). Phosphorous acid (oxidation number $+3$) is diprotic, not triprotic and ionizes two protons in aquatic systems resulting in hydrogen phosphonate (or hydrogen phosphite) $[H_2PO_3]^-$ and phosphonate (or phosphite) $[HPO_3]^{2-}$ [104]. The last hydrogen atom is bonded directly to the phosphorus atom and is not ionizable. Phosphorous acid is an intermediate in the preparation of other phosphorus compounds.

Orthophosphoric acid (oxidation number $+5$) is a triprotic acid. Its three derived oxyanions dihydrogen phosphate $[H_2PO_4]^-$, hydrogen phosphate $[HPO_4]^{2-}$ and orthophosphate $[PO_4]^{3-}$ coexist according to the dissociation and recombination equilibria. Premised on the observations that orthophosphoric acid and its oxyanions could be transformed into various forms, the aqua forms of phosphorus were classified into three types, viz. orthophosphates, pyrophosphates $[P_2O_7]^{4-}$, triphosphate $[P_3O_{10}]^{5-}$ and metaphosphates, which are usually long linear polymers [2]. The condensed phosphates, which include pyrophosphates and metaphosphates, are polymers of phosphate connected through a phosphoanhydride bond and serves many biological roles, including storage of phosphate, a source of energy in anoxic environments, and the sequestration of multivalent cations [105]. In some cases, during treatment, water supplies are dosed with low concentration of condensed phosphates (polyphosphates and polyphosphosilicates) pigments to prevent corrosion [106]. Much attention has been given to phosphate antirust pigments (mainly aluminium tripolyphosphate and zinc phosphate) [107]. Phosphate coatings are used to resist corrosion on steel, aluminium, zinc, cadmium, silver and tin parts and can also be found in coatings or painting [108].

This same treatment is also applicable in the treatment of boiler waters, for industrial purposes. An important ingredient of laundry detergents is phosphate in the form of sodium tripolyphosphate [109]. One of the components of commercial fertilizer is orthophosphates [110]. Contrary to the highly mobile nitrogen in the soil matrix, phosphate is immobilized in the soil by a combination of biological and chemical (i.e., absorption and mineralization) actions. Losses of phosphates in agricultural runoff water can be a result of erosive processes under the land uses, where extreme rainfall events contribute disproportionately to such losses. Eutrophication due to agricultural phosphorus sources affects a relatively high proportion of rivers, lakes, and reservoirs [111]. Phosphate concentration that exceeds about 2 μM in water bodies is known to stimulate eutrophication. Eutrophication leads to significant deterioration of water quality by reducing the dissolved oxygen, which not only kills the aquatic life but also disrupts the natural food chain web. Excessive amounts of phosphates discharged from wastewater treatment plants effluents into aquatic environments have become one of the main causes of water eutrophication [112]. In water treatment plant, removal technologies are chemical precipitation of phosphate with iron(III) or Al(III) salts and enhanced biological phosphorus removal [93, 113]. The presence of phosphates in drinking water has not been reported to be toxic to both human and animal except when present in very high concentrations, which could induce digestion problems. Therefore, there are no limit values for phosphate in drinking water from WHO, US EPA, and EU.

In plants and animals, the soluble or bio-available phosphate is incorporated into the biological system, and these key biological systems include adenosine triphosphate (ATP), deoxyribonucleic acid (DNA), and ribonucleic acid (RNA). These biological systems are important in the storage and use of energy and a key stage in the Kreb's Cycle, forming the backbones of life (i.e., via genetics) and a key factor controlling photosynthesis. In order to meet the EU Drinking Water Directive Standards in the UK, domestic water is dosed with phosphate to reduce the possible leaching of lead from the plumbing systems during water distribution. It has been reported [114] that the presence of phosphate in the aqua matrix significantly reduces the amount of copper in domestic sewage. The dosage of water supplies with phosphate greatly retarded the leaching of lead compounds from plumbing pipes, and this procedure has been in use since the 1980s.

1.3.11 Lead (Plumbum) Oxyanions

Lead (plumbum; Pb) has several oxyanions which are in the oxidation states $+4$ (plumbates) or $+2$ (plumbites). Plumbates are formed by the reaction of lead oxide PbO_2 with alkali. The resulting plumbate salts are the hydrated plumbate anion, $[Pb(OH)_6]^{2-}$ or the anhydrous anions meta-plumbate $[PbO_3]^{2-}$ or orthoplumbate $[PbO_4]^{4-}$ [7]. The plumbate (IV) salts are very strong oxidising agents. Some hydrated plumbate (IV) salts decompose upon dehydration and by carbon dioxide [115]. Lead trioxide Pb_3O_4 and lead sesquioxide Pb_2O_3 are thought of as lead(II) orthoplumbate(IV) $[Pb^{2+}]_2[PbO_4]^{4-}$ and lead(II) meta-plumbate(IV) $[Pb^{2+}][PbO_3]^{2-}$, respectively [7]. Plumbite $[PbO_2]^{2-}$ and the hydrated anion $[HPbO_2]^-$ result as oxyanions of lead(II) oxide PbO, when PbO is dissolved in alkali medium. Plumbite is a weak reducing agent; when it functions as one, it is oxidized to plumbate [115].

Lead is one of the naturally occurring metals found in the Earth's crust, and the widespread use has resulted in extensive environmental contamination, human exposure, and significant public health problems in many parts of the world [116]. The use of lead for water pipes is known to be a problem in areas with soft or acidic water [116]. There are two reasons for lead contamination of drinking water pipes: If carbon dioxide CO_2 is dissolved in the water phase, it may result in the formation of soluble lead bicarbonate $Pb(HCO_3)_2$. If water has oxygen dissolved it may dissolve lead as lead(II) hydroxide $Pb(OH)_2$. Both are not oxyanions, but result in toxicity [116].

Lead poisoning or plumbism occurs when there is prolonged (usually over a few months and even years) accumulation of lead in the body [117]. The accumulation of lead occurs when an individual has absorbed excessive lead by either swallowing a lead containing substance or breathing it, such as dust, water, paint, or food. Lead accumulation, even at low concentration, causes damage to almost every organ in the body system. Consumption of water containing lead has been linked to reduced cognitive development in children and an increased risk of heart disease. It has been observed that children are more vulnerable to lead poisoning than adults and thus suffer profound and permanent adverse health effects. The effects of lead poisoning (e.g., risk of high blood pressure and kidney damage) in adults also manifest as a long-term harm. Exposure of pregnant women to high levels of lead has been to the cause of miscarriage, stillbirth, premature birth and low birth weight, and minor malformations.

While EPA sets the action level to 15 $\mu g/L$, the maximum contaminant goal was 0 $\mu g/L$ [10]. The EU Council Directive and the WHO recommended a maximum concentration of 10 $\mu g/L$ for total lead in drinking waters [11, 12].

1.3.12 Sulfur Oxyanions

Sulfur (S) forms sulfur oxoacids in water, some of which cannot be isolated and are only known through their salts. Sulfites $[SO_3]^{2-}$ as oxyanions in the oxidation state +4 are related to the unstable sulfurous acid (H_2SO_3) while sulfates $[SO_4]^{2-}$ have an oxidation state of +6 and are related to sulfuric acid (H_2SO_4). The salts of thiosulfates $[S_2O_3]^{2-}$ are used in photographic fixing and as reducing agents, feature sulfur in two oxidation states. Sodium dithionite $(Na_2S_2O_4)$ contains the more highly reducing dithionite anion $[S_2O_4]^{2-}$.

A large number of ionic sulfate salts (e.g., sodium, potassium and magnesium sulfates) which are readily soluble in water, abound. The occurrence of sulfates in a large number of naturally occurring minerals (e.g., barite $(BaSO_4)$, epsomite $(MgSO_4 \cdot 7H_2O)$, and gypsum $(CaSO_4 \cdot 2H_2O)$ has been reported [79]. The dissolution of sulfate containing minerals in water contribute to the mineral content of many drinking waters. The reported taste threshold concentrations of sulfates in drinking water are diverse and a function of the type of sulfate salt [118]. The salts, acid derivatives, and peroxides of sulfate are important industrial chemicals. Sulfates and sulfuric acid derivative are used in the production of a large number of industrial products (e.g., fertilizers, chemicals, dyes, glass, paper, soaps, textiles, fungicides, insecticides, astringents and emetics). They have also found wide applications in the mining, wood pulp, metal and plating industries, sewage and drinking treatment and leather processing [79].

Effluents containing sulfates are discharged into water from a large number of industries (e.g., mines and smelters, kraft pulp and paper mills, textile mills and tanneries) [118]. The combustion of fossil fuels and metallurgical roasting processes are major contributors to the atmospheric sulfur dioxide, which in turn contribute to the sulfate content of rainwater (acid rain) as well as of surface waters [119, 120]. High sulfate concentrations in contact with concrete can cause chemical changes to the cement, which can cause significant microstructural effects leading to the weakening of the cement binder (chemical sulfate attack) [121]. This is relevant when industrial wastewater will be transported by concrete sewer systems. Premised on the report by a global network of water monitoring stations (GEMS/Water), typical sulfate concentration in fresh water is around 20 mg/L, river water ranges between 0 and 630 mg/L, lakes range between 2 and 250 mg/L and groundwater ranges between 0 and 230 mg/L [122].

Drinking of water containing sulfate in concentrations exceeding 600 mg/L has been attributed to the cause of catharsis in adult but humans can adapt to higher concentrations with time [123]. Dehydration has also been reported as a common side effect following the ingestion of large amounts of magnesium or sodium sulfate [124]. It has also been reported that most people experienced a laxative effect when they drank water containing >1000 mg/L of sulfate [125]. The presence of sulfate in drinking water can also result in a noticeable taste, and the lowest taste threshold concentration was fixed at approximately 250 mg/L as the sodium salt. Sulfate may also contribute to the corrosion of distribution systems [118]. Consequent upon the inability to really ascertain the health implications of sulfate in drinking water, no health-based guideline value has been proposed [118].

1.3.13 Selenium Oxyanions

Selenite $[SeO_3]^{2-}$ and selenate $[SeO_4]^{2-}$, the oxyanion of the element selenium (Se) in the oxidation states $+4$ and $+6$, respectively, are highly soluble in aqueous solutions at ambient temperatures and are analogues of sulfates. Despite the proximity in the chemistry of these anionic species, unlike sulfate, selenates exhibit very high oxidizing ability and get easily reduced to selenite or selenium. Selenium itself exhibits variable oxidation state and the oxidation state determine the toxicity. The oxyanion selenate is required by organisms that need selenium as a micronutrient. These organisms have the ability to acquire, metabolize, and excrete selenium. The level at which selenium becomes toxic varies from species to species and is related to other environmental factors like pH and alkalinity that influence the concentration of selenite over selenate.

The formation of selenate from selenite is slow, and both forms exist together in solution [126]. In acidic and reducing milieu, inorganic selenites get reduced to elemental selenium, while in alkaline and oxidizing milieu the formation of selenates is favored. Consequent upon the high solubility of selenites and selenates in water, selenium is leached from well-aerated alkaline soils that favor its oxidation. Although elemental selenium and many selenides have garlicky odors, the dominant forms of selenium found in water (i.e., selenites and selenates) are odorless [127]. The least reduced oxidation state of selenium is the oxyanion (i.e., selenates), and it is the form that serves as micronutrients in biological systems. Although the biological system can acquire, metabolize, and excrete selenium, the toxicity threshold greatly depends on the particular biological specie and other environmental factors (e.g., pH and alkalinity) that influence the concentration of selenite over selenate. Selenium derivatives, including selenate, are abundant in locations where ancient seas have evaporated. Selenium concentrations that ranged between 0.06 and 400 µg/L have been reported in groundwater and surface water [128–130]. Selenium concentration values, as high as 6000 µg/L, have been reported in groundwater in some localities [131]. At the global level, the selenium concentrations in potable water samples from public water supplies are usually much less than 10 µg/L but may exceed 50 µg/L in drinking water from a high soil selenium area [132]. Since drinking water often contains selenium concentrations that are much lower than 10 µg/L, except in certain seleniferous areas, drinking water rarely contribute significantly to total selenium intake in humans [127]. Very low selenium status in humans has been associated with a juvenile, multifocal myocarditis called Keshan disease and a chondrodystrophy called Kashin–Beck disease [133–135].

Selenium poisoning of natural waters may result whenever new agricultural runoff courses through undeveloped lands. This process leaches natural available soluble selenium compounds including selenates into the water. Selenium pollution of waterways also occurs when selenium is leached from coal flue ash, mining and metal smelting, crude oil processing, and landfill [136].

A major health focus of selenium has been its putative role in anticarcinogenesis. Studies have shown that the blood selenium levels are inversely linked with the prevalence of several types of cancer [137–140]. The Nutritional Prevention of Cancer Trial [141] has shown that supplemental selenium (200 µg/day as high-selenium yeast) reduced risks for total cancers and prostate and colorectal carcinomas. The symptoms that have been associated with high dietary intakes of selenium include gastrointestinal disturbances, discoloration of the skin, and decayed teeth [142]. Children living in a seleniferous area exhibited more pathological nail changes, loss of hair, and dermatitis in Venezuela [129]. High morbidity has been reported in china, where endemic selenium intoxication occurred [143], and the main symptoms exhibited were brittle hair with intact follicles, lack of pigment in new hair, thickened and brittle nails and skin lesions. Symptoms of neurological disturbances have also been reported among inhabitants of a heavily affected village; and those affected recovered as soon as the diets were changed. Subsequently, they were evacuated from the village. The average dietary intake that is associated with selenosis (hair or nail loss, nail abnormalities, mottled teeth, skin lesions and changes in peripheral nerves) is in excess of 900 µg/day [144, 145].

For drinking water, the WHO sets a guideline value of 40 µg/L for Se [11], the EU Council Directive [12] sets a value of 10 µg/L for Se, and the US EPA [10] recommended a maximum contaminant level of 50 µg/L for Se.

1.3.14 Silicon Oxyanions

Silicon (Si) has a great number of oxyanions, which are called silicates with the general formula $[SiO_{4-x}]^{(4-2x)-}$, where $0 \leq x < 2$. The group of oxyanions includes orthosilicate $[SiO_4]^{4-}$ metasilicate $[SiO_3]^{2-}$, and pyrosilicate $[Si_2O_7]^{6-}$. Silicate anions are often large polymeric molecules with an extensive variety of structures, including chains and rings (as in polymeric metasilicate $[[SiO_3]^{2-}]_n$, double chains $[[Si_2O_5]^{2-}]_n$, and sheets (e.g., $[[Si_2O_5]^{2-}]_n$) [79].

Silicate oxyanions are formally the conjugate bases of silicic acids, e.g., orthosilicate is the oxyanion of the deprotonated orthosilicic acid $Si(OH)_4$. Silicic acids are generally very weak and cannot be isolated in pure form. They exist in water solution, as mixtures of condensed and partially protonated anions, in a dynamic equilibrium [146].

Solid silicates are generally stable and are fairly soluble in water as sodium or potassium silicates. They form several solid hydrates when crystallized from solution. Soluble sodium silicates and their mixtures (water glass) are important industrial and household chemicals. Silicates of non-alkali cations, or with sheet and tridimensional polymeric anions, generally have negligible solubility in water at normal conditions [7]. Dissolved silicates, which are referred as silicic acid in water, are present in surface waters in concentrations of 1–100 mg/L [147]. Few studies indicate a No Observed Adverse Effects Level (NOAEL) of 50,000 mg/L for dietary silica [Martin, 2007]. It was reported [147] that many forms of silicon and its oxyanions exist in nature, which have beneficial effects in water. There are no limit values for drinking water.

1.4 Choosing Being Friend or Foe-The Determinants

An overview of the features, mode of occurrence, and health implications of the presence of oxyanions in aqua stream showed that they are multifunctional in their roles. In a particular instance, the role exhibited by an oxyanion present in an aqua system is a function of a number of variables. Premised on the critical review of literatures on the role of the presence or the absence of an array of oxyanions that are commonly found in aqua system (i.e., both water and wastewater), the determinants of the role adopted by these oxyanions, either as a beneficial (i.e., regarded as a friend) or a deleterious (i.e., regarded as a foe) chemical specie, were identified and delineated below.

1.4.1 Aqua Phase Concentration Range

Oftentimes, the impact of any chemical specie on living beings hinges on the available aqua phase concentration. Majority of the oxyanions found in aqueous system are required basic micronutrients (e.g., $[SeO_4]^{2-}$, $[PO_4]^{3-}$, $[NO_3]^-$) for the sustenance of both plants and animals. For example, phosphate and nitrate are desirable essential nutrients that are needed for plant growth at minimal concentration range but at elevated concentration range (e.g., values >2 μM for phosphorus) in the aqua matrix, they become undesirable because of the critical and prominent role they play in the onset of eutrophication.

Water alkalinity is attributed to the presence of carbonates and bicarbonates of some alkali and alkaline earth metals in the water matrix. Alkalinity in aqua system has both, the positive and negative side, and the side it displays is strongly dependent on the aquatic phase concentration range. Generally, alkalinity in water has a buffer effect (i.e., it stabilizes and prevents fluctuations in the water pH value); thus, at low values (<80 mg/L), enhances the corrosive tendencies of the water system. In anaerobic digestion of wastewater, sufficient alkalinity (i.e., high values that ranged between 1000 and 5000 mg/L as calcium carbonate) is required to ensure the optimal performance of the methane bacteria. In some water and wastewater treatment operations (e.g., coagulation flocculation protocol, biological nutrient removal, and ammonia removal by air stripping), sufficient water alkalinity is required for an optimal performance. Despite the identified benefits, at higher concentrations, water alkalinity (i.e., the presence of carbonates, bicarbonates, chlorides, and sulphates of calcium and magnesium in water) has also been attributed to the cause of a global water problem known as water hardness. Water hardness has been fingered in the reduction of the effectiveness of soaps and detergents during washing operations (at both domestic and industrial levels) and contributes to scale formation in pipes and boilers.

The presence of molybdenum oxyanions at concentration ≥ 10 mg/L is known to impart a slightly astringent taste to water. At certain concentrations, Mo is an essential trace element (i.e., micronutrient) that is required for the sustenance of both plants and animals. Safe and adequate Mo intake (i.e., concentration range of 0.015–0.04 mg/day for infants, 0.025–0.15 mg/day for children and 0.075–0.25 mg/day for persons above 10 years) has been prescribed for the different categories of humans, based on body mass. Some ailments and abnormalities (e.g., neurological disorder, severe headache, night blindness, nausea, edema) have been traced to the Mo deficiency in humans.

It has been reported [106–108] that the deficiency of selenium oxyanion in humans is responsible for some specific and identifiable health challenges (e.g., juvenile multifocal myocarditis and chondrodystrophy). Its putative role in anticarcinogenesis has also been reported [109–112] and the Nutritional Prevention of Cancer Trial [113] has shown that supplemental selenium (200 μg/day as high-selenium yeast) reduced risks for total cancers and prostate and colorectal carcinomas. Contrastingly, high dietary intake of Se (i.e., excess of 900 μg/day) has been associated with a number of health challenges dubbed as selenosis (i.e., hair or nail loss, nail abnormalities, mottled teeth, skin lesions and changes in peripheral nerves) [116].

The influence of the oxyanions of boron in aqua system is also determined by the aqua phase concentration range. Lower concentration (<2 mg/L) of borates in soils serve as nutrients to some plants while higher values (≥ 2 mg/L) portends danger for crops. Although the higher concentration of borates is toxic, the toxicity nature has also been harnessed as active ingredients in the formulations of bactericidal, fungicidal, and herbicidal.

1.4.2 pH Value of the Aqua Phase

In aqua system, the pH value determines the mode of occurrence (i.e., the speciation) and reactivity of the chemical species within the system. Oxyanions, like any inorganic species, are very sensitive to variation in the pH value, and the pH value determines the species of the available oxyanions. A peep at the hydrochemistry of carbonates, the oxyanion that has been ascribed to the alkalinity of an aqueous system shows that the available specie is highly dependent on the aqua phase pH value. In any aqueous system, the presence of the carbonate species (i.e., carbonate, bicarbonate, carbon dioxide, and carbonic acid) exist in dynamic equilibrium. In strongly basic conditions, carbonate ion dominates, in weakly basic conditions, bicarbonate dominate and in more acid medium carbon dioxide ($CO_{2(aq)}$) is the dominant specie. In the strongly acidic medium, the aqueous phase carbon dioxide is in equilibrium with carbonic acid, but the equilibrium favors carbon dioxide. This shows that at low pH value, the carbonate species that is present enhances the corrosive tendency of the aqua system while at higher pH value, where carbonate and bicarbonate species are present, the possibility of water hardness is promoted.

The effect of aqueous phase pH value on the role of oxyanions is also prominent in the aqueous phase distribution of the molybdenum species. When molybdenum trioxide (MoO_3) is dissolved in basic solution, it produces the simplest molybdate oxyanion (i.e., $[MoO_4]^{2-}$), but in acidic medium, heptamolybdate $[Mo_7O_{24}]^{6-}$ is the first species to be formed and at lower pH, octamolybdate $[Mo_8O_{26}]^{4-}$) and other molybdate oxyanions with, probably sixteen to eighteen (16–18) Mo atoms are formed.

The dependence of the oxyanion of arsenic on aqua phase pH value is also a good example of the influence of pH value on the available species of oxyanion in aqua system. Under oxidic condition, arsenic acid (H_3AsO_4) dominates at pH value <2 and within a pH value range of 2–11, $[H_2AsO_4]^-$, and $[HAsO_4]^{2-}$ subsists. Under mildly reduced conditions and low pH value, arsenious acid (H_3AsO_3) is formed but as the pH value increases, it is replaced by $[H_2AsO_3]^-$. When the pH value of the aqua system exceeds 12, $[HAsO_3]^{2-}$ forms [17]. Since it has been confirmed that all oxyanions of arsenic are extremely toxic, it shows that despite the ease of speciation and the dependence of the speciation on the aqua phase pH values, it does not enjoy the benefits of choosing between being a friend or a foe, instead it remains a foe, irrespective of the specie(s) present at a particular pH value.

The available specie of borates is also strongly dependent on the pH value. Within the acidic pH range of water, boric acid (H_3BO_3 or $B(OH)_3$), is the available specie, but it transforms easily to tetrahydroxyborate complex $[B(OH)_4]^-$, via autopyrolysis of water. Within the basic pH range and the aqua phase concentration that is higher than 0.025 mol/L, the polyhydroxoborates are formed. Although borates speciate with changes in the pH value of the aqua system, all the forms are equally toxic.

1.4.3 Valency/Oxidation State

Chemical species with variable oxidation states are endowed with rich chemistries. Information about the oxidation state of any chemical species provide the basis for the prediction of the possible activities in any medium. Consequent upon the fact that oxyanions of a particular element can exist in more than one oxidation state, they certainly possess rich and interesting chemistries in aqueous system. An overview of the hydrochemistry of oxyanions showed that non-metals exhibit more than one oxidation state which enabled them to form oxyanions with variable number of oxygen atom attached to the central element. Therefore, oxyanions of an element with different number of oxygen atom differ in the degree of basicity and the basicity reduces with increasing number of oxygen atom in the oxyanion.

Aside the issue of basicity, the variable oxidation state of oxyanions is also a big factor in the choice it adopts, either as a friend or foe to plants and animals. A case that readily comes to mind in this respect is the differences in the role of Cr(VI) and Cr(III) to both plants and animals. Cr(III) has been identified as one of the essential plant and animal micronutrients for the maintenance of good health and the metabolism of glucose, cholesterol, and fat in human bodies. Unlike Cr(III), Cr(VI) is highly mobile and toxic and different forms of cancer diseases have been ascribed to its accumulation in humans.

It has also been identified that one of the features of the oxyanions of the element antimony that determines its influence on humans is the oxidation state. Antimony(III) has been found to be more toxic than antimony(V). The toxicity associated with antimony (III) includes gastrointestinal mucosa, abdominal cramps, diarrhea and cardiac toxicity, optic nerve destruction, uveitides and retinal bleeding, headache, coughing, anorexia, troubled sleep and vertigo.

Nitrate poisoning has been attributed to the cause of methaemoglobinaemia, which manifest as cyanosis and asphyxia at high concentrations, respectively. A critical analysis of the nitrate (NO_3^-) poisoning showed that nitrate is not the actual culprit but the nitrogen oxyanion in another state of oxidation (i.e., nitrite (NO_2^-)).

1.5 Conclusion

Albeit, oxyanions are important and pervasive components of aqua systems (i.e., both water and wastewater), but they all serve different purposes when they find their ways into the food chain. The presence of oxyanions in water could either be beneficial (i.e., as a friend) or deleterious. On this premise, the presence of oxyanions in aqua system is likened to a double-sided coin and the side shown by the coin depends on certain factors. Using experimental evidences (i.e., laboratory and clinical reports) that have been reported as a basis, some of the factors that have been identified to influence the adopted role of aqua phase oxyanions include aqua phase concentration, aqua phase pH values, and the valence state or the oxidation state of the oxyanions in the aqua system.

References

1. Helmreich B, Metzger S (2017) Post-treatment for micropollutants removal. In: Lema JM, Suarez Martinez S (eds) Innovative wastewater treatment & resource recovery technologies: impacts on energy, economy and environment. IWA-Publishing, ISBN 13; 9781780407869, pp 214–232
2. Bowen HJM (1979) Environmental chemistry of the elements. Academic Press, London
3. Mueller U (1993) Inorganic structural chemistry. Wiley. ISBN 0-471-93717-7
4. Cotton FA, Wilkinson G, Murillo CA, Bochmann M (1999) Advanced inorganic chemistry, 6th edn. NY, Wiley, New York
5. Butterman C, Carlin JF (2004) Mineral commodity profiles: antimony. United States Geological Survey. Open file report 03-019
6. Nriagu JO, Pacyna JM (1988) Quantitative assessment of worldwide contamination of air, water and soils by trace metals. Nature 333:134–139
7. Holleman AF (2001) Inorganic chemistry, continued by Egon Wiberg (trans: Eagleson M, Brewer W); revised by Bernhard J. Holleman-Wiberg's, 1st English edition: Wiberg N. Academic Press, San Diego, California, Berlin, p 454. W. de Gruyter. ISBN 0123526515
8. Gebel T (1999) Arsenic and drinking water contamination [letter]. Science 283:1458–1459
9. Crecelius EA, Bothner MH, Carpenter R (1975) Geochemistries of arsenic, antimony, mercury, and related elements in sediments of Puget Sound. Environ Sci Technol 9:325–333
10. US EPA: United States Environmental Protection Agency (2009) National primary drinking water regulations. EPA 816-F-09-004
11. WHO (2017) Worlds Health Organisation: Guidelines for drinking water quality, 4th edition, incorporating the 1st addendum. ISBN: 978-92-4-154995-0
12. Council of the European Union (1998) Council Directive 98/83/EC of 3 November 1998 on the quality of water intended for human consumption. Official Journal L 330, 05/12/1998, pp 32–54, latest consolidation version October 2015
13. World Health Organization Antimony in Drinking water, Background document for development of WHO Guidelines for Drinking water Quality, WHO/SDE/WSH/03.04/74 (2003)
14. Andreae MO, Asmode JF, Foster P, Vantdack L (1981) Determination of antimony (III), antimony (V), and methylantimony species in natural waters by atomic absorption spectrometry with hydride generation. Anal Chem 53:1766–1771
15. Willis SS, Haque SE, Johannesson KH (2011) Arsenic and antimony in groundwater flow systems: a comparative study. Aquat Geochem 17:775–807
16. Enders R, Jekel M (1994) Entfernung von Antimon(V) und Antimon(III) aus wässrigen Lösungen. Teil I: Mitfällung und Adsorption bei der Flockung mit Eisen(III)-Salzen. [Elimination of Sb(V) and Sb(III) from aqueous solutions. Part I: coprecipitation and adsorption during flocculation with Fe(III)salts.] Wasser Abwasser 135:632–641
17. Longtin JP (1985) Status report—national inorganics and radionuclides survey. Cincinnati, OH, US Environmental Protection Agency, Office of Drinking Water
18. US Epa Antimony (1984) An environmental and health effects assessment. DC, US Environmental Protection Agency, Office of Drinking Water, Washington
19. Fowler BA, Goering PL (1991) Antimony. In: Merian E (ed) Metals and their compounds in the environment: occurrence, analysis, and biological relevance. VCH, Weinheim, pp 743–750
20. Stemmer KL (1976) Pharmacology and toxicology of heavy metals: antimony. Pharmacol Ther Part A 1:157–160

21. Elinder CG, Friberg L (1986) Antimony. In: Friberg L, Nordberg GF, Vouk VB (eds) Handbook on the toxicology of metals. Amsterdam Elsevier, pp 26–42
22. Francesconi KA, Edmonds JS (1994) Determination of arsenic species in marine environmental samples. In: Nriagu JO (ed) Arsenic in the environment. Part I: cycling and characterization. Wiley, New York
23. Report #03–07 Arsenic and chromium speciation of leachates from CCA-treated wood, State University System of Florida Florida Center for Solid and Hazardous Waste Management (2004)
24. Squibb KS, Fowler BA (1983) The toxicity of arsenic and its compounds. In: Fowler BA (ed) Biological and environmental effects of arsenic. Elsevier, Amsterdam, pp 233–263
25. Ravenscroft P, Brammer H, Richards K (2009) Arsenic pollution: a global synthesis. Wiley–Blackwell
26. Cullen WR, Reimer KJ (1989) Arsenic speciation in the environment. Chem Rev 89:713–764
27. Thirunavukkarasu OS, Viraraghavan T, Subramanian KS, Tanjore S (2002) Organic arsenic removal from drinking water. Urban Water 4:415–421
28. Wiberg E (2001) Arnold Frederick Holleman inorganic chemistry. Elsevier. ISBN 0-12-352651-5
29. Atkins PW, Overton T, Rourke JP, Weller M, Armstrong FA (2010) Inorganic chemistry, 5th edn. Oxford University Press, p 334. ISBN 9780199236176
30. Coughlin JR (1998) Sources of human exposure: overview of water supplies as sources of boron. Biol Trace Elem Res 66(1–3):87–100
31. Flores HR, Mattenella LE, Kwok LH (2006) Slow release boron micronutrients from pelletized borates of the northwest of Argentina. Miner Eng 19(4):364–367
32. Ozturk N, Kavak D (2005) Adsorption of boron from aqueous solutions using fly ash: batch and column studies. J Hazard Mater B 127:81–88
33. Peak D, Luther GW, Sparks DL (2003) ATR-FTIR spectroscopic studies of boric acid adsorption on hydrous ferric oxide. Geochim Cosmochim Acta 67:2551–2560
34. Yılmaz AE, Boncukcuogl R, Yılmaz MT, Kocakerim MM (2005) Adsorption of boron from boron-containing wastewaters by ion exchange in a continuous reactor. J Hazard Mater B 117:221–226
35. Sahin S (2002) A mathematical relationship for the explanation of ion exchange for boron adsorption. Desalination 143:35–43
36. Mack RB (1988) Round up the usual suspects. Potassium bromate poisoning. N C Med J 49:243–245
37. IARC (1999) Some chemicals that cause tumours of the kidney or urinary bladder in rodents and some other substances. Lyon, international agency for research on cancer, IARC monographs on the evaluation of carcinogenic risks to humans, vol 73, pp 481–496
38. US FDA (1994) Code of Federal regulations, vol 21. Food and Drugs, 1994, Part 136, 137 and 172.730. Washington, DC, US Food and Drug Administration; Office of the Federal Register, National Archives and Records Administration
39. Ministry of Health and Welfare (1979) The Japanese standards of food additives, 4th edn. Tokyo, p 367
40. Haag WR (1983) Hoigné J Ozonation of bromide-containing water: kinetics of formation of hypobromous acid and bromate. Environ Sci Technol 17:261–267
41. von Gunthen U (2003) Ozonation of drinking water: part II. Disinfection and by-product formation in presence of bromide, iodide or chlorine. Water Res 37:1469–1487
42. IPCS (2000) Disinfectants and disinfectant by-products. Geneva, World Health Organization, International Programme on Chemical Safety (Environmental Health Criteria 216)
43. AWWARF Disinfection by-products database and model project (1991) Denver. CO, American Water Works Association Research Foundation
44. Siddiqui MS, Amy G (1993) Factors affecting DBP formation during ozone–bromide reactions. J Am Water Works Assoc 85(1):63–72
45. Siddiqui MS, Amy GL, Rice RG (1995) Bromate ion formation: a critical review. J Am Water Works Assoc 87(10):58–70

46. Fielding M, Hutchison J (1993) Formation of bromate and other ozonation by-products in water treatment. In: Proceedings of the IWSA international workshop on bromate and water treatment. International Water Supply Association, Paris London, pp 81–84
47. Canada H (1999) Bromate: guidelines for Canadian drinking water quality—supporting document. Ontario, Health Canada, Environmental Health Directorate, Health Protection Branch, Ottawa
48. McGuire MJ, Krasner SW, Gramith JT (1990) Comments on bromide levels in state project water and impacts on control of disinfectant by-products. Los Angeles, CA, Metropolitan Water District of Southern California
49. World Health Organization (2005) Bromate in Drinking water Background document for development of WHO Guidelines for Drinking water Quality. WHO/SDE/WSH/05.08/78
50. Quick CA, Chole RA, Mauer SM (1975) Deafness and renal failure due to potassium bromate poisoning. Arch Otolaryngol 101:494–495
51. Stumm W, Morgan JJ (1995) Aquatic chemistry: chemical equilibria and rates in natural waters. Wiley-Interscience; 3rd edn. ISBN-13: 978-0471511854
52. Chisholm H (1911) Carbonates. In: Encyclopaedia Britannica, 11th edn. Cambridge University Press
53. World Health Organization: Hardness in Drinking water, 2003WHO/HSE/WSH/10.01/10/Rev/1 (2003)
54. Biology.arizona.edu. (2006) Clinical correlates of pH levels: bicarbonate as a buffer. Accessed on Oct 2006
55. Spellman FR (2003) Handbook of water and wastewater treatment plant operations. A CRC Press Company, Lewis Pubishers
56. Vogt H, Balej J, Bennett JE, Wintzer P, Sheikh SA, Gallone P (2002) Chlorine oxides and chlorine oxygen acids. In: Ullmann's encyclopedia of industrial chemistry. Wiley-VCH
57. Jenica McMullen, Akhgar G, Brenda K, Leonardo T (2017) Identifying subpopulations vulnerable to the thyroid-blocking effects of perchlorate and thiocyanate. J Clin Endocrinol Metab 102(7):2637–2645
58. Brandhuber P, Clark S, Morley K (2009) A review of perchlorate occurrence in public drinking water systems. J Am Water Works Assoc 101(11):63–73
59. DuBois JL, Ojha S (2015) Chapter 3, Section 2.2 natural abundance of perchlorate on earth. In: Kroneck PMH, Torres mES (eds) Sustaining life on planet earth: metalloenzymes mastering dioxygen and other chewy gases. Metal Ions in Life Sciences. vol 15. Springer. Berlin, p 49
60. Susarla S, Collette CW, Garrison AW, Wolfe NL, McCutcheon SC (1999) Perchlorate identification in fertilizers. Environ Sci Technol 33:3469–3472
61. Urbansky T, Brown SK, Magnuson ML, Kelty CA (2001) Perchlorate levels in samples of sodium nitrate fertilizer derived from Chilean caliche. Environ Pollut 112:299–302
62. Agency for Toxic Substances and Disease Registry (ATSDR) (2008) Toxicological profile for perchlorates. www.atsdr.cdc.gov/toxprofiles/tp162.pdf
63. Block SS (2001) Disinfection, sterilization, and preservation, 5th edn. Lippincott, Williams & Wilkins. p 215. ISBN 0-683-30740-1
64. Rao B, Hatzinger PB, Böhlke JK, Sturchio NC, Andraski BJ, Eckardt FD, Jackson W (2010) Natural chlorate in the environment: application of a New IC-ESI/MS/MS method with a $Cl^{18}O_3{}^-$ internal standard. Environ Sci Technol 44:8429–8434. https://doi.org/10.1021/es1024228.PMID20968289
65. NAS (1987) Drinking water and health, vol 7. Washington, DC, National Academy of Sciences, National Academy Press
66. Meister R (ed) (1989) Farm chemicals handbook. OH, Meister Publishing Co, Willoughby
67. Budavari S, O'Neill M, Smith A (eds) (1989)The Merck index. In: An encyclopedia of chemicals, drugs and biologicals, 11th ed. Rahway, NJ, Merck
68. IARC (1991) Chlorinated drinking water; chlorination by-products; some other halogenated compounds; cobalt and cobalt compounds. Lyon, International Agency for Research on Cancer, (IARC Monographs on the Evaluation of Carcinogenic Risks to Humans, vol 52, pp. 45–359

69. Harrington RM, Romano RR, Gates D (1995) Subchronic toxicity of sodium chlorite in the rat. J Am Coll Toxicol 14:21–33
70. Heffernan WP, Guion C, Bull RJ (1979) Oxidative damage to the erythrocyte induced by sodium chlorite, in vivo. J Environ Pathol Toxicol 2:1487–1499
71. WHO (2005) Guidelines for Drinking water Quality, Chlorite and Chlorate in Drinking water, Background document for development of, WHO/SDE/WSH/05.08/86
72. Barnowski C, Jakubowski N, Stuewer D, Broekaert JAC (1997) Speciation of chromium by direct coupling of ion exchange chromatography with ICP-MS. At Spectrom 1155(12):1155–1161. https://doi.org/10.1039/a702120h
73. Gil RA, Cerutti S, Gásquez JA, Olsina RA, Martinez LD (2006) Preconcentration and speciation of chromium in drinking water samples by coupling of online sorption on activated carbon to ETAAS determination. Talanta 1065(68):1065–1070
74. Das AK (2004) Micellar effect on the kinetics and mechanism of chromium(VI) oxidation of organic substrates. Coord Chem Rev 248:81–89
75. Katz F, Slem H (1994) The biological and environmental chemistry of chromium. VCH, New York, pp 51–58
76. Kotas J, Stasicka Z (2000) Chromium occurrence in the environment and methods of its speciation. Environ Pollut 107(3):263–283
77. Venitt S, Levy LS (1974) Mutagenicity of chromates in bacteria and its relevances to chromate carcinogenesis. Nature 250(5466):493–495
78. Kimbrough DE, Cohen Y, Winer AM, Creelman L, Mabuni CA (1999) Critical assessment of chromium in the environment critical reviews in environmental science and technology. Crit Rev Environ Sci Technol 29(1):1–46
79. Pope MT, Müller A (1997) Polyoxometalate chemistry: an old field with new dimensions in several disciplines. Angew Chem Int Ed 30:34–48
80. Greenwood NN, Earnshaw A (1997) Chemistry of the elements, 2nd edn. Butterworth-Heinemann (1997). ISBN 0-08-037941-9
81. Baird RB, Eaton AD, Rice EW (2017) Standard methods for the examination of water and wastewater, 23rd edn. American Public Health Association, Washington, DC
82. Asmanguljan TA (1965) Determination of the maximum permissible concentration of molybdenum in open bodies of water. Gigiena I Sanitarija 30:1–5 (in Russian)
83. Weast RC (ed) (1986) Handbook of chemistry and physics, 67th edn. OH, CRC Press, Cleveland
84. National Academy of Sciences (1977) Drinking water and health. DC, Washington, pp 279–285
85. Kehoe RA, Chalak J, Largent EJ (1944) The concentration of certain trace metals in drinking waters. J Am Water Works Assoc 36:637–644
86. Durfor CN, Becker E (1964) Public water supplies of the 100 largest cities in the United States. Washington, DC, United States Geological Survey Water Survey (Water Supply Paper No. 1812)
87. Greathouse DG, Osborne RH (1980) Preliminary report on nationwide study of drinking water and cardiovascular diseases. J Environ Pathol Toxicol 4(2–3):65–76
88. World Health Organization. Molybdenum in Drinking water, Background document for development of WHO Guidelines for Drinking water Quality, WHO/SDE/WSH/03.04/11/Rev/1(2011)
89. Fairhall LT, Dunn RC, Sharpless NE, Pritchard EA (1945) The toxicity of molybdenum. Washington, DC, United States Government Printing Office, (Public Health Service Bulletin No. 293), pp 1–35
90. Robinson MF, McKenzie JM, van Thomson CD, Rij AL (1973) Metabolic balance of zinc, copper, cadmium, iron, molybdenum and selenium in young New Zealand women. Br J Nutr 30:195–205
91. National Academy of Sciences (1989) Recommended dietary allowances, 10th edn. Washington, DC, National Academy Press, pp 243–246, 284

92. Johnson JL, Waud WR, Rajagopalan KV, Duran M, Beemer FA, Wadman SK (1980) Inborn errors of molybdenum metabolism: combined deficiencies of sulfite oxidase and xanthine dehydrogenase in a patient lacking the molybdenum cofactor. Proc Natl Acad Sci 77:3715–3719

93. Abumrad NN, Schneider AJ, Steel D, Rogers LS (1981) Amino acid intolerance during prolonged total parenteral nutrition reversed by molybdate therapy. Am J Clin Nutr 34:2551–2559

94. Metcalf and Eddy (2013) Wastewater Engineering: treatment and reuse. 5th edn. McGraw-Hill Education. ISBN-13: 978–1259010798

95. ICAIR Life Systems, Inc. (1987) Drinking water criteria document on nitrate/nitrite. Washington, DC, United States Environmental Protection Agency, Office of Drinking Water

96. Laue W, Thiemann M, Scheibler E, Wiegand KW (2006) Nitrates and nitrites. In: Ullmann's encyclopaedia of industrial chemistry. Wiley-VCH, Weinheim

97. World Health Organization (2011) Nitrate and nitrite in drinking water, Background document for development of WHO Guidelines for Drinking water Quality, WHO/SDE/WSH/07.01/16/Rev/1

98. AWWARF (1995) Nitrification occurrence and control in chloraminated water systems. Denver, CO, American Water Works Association Research Foundation

99. van Duijvenboden W, Loch JPG (1983) Nitrate in the Netherlands: a serious threat to groundwater. Aqua 2:59–60

100. USEPA (1987) Estimated national occurrence and exposure to nitrate and nitrite in public drinking water supplies. Washington, DC, United States Environmental Protection Agency, Office of Drinking Water

101. FAO/WHO (1996) Toxicological evaluation of certain food additives and contaminants. Geneva, World Health Organization, Joint FAO/WHO Expert Committee on Food Additives (WHO Food Additives Series No. 35)

102. Manassaram DM, Baker LC, Moll DM (2007) A review of nitrates in drinking water: maternal exposure and adverse reproductive and developmental outcomes. Cien Saude Colet 12(1):153–163

103. WHO (1985) Health hazards from nitrate in drinking water. Report on a WHO meeting, Copenhagen, 5–9 March 1984. Copenhagen, WHO Regional Office for Europe (Environmental Health Series No. 1)

104. Kuper F, Til HP (1995) Subchronic toxicity experiments with potassium nitrite in rats. In: Health aspects of nitrate and its metabolites (particularly nitrite). Proceedings of an international workshop, Bilthoven (Netherlands). Strasbourg, Council of Europe Press, pp 195–212

105. Larson JW, Margaret P (1989) Thermodynamics of ionization of hypophosphorous and phosphorous acids. Substituent effects on second row oxy acids. Polyhedron 8:527–530

106. Sidney JO, Grynpas MD (2008) Relationships between polyphosphate chemistry, biochemistry and apatite biomineralization. Chem Rev 108:4694–4715

107. Kalendova A (2003) Comparison of the anticorrosion efficiencies of pigments based on condensed phosphates and polyphosphosilicates Anti-Corros Methods Mater 50(2):82–90

108. Chromy L, Kamińska E (1990) Non-toxic anticorrosive pigments. Prog Organ Coat 18:319–324

109. Edwards J (1997) Coating and surface treatment systems for metals. Finishing Publications Ltd. and ASM International, pp 214–217

110. Köhler J (2006) Detergent phosphates: an EU policy assessment. J Bus Chem 03:15–30

111. Richardson AE (2007) Making microorganisms mobilize soil phosphorus. In: Velázquez E, Rodríguez-Barrueco C (eds) First international meeting on microbial phosphate solubilization. Developments in plant and soil sciences, vol 102. Springer, Dordrecht

112. Torrent J, Barberis E, Gil-Sotres F (2007) Agriculture as a source of phosphorus for eutrophication in southern Europe. Soil Use Manag 23(s1):25–35

113. Peng L, Dai H, Wu Y et al (2018) A comprehensive review of the available media and approaches for phosphorus recovery from wastewater. Water Air Soil Pollut. 229:115

114. Bunce JT, Ndam E, Ofiteru ID, Moore A, Graham DW (2018) A review of phosphorus removal technologies and their applicability to small-scale domestic wastewater treatment systems. Front Environ Sci
115. Comber S, Cassé F, Brown B et al (2011) Phosphate treatment to reduce plumbosolvency of drinking water also reduces discharges of copper into environmental surface waters. Water Environ J 25:266–270
116. Arora A (2005) Text book of inorganic chemistry. Discovery Publishing House, pp 450–452
117. Thornton I, Rautiu R, Brush SM (2001) Lead: The facts (PDF). International Lead Association
118. The Editors of Encyclopedia Britannica, February 2020, Lead poisoning pathology, Encyclopaedia Britannica. Cambridge University Press
119. World Health Organization, Sulfate in Drinking water, Background document for development of WHO Guidelines for Drinking water Quality, WHO/SDE/WSH/03.04/114 (2004)
120. Delisle CE, Schmidt JW (1977) The effects of sulphur on water and aquatic life in Canada. In: Sulphur and its inorganic derivatives in the Canadian environment. Ottawa, Ontario, National Research Council of Canada (NRCC No. 15015)
121. Keller W, Pitblade JR (1986) Water quality changes in Sudbury area lakes: a comparison of synoptic surveys in 1974–1976 and in 1981–1983. Water Air Soil Pollut 29:285
122. Tian B, Cohen MD (2000) Does gypsum formation during sulfate attack on concrete lead to expansion? Cem Concr Res 30(1):117–123
123. UNEP (1990) GEMS/Water data summary 1985–1987. Burlington, Ontario, Canada Centre for Inland Waters; United Nations environment programme, global environment monitoring system, GEMS/Water Programme Office
124. US EPA (1985) National primary drinking water regulations; synthetic organic chemicals, inorganic chemicals and microorganisms; proposed rule. US Environmental Protection Agency. Fed Reg 50(219):46936
125. Fingl E (1980) Laxatives and cathartics. In: Gilman AG et al (eds) Pharmacological basis of therapeutics. MacMillan Publishing, New York
126. US EPA (1999) Health effects from exposure to high levels of sulfate in drinking water workshop. Washington, DC, US Environmental Protection Agency, Office of Water (EPA 815-R-99-002)
127. Sorg TJ, Logsdon GS (1978) Treatment technology to meet the interim primary drinking water regulations for inorganics: part 2. J Am Water Works Assoc 70(7):379–393
128. World Health Organization Sulfate in Drinking water, Background document for development of WHO Guidelines for Drinking water Quality, WHO/HSE/WSH/10.01/14 (2004)
129. Smith MJ, Westfall BB (1937) Further field studies on the selenium problem in relation to public health. Uni States Public Health Rep 52:1375–1384
130. Scott RC, Voegeli Jr PT (1961) Radiochemical analysis of ground and surface water in Colorado. Colorado water conservation board (Basic Data Report 7)
131. Lindberg P (1968) Selenium determination in plant and animal material, and in water. A methodological study. Acta Veterinaria Scandinavica, Suppl. 23:1–48
132. Cannon HG (1964) Geochemistry of rocks and related soils and vegetation in the Yellow Cat area, Grand County, Utah. Washington, DC, United States Geological Survey (Bulletin No. 1176)
133. Gore F, Fawell J, Bartram J (2010) Too much or too little? A review of the conundrum of selenium. J Water Health 8(3):405–416
134. Högberg J, Alexander J (1986) Selenium. In: Friberg L, Nordberg GF, Vouk VB, (eds) Handbook on the toxicology of metals, 2nd edn. Vol, 2. Amsterdam, Elsevier, pp 482–520
135. IPCS (1987) Selenium. Geneva, World Health Organization, International Programme on Chemical Safety (Environmental Health Criteria, No. 58)
136. FAO/WHO (2004) Vitamin and mineral requirements in human nutrition, 2nd edn. Report of a Joint FAO/WHO Expert Consultation, Bangkok, Thailand, 21–30 Sept 1998. Geneva, World Health Organization
137. Lemly D (2004) Aquatic selenium pollution is a global environmental safety issue. Ecotoxicol Environ Saf 59(1):44–56

138. Combs G, Gray W (1998) Chemopreventive agents: selenium. Pharmacol Ther 3:179–198
139. Rayman M (2005) Selenium in cancer prevention: a review of the evidence and mechanism of action. Proc Nutrition Soc 64:527–542
140. Combs G, Lü J (2006) Selenium as a cancer preventive agent. In: Hatfield D (ed) Selenium: molecular biology and role in health, 2nd edn. NY, Kluwer Academic, New York, pp 205–218
141. Gromadzińsk J, Reszka E, Bruzelius K, Wąsowicz W, Åkesson B (2008) Selenium and cancer: biomarkers of selenium status and molecular action of supplements. Eur J Nutr 47:29–50
142. Clark LC, Combs Jr GF, Turnbull BW et al (1996) Effects of selenium supplementation for cancer prevention in patients with carcinoma of the skin. J Am Med Assoc 276:1957–1963
143. Jaffe WG (1976) Effect of selenium intake in humans and in rats. In: Proceedings of the symposium on selenium and tellurium in the environment. Pittsburgh, PA, Industrial Health Foundation, pp 88–193
144. Yang GQ, Wang SZ, Zhou RH, Sun SZ (1983) Endemic selenium intoxication of humans in China. Am J Clin Nutr 37:872–881
145. Yang G, Yin S, Zhou R, Gu L, Yan B, Liu Y, Liu Y (1989) Studies of safe maximal daily selenium intake in a seleniferous area in China. Part II. J Trace Elem Electrolytes Health Dis 3:123–130
146. ATSDR (2003) Toxicological profile for selenium. Atlanta, GA, Agency for Toxic Substances and Disease Registry. https://www.atsdr.cdc.gov/ToxProfiles/tp.asp?id=153&tid=28
147. Knight CTG, Balec RJ, Kinrade SD (2007) The structure of silicate anions in aqueous alkaline solutions. Angew Chem Int Ed 46:8148–8152

Chapter 2
Ecological Impacts of Oxyanion in Aqua Systems

Y. I. Bulu and T. D. Saliu

Abstract The growth and development of organisms in aquatic ecosystem are affected by the presence of oxyanions which are toxic even at low concentrations. The presence of undesirable types of oxyanions at levels above stipulated threshold concentrations, cause ecological imbalance and the ultimate destruction of the ecosystem. This chapter is focused on the ecological impact of oxyanions in aquatic system. Aside their bioaccumulation in the living organism, their direct impacts include significant reduction in the water quality, alteration of water pH value, which may be intolerable to plants and animals and reduction in the abundance and fitness of biota. Although some oxyanions, such as selenium, nitrogen and phosphorus, are required as nutrients in biological systems, they become toxic with increasing concentrations. Their toxicity is associated with their ability to substitute some important oxyanions in biological systems due to their structural similarities, their oxidation state, mobility and the pH value of the medium in which they occur. However, the oxyanions of arsenic, chromium and selenium are harmful because of their nonbiodegradability, higher solubility in water and deleterious effects. The oxyanions of nitrogen and phosphorus are associated with excessive growth of algae and macrophytes, with the attendant effects on the structure and functioning of the aquatic system. This study posits that the dearth of organisms (due to the toxicity of oxyanions) or their abundance (due to nutrient enrichment by oxyanions) impact negatively on the food web and stability of the ecosystem.

Keywords Oxyanions · Toxicity · Ecological Impact · Water Pollution · Transport

Y. I. Bulu (✉)
Department of Plant Science and Biotechnology, Adekunle Ajasin University, Akungba-Akoko, Nigeria
e-mail: yetunde.bulu@aaua.edu.ng

T. D. Saliu
Hydrochemistry Research Laboratory, Department of Chemical Sciences, Adekunle Ajasin University, Akungba-Akoko, Nigeria

© Springer Nature Switzerland AG 2021   33
N. A. Oladoja and E. I. Unuabonah (eds.), *Progress and Prospects in the Management of Oxyanion Polluted Aqua Systems*, Environmental Contamination Remediation and Management, https://doi.org/10.1007/978-3-030-70757-6_2

2.1 Introduction

Pollutants released into the environments can have adverse effects on the quality and biological communities of an aquatic ecosystem, creating ecological imbalance. Long et al. [1] opined that the predictive criteria to evaluate risk and establish permissible levels of contamination are based on species responses to the contaminants. Toxic substances in aquatic environment can reduce organism abundance by causing damage to their reproductive system and decrease their viability, or change the physical properties of their environment [2]. Unlike most organic pollutants which are biodegradable, inorganic chemicals, including oxyanions, are continuously accumulated in the environment [3]. They pose threat to life, due to their potential risk of entry into the food chain. He et al. [4] noted that industrial activities, such as electroplating, metal smelting, chemical industries, and manufacturing processes, are some of the anthropogenic sources of heavy metals in water. Poorly treated domestic, industrial, and agricultural wastewater also contain high concentrations of metals, which are often discharged into the environment [5].

Oxyanions/oxoanions are polyatomic negatively charged ions [an anion] containing oxygen, with the general formula of $A_xO_y^{z-}$, where A is an element symbol, O is an oxygen atom, and x, y, and, z are integer values. Oxyanions typically have increased solubility and mobility with increasing pH value, in contrast to metals, which show the opposite behavior. In solution, many chemical elements form oxyanions at specific oxidation states, which may be toxic even at low concentrations. They find their way into living organisms via ingestion, inhalation, and skin adsorption, causing irrevocable effects [6]. Their toxicity has been linked to their structural similarities with some important oxyanions in biological systems [e.g., phosphate and sulfate] and their mobility in soils and groundwater [7]. The competitive selection of such structurally similar oxyanion rather than the endogenous ones, in biological reaction, is one of the bases for their toxicity. The mobility and retention of oxyanions in groundwater are regularly controlled by adsorption reactions with the solid surface, particularly in oxic groundwater system, which contain significant amount of iron hydroxides [8].

Although some of these oxyanions are essential in trace amounts in biological systems, yet they become toxic beyond certain threshold concentrations. Implicated in this scenario include the oxyanions of arsenic (As), antimony (Sb), chromium (Cr), molybdenum (Mo), and boron (B). The oxyanions of these elements are classified as harmful species because of their nonbiodegradability [6, 9] and higher solubility in water. These species equally bioaccumulate in the food chain and environment, thus increasing their toxicity [6]. Other oxyanions implicated as water contaminants include sulfate, carbonates, chlorates, nitrate, and phosphates. Phosphates and nitrates can cause serious environmental disturbances such as increasing the occurrence of algae bloom in aquatic ecosystem.

Oxyanions released into the environment often accumulate most rapidly in aquatic habitats, where they enter the biota and are subsequently transferred to higher trophic levels and, in many cases, eventually to humans. Aside bioaccumulating in living organism via the food chain and web, direct effects of toxic chemicals in the aquatic environment include significant reduction in the quality of surface water and drinking water, lowering or raising of water pH which may be intolerable to plants and animals, reduction in organism abundance and fitness. Although some organisms are tolerant to the toxic effects of pollutants when exposed to it, some pollutants may directly affect source species thereby leading to changes in the abundance of associated species [10].

2.2   The Ecological System

An ecosystem is a dynamic entity composed of a biological community and its inter-action with the associated abiotic environment. The two major forces identified as linking the ecosystem components [biotic and abiotic] together are the flow of energy through the ecosystem and the cycling of nutrients within the system [11]. Often the dynamic interactions that occur within an ecosystem are numerous and complex. The balance between these interactions is essential for the survival, existence, and stability of the ecosystem. Survival of all organisms is actualized by the ecological balance that exist within various species as a result of the different relationship that exist between them.

An ecosystem is said to be stable if following a disturbance, it returns to its equilibrium state. Stability of an ecosystem has therefore been characterized based on its resistance to disturbance and its resilience (recovery) in the event of a disturbance. While the ability of an ecosystem to resist disturbance as a result of its accumulated structure is termed resistance, resilience infers the ability of an ecosystem to return to a normal state, following displacement [12]. Holing [13] opined that although a natural system has high ability to withstand different disturbances or perturbations without modifications, the system may alter to another, in the face of excessive disturbances. An example of this scenario is the rapid increase in the population of algae in the aquatic ecosystem as a result of introduction of excessive nitrogen and phosphate nutrient or oil spillage which may lead to death of hundreds of thousands of seabirds, otters, seals, and whales [See Exxon Valdex spill in Alaska, 1989 and Deep water horizon spill in the gulf of Mexico, 2010].

In order to assess the full range of the effects of disturbance on ecosystems, an understanding of ecosystem-level interactions is essential. Disruption of the natural patterns and processes in an ecosystem may have great impact because species adaptation to these disturbances may be delayed. Ecologists have suggested that ecosystem stability is linked to nutrient cycling characteristics and this position was supported by Odum [14], Jordan et al. [15] and Waide et al. [16].

Energy flow in the ecosystem occur through the food chain and the food web [17] (Fig. 2.1). Autotrophic organisms are able to convert inorganic compounds to organic by using energy from sunlight. This autotrophic organism/producer occupies the lowest level in the food chain. Energy stored in the producers is transferred to the primary, secondary, and tertiary consumer in the food chain, through successive

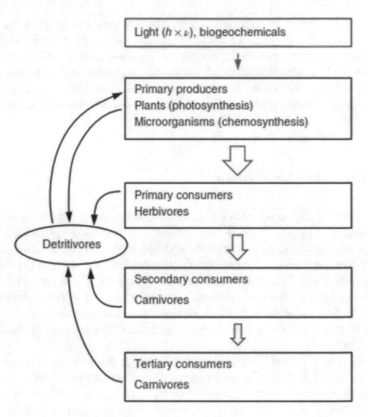

Fig. 2.1 Energy flow through different trophic levels [17]

feeding. In the food chain, the survival of one organism depends on the other, such that the removal of any organism results in imbalance in the ecological system.

The food web is embedded in a natural ecosystem that consists of interwoven food chains, in which individual animal or plant species function as either a predator or prey. Apart from observable changes in the ecosystem as a result of disturbance, an essential effect of disturbance on food chain is the removal of the keystone species. According to Kotliar [18], Delibes-Mateos et al. [19] and Hale and Koprowski [20], keystone species define an entire ecosystem by maintaining its organization, stability, and functioning. Notable examples of keystone species are sharks, African elephant, gray wolves, otters, etc. As described by Mclaren and Peterson [21], the removal of wolves (*Canis lupus* L.) from the food chain in a three-trophic-level system comprising of Balsam fir (*Abies balsamea* (L.) Mill), moose (*Alces alces* L.) and the wolves, allowed for over browsing of the fir tree population by elevated moose densities.

The structure and dynamics of the food web in an ecosystem are affected by nutrients availability [17]. Growth of organisms in aquatic ecosystem is reportedly hindered by one or more macronutrients such as phosphorus and nitrogen and some-times by micronutrients such as molybdenum. Considering the need to maintain ecological balance in an ecosystem, the presence of oxyanions, of an undesirable type or in a detestable concentration, is bound to cause ecological imbalance and the ultimate destruction of the ecosystem.

2.3 Sources and Pathways of Contamination

Formation of oxyanions is normally assumed to take place in an alkaline aquatic environment. However, there is evidence that the formation of oxyanions salts can also occur in a solid-phase high-temperature environment [22]. At specific oxidation states, elements form oxyanions, which could exist in monomeric or polymeric form. Monomeric oxyanions have the general formulae, AO_n^{m-}, which is dictated by the oxidation state of element A and its position in the periodic table. In period two of the periodic table, carbon and nitrogen form trigonal planar structured oxyanions nitrate (NO_3^-) and carbonate (CO_3^{2-}) when covalently bonded with oxygen. The π bond formed between the central atoms and the oxygen atoms is favored by the similarities between the bonding atoms. The oxyanions formed by metals of period three elements include olivine minerals $(Mg,Fe)SiO_4$, Phosphate (PO_4^{3-}), sulfate (SO_4^{2-}), and perchlorate (ClO_4^-). Phosphorus can exist as phosphate (PO_4^{3-}), hypophosphite (PO_2^{3-}), and phosphite (PO_3^{3-}). Arsenic species in ground and surface waters exist primarily as oxyanions with varying oxidation states, which are arsenate (AsO_4^{3-}) and arsenite (AsO^{2-} or AsO_3^{3-}), [23] and selenium exists in the following oxidation state: -2; 0; $+2$; $+4$; $+6$.

Majority of oxyanions are highly water soluble, toxic [6, 9, 24, 25], and nonbiodegradable [6, 9]. Their high solubility makes them extremely mobile and bioaccumulate [26–29]. Sources of these toxic oxyanions include: alkaline wastes emanating from high-temperature operations such as thermal treatment of waste, fossil fuel combustion, metal smelting and metal finishing, electroplating, microelectronics, battery manufacturing, tannery, fertilizer industries, agricultural processes, and municipal wastewater [30].

Review of several studies revealed that municipal wastewater releases 100–1400 mg/L nitrate, 15–273 mg/L phosphate, and low concentrations of metal oxyanions into the environment [31–33]. Daily industrial activities release, from 100 to 4535 mg/L nitrate, 2.5–3490 mg/L phosphate into the environment [34–38]. Examples of oxyanion contamination in groundwater include: perchlorate, borate, arsenate, selenate, chromate, molybdate, nitrates, sulfates, carbonates, and phosphates. Most of these oxyanions are released into the aquatic ecosystem through industrial activities.

2.4 Ecological Impact of Selected Oxyanions

2.4.1 Oxyanion of Selenium

Selenium [Se] is a metalloid required in trace amount in the nutrition of organisms. Since it shares many similar properties with sulfur [39], it is considered to be the analogue [40, 41]. Selenium enters the aquatic environment through natural and anthropogenic processes [42]. Natural sources of selenium have been attributed to weathering of seleniferous rocks, shales, and soils [43], while anthropogenic sources include emission from burning of fossil fuels [44], as by-products of several activities including electricity generation, coal ash leakages, mining [coal, copper], industrial wastewater, as well as agricultural drainage water from irrigation [45].

Soluble inorganic Se oxyanions includes selenite (SeO_3^{2-}) and selenate (SeO_4^{2-}). They account for the majority of the total Se concentration in water, with their proportion determined by the redox potential and pH of the environment. While selenate occurs more in an oxidizing and alkaline environment, selenite is stable under a more reducing condition [46] (Fig. 2.2). The availability of selenate and selenite in aquatic system is likewise affected by the kinetics of their adsorption property. Selenite is readily adsorbed by particulate organic matter [47], oxides of Al and Fe, clay minerals and calcite.

Many enzymes and proteins require Se for their activity. It is readily incorporated into protein as the amino acid selenocysteine (Sec) [49, 50]. Glutathione peroxidases,

Fig. 2.2 Selenium speciation in an aqueous system as a function of pH and redox potential [48]

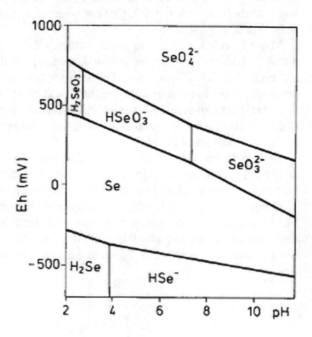

thioredoxin reductases, and iodothyronine deiodinases are examples of selenoproteins described in humans [51], while formate dehydrogenase [52, 53], selenophosphate, and peroxiredoxin [53] have been found in bacteria. Selenoprotein functions as antioxidant defenses, [54], redox signaling, and immune responses in many species. Hence, its role as cancer chemopreventive agent [55] and tendency to reduce the expression of viral infection [56] has been suggested.

According to Ogle et al. [57], selenate is taken by active uptake process via the sulfate membrane pathway in living aquatic organisms or plants as shown in Fig. 2.3, while selenite has an independent uptake mechanism that is specific to it. They further reported that selenite accumulates to greater extent and it is easily metabolized to form organic selenium compounds, notably selenoamino acids. Two processes of bioaccumulation of Se in the aquatic system have been pointed out by Ogle et al. [57] as bioconcentration and biomagnification.

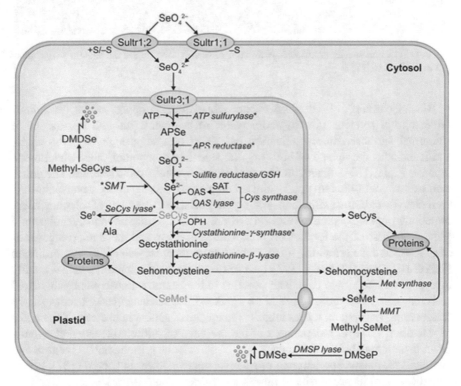

Fig. 2.3 Sulfate/selenate assimilation pathway in plant Red text and arrows indicate Se hyperaccumulator processes. Asterisks indicate enzymes overexpressed via genetic engineering. Sultr, sulfate/selenate cotransporter; APSe, adenosine phosphoselenate; APS, adenosine phosphosulfate; GSH, glutathione; SAT, serine acetyltransferase; OAS, O-acetylserine; (Se) Cys, (seleno)cysteine; OPH, Ophosphohomoserine; (Se)Met, (seleno) methionine; MMT, methylmethionine methyltransferase; DMSeP, dimethylselenoproprionate; DM(D)Se, dimethyl(di)selenide (volatile); SMT, selenocysteine methyltransferase [58]

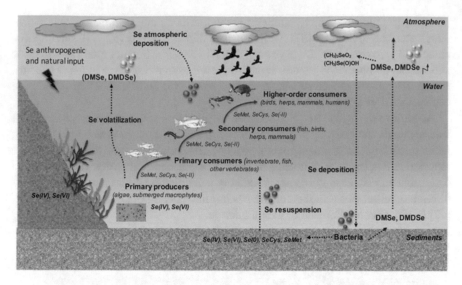

Fig. 2.4 Selenium species associated with major processes occurring in the aquatic environment and in the food web [67]

Bioconcentration of selenium occurs when there is uptake of its oxyanions, selenate, and selenite, directly from water or sediments through respiratory or epidermal surfaces. Biomagnification occurs as uptake through the food chain, and it includes the uptake of both organic selenium compound, selenomethionine (SeMet) ($C_5H_{11}NO_2Se$), and the inorganic species, selenate and selenite. The dominant pathway for selenium uptake and accumulation reported in organism is through their diet, hence biomagnification is of high importance in the study of selenium accumulation in aquatic organism (Fig. 2.4). Microbes such as diatoms and cyanobacteria exist at the base of the food chain in aquatic ecosystem and these microorganisms can bioconcentrate selenium, up to tenfold, from ambient water [59]. According to Foda et al. [60] and Ogle et al. [57], bacteria rapidly take in selenium which is readily incorporated into amino acids and proteins in the bacteria. Subsequently, bacterial cells may serve as sources of selenium uptake. Since selenate uptake occurs via the sulfate transport system, it thus inhibits the uptake of sulfate in bacteria. The uptake of selenite is however independent of the presence of sulfite. Selenate and selenite differ in their metabolism when taken up. While selenite is immediately reduced and metabolized, the metabolism rate of selenate appears lower in bacteria [57], giving an indication of its propensity to inhibit the uptake of sulfate [61].

In the aqueous system, selenate competes with sulfate for uptake in algae. Due to the adsorption of selenite to mineral and organic matter surfaces, selenate compounds are normally more readily bioavailable to marine and freshwater algae than selenite compounds [62]. However, the absorption and accumulation of selenite in microalgae are suggested to be more rapid, compared to selenate [63, 64]. Freshwater green algae take up selenite, selenate, and selenomethionine, with selenomethionine uptake being higher than the inorganic seleniums. Kiffney and Knight [65] reported the accumulation of selenate, selenite, and selenomethionine in the cyanobacterium *Anabaena flosaquae* (Lyngb.) Breb. Selenium bioconcentration in the algae was in the order of Seleno-L-methionine > selenite > selenate.

Immediately after absorption, selenium in the algae is reduced to selenide via the sulfur reductive assimilation pathway, which is further used as a substrate for the synthesis of the amino acids selenocysteine (SeCys) ($C_3H_7NO_2Se$) and selenomethionine (SeMet) ($C_5H_{11}NO_2Se$) (Fig. 2.4) [66, 67].

While some microalgae can efficiently methylate the selenoamino acid produced, releasing volatile compounds (i.e., dimethyl selenide, dimethyl diselenide, and dimethyl selenyl sulfide), others may accumulate selenium in the form of Se^0 [66, 67]. For example, Luxem et al. [68] reported that in the marine algae *Emiliania huxleyi* (Lohm.) Hay and Mohler, the uptake of selenite was higher, when compared with selenate. Similar result was also observed in the *Chlamydomonas reinhardtii* P. A. Dangeard, a freshwater algae, by Vriens et al. [69]. They discovered that selenite was accumulated more than ten times selenate and the production of methylated volatile compounds in the freshwater algae studied. Other studies that have reported the accumulation ability of selenium by algae include *Chlorella vulgaris* Beijerinck [66], *Chara canescens* Loiseleur [70], *Cladophora hutchinsiae* (Dillwyn) Kützing [71], *Fucus vesciculosus* L., and *Fucus ceranoides* L. [72]. These findings provided indication that selenium methylation and bioaccumulation by microalgae could contribute immensely to its environmental cycling.

Selenium promotes the rate of growth and photosynthesis of algae at low concentration [73]. And in line with its function as an antioxidant [54], selenite facilitated the activities of the antioxidant enzymes, guaiacol peroxidase (GPX), catalase (CAT), and superoxide dismutase (SOD) in *C. vulgaris*, at concentration lower than or at 75 mg/L [73]. However, high concentration of selenium decreases algal growth in an aquatic ecosystem [74] and this is linked to damages in photosynthetic apparatus. Other toxic effects associated with elevated concentration of selenium in algae include inhibition of cell division and formation of malformed proteins [75]. Some plants are tolerant to the toxic effects of selenium, by accumulating and sequestering it in non-protein selenoamino acids. Non-selenium accumulating plants synthesizes selenocysteine and selenomethionine from selenate uptake and the selenocysteine formed is readily incorporated into proteins.

Selenium biomagnification in aquatic food chain was observed by Bennett et al. [76], in algae, rotifer, and larval fish, and this was ascribed to the availability of food and time. Selenium concentrations in the rotifers (*Brachionus calyciflorus* Pallas) ranged from 46 to 91 μg Se g^{-1} dry weight, after 5 h of feeding on selenium contaminated algae (*Chlorella pyrenoidosa* H. Chick). The feeding of the Se-contaminated rotifers to larval fish (*Pirnephales prornelas* Rafinesque) resulted in its accumulation of Se to 61.1 μg Se g^{-1} dry weight in the 9-day-old larvae and 51.7 μg Se g^{-1} in 17-day-old larvae, after 7 and 9 days, respectively.

Egg-laying vertebrates at the top of aquatic food chains are most at risk in environments with elevated aqueous selenium concentrations. Janz et al. [77] reported that the ability of selenium to induce embryo toxicity and teratogenicity could be linked to its substitution of sulfur in biota during protein synthesis. Sunde [78] pointed out that the substitution of sulfur in protein synthesis, by selenium, could result in dysfunctional proteins that causes deformities. Study on the bioaccumulation of selenium in fish revealed high concentration level of selenium in the spleen, liver, and kidney and low concentration in the muscle tissue [57].

2.4.2 Oxyanions of Arsenic

Relative to the other oxyanion-forming elements, arsenic (As) is among the most problematic in the environment because of its relative high mobility over a wide range of redox conditions. Worldwide, water pollution by arsenic is one of the most common environmental issues, resulting in high incidence of arsenicosis (Fig. 2.5) across the world [79, 80]. Arsenic has been reported to induce black foot disease, hyperglycemia, hyperkeratosis, as well as immune system dysfunction as a result of its bioaccumulation in the liver and kidney with the potential of causing cancer of the bladder, lung, kidney, and liver [81].

Several studies have shown the presence of the inorganic arsenic in large aquifers in various parts of the world, at concentrations higher than the World Health Organization (WHO) recommendation of 10 µg/L. Notable records of As occurrence have been reported in Argentina, Bangladesh, Chile, China, Hungary, India (West Bengal), Mexico, Romania, Taiwan, Vietnam, and many parts of the USA (Fig. 2.6) [83].

Arsenic is used as a doping agent in semiconductors (gallium arsenide) for solid-state devices such as semiconductors, light-emitting diodes, lasers, and a variety of transistors [84]. In industry, it is used to manufacture paints, fungicides, insecticides,

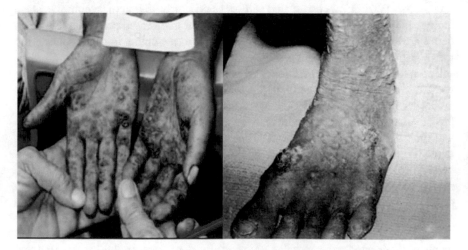

Fig. 2.5 Arsenical nodular keratosis disease (Indication of arsenic poisoning) [82]

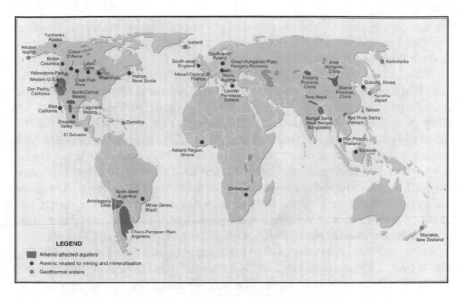

Fig. 2.6 Distribution of arsenic in groundwater in major aquifers in the world [83]

pesticides, herbicides, wood preservatives, and cotton desiccants [84]. It is also used in bronzing, pyrotechnics, and for hardening shot. Today, organoarsenic compounds are added to poultry feed to prevent disease and improve their weight gain [85].

The degree of toxicity of arsenic in the environment was identified to depend on its form [i.e., inorganic or organic form] and its oxidation state [86]. The predominant form of inorganic arsenics are its oxyanions: arsenate (As (V)), as $H_2AsO_3^-$ and $HAsO_4^{2-}$ in oxic environment, and arsenite (As (III)), as $H_3AsO_3^0$ and $H_2AsO_3^-$ under anoxic conditions [87]. Organic arsenic species available in contaminated surface and groundwater are monomethylarsenate (MMA) (CH_4AsO_4) and dimethylarsenate (DMA) ($C_2H_7AsO_4$) [88]. According to Smedley and Kinniburgh [83], the most important factor controlling the inorganic arsenic species formation is redox potential and pH of the environment. Such that in an oxidizing environment with low pH value, $H_2AsO_4^-$ dominates, while $HAsO_4^{2-}$ dominates at high pH value. At reducing condition however, arsenite $H_3AsO_3^0$ predominates (Fig. 2.7) [83].

Both arsenic oxyanions are reported to have different mobility levels in aquatic environment, due to their adsorption behaviors. While arsenite (As III) is relatively mobile, the arsenate (As V) species is strongly adsorbed onto aquifer material [89, 90], such as ferrihydrite and alumina. The adsorption vulnerability of arsenate infers that it will pose little risk to the aquatic environment [91].

Due to its structural similarity to phosphate [83], arsenate is taken up by the phosphate transport system in phytoplanktons and plants, leading to diminution in phosphate uptake in a competitive way [92]. As described by Levy et al. [93], the uptake of arsenate by these organisms may lead to the disruption of phosphorus metabolism, because it (arsenate) is incorporated into phosphorylated compounds that are vital for ATP cycling. Arsenite (As III) moves across plasma membrane via aquaglyceroporins (AQP) and hexose permeases causing toxicity by blocking sulfhydryl groups [94, 95]. This may result in membrane degradation and cell death through the production of reactive oxygen species.

Arsenate is readily taken up by plants because it is an analogue of phosphates. Lee et al. [96] reported that the uptake of arsenic in the aquatic plant *Hydrilla verticillata* (L. f.) Royle was inhibited by the presence of high concentration of phosphate. The study of Sanders et al. [97] elucidated the pathway of arsenic uptake and incorporation in aquatic organisms. Their findings provided evidence that phytoplankton in estuaries readily uptake and incorporate inorganic arsenic, resulting in growth reduction, even at low concentration. Uptake of inorganic arsenic into the soft tissues of copepod *Eurytemora affinis* Poppe, the barnacle *Balanus improvises* Darwin, and the oyster *Crassostrea virginica* Gmelin do not occur from seawater but via food chain, by feeding on arsenic contaminated phytoplankton. Their study revealed that arsenate is accumulated in calcareous shell of the barnacles and oysters [97].

Arsenic accumulates in the tissue of organism as organo-arsenic compounds [98]. The uptake of arsenic by invertebrate is done readily in the form of arseno-sugars, synthesized from arsenate by algae. The organo-arsenic taken in by animals is not incorporated into their tissues but eliminated unmetabolized as the organic arsenobetaine and other complex organoarsenic compounds [99]. Various studies has however affirmed that the lower trophic marine organisms are able to accumulate high level of arsenic in their tissues, while higher animals rapidly get rid of arseno-arsenics through urine [100].

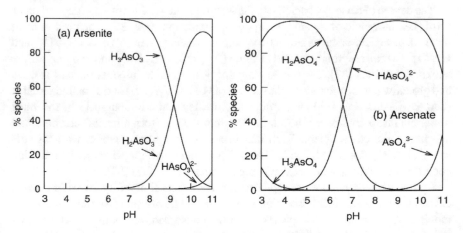

Fig. 2.7 **a** Arsenite and **b** arsenate speciation as a function of pH (ionic strength of about 0.01 M) [83]

2.4.3 Oxyanion of Chromium

The common chromium (Cr) containing mineral is chromite ($FeCr_2O_4$), with chromium occurring in the +3-oxidation state [101]. Major deposits of chromite occur in South Africa, Kazakhstan, Turkey, India, and the western hemisphere (i.e., Brazil and Cuba). Chromium occurs in several oxidation states from Cr (II) to Cr (VI), with trivalent and hexavalent states being the most stable [102]. The trivalent form, in trace amount, is essential in human nutrition, but toxic at high concentration [103]. It forms strong complexes with various ligands, with special affinity for ligands containing oxygen, nitrogen, and sulfur [104]. Its solubility is limited by the formation of highly insoluble oxide, hydroxide, and phosphates.

Chromium enters natural waters by weathering of chromium containing rocks, direct discharge from industrial operations, wet and dry depositions and leaching from soil. Chromium (III) and chromium (VI) are regulated in different ways because of their different toxicity. Chromium (III), in organic form, serves as a micronutrient essential for the growth of plant [105, 106]. In animals, including humans, it is required for sugar and fat metabolism, with estimated safe and adequate daily dietary intake as 50–200 µg [107]. Symptoms of deficiency in humans are reported by Anderson [108] and are similar to the symptoms reported for non-insulin-dependent diabetes mellitus and cardiovascular diseases. Chromium prevents and reverses atherosclerosis [109, 110], causes decrease in total cholesterol, and helps in controlling hypertension in rats [111].

Recently, the importance of chromium as an essential element is generating a lot of controversies. No adverse effects on body composition, glucose metabolism, or insulin sensitivity were observed in rats exposed to low dosage of Cr (III) when compared with rats exposed to sufficient dosage of Cr (III) [112]. Likewise, Cr (III) complexes accumulating in the body are reported to be potentially genotoxic [113]. The signs and symptoms of Cr (III) in man include: impaired glucose tolerance, glycosuria, hypoglycemia, elevated circulating insulin, decrease in insulin binding, peripheral neuropathy, encephalopathy, and low respiratory quotient [107, 108].

In the presence of excessive oxygen, it is oxidized to Cr (VI) [114], which is soluble and very mobile in water [115]. When inhaled or ingested, the hexavalent form is hematoxic, genotoxic, and carcinogenic [116]. The oxyanions of chromium in the +6 oxidation state are hydrochromate $HCrO_4^-$, chromate CrO_4^{2-}, and dichromate CrO_7^{2-}, which are not readily adsorbed to surfaces. The main oxyanions, chromate and dichromate, are involved in reversible transformation reaction [117].

$$2CrO_4^{2-} + 2H_3O^+ \leftrightarrow CrO_7^{2-} + 3H_2O$$

Chromium speciation is determined by the redox potential and pH condition of the environment where they occur. In an aquatic environment, Cr (VI) predominates under oxidizing conditions and Cr (III) predominates under more reducing conditions [118]. The species distribution of Cr (III) and Cr (VI) in an aqueous system is presented in Fig. 2.8 [119].

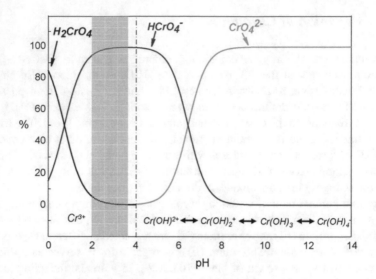

Fig. 2.8 Species distribution of Cr (III) and Cr (IV) in an aqueous system [119]

While Cr (III) is used in leather tanning because it forms stable complexes with amino groups in organic material, Cr (VI) chemicals are extensively used for metal plating, dyes, and paint pigments [120, 121]. Chromium (VI) is toxic to many plants and aquatic animals [122], severely damaging drinking water resources and labeled as human carcinogen [123]. Chromate (VI) species (i.e., chromate (CrO_4^{2-}) and dichromate ions (CrO_7^{2-}) are water soluble at all pH and are more mobile in the aquatic environment [116]. Both chromium species are adsorbed on iron and aluminum [hydro] oxides, organic substance and clay minerals [124]. A decrease in solution pH results in an increase in Cr (VI) and a decrease in Cr (III) adsorption [115].

Chromium (III) enters the cell of biotic organism through diffusion while chromium (VI) easily moves across the cell membrane. Other chromate anions are transported via the phosphate–sulfate carrier pathway [125]. Algae and plant species are negatively affected by chromium at high concentration with the toxicity depending on the type of algae and plant species. For instance, toxicity is reported at 1 μg/L in the diatom *Thalassiosira pseudonana* Cleve [126], 20 μg/L for *Chlorella pyrenoidosa*, 150 μg/L for *Ulothrix fimbriata* Bold up to 980 μg/L for *Skeletonema costatum*(Greville) Cleve [127]. The accumulation of Cr (VI) by the diatoms *Phaeodactylum tricornutum* Bohlin and *Navicula pelliculosa* (Kützing) Hilse was investigated by Hedayatkhah et al. [127]. The diatoms were tolerant to Cr (VI) at concentration up to 1 mg/L with significant decrease in growth yield at a concentration of 5 and 10 mg/L. Husien et al. [128] opined that *Chlorella sorokiniana* Shihira and R. W. Krauss is a good biosorbent material of Cr (VI) from solution with an efficiency of approximately 99.69% achieved at 100 ppm. At higher concentrations of Cr (VI) and after three days of exposure, a reduction in chlorophyll content followed by death of the *C. sorokiniana* was observed.

Although toxicity is reported by plants when exposed to chromite and chromate, bioaccumulation occurs with highest concentration in the root followed by the shoot [129, 130]. Macrophytes in aquatic habitat have been shown to accumulate Cr (VI) with highest accumulation occurring in the root. Studies on chromium accumulation in aquatic plant show that the roots of submerged plant species accumulate more chromium than the emerged plant species [130]. For example, Choo et al. [131] reported that water lilies (*Nymphaea spontanea* L.) are capable of accumulating Cr (VI) up to 2.119 mg/g from a 10 mg/L solution. Accumulation was highest at the root region followed by the leaves and petioles. Jana [132] observed that out of the estimated 19.23×10^{-6} M of Cr concentration in a solution, the roots of the floating water hyacinth (*Eichhornia crassipes* (Mart.) Solms) showed as high as 18.92 μmol (g dry tissue/wt) Cr accumulation, which was about two times higher than that in its shoots (1.5 μmol (g dry tissue/wt)). *Hydrilla verticillata* (L. f.) Royle, a submerged aquatic plant, is exposed to the same concentration of chromium, however accumulated lower concentration (9 μmol (g dry tissue/wt)) is in its roots.

While morphological stress, such as yellowing and chlorosis of the leaves, was observed in *N. spontanea* and correspondent reduction in chlorophyll, protein, and sugar contents [131], *E. crassipes* was able to accumulate Cr without any significant changes in the physiological and biochemical parameters such as Hill activity, protein, and chlorophyll content [132].

Similarly, in the emergent macrophyte species, *Bacopa monnieri* L. and *Scripus lacustris* L., Gupta et al. [5] found more uptake of Cr in the root of *B. monnieri* (1600 μg/g dw) than in *S. lacustris* L. (739 μg/g dw) at the same concentration and treatment duration. The translocation of chromium from the root to the shoot increased with an increase in total accumulation in the root. The study reported that at high concentration, the two test plants showed decrease in malondialdehyde and chlorophyll content.

Several studies have shown that dissolved chromium moved readily through the gill membrane of fish and accumulates in other organs of the fish. Fish has been reported to have erratic behavior when exposed to chromium-contaminated environment. For example, Rohu fingerlings (*Labio rohita* Hamilton) exhibited restlessness, decreased body balance, and higher rate of mucus secretions when exposed to 56.59 mg/L concentration of chromium [117]. This finding was also reported by Nisha et al. [133], where it was discovered that exposure of the zebra fish (*Danio rerio* Hamilton) to chromium resulted in mucus secretion, erratic swimming, and jerky movements by the fish.

Difficulty exists in distinguishing between the effects generated by Cr (VI) and Cr (III) since the former is rapidly reduced to the latter after penetration of biological membranes [134]. Since Cr (III) is not mobile, the reduction of Cr (VI) to Cr (III) in cells may be non-toxic [135]. However, intermediary products released during the reduction reaction may interrupt integrity and functioning of the cells [136]. The reduction of Cr (VI) to Cr (III) inside of cells may be an important mechanism for the toxicity of Cr compounds, whereas the reduction of Cr (VI) to Cr (III) outside of cells is a major mechanism of protection.

2.4.4 Oxyanion of Molybdenum

The oxyanions of molybdenum [Mo] are the molybdate anions and include MoO_4^{2-}, $Mo_2O_7^{2-}$ [137–142] and $Mo_8O_{26}^{4-}$ [143]. The oxidation state $+V$ and $+VI$ are prevalent in natural waters depending on the pH of the environment. According to Weidner and Ciesielczyk [30] at pH 5–6, MoO_4^{2-} is the dominant anion with concentration ranging from approximately 1 to 100 nM [144]. At lower pH, $HMo\ O_4^-$ dominates. At further lower pH and high molybdenum concentration, $Mo_7O_{24}^{6-}$ or $Mo_8O_{26}^{4-}$ occur (Fig. 2.9).

In the molybdate MoO_4^{2-} form, molybdenum is required as a cofactor in the synthesis of nitrogenase enzyme in nitrogen fixation and in nitrate reductase system.

Molybdenum (Mo) is a low toxic element, used as alloy material in steel production and as an inhibitor for steel corrosion [145, 146]. The two main sources of molybdenum are through mining and its occurrence as a by-product of copper mining [147]. China is the largest producer of molybdenum, producing about 130,000 metric tons, which is equivalent to almost 45% of the global output by volume in 2017 [148]. Other important producing countries are Chile, USA, Peru, and Mexico [149]. Detailed overall end-use statistics of molybdenum by the European Union is provided in Fig. 2.10. Molybdenum oxidation states ranges from -2 to $+6$, but the $+4$, $+5$, and $+6$ are the most important in the environment. The variable oxidation states are the reason for its involvement in a variety of redox reactions. It is required in trace amount for the growth and development of plants, animal, and humans. In mammals, it forms part of a complex called molybdenum cofactor in the flavoprotein xanthine oxidase and aldehyde oxidase enzymes and the heme protein sulfite oxidase enzyme [150].

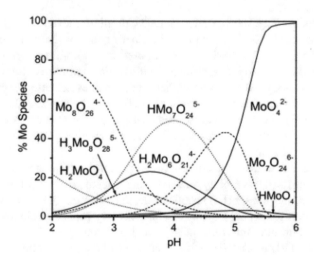

Fig. 2.9 Mo speciation as a function of pH at initial [Mo] of 10 nM [30]

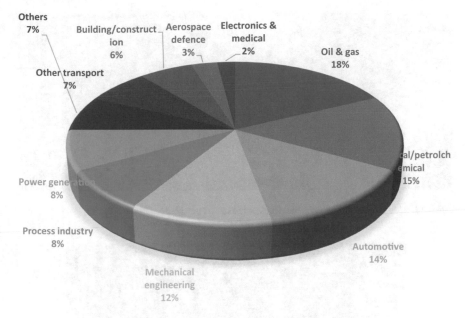

Fig. 2.10 Detailed overall end-use statistic of molybdenum [149]

While xanthine oxidase is involved in purine metabolism, by catalyzing the oxidative hydroxylation of purine substrates and a subsequent increase in the formation of uric acid or urate, aldehyde oxidase oxidizes various aldehydes, purines, and pyrimidines variety of heterocyclic compounds and xenobiotics [150]. Sulfite oxidase enzyme is involved in the metabolism of sulfur-containing amino acids by converting sulfite to inorganic sulfate [151].

Although molybdenum deficiency has not been identified in free living human or animal species, deficiency symptoms observed in patients receiving total parenteral nutrition include tachycardia, tachypnea, severe bifrontal headache, night blindness, nausea, vomiting, central scotomas, periods of lethargy, disorientation, and coma [152]. Its deficiency had also been linked to increase incidence of esophageal cancer [153] and dental caries [154] cases in humans. High intake of molybdenum has been associated with anemia, liver and kidney abnormalities, sterility, bone and joint deformities in human. Also, implicated is the increase in level of uric acid in blood and urine due to the action of the enzyme xanthine oxidase with high incidence of gout as shown in Fig. 2.11 [155].

High dietary intake in combination with low copper is associated with molybdenum poisoning [156, 157] [molybdenosis or teart], a condition recognized in grazing cattle (Fig. 2.12).

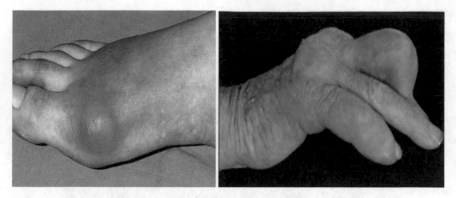

Fig. 2.11 Gout formation in man due to elevated levels of uric acid in blood [155]

Fig. 2.12 Molybdenum toxicity: depigmented coat, alopecia, and periocular gray hair [157]

This condition includes a variety of symptoms and signs including bone disorders, diarrhea, rough hair coat, hair color change, and in extreme cases, death [157]. Molybdenosis could however be controlled by treatment with copper sulfate [157].

Two competitive inhibitors of the transport and uptake of $MoO_4{}^{2-}$ across cells are sulfate and tungstate that are most similar to it in structure. Shah et al. [158] and Cole et al. [159] reported a mole-for-mole inhibition of molybdate by tungstate. Hence, the availability of molybdenum is affected by the presence of sulfate. The concentration of molybdenum reported in freshwater is approximately 0.001 µg/ml [160]. Eisler [161] stated that molybdenum concentration in surface water do not exceed 0.020 µg/ml.

Molybdenum acts as a cofactor for nitrogenase enzyme responsible for catalyzing nitrogen fixation, i.e., the reduction of nitrogen (N_2) to ammonia (NH_3) [162] and nitrate reductase enzyme responsible for NO_3^- assimilation. Studies have shown that N_2 fixation requires more Mo than NO_3^- assimilation, while NH_4^+ do not require molybdenum for assimilation [163, 164]. The uptake of molybdenum in oxic aquatic environment is inhibited by the presence of sulfate and tungstate. According to Howarth and Cole [165], SO_4^{2-} is a competitive inhibitor of MoO_4^{2-} transport, thus lowering molybdenum availability to N_2 fixing organisms.

Molybdenum concentrations increase with salinity, and therefore, its concentrations are low in most freshwaters [<20 nM], when compared with seawater (107 nM) [166]. This explains the high affinity for molybdenum by algae in freshwater ecosystem. For example, the algae Anabaena had optimal growth when exposed to molybdenum at concentration in the range of 50–2000 nM. Also, when exposed to molybdenum at concentration higher than the ambient <20 nM of freshwater, heterocystous cyanobacteria are able to accumulate molybdenum to >100 μmol/mol without observable adverse effect [164].

The presence or absence of a molybdenum transport system in algae is reported to have effect on its ability to assimilate molybdenum. *Chlamydomonas reinhardtii* with the MoO_4^{2-} uptake system is less susceptible to molybdenum limitation compared to *Navicula pelliculosa* in aquatic system [167]. The presence of such transport system enables organisms to cope with molybdenum limitation for longer period of time.

Toxicity effects on aquatic organisms as a result of exposure to molybdenum have not been reported. The exposure of the marine microalgae *Isochysis galbana* Parke, to dissolved molybdenum of concentration up to 9500 μg/L, did not show any observable toxicity effect [168]. In recent studies, no adverse effect was reported when the larvae of the barnacle *Amphibalanus amphitrite* Darwin were exposed to high concentration of molybdenum (E_rC_{10} >10 mg Mo/L) [169]. Further study with tropical marine snail *Nassarius dorsatus* Röding showed that at concentration of ≤7000 μg/L Mo exposure, the survival and growth rate of the larval of the snail was not affected adversely [170].

In generating chronic ecotoxicity data for molybdenum for freshwater and marine species, the freshwater amphipod *Hyalella azteca* Saussure and marine atherinid fish *Menida beryllina* Cope were exposed to high concentration of molybdenum, using sodium molybdate dihydrate as test substance [171]. No significant adverse effects were observed on the survival of *H. azteca* at ≥245.1 mg Mo/L. The total weight of *H. azteca* was also not affected at concentration up to 103.6 mg Mo/L. A statistically significant effect was however observed in adult reproductive performance at highest test concentration (1070 mg Mo/L). In the test organism *Menida beryllina*, no statistically significant differences on survival were recorded when compared with the control at molybdenum concentration up to 1070 mg Mo/L. While overall hatching success in the control was 93%, it ranged from 95 to 99% in the test substance treatments. A significant reduction in standard length and blotted wet weight of *M. beryllina* exposed to 265, 532, and 1070 mg Mo/L was recorded when compared to the control ($p < 0.05$).

In young Chinook salmon (*Oncorhynchus tshawytscha* Walbaum in Artedi), molybdenum was not detected in fish exposed to concentrations as high as 193 μg molybdenum/L for 90 d in well water or freshwater. These results were similar to the findings of Ohlendorf et al. [172] and Saiki and May [173], who reported that molybdenum was not detected in fish from Kesterson Reservoir or Salt Slough. These results suggest that molybdenum seem not to bioaccumulate in fish, therefore may not pose adverse effects on its populations [174].

In the observation of Saiki et al. [175], concentrations of molybdenum were highest in detritus and lesser amounts in filamentous algae. Sediment, invertebrates, and fishes were reported to contain uniformly low concentrations of Mo, with lowest measurable concentrations of molybdenum occurring in water [175]. Hence, it was postulated that molybdenum does not bioaccumulate.

2.4.5 Oxyanions of Nitrogen

The common oxyanions of nitrogen found in the environment include nitrates (NO_3^-) and nitrite (NO_2^-), containing a central nitrogen atom enclosed by three identically bonded oxygen atoms in a trigonal planar organization with a net charge of -1. As shown in the Pourbaix diagram (Fig. 2.13) [176], pH affects the forms of nitrogen

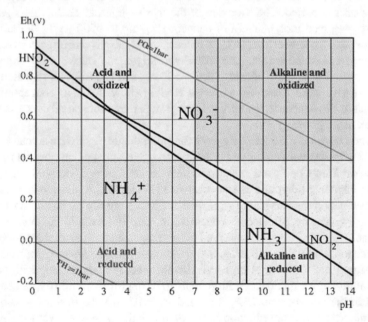

Fig. 2.13 Pourbaix diagram of nitrogen (N) representing the various forms of N in a 100 μM solution at 25 °C as a function of Eh (in V) and pH. Diagram adapted from MEDUSA Software. Puigdomenech 2009–2011 [176]

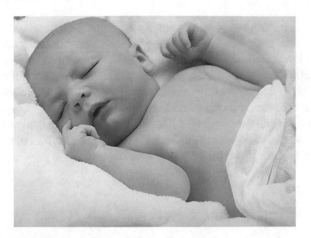

Fig. 2.14 Methemoglobinemia in human [178]

that predominates in an aquatic medium. In oxidized conditions (Eh >500 mV at pH 7), the thermodynamically stable species of nitrogen is NO_3^-, whereas under reduced or moderately oxidized conditions (Eh <400 mV at pH 7). As NO_3^- is soluble, redox potential and pH mainly influence the form under which N is assimilated by plants.

Nitrate contamination in natural water bodies has become an increasingly serious environmental problem around the world, mainly because of the extensive use of chemical fertilizers, improper treatment of wastewater from industrial and municipal sites, landfills, and animal wastes (particularly from animal farms). High levels of nitrite and nitrate contaminants in drinking water have been linked with the blue baby syndrome (methemoglobinemia) in children (Fig. 2.14) [177, 178].

The mechanism of the process is based on the fact that nitrate binds with hemoglobin to form methemoglobin (MHb), a substance that does not enable oxygen transport to tissues, thereby causing asphyxia (lack of oxygen) and cyanosis of body tissues. A Colombian hospital reported a case of methemoglobinemia in 1998 as a result of preparing a baby's food with well water contaminated with 22.9 mg/L nitrate-N [177, 179]. Ayebo et al. [180] found methemoglobinemia incidence rates ranging from 24 to 363 cases per 100,000 live births in the Transylvania region of Romania between 1990 and 1994. In Poland, 90% of 239 investigated cases of infant methemoglobinemia were reported to be associated with ingestion of nitrate-N contaminated water [181]. Studies conducted in China [182], Spain [183] and Taiwan [184] showed that long-term ingestion of nitrate-contaminated water can increase the risk of gastric cancer.

The proposed mechanism was said to involve the conversion of nitrate to nitrite, which is followed by the transformation of nitrite to nitrosamines. The European studies recommended a maximum nitrate NO_3^- and nitrite NO_2^- concentration of 10 and 0.03 mg/L in aquatic system, respectively.

Apart from the toxic effect on human, excess nitrate concentration in aquatic system causes algae boom. Mainly, excess nitrate in the aquatic system serves as source of fertilizer for aquatic plants and algae. In many circumstances, the amount of nitrate in the water is what determine the extent of plant and algae growth. If there is a superfluous level of nitrates, plants and algae will grow extremely well. This process reduces the dissolved oxygen in aqua system which creates stressful conditions for aquatic life. High densities of algae also create a condition where sunlight cannot penetrate into the water. Since aquatic lives require some sunlight, aquatic lives not receiving enough sunlight will die off. These dead aquatic lives increase the activities of microorganisms and organic matter in the system. The activities of microorganism in the system further reduce the water dissolve oxygen.

2.4.6 Oxyanions of Phosphorus

Phosphate $(PO_4)^{3-}$, an oxyanion of phosphorus, contains a central phosphorus atom enclosed by four oxygen atoms in a tetrahedral arrangement. Phosphate is an important plant nutrient and has found use in fertilizer production. It is also used as precursor in production of many industrial products including detergents, animal feed, beverages, food, surface treatment (metal coating and cleaning), water treatment, tooth pastes, fire extinguishers among many others. It has been observed that large chunk of the applied fertilizers is lost into water bodies, through runoffs [185]. In aquatic system, phosphate exists in different forms as a result of the variation in the pH value of the medium [186].

As described in Fig. 2.15, $H_2PO_4^-$ and HPO_4^{2-} species occurred at pH values that ranged between 5 and 10. The concentration of $H_2PO_4^-$ species is higher for pH value below 7, while HPO_4^{2-} species predominate over other species at pH values that ranged between 7 and 10. At pH values that ranged between 10 and 12, HPO_4^{2-} dominate over PO_4^{3-} species, while at pH values higher than 12.5, the concentration of PO_4^{3-} was more copious. Phosphate availability for plant uptake in aqua system depends on the species of phosphorus in abundant in the system.

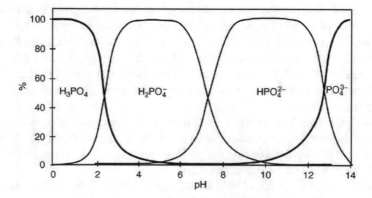

Fig. 2.15 Relative distribution of phosphate species as a function of pH [187]

Fig. 2.16 Effects of phosphorus pollution on aquatic ecosystem [190, 191]

A study conducted by Mekonnen and Hoekstra [188] observed that human activity discharges 1.47 Tg (1.62 million US tons) of phosphorus into the world's freshwater bodies annually. China was found to contribute 30% of the freshwater phosphorus load, followed by India at 8% and the USA at 7%. The largest contribution to the global phosphorus pollution was observed to come from domestic sewage, at 54%, followed by agriculture, at 38%, and industry, at 8%.

Excess phosphates in water results in an enormous growth of macrophytes and algae, which are aquatic plants that include several single-celled, free-floating plants [189]. Extreme amounts of algae cloud the water in an effect called algal bloom (eutrophication) (Fig. 2.16) [190, 191] which reduces the sunlight penetration into the aquatic system and also reduce dissolved oxygen. The reduction in the available dissolved oxygen affects aquatic lives and in severe cases, the marine lives die. When the algae and marine lives die, the bacteria that break them down use up the residual dissolved oxygen in the water, depriving and suffocating other aquatic life [192, 193].

2.4.7   Vanadium Oxyanions

Vanadium is a transition metal that can exist in different oxidation states that ranged from $+1$ to $+5$, with the most common being $+3$, $+4$, and $+5$ due to it redox-sensitive nature [194]. The most stable oxidation state is vanadyl salts (VO^{2+}). Trivalent vanadium (V_2O_3) is a strong reducing agent that dissolves in acid to form a green hexa-aqua ion. Other forms of vanadium oxyanion include metavanadate (VO_3^-), orthovanadate ($H_2VO_4^-$), and pyrovanadate ($V_2O_7^{-4}$). Vanadium exists as vanadyl (VO^{2+}), at pH below 3.5 while in basic medium, it occurs as orthovanadate (VO_4^{3-}), which is chemically similar to phosphates (PO_4^{3-}) [195]. Vanadium occurs as $H_2VO_4^-$ in neutral solutions.

It has been observed that vanadium exists in different hydrolyzed species, subject to the pH of the system [196]. In an aquatic environment, vanadium occurs mostly in the presence of oxygen, creating several oxyanions of different oxidation states [197]. Under acidic conditions (pH < 3), vanadium (V) exists in cationic form as VO_2^+, whereas in the neutral–alkaline (pH = 4–11) they occur in neutral ($VO(OH)_3$) and anionic forms, including decavanadate species ($V_{10}O_{26}(OH)_2^{4-}$, $V_{10}O_{27}(OH)^{5-}$ and $V_{10}O_{28}^{6-}$) and mono- or polyvanadate species (e.g., ($VO_2(OH)_2^-$, $VO_3(OH)^{2-}$, VO_4^{3-}, and $V_2O_6(OH)^{3-}$, $V_2O_7^{4-}$, $V_3O_9^{3-}$ $V_4O_{12}^{4-}$) system (Fig. 2.17) [196].

Commercial vanadium-containing ores include varnotite (potassium uranyl vanadate), patronite (impure vanadium sussslfide), roscoelite, and vandinite as well as phosphate rock [30].

Vanadium has found use in industries like glass, photography, rubber, textile, ceramic, mining, pigments, metallurgy, automobile, oil refining, and in the production of inorganic chemicals [198, 199]. This vast area of applications results in the production of huge amounts of vanadium polluted effluent from industries, which are discharged into aqua systems without proper treatment. Vanadium has been recognized as a potentially dangerous pollutant in the same class as lead, mercury, and arsenic [200]. The level of vanadium in freshwater hinge on geographical difference in leachates and effluents from both natural and anthropogenic sources. Concentrations up to 220 µg V/L has been detected in rivers in the Colorado Plateau, which contains vanadium associated with naturally occurring uranium ores [201]. Some

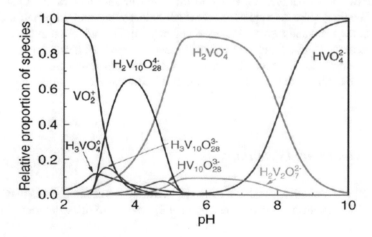

Fig. 2.17 Distribution of vanadium (V) species as a function of pH [196]

water with high mineral content may contain concentrations up to 10 μg V/L, in conjunction with other elements, such as fluorides and arsenic [201].

The only obvious documented effect of vanadium toxicity is the upper respiratory tract irritation, characterized by nasal hemorrhage, rhinitis, cough, wheezing, conjunctivitis, sore throat, and chest pain [202]. These symptoms, particularly cough and rhinitis, are noticeable in boilermakers, who are exposed to high concentrations of V_2O_5 fumes during the cleaning of the boilers; hence, the name "boilermakers" bronchitis [202]. Eight (8) case studies suggested the likelihood of asthma being developed after heavy exposure to vanadium compounds [203].

Vanadium has been found to be weakly mutagenic [204] and there is a possibility of carcinogenicity because vanadium interferes with mitosis and chromosome distribution. In vitro studies indicated that vanadium pentoxide is not cytotoxic, but this compound induces rapidly repairable DNA damage (single-strand breaks) [205].

It has been observed that vanadate (VO_3^-), vanadyl (VO^{2+}), can have a large effect on the function of a range of enzymes, either as an activator or inhibitor of the enzyme functions (206). Pentavalent vanadium (V_5O_{14}) is especially harmful to human health—it causes damage to the respiratory, gastrointestinal, and central nervous systems and disturbs metabolism [197]. The International Agency for Research on Cancer had classified vanadium pentoxide as a possible carcinogen. Presently, vanadium is on the United States Environmental Protection Agency (USEPA), Drinking Water Contaminant Candidate List (CCL3), due to its potential carcinogenic effects [206]. Maximum concentrations of vanadium in drinking water ranged from about 0.2 to 100 μ g/L, with typical values ranging from 1 to 6 μ g/L [196].

2.5 Conclusion

Ecological imbalance created by the introduction and accumulation of oxyanions in the ecosystem is colossal. Although some are beneficiary at low concentration, the predisposition to be toxic at higher concentration is high. The effects of oxyanions on the stability of the ecological community are predicated mostly on its movement through the food chain and the food web, accentuated by its bioaccumulation in the biological organism. The toxicity of most oxyanions is linked to the form in which they occur and their oxidation states. Another important factor that determines their toxicity is the pH value of the medium in which they occur.

Removal of organisms, due to the deleterious effects of oxyanions, or addition of organism, due to nutrient enrichment, has serious consequences on the food web and stability of the ecosystem. Disruption and alteration of many life processes and functioning in aquatic organisms have been linked to the toxic effects of oxyanions at higher concentrations. The accepted level of different contaminants in aquatic ecosystem has been published by different regulatory agencies but both natural and human interferences tend to increase the level of oxyanions pollutants, inadvertently in aquatic ecosystems.

References

1. Long ER, Macdonald DD, Smith SL, Calder FD (1995) Incidence of adverse biological effects within ranges of chemical concentrations in marine and estuarine sediments. Environ Manage 19(1):81–97
2. Loya Y (1975) Possible effects of water pollution on the community structure of red sea corals. Mar Biol 29:177–185
3. Sarwar N, Imran M, Shaheen MR, Kamran MA, Matloob A, Rehim A, Ishaque W, Hussain S (2017) Phytoremediation strategies for soils contaminated with heavy metals: modifications and future perspectives. Chemosphere 171:710–721
4. He TR, Feng XB, Guo YN, Qiu GL (2008) The impact of eutrophication on the biogeochemical cycling of mercury species in a reservoir: a case study from Hongfeng Reservoir, Guizhou, China. Environ Pollut 154:56–67
5. Gupta M, Sinha S, Chandra P (1994) Uptake and toxicity of metal in *Scirpus lacustris* L. and *Bacopa monnieri* L. J Environ Sci Health [A 29] 10:2185–2202
6. Hajji S, Montes-Hernandez G, Sarret G, Tordo A, Morin G, Ona-nguema G, Bureau S, Turkia T, Mzoughi N (2019) Arsenite and chromate sequestration onto ferrihydrite, siderite and goethite nanostructured minerals: Isotherms from flow-through reactor experiments and XAS measurements. J Hazard Mater 362:358–367
7. Razzaque MS (2011) Phosphate toxicity: new insights into an old problem. Clin Sci 120:91–97
8. Raju NJ, Schroeter MI, Kofod M (2007) Mobility and retention of oxoanions in iron hydroxide sandy aquifers—batch and column tests. Indian J Geochem 22(2):257–273
9. Wang Y, Ding S, Shi L, Gong M, Xu S, Zhang C (2017) Simultaneous measurements of cations and anions using diffusive gradients in thin films with a ZrO-Chelex mixed binding layer. Anal Chim Acta 972:1–11
10. Bruno JF, Bertness MD (2001) Habitat modification and facilitation in benthic marine communities. In: Bertness M, Gaines S, Hay M (eds) Marine community ecology. Sinauer Associates, Massachusetts, USA, pp 201–218
11. Loreau M (2010) From populations to ecosystems: theoretical foundations for a new ecological synthesis, No. 46. In: Monographs in population biology. Princeton University Press, Princeton, NJ, oCLC: 699867402
12. Webster JR, Waide JB, Patten BC (1945) Nutrient cycling and the stability of ecosystems. In: Howell FG, Gentry JB, Smith MH (eds), Mineral cycling in Southeastern ecosystems. ERDA conference 740513, National Technical Information Service, US Department of Commerce, Springfield, VA, pp 1–27
13. Holing CS (1973) Resilience and stability of ecological systems. Annu Rev Ecol Syst 4:1–23
14. Odum EP (1969) The strategy of ecosystem development. Science 164:262–270
15. Jordan F, Kline JR, Sasscer DS (1972) Relative stability of mineral cycles in forest ecosystems. Am Nat 106:237–253
16. Waide JB, Krebs JE, Clarkson SP, Setzler EM (1974) A linear systems analysis of the calcium cycle in a forested watershed ecosystem. In: Rosen R, Snell PM (eds) Progress in theoretical biology, vol 3. Academic Press, Inc., New York, pp 261–345
17. Attayde JL, Ripa J (2008) The coupling between grazing and detritus food chains and the strength of trophic cascades across a gradient of nutrient enrichment. Ecosystems 11:980–990
18. Kotliar NB (2000) Application of the new keystone-species concept to prairie dogs: how well does it work? Conserv Biol 14:1715–1721
19. Delibes-Mateos M, Smith AT, Slobodchikoff CN, Swenson JE (2011) The paradox of keystone species persecuted as pests: a call for the conservation of abundant small mammals in their native range. Biol Cons 144:1335–1346
20. Hale SL, Koprowski JL (2018) Ecosystem-level effects of keystone species reintroduction: a literature review. Restor Ecol 26:439–445
21. McLaren BE, Peterson RO (1994) Wolves, moose and tree rings on Isle Royale. Science 266:1555–1558

22. Kersten M, Vlasova N (2013) The influence of temperature on selenite adsorption by goethite. Radiochim Acta 101:413–419
23. Peshut PJ, Morrison RJ, Brooks BA (2008) Arsenic speciation in marine fish and shellfish from American Samoa. Chemosphere 71(3):484–492
24. Verbinnen B, Block C, Van Caneghem J, Vandecasteele C (2015) Recycling of spent adsorbents for oxyanions and heavy metal ions in the production of ceramics. Waste Manage 45:407–411
25. Mon M, Bruno R, Ferrando-Soria J, Armentano D, Pardo E (2018) Metal-organic framework technologies for water remediation: towards a sustainable ecosystem. J Mater Chem A 6:4912–4947
26. Ungureanu G, Filote C, Santos SCR, Boaventura RAR, Volf I, Botelho CMS (2016) Antimony oxyanions uptake by green marine macroalgae. J Environ Chem Eng 4:3441–3450
27. Gupta VK, Fakhri A, Kumar A, Agarwal S, Naji M (2017) Optimization by response surface methodology for vanadium [V] removal from aqueous solutions using PdO-MWCNTs nanocomposites. J Mol Liq 234:117–123
28. Hristovski KD, Markovski J (2017) Science of the total environment engineering metal [hydr]oxide sorbents for removal of arsenate and similar weak-acid oxyanion contaminants: a critical review with emphasis on factors governing sorption processes. Sci Total Environ 598:258–271
29. Kołodyńska D, Budnyak TM, Hubicki Z, Tertykh VA (2017) Sol–gel derived organic—Inorganic hybrid ceramic materials for heavy metal removal. In: Mishra AK (ed) Sol-Gel based nanoceramic materials: preparation, properties and applications. Springer International Publishing, Berlin, Germany, pp 253–274
30. Weidner E, Ciesielczyk F (2019) Removal of hazardous oxyanions from the environment using metal-oxide-based materials, review. Materials 12:927
31. Latifian M, Holst O, Liu J (2013) Nitrogen and phosphorus removal from urine by sequential struvite formation and recycling process. Clean-Soil Air Water 41:1–5
32. Xu H, He P, Gu W, Wang G, Shau L (2012) Recovery of phosphorus as struvite from sewage sludge ash. J Environ Sci 24:1533–1538
33. Mehta CM, Hunter MN, Leong G, Batstone DJ (2018) The value of wastewater derived struvite as a source of phosphorus fertilizer, WILEY-VCH Verlag GmbH & Co. KGaA, Weinheim. https://doi.org/10.1002/clen.201700027
34. Huang H, Song Q, Wang W, Wu S, Dai J (2012) Treatment of anaerobic digester effluents of nylon wastewater through chemical precipitation and a sequencing batch reactor process. J Environ Manage 30:68–74
35. Khai NM, Tang HTQ (2012) Chemical precipitation of ammonia and phosphate from Nam son landfill leachate. Hanoi Iran J Energy Environ 3:32–36
36. Kumar R, Pal P (2015) Assessing the feasibility of N and P recovery by struvite precipitation from nutrient-rich wastewater: a review. Environ Sci Pollut Res 22(22):17453–17464
37. Foletto EL, dos Santos WRB, Mazutti MA, Jahn SL, Gundel A (2013) Production of struvite from beverage waste as phosphorus source. Mater Res 16:242–245
38. Pradhan SK, Mikola A, Vahala R (2017) Nitrogen and phosphorus harvesting from human urine using a stripping, absorption, and precipitation process. Environ Sci Technol 51:5165–5171
39. Lauchli A (1993) Selenium in plants: uptake, functions, and environmental toxicity Botanica Acta 106:455–468
40. Sunde RA, Selenium. In: O'Dell BL, Sunde RA (ed) Handbook of nutritionally essential mineral elements. Marcel Dekker, New York, pp 493–556
41. Unrine JM, Jackson BP, Hopkins WA (2007) Selenomethionine biotransformation and incorporation into proteins along a simulated terrestrial food chain. Environ Sci Technol 41:3601–3606
42. Haygarth PM (1994) Global importance and global cycling of selenium. In: Frankenberger Jr WT, Benson S (eds) Selenium in the environment. Marcel Dekker, New York, pp 1–27

43. Rosenfeld I, Beath OA (1964) Selenium: geobotany, biochemistry, toxicology and nutrition. Academic Press, New York
44. Health and Welfare Canada Selenium (1980) Guidelines for Canadian drinking water quality 1978: supporting documentation. Canada, Hull, Supply & Services, pp 541–554
45. Lemly AD (2004) Aquatic selenium pollution is a global environmental safety issue. Ecotoxicol Environ Saf 59:44–56
46. Geering HR, Cary EE, Jones LHP, Allaway WH (1968) Solubility and redox criteria for the possible forms or' selenium in soils. Soil Sci Soc Am Proc 32:35–40
47. Cohen R, Schuhmann D, Sinan F, Vanel P (1992) The role of organic coatings in the enrichment of marine particles with selenium. The fixation of selenite on an adsorbed amino acids. Mar Chem 40:249–271
48. Mikkelsen RL, Page AL, Bingham FT (1989) Factors affecting selenium accumulation by agricultural crops. In: Jacobs LW (ed) Selenium in agriculture and the environment. American Society Agronomy, Madison, Wisconsin, pp 65–94
49. Hatfield DL, Gladyshev VN (2002) How selenium has altered our understanding of the genetic code. Mol Cell Biol 22(11):3565–3576
50. Bulteau AL, Chavatte L (2015) Update on selenoprotein biosynthesis. Antioxid Redox Signal 23:775–794
51. Lu J, Holmgren A (2008) Selenoproteins. J Biol Chem 284:723–727
52. Andreesen JR, Ljungdahl L (1973) Formate dehydrogenase of Clostridium thermoaceticum:incorporation of selenium-75, and the effect of selenite, molybdate and tungstate on the enzyme. J Bacteriol 116:867–873
53. Zhang Y, Romero H, Salinas G, Gladyshev VN (2006) Dynamic evolution of selenocysteine utilization in bacteria: a balance between selenoprotein loss and evolution of selenocysteine from redox active cysteine residues. Genome Biol 7(10)
54. Pappas AC, Zoidis E, Surai PF, Zervas G (2008) Selenoproteins and maternal nutrition. Comp Biochem Physiol B: Biochem Mol Biol 151:61–372
55. Combs GF, Lu L (2001) Selenium as a cancer preventive agent. In: Hatfield DL (ed) Selenium: its molecular biology and role in human health. Kluwer Academic Publishers, Norwell, Mass, pp 205–217
56. Beck MA (2001) Selenium as an antiviral agent. In Hatfield DL (ed) Selenium: its molecular biology and role in human health. Kluwer Academic Publishers, Norwell, Mass, pp 235–245
57. Ogle RS, Maier KJ, Kiffney P, Williams MJ, Brasher A, Melton LA, Knight AW (1988) Bioaccumulation of selenium in aquatic ecosystems. Lake Reservoir Manage 4(2):165–173
58. Schiavon M, Pilon-Smits EAH (2016) The fascinating facets of plant selenium accumulation—biochemistry, physiology, evolution and ecology. New Phytol 213(4):1582–1596
59. Baines SB, Fisher NS (2001) Interspecific differences in the bioconcentration of selenite by phytoplankton and their ecological implications. Mar Ecol Prog Ser 213:1–12
60. Foda A, Vandermeulen J, Wrench JJ (1983) Uptake and conversion of selenium by a marine bacterium. Can J Fish Aquat Sci 40(supplementary 2):215–220
61. Postgate J (1949) Competitive inhibition of sulfate reduction by selenate. Nature 164:670–671
62. Chapman PM, Adams WJ, Brooks ML, Delos CG, Luoma SN, Maher WA, Ohlendorf HM, Presser TS, Shaw DP (2010) Ecological assessment of selenium in the aquatic environment. Pensacola, FL, SETAC, p 339
63. Boisson F, Gnassia BM, Romeo M (1995) Toxicity and accumulation of selenite and selenate in the unicellular marine alga *Cricosphaera elongata*. Arch Environ Contam Toxicol 28:487–493
64. Vriens B, Behra R, Voegelin A, Zupanic A, Winkel LHE (2016) Selenium uptake and methylation by the microalga *Chlamydomonas Reinhardtii*. Environ Sci Technol 50:711–720
65. Kiffney P, Knight A (1990) The toxicity and bioaccumulation of selenate, selenite and selenoi-methionine in the cyanobacterium *Anabaena flos-aquae*. Arch Environ Contam Toxicol 19:488–494
66. Neumann PM, De Souza MP, Pickering IJ, Terry N (2003) Rapid microalgal metabolism of selenate to volatile dimethylselenide. Plant Cell Environ 26:897–905

67. Schiavon M, Ertani A, Parrasia S, Vecchia FD (2017) Selenium accumulation and metabolism in algae. Aquat Toxicol 189:1–8
68. Luxem KE, Vriens B, Wagner B, Behra R, Winkel LHE (2015) Selenium uptake and volatilization by marine algae. In: EGU general assembly conference abstract 17 No. EGU 2015–6613
69. Vriens B, Mathis M, Winkel LHE, Berg M (2015) Quantification of volatile alkylated selenium and sulfur in complex aqueous media using solid-phase microextraction. J Chromatogr A 1407:11–20
70. Lin ZQ, De Souza M, Pickering IJ, Terry N (2002) Evaluation of the macroalga, muskgrass, for the phytoremediation of selenium-contaminated agricultural drainage water by microcosms. J Environ Qual 31:2104–2110
71. Tuzen M, Sari A (2010) Biosorption of selenium from aqueous solution by green algae [*Cladophora hutchinsiae*] biomass: equilibrium, thermodynamic and kinetic studies. Chem Eng J 158:200–206
72. Turner A (2013) Selenium in sediments and biota from estuaries of southwest England. Mar Pollut Bull 73:192–198
73. Sun X, Zhong Y, Huang Z, Yang Y (2014) Selenium accumulation in unicellular green alga *Chlorella vulgaris* and its effects on antioxidant enzymes and content of photosynthetic pigments. PLoS ONE 9:e112270
74. Zheng Y, Li Z, Tao M, Li J, Hu Z (2017) Effects of selenite on green microalga *Haematococcus pluvialis*: bioaccumulation of selenium and enhancement of astaxanthin production. Aquat Toxicol 183:21–27
75. Vallentine P, Hung CY, Xie J, Van Hoewyk D (2014) The ubiquitin-proteasome pathway protects *Chlamydomonas reinhardtii* against selenite toxicity, but is impaired as reactive oxygen species accumulate. AOB Plants 6:1–11
76. Bennett WN, Brooks AS, Boraas ME (1986) Selenium uptake and transfer in an aquatic food chain and its effects on fathead minnow larvae. Arch Environ Contam Toxicol 15:513–517
77. Janz DM, DeForest DK, Brooks ML, Chapman PM, Gilron G, Hoff D, Hopkins WD, McIntyre DO, Mebane CA, Palace VP, Skorupa JP, Wayland M (2010) Selenium toxicity to aquatic organisms. In: Chapman PM, Adams WJ, Brooks ML, Delos CG, Luoma SN, Maher WA, Ohlendorf HM, Presser TS, Shaw DP (eds) Ecological assessment of selenium in the aquatic environment. CRC Press, Boca Raton, FL, pp 141–231
78. Sunde RA (1984) The biochemistry of selenoproteins. J Am Org Chem 61:1891–1900
79. Mandal BK, Suzuki KT (2002) Arsenic round the world: a review. Talanta 58(1):201–235
80. Paikaray S (2012) Environmental hazards of arsenic associated with black shales: a review on geochemistry, enrichment and leaching mechanism. Rev Environ Sci Bio/Technol 11(3):289–303
81. Smith AH, Hopenhayn-Rich C, Bates MN, Goeden HM, Hertz-Picciotto I, Duggan IIM, Wood R, Kosnett MJ, Smith MT (1992) Cancer risks from arsenic in drinking water. Environ Health Perspect 97:259–267
82. Mazumder DNG (2008) Chronic arsenic toxicity and human health. Indian J Med Res 128:436–447
83. Smedley PL, Kinniburgh DG (2002) A review of the source, behavior and distribution of arsenic in natural waters. Appl Geochem 17:517–568
84. Ratnaike RN (2003) Acute and chronic arsenic toxicity. Postgrad Med J 79:391–396
85. Mangalgiri KP, Adak A, Blaney L (2015) Organoarsenicals in poultry litter: detection, fate, and toxicity. Environ Int 75:68–80
86. Agusa T, Takagi K, Kubota R, Anan Y, Iwata H, Tanabe S (2008) Specific accumulation of arsenic compounds in green turtles [*Chelonia mydas*] and hawksbill turtles [*Eretmochelys imbricata*] from Ishigaki Island, Japan. Environ Pollut 153:(127–136)
87. Masscheleyn PH, DeLaune RD, Patrick WH Jr (1991) Arsenic speciation and solubility in a contaminated soil. Environ Sci Technol 25:1414–1419
88. Mondal P, Majumder CB, Mohanty B (2006) Laboratory based approaches for arsenic remediation from contaminated water: recent developments. J Hazard Mater 137(1):464–479

89. Gulens J, Champ DR, Jackson RE (1979) Influence of redox environments on the mobility of arsenic in ground water. In: Jenne EA (ed) Chemical modelling in aqueous systems. American Chemical Society, pp 81–95

90. Xu H, Allard B, Grimvall A (1991) Effects of acidification and natural organic materials on the mobility of arsenic in the environment. Water Air Soil Pollut 57–8:269–278

91. Cervantes C, Ji G, Ramírez JL, Silver S (1994) Resistance to arsenic compounds in microorganisms. FEMS Microbiol Rev 15:355–367

92. Meharg AA, Macnair MR (1990) An altered phosphate uptake system in arsenate-tolerant Holcus lanatus L. New Phytol 115(1):29–35

93. Levy JL, Stauber JL, Adams MS, Maher WA, Kirby JK, Jolley DF (2005) Toxicity, biotransformation, and mode of action of arsenic in two freshwater microalgae [*Chlorella sp.* and *Monoraphidium arcuatum*]. Environ Toxicol Chem 24:2630–2639

94. Lloyd JR, Oremland RS (2006) Microbial transformations of arsenic in the environment: from soda lakes to aquifers. Elements 2:85–90

95. Zhang SY, Rensing C, Zhu YG (2014) Cyanobacteria-mediated arsenic redox dynamics is regulated by phosphate in aquatic environments. Environ Sci Technol 489:994–1000

96. Lee CK, Low KS, Hew NS (1991) Accumulation of arsenic by aquatic plants. Sci Total Environ 103:215–227

97. Sanders JG, Osman RW, Riedel GF (1989) Pathways of arsenic uptake and incorporation, in estuarine phytoplankton and the filter-feeding invertebrates *Eurytemora affinis*, *Balanus improvises* and *Crassostrea virginica*. Mar Biol 103:319–325

98. Sanders JG, Windom HL (1980) The uptake and reduction of arsenic species by marine algae. Estua Cstl Mar Sci 10:555–567

99. Lindsay DM, Sanders JG (1990) Arsenic uptake and transfer in a simplified estuarine food chain. Environ Toxicol Chem 9(3):391–395

100. Langlois C, Langis R (1995) Presence of airborne contaminants in the wildlife of northern Québec. Sci Total Environ 160–161:391–402

101. Nriagu JO (1988) Production and uses of chromium. In Nriagu JO, Nieboer E (eds) Chromium in the natural and human environments, vol 20, pp 81–103. Wiley Interscience, New York, NY

102. Zayed AM, Terry N (2003) Chromium in the environment: factors affecting biological remediation. Plant Soil 249:139–156

103. Dazy M, Béraud E, Cotelle S, Meux E, Masfaraud J, Férard J (2008) Antioxidant enzyme activities as affected by trivalent and hexavalent chromium species in *Fontinalis antipyretica* Hedw. Chemosphere 73:281–290

104. Nakayama E, Tokoro H, Kuwamoto T, Fujinaga T (1981) Dissolved state of chromium in seawater. Nature 290:768–770

105. Anderson RA, Polansky MM, Brantner JH, Roginski EE (1977) Chemical and biological properties of biologically active chromium. In: International symposium on trace element metabolism in man and animals, Freising, Germany, pp 269–271

106. Nieboer E, Jusys AA (1988) Biologic chemistry of chromium. In: Nriagu JO, Nieboer E (eds) Chromium in the natural and human environments. Wiley, New York, pp 21–79

107. Anderson RA (1997) Chromium as an essential nutrient for humans. Regul Toxicol Pharmacol 26(1):S35–S41

108. Anderson RA (1994) Stress effects on chromium nutrition of humans and farm animals. In: Biotechnology in the feed industry. Proceedings of Alltech's tenth symposium. University Press, Nottingham, England, pp 267–274

109. Abraham AS, Sonnenblick M, Eini M (1982) The action of chromium on serum lipids and on atherosclerosis in cholesterol-fed rabbits. Atherosclerosis 42:115–195

110. Abraham AS, Brooks BA, Eylath U (1991) Chromium and cholesterol-induced atherosclerosis in rabbits. Ann Nutr Metab 35:203–207

111. Preuss HG, Gondal JA, Bustos E, Bushehri N, Lieberman S, Bryden NA, Polansky MM, Anderson RA (1995) Effects of chromium and guar on sugar-induced hypertension. Clin Nephrol 44:170–177

112. Di Bona KR, Love S, Rhodes NR et al (2011) Chromium is not an essential trace element for mammals: effects of a 'low-chromium' diet. J Biol Inorg Chem 16:381–390
113. Levina A, Codd R, Dillon CT, Lay PA (2003) Chromium in biology: toxicology and nutritional aspects. Prog Inorg Chem 51:145–250
114. Gauglhofer J, Bianchi V (1991) Chromium. In: Merian E (ed), Metals and their compounds in the environment. VCH, Weinheim, Germany, New York, pp 853–861
115. Rai D, Eary LE, Zachara JM (1989) Environmental chemistry of chromium. Sci Total Environ 86:15–23
116. Barceloux DG, Barceloux D (1999) Chromium. J Toxicol Clin Toxicol 37:173–194
117. Bakshi A, Panigrahi AK (2018) A comprehensive review on chromium induced alterations in fresh water fishes. Toxicol Rep 5:440–447
118. Guertin J, Jacobs JA, Avakian CP (2005) Chromium [VI] handbook. Independent Environmental Technical Evaluation Group, Point Richmond, CA
119. Hagendorfer H, Goessler W (2008) Separation of chromium [III] and chromium [VI] by ion chromatography and an inductively coupled plasma mass spectrometer as element-selective detector. Talanta 76:656–661
120. Westbrook J (1983) Chromium and chromium alloys. In: Grayson M (ed) Kirk-Othmer encyclopedia of chemical technology, vol 6, 3rd edn. Wiley-Interscience, New York, pp 54–82
121. Hartford W (1983) Chromium chemicals. In: Grayson M (ed) Kirk-Othmer encyclopedia of chemical technology, vol 6, 3rd edn. Wiley-Interscience, New York, pp 83–120
122. National Academy of Sciences [NAS] (1974) Medical and biological effects of environmental pollutants: chromium. National Academy Press, Washington, DC
123. The biological and environmental chemistry of chromium. VCH Publishers, New York
124. Manceau A, Charlet L (1992) X-ray absorption spectroscopic study of the sorption of Cr [III] at the oxide–water interface: I. Molecular mechanism of Cr [III] oxidation on Mn oxides. J Colloid Interface Sci 148(2):425–442
125. Riedel GF (1989) Interspecific and geographical variation of the chromium sensitivity of algae. In: Suter II GW, Lewis MA (eds) Aquatic toxicology and environmental fate, vol 11. ASTM, Philadelphia, pp 537–548
126. Riedel GF (1984) Influence of salinity and sulfate on the toxicity of chromium [vi] to the estuarine diatom *Thalassiosira pseudonana*. J Phycol 20(4):496–500. https://doi.org/10.1111/j.0022-3646.1984.00496.x
127. Hedayatkhah A, Cretoiu MS, Emtiazi G, Bolhuis H, Stal LJ (2018) Bioremediation of chromium contaminated water by diatoms with concomitant lipid accumulation for biofuel production. J Environ Manage 227:313–320
128. Husien Sh, Labena A, El-Belely EF, Mahmoud HM, Hamouda AS (2019) Absorption of hexavalent chromium by green micro algae *Chlorella sorokiniana*: live planktonic cells. Water Pract Technol 14(3):515–529
129. Kleiman ID, Cogliatti DH (1998) Chromium removal from aqueous solutions by different plant species. Environ Technol 19(11):1127–1132
130. Chandra P, Kulshreshtha K (2004) Chromium accumulation and toxicity in aquatic vascular plants. Botan Rev 70(3):313–327
131. Choo TP, Lee CK, Low KS, Hishamuddin O (2006) Accumulation of chromium [VI] from aqueous solutions using water lilies [*Nymphaea spontanea*] Chemosphere 62:961–967
132. Jana S (1988) Accumulation of Hg and Cr by three aquatic species and subsequent changes in several physiological and biochemical plant parameters. Water Air Soil Pollut 38:105–109. https://doi.org/10.1007/BF00279589
133. Nisha JC, Raja R, Sekar J, Chandran R (2016) Acute effect of chromium toxicity on the behavioral response of zebra fish *Danio rerio*. Int J Plant Anim Environ Sci 6(2):6–14
134. Petrilli FL, Rossi GA, Camoirano A, Romano M, Serra D, Bennicelli C, De Flora A, De Flora S (1986) Metabolic reduction of chromium by alveolar macrophages and its relationships to cigarette smoke. J Clin Investig 77:1917–1924
135. Ilias M, Rafiqullah IM, Debnath BC, Mannan KS, Hoq M (2011) Isolation and characterization of chromium [VI]-reducing bacteria from tannery effluents. Indian J Microbiol 5(1)76–81

136. O'Brien TJ, Ceryak S, Patierno SR (2003) Complexities of chromium carcinogenesis: role of cellular response, repair and recovery mechanisms. Mutat Res Fundam Mol Mech Mutagen 533(1–2):3–36
137. Day VW, Fredrich MF, Klemperer WG, Shum W (1977) Synthesis and characterization of the dimolybdate ion, $Mo_2O^{2-}_7$. J Am Chem Soc 99(18):6146
138. Guillou N, Ferey G (1997) Hydrothermal synthesis and crystal structure of anhydrous Ethylenediamine Trimolybdate $[C_2H_{10}N_2][Mo_3O_{10}]$. J. Solid State Chem 132(1): 224–227(4)
139. Gatehouse BM, Leverett P (1971) Crystal structure of potassium tetramolybdate, $K_2Mo_4O_{13}$, and its relationship to the structures of other univalent metal polymolybdates. J Chem Soc A 2107–2112
140. Lasocha W, Schenk H (1997) Crystal structure of Anilinium Pentamolybdate from powder diffraction data. The solution of the crystal structure by direct methods package POWSIM. J Appl Crystallogr 30(6):909–913
141. Ghammami S (2003) The crystal and molecular structure of bis[tetramethylammonium] hexamolybdate[VI]. Cryst Res Technol 38(913):913–917
142. Evans Jr HT, Gatehouse BM, Leverett P (1975) Crystal structure of the heptamolybdate[VI][paramolybdate] ion, $[Mo_7O_{24}]^{6-}$, in the ammonium and potassium tetrahydrate salts. J Chem Soc Dalton Trans 6:505–514
143. Greenwood NN, Earnshaw A (1997) Chemistry of the elements, 2nd edn. Butterworth-Heinemann. ISBN 978-0-08-037941-8
144. Marino R, Howarth RW, Shamess J, Prepas E (1990) Molybdenum and sulfate as controls on the abundance of nitrogen-fixing cyanobacteria in saline lakes in Alberta. Limnol Oceanogr 35:245–259
145. Morrison SJ, Mushovic PS, Niesen PL (2006) Early breakthrough of molybdenum and uranium in a permeable reactive barrier. Environ Sci Technol 40:2018–2024
146. Atia AA, Donia AM, Awed HA (2008) Synthesis of magnetic chelating resins functionalized with tetraethylenepentamine for adsorption of molybdate anions from aqueous solutions. J Hazard Mater 155(1–2):100–108
147. Roskill, Global molybdenum market outlook, Presentation, Roskill Information Services, London (2010)
148. Market Research Reports (2018) Molybdenum market: current scenario & future outlook. Accessed online on 20 Apr 2020
149. Moss RL, Tzimas E, Willis P, Arendorf J, Thompson P, Chapman A, Morley N, Sims E, Bryson R, Pearson J, Espinoza LT, Marscheider-Weidemann F, Soulier M, Lüllmann A, Sartorius C, Ostertag K (2013) Critical metals in the path towards the decarbonisation of the EU energy sector-assessing rare metals as supply-chain bottleneck in low-carbon energy technologies. JRC scientific and policy reports. Publications Office of the European Union, Luxembourg, p 246
150. Sardesai VM (1993) Molybdenum: an essential trace element. Nutr Clin Prac 8(6):277–281
151. Gunnison AF, Farruggella TJ, Chiang G, Dulak L, Zaccardi J, Birkneret J (1981) A sulfite oxidase-deficient rat model: metabolic characterization. Food Cosmet Toxicol 19:209–220
152. Abumrad NN, Schneider AJ, Steet D, Rogers LS (1981) Amino acid intolerance during prolonged total parenteral nutrition reversed by molybdenum therapy. Am J Clin Nutr 34:2551–2559
153. Yang CS (1980) Research on esophageal cancer in China: a review. Can Res 40:2633–2644
154. Losee FL, Adkins BL (1968) Anticariogenic effect of minerals in food and water. Nature 219:630–631
155. Grassi W, De Angelis R (2011) Clinical features of gout. Reumatismo 63(4):238–45. https://doi.org/10.4081/reumatismo.2011.238
156. Erdman JA, Ebens RJ, Case AA (1978) Molybdenosis: a potential problem in ruminants grazing on coal mine spoils. J Range Manage 31(l):34–36
157. Blowey R, Weaver A (2011) Color atlas of diseases and disorders of cattle, 3rd edn. Mosby
158. Shah VK, Ugalde RA, Imperial J, Brill WJ (1984) Molybdenum in nitrogenase. Annu Rev Biochem 53:231–257

159. Cole JJ, Lane JM, Marino R, Howarth RW (1993) Molybdenum assimilation by cyanobacteria and phytoplankton in freshwater and salt water. Limnol Oceanogr 38(1):25–35
160. Forstner U, Wittmann GTW (1979) Metal pollution in the aquatic environment. Springer, Berlin
161. Eisler R (1989) Molybdenum Hazards to fish, wildlife, and invertebrates. Washington: a synoptic review. In: Fish and wildlife service
162. Mendel RR, Hänsch R (2002) Molybdoenzymes and molybdenum cofactor in plants. J Exp Bot 53:1689–1698
163. Walker J (1953) Inorganic micronutrient requirements of Chlorella. I. Requirements for calcium [or strontium], copper, and molybdenum. Arch Biochem Biophys 46:1–11
164. Glass JB, Wolfe-Simon F, Elser JJ, Anbar AD (2010) Molybdenum-nitrogen co-limitation in freshwater and coastal heterocystous cyanobacteria. Limnol Oceanogr 55:667–676
165. Howarth RW, Cole JJ (1985) Molybdenum availability, nitrogen limitation and phytoplankton growth in natural waters. Science 229:653–655
166. Collier RW (1985) Molybdenum in the northeast Pacific Ocean. Limnol Oceanogr 30:1351–1354
167. Wallen DG, Cartier LD (1975) Molybdenum dependence, nitrate uptake and photosynthesis of freshwater plankton algae. J Phycol 11:345–349
168. Trenfield MA, van Dam JW, Harford AJ, Parry D, Streten C, Gibb K, van Dam RA (2015) Aluminium, gallium, and molybdenum toxicity to the tropical marine microalga *Isochrysis galbana*. Environ Toxicol Chem 34(8):1833–1840
169. Van Dam JW, Trenfield MA, Harries SJ, Streten C, Harford AJ, Parry D, van Dam RA (2016) A novel bioassay using the barnacle *Amphibalanus amphitrite* to evaluate chronic effects of aluminium, gallium and molybdenum in tropical marine receiving environments. Mar Pollut Bull 112(1–2):427–435
170. Trenfield MA, van Dam JW, Harford AJ, Parry D, Streten C, Gibb K, van Dam RA (2016) A chronic toxicity test for the tropical marine snail *Nassarius dorsatus* to assess the toxicity of copper, aluminium, gallium, and molybdenum. Environ Toxicol 35(7):1788–1795
171. Heijerick DG, Carey S (2017) The toxicity of molybdate to freshwater and marine organisms. III. Generating additional chronic toxicity data for the refinement of safe environmental exposure concentrations in the US and Europe. Sci Total Environ 609:420–428
172. Ohlendorf HM, Hoffman DJ, Saiki MK, Aldrich TW (1986) Embryonic mortality and abnormalities of aquatic birds: apparent impacts of selenium from irrigation drainwater. Sci Total Environ 52:49–63
173. Saiki MK, May TW (1988) Trace element residues in bluegills and common carp from the lower San Joaquin River, California, and its tributaries. Sci Total Environ 74:199–217
174. Hamilton SJ, Wiedmeyer RH (1990) Concentrations of boron, molybdenum, and selenium in Chinook salmon. Am Fish Soc 119(3):500–510
175. Saiki MK, Jennings MR, Brumbaugh WG (1993) Boron, molybdenum, and selenium in aquatic food chains from the Lower San Joaquin River and its tributaries, California. Arch Environ Contam Toxicol 24:307–319
176. Husson O (2013) Redox potential [Eh] and pH as drivers of soil/plant/microorganism systems: a transdisciplinary overview pointing to integrative opportunities for agronomy Plant Soil 362:389–417. https://doi.org/10.1007/s11104-012-1429-7
177. Knobeloch L, Salna B, Hogan A, Postle J, Anderson H (2000) Blue babies and nitrate-contaminated well water. Environ Health Perspect 108(7):675–678
178. Amritha K (2019) Blue baby syndrome: causes, symptoms, treatment and prevention. (online) https://www.boldsky.com. Available at: https://www.boldsky.com/pregnancy-parent ing/baby/2019/blue-baby-syndrome-causes-symptoms-treatment-prevention-128919.html. Accessed 19 June 2020
179. Hafshejani DL, Naseri AA, Hooshmand AR, Abbasi F, Mohammadi SA (2016) Removal of nitrate ions from aqueous solution by modified sugarcane bagasse vermicompost. Nat Sci 14(3):16–20

180. Ayebo A, Kross BC, Vlad M, Sinca A (1997) Infant methemoglobinemia in the Transylvania region of Romania. Int J Occup Environ Health 3(1):20–29
181. Lutynski R, Steczek-Wojdyla Z, Kroch S (1996) The concentrations of nitrates and nitrites in food products and environment and the occurrence of acute toxic methemoglobinemias. Przegl Lek 53(4):351–355
182. Xu G, Song P, Reed PI (1992) The relationship between gastric mucosal changes and nitrate intake via drinking water in a high-risk population for gastric cancer in Moping County, China. Eur J Cancer Prevent 1(6):437–443
183. Morales-Suarez-Varela MM, Llopis-Gonzales A, Tejerizo-Perez ML (1995) Impact of nitrates in drinking water on cancer mortality in Valencia, Spain. Eur J Epidemiol 11(1):15–21
184. Yang CY, Cheng MF, Tsai SS, Hsieh YL (1998) Calcium, magnesium, and nitrate in drinking water and gastric cancer mortality. Jpn J Cancer Res 89(2):124–130
185. Schoumans OF, Bouraoui F, Kabbe C, Oenema O, van Dijk KC (2015) Phosphorus management in Europe in a changing world. Ambio 44:180–192
186. Dai L, Pan G (2014) The effects of red soil in removing phosphorus from water column and reducing phosphorus release from sediment in Lake Taihu. Water Sci Technol 69(5)
187. Reddy KR, Delaune RD (2008) Biogeochemistry of Wetlands: science and applications, 1st edn. CRC Press, Boca Raton
188. Mekonnen MM, Hoekstra AY (2017) Global anthropogenic phosphorus loads to freshwater and associated grey water footprints and water pollution levels: a high-resolution global study. Water Resour Res 51(1):345–358
189. Stapanian MA, Schumacher W, Gara B, Monteith SE (2016) Negative effects of excessive soil phosphorus on floristic quality in Ohio wetlands. Sci Total Environ 551:556–562
190. Scannone F (2018) What is eutrophication? Causes, effects and control—Eniscuola. (Online) Eniscuola. Available at: https://www.eniscuola.net/en/2016/11/03/what-is-eutrophication-causes-effects-and-control/. Accessed 21 June 2020
191. US EPA (n.d.) The effects: environment|US EPA. (online) Available at: https://www.epa.gov/nutrientpollution/effects-environment. Accessed 22 June 2020
192. Waajen G, van Oosterhout F, Douglas G, Lürling M (2016) Management of eutrophication in Lake De Kuil [The Netherlands] using combined flocculant—Lanthanum modified bentonite treatment. Water Res 97:83–95
193. Lee J, Rai PK, Jeon YJ, Kim KH, Kwon EE (2017) The role of algae and cyanobacteria in the production and release of odorants in water. Environ Pollut 227:252–262
194. Wright MT, Stollenwerk KG, Belitz K (2014) Assessing the solubility controls on vanadium in groundwater, Northeastern San Joaquin Valley, CA. Appl Geochem 48:41–52
195. Harland BF, Harden-Williams BA (1994) Is vanadium of human nutritional importance yet? J Am Diet Assoc 94:891–894
196. Huang J, Huang F, Evans L, Glasauer S (2015) Vanadium: global [bio] geochemistry. Chem Geol 417:68–89
197. Leiviskä T, Khalid MK, Sarpola A, Tanskanen J (2017) Removal of vanadium from industrial wastewater using iron sorbents in batch and continuous flow pilot systems. J Environ Manage 190:231–242
198. Omidinasab M, Rahbar N, Ahmadi M, Kakavandi B, Ghanbari F, Kyzas GZ, Martinez SS, Jaafarzadeh N (2018) Removal of vanadium and palladium ions by adsorption onto magnetic chitosan nanoparticles. Environ Sci Pollut Res 25:34262–34276
199. Sharififard H, Soleimani M, Zokaee Ashtiani F (2016) Application of nanoscale iron oxide-hydroxide-impregnated activated carbon [Fe-AC] as an adsorbent for vanadium recovery from aqueous solutions. Desalin Water Treatment 57:15714–15723
200. Larsson MA, Hadialhejazi G, Petter J (2017) Vanadium sorption by mineral soils: development of a predictive model. Chemosphere 168:925–932
201. Barceloux DG (1999) Vanadium. Clin Toxicol 37(2):265–278
202. Levy BS, Hoffman L, Gottsegen S (1984) Boilermakers' bronchitis respiratory tract irritation associated with vanadium pentoxide exposure during oil-to-coal conversion of a power plant. J Occup Environ Med 26:567–570

203. Musk AW, Tees JG (1982) Asthma caused by occupational exposure to vanadium compounds. Med J Aust 1:183–184
204. Rojas E, Valverde M, Herrera LA, Altamirano-Lozano M, Ostrosky-Wegman P (1996) Genotoxicity of vanadium pentoxide evaluate by the single cell gel electrophoresis assay in human lymphocytes. Mutat Res 359:77–84
205. Leonard A, Gerber GB (1994) Mutagenicity, carcinogenicity and teratogenicity of vanadium compounds. Mutat Res 317:81–88
206. Vega ED, Pedregosa JC, Narda GE, Morando PJ (2003) Removal of oxovanadium[IV] from aqueous solutions by using commercial crystalline calcium hydroxyapatite. Water Res 37:1776–1782
207. Ghazvini MPT, Ghorbanzadeh SG (2009) Bioresource technology effect of salinity on vanadate biosorption by Halomonas sp. GT-83: preliminary investigation on biosorption by micro-PIXE technique. Bioresour Technol 100:2361–2368

Chapter 3
Oxyanions in Groundwater System—Prevalence, Dynamics, and Management Strategies

Eric T. Anthony and Nurudeen A. Oladoja

Abstract Groundwater (GW) is considered one of the most critical drinking water sources in many parts of the world. Thus, GW contributes immensely to the United Nations clean water project. However, due to GW proximity to the ecosystem, it is easily exposed to natural and anthropogenic pollutants. Among the pollutants that are ubiquitous in the GW system is the oxyanions (also referred to as oxoanions). The oxyanions contamination of the GW could be intrinsic (e.g., flooding, volcanic activities, and geogenic processes) or extrinsic (e.g., GW mining, industrial, and agricultural release). A significant number of elements form stable and mobile oxyanions in the GW system, resulting in an impaired health condition when consumed. The balance between the negatively charged oxyanions and GW's physicochemical properties provides oxyanions with the ability to be easily transported effortlessly, in the GW system. Therefore, several techniques, which have shown tremendous capacity, have been employed for the removal of potentially harmful oxyanions from GW before consumption. These techniques involve the alteration of the oxidation state of the oxyanions into a more tolerable oxidation state, inducing the formation of separable insoluble solids, and mass transfer onto another substrate. The development or modification of nanomaterials, with high potentials for obliterating the negative impacts of oxyanions in the GW system, is necessary to obtain high-quality potable GW.

E. T. Anthony
Department of Chemistry, University of Fort Hare, Alice, South Africa

N. A. Oladoja (✉)
Hydrochemistry Research Laboratory, Department of Chemical Sciences,
Adekunle Ajasin University, Akungba Akoko, Nigeria
e-mail: nurudeen.oladoja@aaua.edu.ng

© Springer Nature Switzerland AG 2021
N. A. Oladoja and E. I. Unuabonah (eds.), *Progress and Prospects in the Management of Oxyanion Polluted Aqua Systems*, Environmental Contamination Remediation and Management, https://doi.org/10.1007/978-3-030-70757-6_3

3.1 Introduction

Groundwater (GW) is a delicate natural resource that is resident in a permeable geological unit called aquifer. It is a crucial water reservoir for drinking water, irrigation [1] industrial applications [2], and a determining factor for food production, water quality/quantity, and flooding [3]. Many communities rely on the use of GW for domestic purposes without prior disinfection or treatment [4]. The availability of GW throughout the year and the buffering potential to resist contamination, which is lacking in other sources of water, makes it a viable and economic freshwater resource [5, 6]. However, sustaining the pristine state of GW has become a big challenge because of the possible entrant of both natural and human-made contaminants into the matrix [7]. This challenge is more precarious in the arid and semi-arid climate [8], because of unfettered GW abstraction [9] and drilling of deeper wells [10], which has not only impaired its quality but have also made it unsustainable. While GW in some regions is not under immediate threat [11], others are heavily polluted. For example, while the rate of abstraction of GW is high in Northern India, due to extensive agriculture, the monsoon precipitation is effectively recharging the GW in southern India [12]. Several wells and boreholes in Bangladesh [13] and Morrocco [14] that supply water to millions of villagers have been proposed for decommissioning due to extensive pollution.

Premised on the genesis of GW and its travails through the tortuous soil strata, it is usually considered as a suite of chemical compounds. These chemical compounds range from those that are nutritionally essential to those that are water contaminants. The characterization of GW is usually based on the domineering ionic species (e.g., $Ca–HCO_3{}^-$, $HCO_3{}^-$, Cl^-, $SO_4{}^{2-}$, Ca^{2+}, Mg^{2+}, and Na^+), which contribute to the ionic mobilities within the system. GW is also characterized by the oxidizing or reducing ability, which strongly correlates with the concentration of available oxygen (i.e., oxic, semi-oxic, and anoxic) [15]. GW redox condition determines the rate of mineral formation, dissolution, speciation, and movement of various dissolved and undissolved constituents present in it.

Oxyanions (also referred to as oxoonions) are negatively charged polyatomic ions with a general formula $E_x O_y^{z-}$, where E is a chemical element bonded to oxygen. Among the vast collections of chemical species present in GW, oxyanions (or oxoanions) have been a constant challenge because of their diversity and rich hydrochemistry. For example, the GW pH value is usually near neutral to alkaline; thus, diverse oxyanions are present in it at this pH range. However, an increase in the pH value initiates a switch in the oxidation state of oxyanion species [16]. Thus, previously prevalent oxyanion may be absent as GW moves from one aquifer to another.

For many GW resources, the primary source of oxyanion contamination is from geological formations through rock–water interaction. The outcome of rock–water interaction includes sorption, dissolution, and leaching, which are dictated by the nature of GW, pH value, redox potential, depth, and temperature. For example, arsenic-bearing mineral is prone to leaching, forming arsenate in the aqueous system, at alkaline medium and low redox conditions [17]. Adsorption attenuates the concentration and mobility of oxyanions in the GW matrix, but dissolution enhances the concentration and mobility, while leaching influences its distribution along the fluid pathway. In the same manner, precipitation restores GW quantity and quality in aquifers, while evaporation concentrates the oxyanions. The presence of natural gas (methane) in aquifers can also improve GW quality, due to the anaerobic microbial demeaning of methane [18]. The synergic contribution of these factors results in an alarming concentration of oxyanions in GW. For instance, a high level of oxyanions, majorly arsenite, chromate, and nitrate, were confirmed in locations devoid of industrial activities [19], a strong indication of natural contamination. Anthropogenic activities (e.g., mining, exploration, agricultural practices, poor waste management, and sanitation) are secondary sources of oxyanions in GW [20]. Upon release from the source, oxyanions move through surface water and are eventually mixed with GW [21]. Oxyanions may also sink directly into GW or adsorb onto a selective surface or compete with similar oxyanions or ions for transport sites. When reaction conditions favor the latter, geological materials/minerals (volcanic glass, sediments, and rocks) dictate oxyanions bio-availability.

3.2 Prevalence of Oxyanions in GW System

Oxyanions are present in GW as one of the determinants of the intrinsic properties (e.g., the values of conductivity and salinity) and the possible end-use. At low concentrations, they are present as an essential dietary requirement, while at elevated concentrations, they are contaminants. Oxyanions in GW exhibit different properties

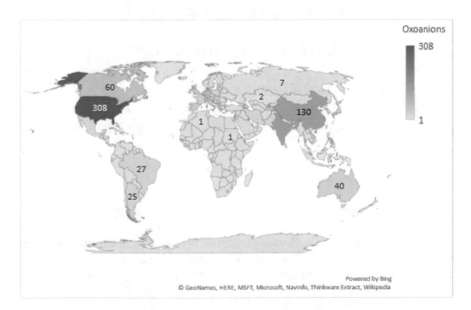

Fig. 3.1 Regional distribution of oxyanions (*Source* Scopus)

that range from serious health implications (e.g., arsenic) to severe environmental threats (e.g., phosphate) and in some cases, both negative ecological and health impacts (e.g., nitrate). The prevalence of oxyanions (Fig. 3.1) in GW stemmed from two broad sources, which are natural and anthropogenic.

3.2.1 Natural Occurrence

Oxyanions are transported from the environment into the GW system; thus, raw GW is replete with different oxyanion species. Arsenic oxyanions (H_3AsO_3 and H_3AsO_4), dominated by arsenite ($HAsO_3^{2-}$) and arsenate ($H_2AsO_4^-$), are primary examples. In natural water systems, arsenic (Eqs. 3.1–3.3) and chromium (Eqs. 3.4–3.9) redox reactions are induced by magnesium, iron, sulfite, and reactive-radical species [22, 23].

$$MnO_2 + H_3AsO_3 \rightarrow 2MnOOH^* + H_3AsO_4 \tag{3.1}$$

$$MnOOH^* + H_3AsO_4 + 4H^+ \rightarrow 2Mn^{2+} + H_3AsO_4 + 3H_2O \tag{3.2}$$

$$2Mn - OH + H_3AsO_4 \rightarrow (MnO_2)AsOOH + 2H_2O \tag{3.3}$$

$$3Mn(III) + Cr(III) \rightarrow 3Mn(II) + Cr(VI) \tag{3.4}$$

$$3Mn(IV) + 2Cr(III) \rightarrow 3Mn(II) + Cr(VI) \tag{3.5}$$

$$3\,OH + Cr(III) \rightarrow 3OH^- + Cr(VI) \tag{3.6}$$

$$3As(III) + 2Cr(VI) \rightarrow 3As(V) + 2Cr(III) \tag{3.7}$$

$$3Fe^{2+} + Cr^{6+} \leftrightarrow 3Fe^{3+} + Cr^{3+} \tag{3.8}$$

$$4HSO_3^- + 2Cr(VI) + 6H^+ \rightarrow 2Cr(III) + 2SO_4^{2-} \tag{3.9}$$

The role of GW flooding in elevating the levels of oxyanions in the GW system has been established [24]. It has also been posited that GW systems with an initial low concentration of arsenic soon experience a concentration surge because of natural weathering and leaching from arsenic-rich aquifer minerals [25, 26].

Volcanic glass, rock, and sediment have been identified as the primary natural sources of arsenic in GW. For instance, arsenopyrite and pyrite are the sources of arsenic oxyanion contamination in the GW of Bolgatanga area of Ghana [27]. In

Peninsula Valdes, Argentina, volcanic shards and iron oxides are the major arsenic-rich geo-materials [28]. During the interaction of GW with geological materials, especially in the presence of $NaHCO_3$, at pH value >6.5, arsenic is released into the GW as oxyanions (i.e., arsenite and arsenate) [19]. In dry climate, loess-type aquifer pyroclastic sediment serves as the source of arsenic in GW. This enrichment is ascribed to the leaching of the volcanic glass that is enhanced at high pH value (9.2) and elevated bicarbonate (HCO_3) level [29]. In a supporting study, Mukherjee et al., (2019) presented more insight into the origin and pathway of aquifer contamination by arsenic oxyanions [30]. Arsenic is contained in the hot metasomatic fluid, and during deep crustal processes, the bulk phase arsenic migrates to the surface. The surgical migration of arsenic is accompanied by extrusion or spread of particulates containing arsenic, and the subsequent leaching into the aquifer. This occurrence is mostly associated with the deep underground system [30].

Similarly, geogenic processes have been ascribed to the distribution of chromium oxyanions in the GW system around North Carolina, USA. The concentration of Cr(VI) anions was associated with the rock lithology, weathering [31], and leaching of rock-bearing chromium [32]. The elevated concentrations of chromium oxyanions detected in different wells in California, USA, were attributed to silicate weathering, which significantly shifted the pH value toward alkalinity, enhanced leaching from rock, and creating an oxic conditions [33]. Leaching of volcanic glass has been ascribed to an elevated concentration of selenium oxyanion in GW system of Poitiers, France. In a batch leaching test, at pH < 2.0, a reduced form of Se (IV) anion was dissolved from apatite and argillaceous minerals. Between pH value of 2.0 and 7.5, Se (VI) oxyanion (selanate) was dominant, while above pH 7.5, Se(IV) oxyanion (selenite) predominated. Generally, leaching was most favorable at pH 11.0 [34].

The ingestion of elevated levels of vanadium oxyanions causes cancer growth and neurological disorder. In shallow GW, the inorganic oxides of iron, aluminum, lead, copper, uranium, and thorium, and other rare earth metals, bearing vanadium and dissolved organic carbon colloidal pool are the main factors contributing to the distribution of vanadium oxyanions [35]. In two separate locations around the USA, Nevada, and Texas, geological mafic phenocrysts were identified as the source of vanadium oxyanions in GW. The oxygen-rich alkaline GW (pH > 7.8) in Nevada favors the dissolution of vanadium as $H_2VO_4^-$ and HVO_4^{2-} from the aquifer rocks. The concentration of vanadium oxyanions in Texas GW decreased due to the formation of insoluble mineral species. In California, vanadium contamination in GW was attributed to the release from the mafic and andesitic rock under favorable geochemical conditions [36].

Molybdenum, though an essential trace element in dietary, can cause diarrhea at elevated concentration. In northern Jordan, molybdenum was leached (up to 1.44 mg/L) from oil–shale rock samples [37]. Studies have shown that the natural occurrence of molybdenum oxyanion in British GW was below 2 μg/L [38]. In Chile, up to 475 μg/L was detected at ore deposit in the Atacama Desert [39] and perchlorate was also in abundant in the same deposit [40].

Aquifers provide a perfect ecosystem for slow-growing bacterial, for example, anaerobic ammonium oxidizing bacterial. This is because of the high residence time, slow rate of water mixing, and low organic carbon [41]. These anaerobic ammonium oxidizing bacteria contribute to the occurrence of nitrate in GW. Nitrate is mainly known to cause methameglobinaemia and carcinogenic nitrosoamine. In Japan, the variations in the concentrations of oxyanions of molybdenum, vanadium, and tungsten were ascribed to the high microbial activities. The microbial activities dominate at the operating pH value, which eventually increased the solubility of vanadium, as HVO_4^{2-}. The concentration of vanadium oxyanions in the shallow section of the GW was 1.6–3.7 times higher than the deep part [42].

3.2.2 Human Activities

Oftentimes, anthropogenic activity's footprints are more disturbing than natural contribution to the oxyanions contamination of GW [43]. Many aquifers have been overexploited above 3.5 times, and an estimated 1.7 billion people live in a location where GW is currently under threat [44]. In a study conducted by Wilkin et al. [45], the minimum value of arsenic concentration detected in abandoned smelting facilities was 2 mg/L [45] (Table 1). Considering that the permissible level published by WHO was 0.01 mg/L, concentration that exceeded 122 mg/L was detected in the GW at the abandoned smelting factories. GW mining, such as well installation, can also disturb the equilibrium state of minerals in the aquifers and the subsequent distribution of oxyanions in the GW matrix. It has been reported that the concentrations of arsenic oxyanions in recently installed wells increased over one year by at least 16% [46]. The mining process, during well installation, leads to the release of arsenic from the weathering of arsenic—rich rocks.

High concentration of chloride in GW (205–3310 mg/L) is an indication of human activities. Chloride enters into groundwater due to mineral reserves; seawater interference; airborne sea spray; agricultural waste; industrial and urban wastewater [47]. Due to inadequacies in the disposal of solid waste from chromate factory in Leon Valley, Mexico, high concentrations of chloride (up to 2000 mg/L) and chromite were detected in the GW [48]. In Southern India, due to the improper disposal of tannery effluent in the open channels, chromium oxyanion with an average concentration of 0.04 mg/L was detected in the GW [49].

In a survey, the prevalence of molybdenum in British GW, above 2 μg/L, signified urban and industrial pollution of the system [38]. The GW was characterized as $CaHCO_3$, $NaHCO_3^-$ and Na_2SO_4, with a mildly reducing condition, containing Fe and Mn minerals. The conditions of the GW did not contribute to the release of the total molybdenum oxyanions. Water–oil–shale interaction induces the leaching of molybdenum into the GW systems and agricultural waste also adds a significant percentage of molybdenum into the GW system. Similarly, due to mining, over-exploitation, and industrialization, elevated levels of molybdenum oxyanions have been reported in the Mongolia, Russia, [50] and Qatar (103 μg/L) GW system.

Although perchlorate (ClO_4^-) occurs naturally in GW system, it is however confined to arid and semi-arid environments. At present, anthropogenic activities

Table 1 The minimum and maximum concentration of some reported oxyanions around the world

Oxyanions	Conc Min	Conc Max (mg/L)	Location	Reference
Cr(VI)	0.012	22.9	North Carolina, USA	[1]
Strontium	0.25	3426	North Carolina, USA	[1]
Arsenic	0.11	0.46	Quebec, Canada	[2]
Perchlorate	-	800	Israel	[3]
Selenium	0.5	4.07	Colorado, USA	[4]
Arsenic	0.01	0.40	Peninsula Valdes, Argentina	[5]
Arsenic	0.001	0.141	Bolgatanga, Ghana	[6]
Selenium Molybdenum Arsenic		0.8 0.457 0.278	Atacama Desert, Chile	[7]
Arsenic	0.0114	1.66	Tucuman, Argentina	[8]
Arsenic	0.0006	0.19	Main Ethiopian Rift, Ethiopian	[9]
Arsenic	0.0013	0.11	Bihar, India	[10]
Arsenic	0.0602	0.13	Pampean, Argentina	[11]
Chromium	0.00	0.0032	California, USA	[12]
Chlorite ClO3-	0.01	38.0	Ferrara Province, Italy	[13]
Arsenic	2.0	112.0	USA	[14]
Selenium	0.005	7.94	USA	[14]
Arsenic	0.0001	2.09	Sindh and Punjab, Pakistan	[15]
Molybdenum	DL	0.10	Northern region, Qatar	[16]
Arsenic	0.01	4.87	Yuncheng Basin, China	[17]
Vanadium	0.003	0.07	California, USA	[18]
Molybdenum	0.0001	0.12	Skagafjordur, Iceland	[19]
Nitrate	0.1	96.0	Yemen	[20]
Arsenic	0.002	0.029	Arizona, USA	[21]
Arsenic	0.0008	0.0086	South East India, India	[22]
Tungsten	0.00027	0.74	Nevada, USA	[23]
Arsenic Vanadium Chromium	0.011 0.037 0.129	1.66 0.30 0.25	Nevada, USA	[24]
Molybdenum	0.07	1.44	Northern Jordan, Jordan	[25]

(continued)

Table 1 (continued)

Oxyanions	Conc Min	Conc Max (mg/L)	Location	Reference
Arsenic	0.00053	0.45	Hetao Basin, Mongolia	[26]
Arsenic	0.01	0.05	Nepal	[27]
Perchlorate	-	800.0	Israel	[28]
Arsenic	0.001	0.0048	South Korea	[27]

contribute to the current significant presence in the GW. Exposure to perchlorate results in birth defects, impaired mental state, diabetes, and thyroid tumors [40]. Perchlorate can compete with the uptake of iodine in the tissue, due to the similarities in the charge and ionic radius of the two chemical species [51]. Perchlorate occurrence in GW around California and Nevada was attributed to industrial discharge [52].

3.3 The Dynamics of Oxyanions in GW System

The complex mode of oxyanion distribution in GW system could be attributed to the diversity of the constituent ionic species in the system. Although both dissolved and undissolved minerals participate in the de/stabilization, solubilization, and distribution of oxyanions within the GW system, pathogen also plays a significant role. Many oxyanions co-exist and interact in the GW systems with other major and minor ionic species. The tendency of oxyanions to act as ligands in the formation of stable geological mineral complex in the GW is directly related to the abundance of these oxyanions in the aqueous phase. The oxyanions in the GW system influence the system pH value, available dissolved oxygen, and electrical conductivity. They also contribute, significantly, to the total carbon, nitrogen, and phosphorous content, thus, imparting color, taste, and overall water acceptability. Therefore, the dynamics of oxyanions in GW are controlled, most importantly, by the biological, physical, and chemical parameters of the aqueous system.

3.3.1 Physical Factors

The physical factors that dictate the dynamics of oxyanions in GW system; include residence time, over-withdrawal/recharge, and flow pathway. These parameters are associated with GW depth, mining, recharge, and water–rock interaction (mixing) [53]. For instance, elevated carbonate content signals–water interaction with calcite and dolomite. Low concentrations of major ions (e.g., Na^+, K^+, Mg^{2+}, Ca^{2+}, Cl^-,

SO_4^{2-}) and total dissolved solids are associated with water flow path in least mineralized zone and HCO_3^- GW type. In contrast, GW containing high concentration of major ions and total dissolved solids flowed through over-saturated mineralized $Ca-SO_4$ water [54]. Saline GW is distinguished by its high Cl^-, Na^+, SO_4^{2-}, Ca^{2+}, and Sr^{2+} content [55], which is a product of the mixing of GW with seawater [56]. The flow of GW influences the availability of oxygen in it and dictates the oxic zone of the GW system. The volume of oxic condition in GW is lower for a steady stream flow, relative to the unsteady flow [57].

The impact of mining and recharge is mostly visible in a long-term study. In the fifteen (15) years of a systematic survey, the concentrations of arsenic and selenium oxyanions reduced over time. The mobile oxyanions were sorbed onto GW minerals, resulting in diminishing oxyanions concentration over the period; since with no fresh contamination occurred. The initial level of arsenic and selenium oxyanions consistently diminished during the 15 years study period [45]. Concerning the GW recharge source and estimated residence time, GW could take over four decades to recharge [58]. Thus, the continuous pumping (usage) of GW has potentially increased the threat posed by arsenic oxyanions in the San Joaquin Valley of California GW. The concentration of arsenic detected is proportional to the rate of mining [59].

The mining of GW involves drilling through the earth's crust—soil, sediment, and rock degradation and disintegration occur. Thus, GW may be encountered at different earth crust layers. Therefore, at a different depth, GW is exposed to specific rock-type and water conditions. Thus, different GW depths may favor the adsorption or desorption of oxyanion from rock-sediment-bearing oxyanions. This an indication that the distribution of oxyanions in GW is also depth-related. The high concentration of oxyanions in shallow GW is associated with a high rate of evaporation, due to the immediate impact of temperature at GW surface. In deep waters, the elevated level of oxyanions is related to the dissolution of volcanic glass [60].

Similarly, in shallow GW, oxyanions contamination may be related to the residuals from anthropogenic activities. For example, the concentration of arsenic oxyanions in GW in the West Bengal, India, was highest at a superficial level. The frequency reduced with depth, due to carbonate mineral dissolution [61]. In the Bengal community, shallow GW has been decommissioned for other GW wells of greater depth (>150 m) due to arsenic contamination. This is because, the deeper GW conformed with the WHO guideline for the concentration of arsenic in GW [62]. The immobilization of arsenic by the GW sediment is responsible for the low level of arsenic-bound oxyanions in some deep well. It is suggested that the abstraction of GW, due to increasing water demand, can also reverse the situation. In any case, GW mining can cause arsenic oxyanions to migrate to more bottomless wells [63].

A contrasting trend has been reported which showed that the concentration of arsenic oxyanions increased with increasing GW depth [64]. This is an indication that the leaching of underground rock contributes to the level of arsenic concentration in GW. At shallow GW, arsenic oxyanions are unavailable due to the oxidation of arsenite to arsenate, and subsequent precipitation as ferric arsenate or sorption onto ferric hydroxide. Beneath 70 m and at pH > 6.0, the oxidation of arsenic-sulfide minerals is favored and sorption onto ferric hydroxide becomes unfavorable.

The local GW temperature is also a defining factor that influences the mobility and solubility of GW bound oxyanions. In the arid region, the local temperature of GW is as high as 40 °C which leads to evaporation and subsequent concentration of the oxyanions. For example, the concentration of molybdenum oxyanions increased linearly with increasing GW temperature (>30 °C), due to enhanced dissolution of plagioclase, pyroxene, and augite minerals [65], while in Nevada, tungsten oxyanion in GW was ascribed to the evaporation process [66].

3.3.2 Chemical Perturbation

The hydrochemistry of GW is a critical factor that controls the prevalence, speciation, and distribution of oxyanion-forming elements. Some of the factors related to chemical perturbation that influences the dynamics of oxyanions in GW systems are discussed below.

3.3.2.1 Bonding

The investigation of the dynamics of GW bound oxyanions has been studied extensively, but knowledge on the influence of the oxyanion-geological mineral bonding strength on the mobility of oxyanions is often overlooked. Meanwhile, the distribution of oxyanions in GW depends on the ease of release from the harboring minerals [61]. Typically, oxyanions bonded to geological minerals via weak forces are prone to leaching at the slightest change in hydrochemistry. Thus, the distribution of such oxyanions occurs via horizontal or vertical transportation. For example, the bonding of molybdenum with oxygen requires little heat of fusion and fission [37]. Thus, Mo-oxyanions are easily released from the minerals [67] into the GW at the slightest pH change. The weak bonded sulfide-bearing arsenic minerals were responsible for the high concentration of arsenic oxyanions detected in Quebec (Canada) GW [68]. Arsenic-bearing minerals (i.e., arsenopyrite and gersdorffite) are susceptible to oxidative reaction, followed by the adsorption of the resulting sulfide-containing arsenic onto Fe–Mn oxyhydroxides and clay materials. The soluble arsenic oxyanion is transported through the fractured geomaterial, and the final distribution of arsenic oxyanions in the GW is achieved via the reduction of Fe–Mn oxyhydroxides [68]. In another study that explored opposite banks (i.e., freshwater), the concentration of arsenic oxyanions was low and high in the west and east, respectively [69]. The sources of arsenic oxyanions in the GW system were ascribed to the silicate weathering in the west and carbonate dissolution in the east. Carbonate minerals are susceptible to a slight change in aqueous anionic character. The east bank displayed a more anoxic behavior, with elevated Cl^-, SO_4^{2-}, and NO_3^-, which possibly suppressed the bonding of arsenic onto the mineral rocks, thereby ensuring the mobility of the arsenic oxyanions in the GW. The stable complex formation between EDTA-acetic acid and arsenic was employed for the preservation of dominant arsenic speciation in GW samples [70] while tungstate adsorbed strongly to hematite [71].

Common ionic species in the GW system may also contribute to the mobilization and the enrichment of arsenic oxyanions, due to the bond formation tendencies. In Yuncheng Basin, China, elevated concentration of arsenic oxyanions (mainly, $HAsO_4^{2-}$) in shallow and deep GW correlated strongly with Na-rich and Ca-poor GW system, especially at a high Na/Ca ratio and a pH (>7.8), which is an indication of desorption from arsenic-bearing hydrous metal oxides and cation exchange [25]. Even though trivalent Al and Ga failed to form a ternary complex with arsenic and selenium oxyanions in the presence of humic acid, trivalent Cr and Fe formed stable ternary complexes [72].

3.3.2.2 Geochemical Process

Several geochemical processes that have been employed to predict the mode of distribution of oxyanions in GW include oxidation/reduction [73, 74], competitive adsorption/desorption [75], infiltration [76], and biogeochemical processes [77]. The redox condition in GW is dictated by the interaction between an electron donor (organic carbon) and electron acceptors (O_2, NO_3^-, SO_4^{2-}, and CO_2) [78]. It can be used to understand water quality and predict microbial activities [79]. In redox influenced GW system, the dominant oxyanions respond to the slightest change in the GW redox potential. For instance, in an abandoned lead and zinc smelting factory, redox potential was the main factor controlling the distribution and speciation of arsenic and selenium oxyanions in the GW [45]. In a hand dug well, arsenic prevailed as arsenite at the slightly reducing condition (i.e., 75 < Eh < 275 mV), pH 6.1–7.9, and low Ca/Na ratio [45]. Under this condition, the solubilities of Fe and Mn arsenic-bound minerals were high. At increased reducing potential (i.e., mild oxidizing condition), in the range 300–490 mV, arsenate dominated and showed a high adsorption rate toward geological minerals. Since the arsenate mobility was restricted, the total concentration of arsenic oxyanions in the GW reduced over the 15 years, due to enhanced adsorption at the oxidizing condition.

The predominant oxyanion in Ethiopian Rift Valley GW was arsenate with ≈80% prevalence [80]. The GW, characterized by positive redox potential, elevated total dissolved solids, dissolved oxygen and HCO_3^-, near neutral-alkaline pH (>6.8), exhibited poor mineral (Fe and Mn) solubility. These features suggested that the GW is in oxidizing condition. In reducing condition, elevated concentration of Fe and Mn was detected in the aquifer and the concentration of ClO_4^- fluctuates through the aquifer as ClO_3^- and ClO_2^-, due to its vulnerability to reducing agents, including NO_3^- [81].

In view of the role of selenium, as a vital essential nutrient, USEPA allowed 50 μg/L [82] in drinking water. Above this concentration, it is most toxic, especially to the liver and lungs. Oxidizing medium is favorable for the distribution of selenium oxyanions in the GW system. The dissolved oxygen and nitrate in GW are two important naturally occurring molecules that influence the oxidation of selenium oxyanion process. Although selenium oxyanions are released from geological minerals in Colorado, USA, high nitrate concentration was one of the major factors that promoted their mobility in the aquifer. Selenium oxyanions were released as $Na_2(SeO_4)_x(SO_4)_x$ and $Mg(SeO_4)_x(SO_4)_x$ in the shale and alluvial materials, even at low dissolved nitrate concentration in non-irrigated soil [83].

The redox chemistry of GW may be classified as oxic and anoxic conditions. The oxic condition is characterized by appreciable dissolved oxygen and low concentration of Fe and Mn, while anoxic condition is distinguished by a minimal or no dissolved oxygen and high Mn and Fe concentration. An establishment of equilibrium between oxic/anoxic conditions is characterized by an elevated level of dissolved oxygen and the presence of Fe and Mn [84]. Therefore, the domineering oxyanion species differs in each water condition. For instance, in oxic states, arsenic is present in GW as H_3AsO_4, $H_2AsO_4^-$, $HAsO_4^{2-}$, and AsO_4^{3-}, which are highly pH dependent [85]. In anoxic GW system, below pH 9.2, arsenic oxyanions are dominant as $H3AsO_3$ [45]. In the environment, inorganic selenium exhibits four oxidation states: selenide (Se(II)), selenium (Se(0)), selenite (Se(IV)), and selenate (Se(VI)). Depending on the redox condition and pH, selenate (SeO_4^{2-}) is dominant at the oxic state in a broad pH range (acidic to alkaline). In oxic/anoxic conditions, selenite (SeO_3^{2-}) and biselenite ($HSeO_3^-$) predominated, depending on the pH of the medium [83].

The dynamics of deep GW were characterized in three hydrogeological conditions, viz: recharge oxic, flow-moderate reducing zone, and flow-reducing region [86]. In the oxic state, the adsorption of arsenic onto Fe mineral prevailed and the concentration of arsenic oxyanions drastically reduced (<0.05 mg/L). In the reducing zone, the desorption of arsenic from Fe/Mn oxyhydroxides/oxides was dominant (0.244 mg/L), recharge, and mixing increased, and the concentration of As-oxyanions was above 0.3 mg/L [86].

During competitive de/sorption, ionic charge plays a significant role. Due to the complex ionic nature of GW, ionic species are displaced and replaced by other species that possess similar ionic radii and character. For example, dissolved natural organic matters are negatively charged species, which are in continuous competition for sorption sites on aquifer minerals and may displace weakly bonded oxyanions, enhanced complex formation, and ultimately influence redox chemistry [87].

Pyrite, a significant deposit of arsenic around the globe, is among the geological materials that contribute to arsenic oxyanion contamination in GW systems. The release of arsenic from aquifer rock (arsenian pyrite) has been reported to be strongly related to bicarbonate ion concentration. Bicarbonate ion (HCO_3^-) could displace sulfur from AsS_3 to form stable arsenic carbonate $As(CO_3)^+/As(CO_3)_2^-$ [88]. Oxyanions that share similar chemistry can co-exist or exhibit absolute domination in GW. Chromium oxyanions co-occurred with vanadium, selenium, and uranium oxyanions and GW rich in nitrate also co-select chromate [33]. Sulfate is a significant interference for the adsorption of inorganic As(V) oxyanion onto GW minerals [85].

Due to chemisorption and desorption, the concentration of vanadium in Texas and Nevada GW varies. In oxic and alkaline conditions (pH > 7.8), GW in Nevada favors the dissolution of vanadium as $H_2VO_4^-$ and HVO_4^{2-} from the aquifer rocks. Thus, the concentration of vanadium oxyanions increased along the GW flow path. In contrast, the level of vanadium oxyanions in Texas GW decreased along the distribution channels, due to the Fe(III) and SO_4^{2-} reduction zones. The oxidized V(V) and V(IV) species are adsorbed onto the Fe(III) and Mn(IV) minerals and the dissolved organic matter (DOM) in the sediments [89]. The prevalence and solubility of vanadium oxyanions in San Joaquin Valley of California GW were redox-dependent. In an oxic condition, vanadium existed as $H2VO_4^-$ and the concentration increased with increasing pH, suggesting increasing solubility as a result of adsorption/desorption reaction. In a sub-oxic condition, vanadium predominated as oxo-cation $(V(OH)_3^+)$, thus, the concentration of oxyanions reduced with increasing pH value. In anoxic GW, vanadium hydroxide $(V(OH)_3)$ is dominated, due to precipitation of V^{3+} or as Fe^{3+} mineral [90].

The ionic charge on the oxyanions enhances their mobility and bond-forming tendencies. For example, oxyanions interact with the natural organic matter, by bridging cations (Fe^{3+}), to form a complex compound. The stability of such a complex is a function of pH, which is used to predict its mobility and speciation [91]. Similarly, tungsten and molybdenum form stable neutral Ca-complexes in alkaline pH and oxyanions in acidic pH region [92]. Thus, in the alkaline medium, the toxicity of molybdenum is minimal.

3.3.3   Biological Activities

In addition to the physicochemical characteristics of GW, lower food-chain micro-organism played a significant role in the bio-availability of major and minor ions. The contribution of micro-organism in the balance of GW nutrients and oxyanions is essential for food cycle. It helps in the transformation of organic and inorganic compounds [93]. For example, *Stygofauna* is capable of degrading natural organic matter and improving GW purity [94] and strain of *pseudomonas* spp., could reduce selenate to selenite in the aqueous system [95].

Leakage from leachate and flooding can cause microbial enrichment of the aquifers [96, 97], resulting in microbial-induced GW condition and pathogenic imbalance. Characterization results of microbial communities in GW system showed that it is depth-dependent. In the near-surface GW (1–2 cm), *Bradyrhizobiaceae* and *Comamonadaceae* were prevalent, while deeper wells were characterized by *Rhodocyclaceae* [98]. Microbial-induced reductive dissolution of the sedimentary iron-oxyhydroxides mineral was identified as the source of high concentration ($>10 \mu g/L$) of As-oxyanions in anoxic GW [99]. Further studies showed that the initial stage of the release of arsenite involves the reduction of arsenate by iron-reducing bacteria, and subsequent reduction by ammonia-producing bacteria [100]. Bio-elements associated with oxyanion-forming elements are mostly affected by micro-organism, since they assimilate oxyanions in the form of an essential nutrient. The reduction in the concentration of oxyanion-forming vanadium toward GW surface was attributed to microbial action [50].

3.4   Management Strategies

The management strategies that have been employed for the treatment of oxyanion contaminated GW systems are diverse and often specific to oxyanion genre. For example, Fe(II)-bound minerals (e.g., chlorite and biotite) are natural reductant for Cr(VI) to Cr(III) in GW [101]. Micro-organism renders potentially toxic oxyanion harmless by switching its oxidation state or enhancing the rate of mobilization on potential substrate [102]. Nanotechnology can recover bulk GW bound oxyanions onto movable substrate or limit the mobility through healthy bond formation or exchange [103]. The discussion in this section shall focus more on bulk removal and redox reaction as a tool for the management of oxonions in GW.

3.4.1   Adsorption

Adsorption-based approaches have been extensively explored for the treatment of GW pollution. As a cheap, sustainable, and non-toxic material, the potential application of clay as adsorbent for oxyanion removal from GW systems has been investigated [104]. The physicochemical characteristics of clays (e.g., sizeable specific surface area and point of ion exchange) enabled them for application in this regard, without prior modification. For example, an anionic clay, layered double hydroxide (LDH), known as hydrotalcite, $[Mg_2Al(OH)_6]_2CO_3.H_2O$ has been used for the attenuation of oxyanions in GW samples. The LDH possesses exchangeable CO_3^{2-} anions within its matrix and a net positively charged surface as a result of the substitution of Al^{3+} for Mg^{2+}. Therefore, oxyanions in the GW matrix can displace CO_3^{2-} in the hydrotalcite lattice. Bare hydrotalcite was used as an adsorbent for the treatment of chromium oxyanion simulated water. The hydrotalcite adsorbent showed poor removal efficiency (<50%) for chromium VI oxyanion at the GW pH value [105].

   In order to enhance the uptake and selectivity of oxyanions from GW, clay materials are usually modified. Modification can be physical or chemical, or a combination of both processes. Calcination is a simple and effective method to cause both physical and chemical disorder within clay material. While physical modification causes surface re-orientation, chemical modification incorporates a guest species into the framework of the adsorbent matrix. For example, the fabrication of hydrocalumite [106] and calcium/iron oxide [107], using a gastropod shell as a precursor produced a highly reactive adsorbent for phosphate removal from water. When the exchangeable anions in a synthesized LDH were exchanged with NO_3 and glycine; it was used for the removal of rapheme, arsenate, and chromate from aqueous solution [108]. Though glycine-LDH displayed larger pore volume, average pore size, and surface area than NO_3-LDH; however, at the GW pH, both adsorbent exhibited similar removal efficiency. Chloride (Cl^-) containing LDH was more effective in the removal borate $B(OH)_4^-$ from aqueous solution than carbonate-LDH [109]. The adsorption capacity of carbonate-LDH and chloride LDH was 10–14 mg/g and <2 mg/g, respectively.

This is because carbonate showed a diminished anionic exchange capacity due to its high affinity toward positively charged molecules.

The tendency of oxyanions and aquifer (rock-sediment) minerals to interact has unlocked another potential class of appropriate adsorbent for the removal of oxyanions from GW matrix. Iron minerals (hematite, goethite, magnetite, and zero-valent iron) are abundant and eco-friendly adsorbent that have been studied for oxyanion removal from aqueous solution. The removal efficiency of goethite, magnetite, and zero-valent iron minerals for arsenate was <20%. The removal efficiency of hematite (Fe_2O_3) displayed 50% removal efficiency [110]. Iron modified peat showed <40% for the recovery of vanadium [111] from simulated GW. Poor adsorption efficiency of many adsorbent has been attributed to competitive adsorption. GW is rich in ions, so during the adsorption process, these ionic species compete with the oxyanions for the adsorbate surface. Phosphate was the primary competing anion, causing a 70% reduction in the removal of arsenic (V), using silica/iron oxide (SiO_2/Fe_3O_4) functionalized adsorbent [112]. Adjusting the operational pH value to neutral, reduced the silica, carbonate, and phosphate interferences [113]. Carbonate is also impeded the adsorption of arsenite and arsenate onto hematite [114]. Increasing the iron loading on silica/iron oxide composite enhanced the removal of arsenic oxyanions containing water at GW pH [115]. The adsorption capacity increased from approximately 70 mg/g to 300 mg/g, at 1 g and 5 g iron content, respectively.

Pure and modified biochar (blended with different metals) has also been receiving increasing attention as sustainable adsorbents for oxyanion removal from GW [116, 117]. Modifications improved were found to improve biochar microporosity and macroporosity, which was evident in its increased adsorption efficiency. $Mg(OH)_2$-biochar and $AlCl_3$-biochar showed high adsorption capacity toward phosphate (250 mg/g) and arsenate (14 mg/g), respectively. Snail shell, a sustainable agricultural waste, was used for the fabrication of nanosized calcium oxide for the removal of chromate from aqueous solution. Both snail shells [118] and the nanosized calcium oxide showed high adsorption capacity (125 mg/g) for the removal of chromate in a broad pH range [119].

Though several kinetic and isotherm equations have been used to model and interpret the process of oxyanion adsorption from aqueous solution. However, the use of requisite analytical instrumentation has been suggested to provide a more valid reaction pathway. During the immobilization of an oxyanion on an adsorbent, the mechanism of the reaction has been shown to be by a combination of several processes, typically, surface adsorption, ion exchange, and precipitation. In a column experiment, for the removal of chromate from stimulated groundwater using acid-washed zero-valent iron [116]. The X-ray photoelectron spectroscopy (XPS) spectra showed that the removal of chromate occurred via reductive precipitation of chromate as chromite on acid-washed zero-valent iron. Fourier transformed infrared (FTIR) and XPS analysis were used to describe the complete removal of chromate onto polyaniline-ethyl cellulose adsorbent [117]. The FTIR adsorption peak of chromate laden-polyaniline-ethyl cellulose adsorbent at 3520 cm^{-1} (–NH) decreased, and the intensity of the peaks at 1580 and 1482 cm^{-1}, reduced with increasing chromate concentration. The XPS analysis showed that chromate was present in a reduced form (i.e., chromite) on the surface of the adsorbent at pH 3. The reduction was ascribed to the presence of amine functional group on the surface of the adsorbent.

Surface adsorption is driven by columbic attraction between oppositely charged species. The removal of chromate using CO_3-LDH was influenced by the electrostatic force of attraction between negatively charged chromate and borate and the abundant cationic species (Mg^{2+} and Al^{3+}) and the exchange of CO_3^{2-} with the oxyanions [120, 121] (Eq. 3.10).

$$Mg_6Al_2CO_3(OH)_{16}4H_2O + Cr_2O_7^{2-} \rightarrow Mg_6Al_2(Cr_2O_7)^{2-}(OH)_{16}4H_2O \quad (3.10)$$

Oxyanions can also be treated as guest anions in the adsorbent lattice, which is peculiar to adsorbent with replaceable anion, and the intercalated oxyanions can be desorbed using an appropriate solvent. Chromate, arsenate, and rapheme in GW systems were guest anions during allophane synthesis [122]. ZrO_2 microsphere was effective in the removal of SeO_3^{2-} and TeO_3^{2-} from aqueous system via ion exchange to form monodentate or bidentate surface chelates [123]. This was confirmed by the enhanced FTIR peaks, which is a representation of Se(IV)–O and Te(IV)–O bond.

The uptake of adsorbate on an adsorbent can proceed via both physical and chemical interaction in a synergistic manner. In the removal of selenite and rapheme from simulated GW, using polymeric engineered material [124], the reaction proceeded by a combination of ligand exchange, between surface hydroxyl group on the polymeric material and selenium oxyanions, and columbic interaction, between positively charged amine on the polymeric material and negatively charged selenium oxyanions. During deprotonation, at alkaline pH range, electrostatic force becomes redundant and the adsorption capacity is lowered. The electrostatic force of attraction was suspected to be the mechanism when Fe_3O_4-chitosan composite was employed for the removal of phosphate from aqueous solution [125]. The removal of phosphate was pH dependent and at pH above 4.0, and the adsorption process was adversely impacted. Competing ions (especially anions) and ionic strength greatly undermined the adsorption process, when it was governed by the electrostatic force of attraction. This is due to competition for active sites on the adsorbent, while ionic strength can impede ionic activities, thus, reducing electrostatic interaction and adsorbent aggregation is induced.

Chitosan and polyethyleneimine were fused with rapheme oxide for the removal of selenium (IV). The composite was rich in –OH and –COOH functionality and the removal efficiency for selenite was pH dependent. The efficiency reduced from pH 3 to pH 8, from approximately 83 to 18% [126]. As shown in Eq. 3.11, selenite was predominant as $HSeO_3^-$ at the operational pH range. At the operational pH range, the surface functional groups (i.e., –OH and –COOH) of the composite got protonated, thereby enhancing the electrostatic force of attraction between adsorbent and adsorbate. The zeta potential analysis of the adsorbent at pH 3–8 showed that the composite displays positive charge surface density.

$$H_2SeO_3 \rightarrow H^+ + HSeO_3^-, pKa = 2.46 \quad (3.11)$$

The uptake of phosphate on zinc-aluminum-zirconium (ZnAlZr) ternary LDH, occurred through outer-sphere complexation, which resulted in the formation of stable crystalline hopeite minerals $Zn_3(PO_4)_3.4H_2O$. The atomic ratio of the LDH was the primary factor that influenced the precipitation of phosphate onto the LDH. The decrease in the atomic ratio of the LDH constituent enhanced surface precipitation [127]. Similarly, the uptake of phosphate using CaO synthesized from gastropod shell formed a mixed stable complexes of Ca_3PO_4, $Ca_5(PO_4)_3OH$, and $Ca_2P_2O_7.2H_2O$ [120] (Eq. 3.12).

$$2Ca^{2+}_{(aq)} + 3PO_4^{3-} + OH^-_{(aq)} \leftrightarrow Ca_5(PO_4)_3OH_{(s)} \tag{3.12}$$

3.4.2   Coagulation

Coagulation is a cost-efficient water management process that has been extensively explored in the treatment of oxyanion contaminated GW. $Fe_2(SO_4)_3$ and $FeCl_3$ were investigated as coagulants for the treatment of molybdate contaminated water [121]. Within the pH value range of 4 and 5, $Fe_2(SO_4)_3$ was more effective than $FcCl_3$, due to the low agglomeration of Fe flocs. The anionic portion (SO_4^{2-}) on $Fe_2(SO_4)_3$ possesses higher charge to compress the electric double layers and neutralize the opposite charge. Similarly, the bi-ionic charge possesses a more attractive force toward negatively charged molybdate. Thus, the iron flocs formed faster and became more stable than $FeCl_3$. However, at pH value between 6 and 9, both coagulants showed similar removal efficiency. The presence of common GW ions influenced the process, thus, increasing concentrations of calcium, silicate, phosphate, and humic acid compete with the molybdate oxyanion. On the other hand, increasing ntimoni ion concentration from 1 to 5 mM at pH 4 enhanced the coagulation process.

The coagulation performance of $FeCl_3$, $AlCl_3$, and polyaluminum chloride (PACl) was compared for the removal of ntimoni and selenite in aqueous solution [128]. At a coagulant dosage of 0.2 and 1.2 mmol/L, the removal efficiency of selenite and ntimoni by $FeCl_3$ was approximately 90% and 98%, and 30% and 98%, respectively. Both ntimonit-based coagulants showed poor removal efficiency. The maximum removal of 78 and 40% was achieved with PACl, at coagulant dosage of 1.2 mmol/L for selenite and ntimoni. The removal efficiency of $AlCl_3$ was generally poor (<45% removal). Competing anions that formed inner-sphere (o-plane) surface complexes, which are barely affected by the change in ionic strength, played a major role in the coagulation process, because they bind strongly with metal hydroxides. Even though phosphate is firmly bonded to metal hydroxide (active sites), ntimoni formed weak bonds with reactive sites [129].

Unlike ntimonite, antimonite was preferably removed by $FeCl_3$ coagulant, at low dosage, wide pH range, and in the presence of phosphate and humic acids [130]. Similarly, arsenic responded differently to $FeCl_3$ coagulation treatment. Increasing coagulant dose enhanced the removal of arsenite, whereas the removal of arsenate depended on the solution pH and zeta potential [131]. The removal of arsenate using Fe(III)-(oxyhydr)oxides as a coagulant, in a complex mixture of silicate, phosphate, natural organic matter, and calcium showed that phosphate and silicate reduced coagulation efficiency by competitively adsorbing onto Fe(III)-precipitates. Whereas, natural organic matter impeded coagulation efficiency is due to the formation of highly mobile and non-floc forming Fe(III)-natural organic matter complex [132].

In the remediation of GW arsenic, arsenite, a more toxic inorganic form of arsenic, is oxidized to arsenate, which is less toxic, prior to coagulation with an in situ generated coagulant species. Ferrate (FeO_4^{2-}) has been described as the right candidate for oxyanion removal from aqueous solution due to its redox potentials, 2.20 and 0.72 V, in acidic and basic solutions, respectively. Fe(VI) is simultaneously reduced to Fe(III) during the oxidation of arsenite to arsenate and the Fe(III) produced, in situ, served as a coagulant for the removal of arsenate from solution [133].

The initial arsenic oxyanion concentration often dictates the optimum coagulant dosage. At a high initial arsenic concentration, the use of high coagulant dosage was detrimental to the removal of arsenate, using ntimonit ntimoni. At lower initial concentration, a high coagulant dosage enhanced the removal efficiency [134]. The use of coagulant aids was employed to enhance the coagulation process [135, 136]. In the presence of ntimonit ntimoni coagulant, the use of polyelectrolyte as coagulant aid was more beneficial than the use of anionic or non-ionic coagulant aid [134].

In the coagulation-flocculation of oxyanions, using inorganic coagulants, three reaction pathways have been highlighted, thus: (1) precipitation and formation of insoluble-oxyanion containing compounds; (2) co-precipitation into growing hydroxide; (3) adsorption of oxyanions onto the formed hydroxide flocs. The removal of selenite from the aqueous phase via coagulation using Fe-based coagulant was governed by precipitation at low concentration and adsorption onto metal hydroxide flocs at higher levels [128]. The removal of ntimoni using ntimonit-based coagulants followed a similar mechanism; adsorption on formed hydroxide flocs [137]. The removal efficiency was dependent on Al speciation, and in situ formed Al_{13} was more effective in the removal of ntimoni than preformed Al_{13}.

Similarly, the mechanism for the removal of arsenic oxyanion in the presence of humic acids depends on arsenic speciation. The removal of arsenate followed its adsorption onto preformed ferric oxides. However, the removal of arsenite is followed precipitation and co-precipitation [131].

Ferric chloride was more effective than alum for the removal of antimony in drinking water. The soluble species antimony was incorporated into the growing hydroxide phase due to adsorption–inner-sphere surface complex model (13–14) [130]. Precipitation and co-precipitation mechanism route was excluded because initial antimony concentration and interfering ion affected the coagulation process, and the result from the adsorption process was similar to the overall reaction. That is, reaction via co-precipitation should be selective toward antimonite or ntimonite.

$$FeOH + Sb(OH)_2 \leftrightarrow FeOSb(OH)_2 + H_2O \qquad (3.13)$$

$$FeOH + Sb(OH)_6^- \leftrightarrow FeOSb(OH)_5^- + H_2O \qquad (3.14)$$

3.4.3 Membrane Separation

Membrane filtration is one of the most practical and sustainable technology for water managements, especially in communities, where end-users are responsible for water treatment. Currently, straws and filters are available and encouraged by the WHO for use. The design of membrane filter varies from organic to inorganic materials. CuO nanoparticles were synthesized via co-precipitation for the removal of arsenic oxyanions in GW [138]. On the filter surface, the CuO nanoparticles simultaneously oxidized arsenite to arsenate and immobilized the residual arsenite. Since the filter composition can interfere with oxyanions during the filtration process, filters could be designed to influence the oxidation state of the oxyanions and subsequently filter the product.

Commercially available nanofilter (NE 90, a negatively charged polyamide filter) has been tested for the removal of arsenic oxyanion from water [139]. In addition to Donnan repulsion due to negative charge oxyanions (arsenite and arsenate), removal of arsenic oxyanions occurred through steric exclusion. Monovalent arsenate was more effectively retained than bivalent arsenite, due to Donnan exclusion. This was enhanced in the presence of Cl^- and deteriorated when SO_4^{2-} was added. Another commercial nanofilter (192-NF 300) was tested for the removal of arsenate from synthetic natural water [140]. The removal of arsenate increased from 93 to 99%, with increasing initial concentration of arsenate from 100 to 382 $\mu g/L$. The removal efficiency was independent of transmembrane pressure, crossflow velocity, and temperature.

The factors that govern the performance of membrane during the filtration of oxyanions in GW, include the molecular weight and size of oxyanion, membrane filter compactness, filter composition, charge density, solution pH, and pressure.

Charged species are retained than neutral species and the change in solution pH can influence the oxyanion speciation. These factors are the determinants of the ionic radii of the oxyanions and the removal efficiencies of oxyanions [141]. For example, the speciation of As(V) changes from monovalent $H2AsO_4^-$ to bivalent $HAsO_4^{2-}$, as the solution pH increases from 4 to 10. Such change can improve the removal efficiencies of arsenic oxyanion in aqueous solutions, due to steric and columbic effects. Similarly, the difference in pH can affect the surface chemistry of the membrane filter. A commercial membrane filter NE 90 contains both amine and carboxylic functional groups. Thus, the surface NE 90 would be positively charged below the isoelectric point, due to the protonation of amine, while it is negatively charged above the isoelectric point, due to the deprotonation of the carboxyl group [139]. A negatively charged membrane surface is repulsive to oxyanions, and an enhanced exclusion efficiency occurs during filtration process.

Polyhedral oligomeric silasesquioxane (POSS) polyamide thin-film nanofilter membrane was fabricated for the removal of arsenic and selenium oxyanions from aqueous solution [142]. The exclusion of common GW ions followed the order: $MgSO_4 > Na_2SO_4 > MgCl_2 > NaCl$. In addition to the average diameter of the nanofilter, and the presence of $–COO^-$ on the surface of the membrane filter resulted in the repulsion of the highly negatively charged ions, SO_4^{2-}, than mono-valence Cl. According to Eqs. 3.15 and 3.16, the removal of selenium oxyanions and arsenate according to their hydrated diameter and charge follows the order: $HAsO_4^{2-} > SeO_4^{2-} > SeO_3^{2-} > HSeO_3^-$ [142].

$$H_2SeO_3 \leftrightarrow HSeO_3^- + H^+ \quad pKa = 2.6 \quad\quad\quad\quad (3.15)$$

$$HSeO_3^- \leftrightarrow SeO_3^{2-} + H^+ \quad pKa = 7.3 \quad\quad\quad\quad (3.16)$$

3.4.4   Oxidation–Reduction

GW remediation via redox-based approach is a robust water management process because it incorporates broad water treatment methods. The process includes the use of chemical reagents, (photo)catalysts and photons, as oxidants and reductants. The simplest form of such reaction involves the use of $KMnO_4$, a potent oxidizing agent. For instance, the Waynesville water system utilizes $KmnO_4$ for the removal of arsenic contamination in GW that is rich in nitrate and iron [143]. When appropriate oxidant dosage is employed, iron/arsenate and Mn are filtered out as insoluble particulates.

In an heterogenous solution, nitrate reduction to gaseous nitrogen was found to depend on the concentration, the reductant, and the acid type [144]. The reduction of nitrate in the presence of the reductants and acid trends was followed, thus: formate > oxalate > citrate and HCl > H_2SO_4 > H_3BO_3 > H3PO$_4$ for organic and inorganic acids, respectively. Citrate and phosphate are strong ligand that can bind strongly to impede reduction reaction.

The Fenton process is also a process that induces the oxidation–reduction reaction of oxyanions through the use of H_2O_2 and iron species. However, due to varying GW physiochemical properties, adjusting the operation parameters is essential to achieve a toxic-free oxyanion GW. Fenton process involves multiple stages that include redox, precipitation, and adsorption. In the treatment of arsenic-contaminated GW, resulting arsenate from the oxidation of arsenite at pH value that ranged between 6 and 8 was coprecipitated as insoluble ferric arsenate [145]. Commercially available micro-scale zero-valent iron was used as Fe(II) source, in the presence of H_2O_2 for the oxidation of As(III) to As(V) in aerated simulated GW [146]. Complete oxidation of arsenite was achieved in 2 h in the presence of micro-scale zero-valent iron. The addition of H2O2, which enhanced the oxidant concentration, increased reaction rate, and the complete oxidation of As(III) was achieved in 8 min. It should be noted that As(III) was preferentially selected for oxidation over Fe(II) [147]. At pH 6–7, the oxidation of As(III) inhibited the formation of lepidocrocite, goethite, and magnetite. Light energy can be incorporated with the Fenton system to promote the reaction rate [148].

It should be noted that the redox species for the reaction can also be generated in situ, using photon energy, and in the presence of semiconductor catalyst [149]. The heat from the photons solely induced a redox reaction, while the photon caused the excitation of electron in the unfilled valence band of the semiconductor. Thus, the excited particle is used for a series of redox reactions [150]. TiO_2 is a model photocatalyst that has been explored in many photons-assisted oxidation of GW oxyanions. In a comparative study, the oxidation of arsenite to arsenate was compared, using UV energy and UV-assisted TiO_2 [151]. The UV-assisted TiO_2 oxidation showed similar oxidation efficiency (97%) with that of photolysis (90%). Natural light (7.5 KJ_{UV}/L) was also employed for the oxidation of arsenite to arsenate at an initial concentration of 1000 µg/L [152] and the oxidation of As(III) was completed in 8 h.

The coupling of semiconductors or metal oxide usually enhances photocatalysis reaction. ZnO/CuO demonstrated improved arsenite oxidation in the presence of irradiation than bare ZnO or CuO. At the initial arsenite concentration of 30 mg/L, photolysis accounted for 8% oxidation in 4 h, photo-oxidation with bare CuO in the presence of irradiation accounted for 20%, while it was 55% for pure ZnO, in the presence of photons. However, when ZnO was coupled with CuO, the reaction was completed at the same reaction time interval [153].

However, since oxidation and reduction reactions occur simultaneously, the in situ generated reactive species are highly responsive and tend to react with each other with the liberation of heat. In order to obviate this shortcoming, hole/electron scavengers are usually introduced to increase the lifetime of redox species. In the reduction of chromate to chromite, humic acid was used as an electron scavenger to accelerate the photoreduction rate and the humic acid was oxidized concurrently [154]. Doping [155] and the creation of homo/hetero-junctions can also be used to achieve a similar purpose [156].

The mechanism of oxidation and reduction reaction solely depends on electron transfer within the reacting medium [157]. Electron transfer is initiated due to entropy change within the system [158]. Energy from photons is used to initiate electron transfer or the addition of catalyst (homogenous and heterogeneous). The oxidation or reduction of oxyanions in GW is initiated with H_2O_2, a homogenous system. It can also be initiated in a mixed system using an iron species, according to Eqs. 3.17 and 3.18;

$$H_2O_{2(aq)} + H_2SeO_{3(aq)} \rightarrow H_2O_{(l)} + H_2SeO_{4(aq)} \tag{3.17}$$

$$NO_{3(aq)}^- + 4Fe_{(s)}^0 + 10H^+ \leftrightarrow NH_{4(aq)}^+ + 4Fe_{(s)}^{2+} + 3H_2O_{(l)} \tag{3.18}$$

The photocatalysis reaction involves the oxidation or reduction of the adsorbed oxyanion by reactive species (electron and hole), which was initially generated on the surface of the photocatalyst. Therefore, some other physiochemical factors, aside from the availability of reactive species, can also predict the reaction pathways [159]. For example, crystallinity, surface area and light source [160].

Redox reaction occurs concurrently, therefore, multicomponent GW impurities can be removed simultaneously. For example, the direct redox reaction of chromate and arsenite is prolonged (Eq. 3.19).

$$2HCr^{6+}O_4^- + 3HAs^{3+}O_2 + 5H^+$$
$$\rightarrow 2Cr^{3+} + 3H_2As^{+5}O_4^- + 2H_2O, \quad \Delta G = -467.95 \text{ kJ/mol} \qquad (3.19)$$

In lieu of this, highly unselective redox species is introduced. Graphitic carbon nitride was used for the conversion of arenite to arsenate and chromate to chromite [161]. The photogenerated e^- and h^+ were the primary reduction and oxidation species for chromate and arsenite, respectively. These species are generated when semiconductor (graphitic carbon nitride) adsorbed energy (photon) that is equal or greater than the bandgap. The light-assisted generated e^- induced the reduction of adsorbed oxygen or water to form superoxide radical ($\cdot O_2^-$) or hydrogen peroxide (H_2O_2) (Eqs. 3.20 and 3.21) and the subsequent oxidation to arsenate (Eq. 3.22). Similarly, the conversion of arsenite can result from the oxidation of adsorbing water molecule by photogenerated h^+ (Eq. 3.23), by direct oxidation (Eq. 3.24) or using the product from Eq. 3.23.

$$O_2 + e^- \rightarrow \cdot O_2^- \qquad (3.20)$$

$$O_2 + 2e^- + 2H^+ \rightarrow H_2O_2 \qquad (3.21)$$

$$As^{3+} + \cdot O_2^- + 2H^+ \rightarrow As^{+5} + H_2O_2 \qquad (3.22)$$

$$H_2O + h^+ \rightarrow \cdot OH + H^+ \qquad (3.23)$$

$$As^{3+} + 2h^+ \rightarrow As^{+5} \qquad (3.24)$$

$$As^{3+} + \cdot OH + H^+ \rightarrow As^{4+} + H_2O \qquad (3.25)$$

$$As^{4+} + \cdot OH + H^+ \rightarrow As^{5+} + H_2O \qquad (3.26)$$

3.4.5 Natural Processes

Natural processes are significant in the remediation of GW contamination. This is viable through adsorption onto minerals and microalgae or precipitation with dissolved nutrient or redox processes. Fe(II)-bearing minerals, chlorite, and biotite were identified as natural reductants in aquifers [101]. These minerals can reduce chromate to chromite. The stability of the Cr(III) precipitate formed $(Cr_xFe_{(1-x)}(OH)_3)$ increased with aging, as it is transformed from the amorphous state to the crystalline state. Biogenic manganese oxide, in the presence of oxidizing bacteria, is efficient for the removal of arsenite/arsenate at low concentration (35/42 µg/L), as insoluble manganese-arsenate complex [162]. The removal mechanism involved the oxidation of Mn(II), Fe(II), and As(III) to Mn(IV), Fe(III), and As(V) by dissolved oxygen and bacteria, respectively. A significant proportion of As(III) was also oxidized by Mn(IV) and the As(V) produced is adsorbed onto Mn(IV).

Green marine algae have also been explored as sustainable biosorbent. *Cladophora sericea* and *Ulva rigida* green algae have shown promise as biosorbent for the removal of antimony oxyanions [163]. The uptake of antimonate was higher than antimonite for both biomasses, with *Cladophora sericea* showing more sorption efficiency than *Ulva rigida* due to the abundant hydroxyl sites. The higher carboxylic group content on *Ulva rigida* is susceptible to deprotonation at pH 5, thereby, impacting negatively on the adsorption process.

Micro-organism can directly reduce high valent oxyanions in GW system. *Dechloromonas* sp. has shown to be a good candidate for the reduction of perchlorate [164]. The addition of an electron donor improves the reduction reaction. In the presence of acetate, *Dechloromonas* sp. completely degraded 100 mg/L perchlorate in 10 h, whereas 22 h were required to achieve a similar result in the absence of acetate.

Although bacterial are capable of expunging toxic oxyanions from the GW system, but the efficiency can be enhanced by the addition of inorganic ions. Selenate reducing bacterial, *Citrobacter braakii*, was used for the removal of selenate as zero-valent selenium or organic selenium [165]. The residual selenate after the reduction experiment, at 1000 µg/L, initial concentration was 550 µg/L during an 8 day experiment. The incorporation of zero-valent iron with *Citrobacter braakii* improved the rate of reaction; thus, the total removal of selenate was completed in less than 3 days.

Perchlorates are prone to reduction in the presence of bacterial; thus, the reduction of perchlorate was achieved (85%) at neutral pH (7–7.5) in the presence of proteobacterial species (*Dechloromonas* spp., *methyloversatilis* spp., and *ferribacterium* spp.). The presence of chloride and sulfate was beneficial to the reduction processes. However, due to the faster reduction rate of nitrate in the presence of the bacterial, the dissolved nitrate negatively impacted the magnitude of perchlorate reduction [166].

3.5 Conclusion

GW is a valuable and abundant natural resource that is threatened by pollutants from human and natural sources. Oxyanions are among the natural and man-made GW pollutants that remained dominant in the GW system, by switching valency state or forming soluble and insoluble species. Consequent upon the environmental, health, and economic impacts of oxyanions, its removal from the GW system is paramount to the production of an eco-friendly, sustainable, and hygienic potable water. Different strategies for the removal of oxyanions have been designed to improve the GW qualities and acceptability. The state of research in water science and technology has shown that complete removal of toxic oxyanions is achievable. The overall removal efficiency of oxyanions in GW can be enhanced by improved material science and engineering and the adjustment of process parameters and reaction variables.

References

1. Madramootoo CA (2012) Sustainable groundwater use in agriculture. Irrig Drain 61:26–33. https://doi.org/10.1002/ird.1658
2. Alagbe SA (2002) Groundwater resources of river Kan Gimi Basin, north-central, Nigeria. Environ Geol 42:404–413. https://doi.org/10.1007/s00254-002-0544-9
3. Qiu J, Zipper SC, Motew M, Booth EG, Kucharik CJ, Loheide SP (2019) Nonlinear groundwater influence on biophysical indicators of ecosystem services. Nature Sustainability 2:475–483. https://doi.org/10.1038/s41893-019-0278-2
4. MacDonald AM, Carlow RC, MacDonald DMJ, Darling WG, Dochartaigh BÉÓ (2009) What impact will climate change have on rural groundwater supplies in Africa? Hydrol Sci J 54:690–703. https://doi.org/10.1623/hysj.54.4.690
5. Taylor RG, Todd MC, Kongola L, Maurice L, Nahozya E, Sanga H, Macdonald AM (2013) Evidence of the dependence of groundwater resources on extreme rainfall in East Africa. Nat Clim Change 3:374–378. https://doi.org/10.1038/nclimate1731
6. Van der Gun J (2012) Groundwater and global change: trends, opportunities and challenges. International Groundwater Resources Assessment Centre, Paris: UNESCO, 2012, France, 2012. https://www.un-igrac.org/resource/groundwater-and-global-change-trends-opportunities-and-challenges.
7. Aeschbach-Hertig W, Gleeson T (2012) Regional strategies for the accelerating global problem of groundwater depletion. Nat Geosci 5:853–861. https://doi.org/10.1038/ngeo1617
8. Gurdak JJ (2017) Climate-induced pumping 1. www.nature.com/naturegeoscience.
9. MacDonald AM, Bonsor HC, Ahmed KM, Burgess WG, Basharat M, Calow RC, Dixit A, Foster SSD, Gopal K, Lapworth DJ, Lark RM, Moench M, Mukherjee A, Rao MS, Shamsudduha M, Smith L, Taylor RG, Tucker J, Van Steenbergen F, Yadav SK (2016) Groundwater quality and depletion in the Indo-Gangetic Basin mapped from in situ observations. Nat Geosci 9:762–766. https://doi.org/10.1038/ngeo2791
10. Perrone D, Jasechko S (2019) Deeper well drilling an unsustainable stopgap to groundwater depletion. Nat Sustain 2:773–782. https://doi.org/10.1038/s41893-019-0325-z
11. Rayne S, Forest K (2011) Temporal trends in groundwater levels from Saskatchewan, Canada. Nat Precedings. https://doi.org/10.1038/npre.2011.6696.1
12. Asoka A, Gleeson T, Wada Y, Mishra V (2017) Relative contribution of monsoon precipitation and pumping to changes in groundwater storage in India. Nat Geosci 10:109–117. https://doi.org/10.1038/ngeo2869
13. Sanchez TR, Levy D, Shahriar MH, Uddin MN, Siddique AB, Graziano JH, Lomax-Luu A, van Geen A, Gamble MV (2016) Provision of well-water treatment units to 600 households in Bangladesh: a longitudinal analysis of urinary arsenic indicates fading utility. Sci Total Environ 563–564(2016):131–137. https://doi.org/10.1016/j.scitotenv.2016.04.112
14. Moyé J, Picard-Lesteven T, Zouhri L, El Amari K, Hibti M, Benkaddour A (2017) Groundwater assessment and environmental impact in the abandoned mine of Kettara (Morocco). Environ Pollut 231(2017):899–907. https://doi.org/10.1016/j.envpol.2017.07.044

15. Tesoriero AJ, Terziotti S, Abrams DB (2015) Predicting redox conditions in groundwater at a regional scale. Environ Sci Technol 49:9657–9664. https://doi.org/10.1021/acs.est.5b01869
16. Zhu M, Zeng X, Jiang Y, Fan X, Chao S, Cao H, Zhang W (2017) Determination of arsenic speciation and the possible source of methylated arsenic in Panax Notoginseng. Chemosphere 168:1677–1683. https://doi.org/10.1016/j.chemosphere.2016.10.093
17. Shafiquzzaman M, Azam MS, Nakajima J, Bari QH (2010) Arsenic leaching characteristics of the sludges from iron based removal process. Desalination 261:41–45. https://doi.org/10.1016/j.desal.2010.05.049
18. Darvari R, Nicot JP, Scanlon BR, Mickler P, Uhlman K (2018) Trace element behavior in methane-rich and methane-free groundwater in north and east texas. Groundwater 56:705–718. https://doi.org/10.1111/gwat.12606
19. del Pilar Alvarez M, Carol E (2019) Geochemical occurrence of arsenic, vanadium and fluoride in groundwater of Patagonia, Argentina: sources and mobilization processes. J South Am Earth Sci 89(2019):1–9. https://doi.org/10.1016/j.jsames.2018.10.006
20. Mukherjee A, Duttagupta S, Chattopadhyay S, Bhanja SN, Bhattacharya A, Chakraborty S, Sarkar S, Ghosh T, Bhattacharya J, Sahu S (2019) Impact of sanitation and socio-economy on groundwater fecal pollution and human health towards achieving sustainable development goals across India from ground-observations and satellite-derived nightlight. Sci Rep 9:1–11. https://doi.org/10.1038/s41598-019-50875-w
21. Telfeyan K, Breaux A, Kim J, Kolker AS, Cable JE, Johannesson KH (2018) Cycling of oxyanion-forming trace elements in groundwaters from a freshwater deltaic marsh. Estuar Coast Shelf Sci 204:236–263. https://doi.org/10.1016/j.ecss.2018.02.024
22. Manning BA, Fendorf SE, Bostick B, Suarez DL (2002) Arsenic(III) oxidation and arsenic(V) adsorption reactions on synthetic birnessite. Environ Sci Technol 36:976–981. https://doi.org/10.1021/es0110170
23. Lin CJ (2002) The chemical transformations of chromium in natural waters—a model study. Water Air Soil Pollut 139:137–158. https://doi.org/10.1023/A:1015870907389
24. MacDonald D, Dixon A, Newell A, Hallaways A (2012) Groundwater flooding within an urbanised flood plain. J Flood Risk Manag 5:68–80. https://doi.org/10.1111/j.1753-318X.2011.01127.x
25. Currell M, Cartwright I, Raveggi M, Han D (2011) Controls on elevated fluoride and arsenic concentrations in groundwater from the Yuncheng Basin China. Appl Geochem 26:540–552. https://doi.org/10.1016/j.apgeochem.2011.01.012
26. Van Geen A, Bostick BC, Trang PTK, Lan VM, Mai NN, Manh PD, Viet PH, Radloff K, Aziz Z, Mey JL, Stahl MO, Harvey CF, Oates P, Weinman B, Stengel C, Frei F, Kipfer R, Berg M (2013) Retardation of arsenic transport through a Pleistocene aquifer. Nature 501(2013):204–207. https://doi.org/10.1038/nature12444
27. Smedley PL (1996) Arsenic in rural groundwater in Ghana. J Afr Earth Sc 22:459–470. https://doi.org/10.1016/0899-5362(96)00023-1
28. Sosa NN, Kulkarni HV, Datta S, Beilinson E, Porfido C, Spagnuolo M, Zárate MA, Surber J (2019) Occurrence and distribution of high arsenic in sediments and groundwater of the Claromecó fluvial basin, southern Pampean plain (Argentina). Sci Total Environ 695:(2019). https://doi.org/10.1016/j.scitotenv.2019.133673
29. Nicolli HB, Bundschuh J, García JW, Falcón CM, Jean JS (2010) Sources and controls for the mobility of arsenic in oxidizing groundwaters from loess-type sediments in arid/semi-arid dry climates—evidence from the Chaco-Pampean plain (Argentina). Water Res 44:5589–5604. https://doi.org/10.1016/j.watres.2010.09.029
30. Mukherjee A, Gupta S, Coomar P, Fryar AE, Guillot S, Verma S, Bhattacharya P, Bundschuh J, Charlet L (2019) Plate tectonics influence on geogenic arsenic cycling: from primary sources to global groundwater enrichment. Sci Total Environ 683:793–807. https://doi.org/10.1016/j.scitotenv.2019.04.255
31. Vengosh A, Coyte R, Karr J, Harkness JS, Kondash AJ, Ruhl LS, Merola RB, Dywer GS (2016) Origin of hexavalent chromium in drinking water wells from the piedmont aquifers of North Carolina. Environ Sci Technol Lett 3:409–414. https://doi.org/10.1021/acs.estlett.6b00342

32. Morrison JM, Goldhaber MB, Mills CT, Breit GN, Hooper RL, Holloway JAM, Diehl SF, Ranville JF (2015) Weathering and transport of chromium and nickel from serpentinite in the Coast Range ophiolite to the Sacramento Valley, California, USA. Appl Geochem 61:72–86. https://doi.org/10.1016/j.apgeochem.2015.05.018
33. Izbicki JA, Wright MT, Seymour WA, McCleskey RB, Fram MS, Belitz K, Esser BK (2015) Cr(VI) occurrence and geochemistry in water from public-supply wells in California. Appl Geochem 63:203–217. https://doi.org/10.1016/j.apgeochem.2015.08.007
34. Bassil J, Naveau A, Bueno M, Razack M, Kazpard V (2018) Leaching behavior of selenium from the karst infillings of the Hydrogeological Experimental Site of Poitiers. Chem Geol 483:141–150. https://doi.org/10.1016/j.chemgeo.2018.02.032
35. Pourret O, Dia A, Gruau G, Davranche M, Bouhnik-Le Coz M (2012) Assessment of vanadium distribution in shallow groundwaters. Chem Geol 294–295:89–102. https://doi.org/10.1016/j.chemgeo.2011.11.033
36. Wright MT, Belitz K (2010) Factors controlling the regional distribution of vanadium in groundwater. Ground Water 48:515–525. https://doi.org/10.1111/j.1745-6584.2009.00666.x
37. Al Kuisi M, Al-Hwaiti M, Mashal K, Abed AM (2015) Spatial distribution patterns of molybdenum (Mo) concentrations in potable groundwater in Northern Jordan. Environ Monit Assess 187:1–26. https://doi.org/10.1007/s10661-015-4264-5
38. Smedley PL, Cooper DM, Ander EL, Milne CJ, Lapworth DJ (2014) Occurrence of molybdenum in British surface water and groundwater: distributions, controls and implications for water supply. Appl Geochem 40:144–154. https://doi.org/10.1016/j.apgeochem.2013.03.014
39. Leybourne MI, Cameron EM (2008) Source, transport, and fate of rhenium, selenium, molybdenum, arsenic, and copper in groundwater associated with porphyry-Cu deposits, Atacama Desert, Chile. Chem Geol 247:208–228. https://doi.org/10.1016/j.chemgeo.2007.10.017
40. Cao F, Jaunat J, Sturchio N, Cancès B, Morvan X, Devos A, Barbin V, Ollivier P (2019) Worldwide occurrence and origin of perchlorate ion in waters: a review. Sci Total Environ 661:737–749. https://doi.org/10.1016/j.scitotenv.2019.01.107
41. Wang S, Zhu G, Zhuang L, Li Y, Liu L, Lavik G, Berg M, Liu S, Long XE, Guo J, Jetten MSM, Kuypers MMM, Li F, Schwark L, Yin C (2019) Anaerobic ammonium oxidation is a major N-sink in aquifer systems around the world. ISME J. https://doi.org/10.1038/s41396-019-0513-x
42. Harita Y, Hori T, Sugiyama M (2005) Release of trace oxyanions from littoral sediments and suspended particles induced by pH increase in the epilimnion of lakes. Limnol Oceanogr 50:636–645. https://doi.org/10.4319/lo.2005.50.2.0636
43. Fendorf S, Benner SG (2016) Indo-Gangetic groundwater threat. Nat Geosci 9:732–733. https://doi.org/10.1038/ngeo2804
44. Gleeson T, Wada Y, Bierkens MFP, Van Beek LPH (2012) Water balance of global aquifers revealed by groundwater footprint. Nature 488:197–200. https://doi.org/10.1038/nature11295
45. Wilkin RT, Lee TR, Beak DG, Anderson R, Burns B (2018) Groundwater co-contaminant behavior of arsenic and selenium at a lead and zinc smelting facility. Appl Geochem 89:255–264. https://doi.org/10.1016/j.apgeochem.2017.12.011
46. Erickson ML, Malenda HF, Berquist EC, Ayotte JD (2019) Arsenic concentrations after drinking water well installation: time-varying effects on arsenic mobilization. Sci Total Environ 678:681–691. https://doi.org/10.1016/j.scitotenv.2019.04.362
47. Brandt MJ, Johnson KM, Elphinston AJ, Ratnayaka DD (2017) Chapter 7—Chemistry, microbiology and biology of water. In: Brandt MJ, Johnson KM, Elphinston AJ, Ratnayaka DD (eds) Butterworth-Heinemann, Boston, pp 235–321. https://doi.org/10.1016/B978-0-08-100 025-0.00007-7
48. Armienta MA, Rodriguez R, Quere A, Juarez F, Ceniceros N, Aguayo A (1993) Groundwater pollution with chromium in Leon valley, Mexico. Int J Environ Anal Chem 54:1–13. https://doi.org/10.1080/03067319308044422
49. Kanagaraj G, Elango L (2019) Chromium and fluoride contamination in groundwater around leather tanning industries in southern India: implications from stable isotopic ratio $\Delta 53Cr/\Delta 52Cr$, geochemical and geostatistical modelling. Chemosphere 220:943–953. https://doi.org/10.1016/j.chemosphere.2018.12.105

50. Mochizuki A, Murata T, Hosoda K, Katano T, Tanaka Y, Mimura T, Mitamura O, Nakano S, Okazaki Y, Sugiyama Y, Satoh Y, Watanabe Y, Dulmaa A, Ayushsuren C, Ganchimeg D, Drucker VV, Fialkov VA, Sugiyama M (2018) Distributions and geochemical behaviors of oxyanion-forming trace elements and uranium in the Hövsgöl–Baikal–Yenisei water system of Mongolia and Russia. J Geochem Explor 188:123–136. https://doi.org/10.1016/j.gexplo.2018.01.009

51. Urbansky ET (2002) Perchlorate as an environmental contaminant. Environ Sci Pollut Res 9:187–192. https://doi.org/10.1007/BF02987487

52. Herman DC, Frankenberger WT (1998) Microbial-mediated reduction of perchlorate in groundwater. J Environ Qual 27:750–754. https://doi.org/10.2134/jeq1998.00472425002700040004x

53. Al-Mikhlafi AS (2010) Groundwater quality of Yemen volcanic terrain and their geological and geochemical controls. Arab J Geosci 3:193–205. https://doi.org/10.1007/s12517-009-0068-7

54. Mora A, Mahlknecht J, Rosales-Lagarde L, Hernández-Antonio A (2017) Assessment of major ions and trace elements in groundwater supplied to the Monterrey metropolitan area, Nuevo León, Mexico. Environ Monit Assess 189. https://doi.org/10.1007/s10661-017-6096-y

55. Giménez-Forcada E, Vega-Alegre M (2015) Arsenic, barium, strontium and uranium geochemistry and their utility as tracers to characterize groundwaters from the Espadán-Calderona Triassic Domain Spain. Sci Total Environ 512–513:599–612. https://doi.org/10.1016/j.scitotenv.2014.12.010

56. Nogueira G, Stigter TY, Zhou Y, Mussa F, Juizo D (2019) Understanding groundwater salinization mechanisms to secure freshwater resources in the water-scarce city of Maputo, Mozambique. Sci Total Environ 661:723–736. https://doi.org/10.1016/j.scitotenv.2018.12.343

57. Galloway J, Fox A, Lewandowski J, Arnon S (2019) The effect of unsteady streamflow and stream-groundwater interactions on oxygen consumption in a sandy streambed. Sci Rep 9:1–11. https://doi.org/10.1038/s41598-019-56289-y

58. Lapworth DJ, MacDonald AM, Krishan G, Rao MS, Gooddy DC, Darling WG (2015) Groundwater recharge and age-depth profiles of intensively exploited groundwater resources in northwest India. Geophys Res Lett 42:7554–7562. https://doi.org/10.1002/2015GL065798

59. Smith R, Knight R, Fendorf S (2018) Overpumping leads to California groundwater arsenic threat. Nat Commun 9:1–6. https://doi.org/10.1038/s41467-018-04475-3

60. Nicolli HB, García JW, Falcón CM, Smedley PL (2012) Mobilization of arsenic and other trace elements of health concern in groundwater from the Salí River Basin, Tucumán Province, Argentina. Environ Geochem Health 34:251–262. https://doi.org/10.1007/s10653-011-9429-8

61. Biswas A, Majumder S, Neidhardt H, Halder D, Bhowmick S, Mukherjee-Goswami A, Kundu A, Saha D, Berner Z, Chatterjee D (2011) Groundwater chemistry and redox processes: depth dependent arsenic release mechanism. Appl Geochem 26:516–525. https://doi.org/10.1016/j.apgeochem.2011.01.010

62. Burgess WG, Hoque MA, Michael HA, Voss CI, Breit GN, Ahmed KM (2010) Vulnerability of deep groundwater in the Bengal aquifer system to contamination by arsenic. Nat Geosci 3:83–87. https://doi.org/10.1038/ngeo750

63. Radloff KA, Zheng Y, Michael HA, Stute M, Bostick BC, Mihajlov I, Bounds M, Huq MR, Choudhury I, Rahman MW, Schlosser P, Ahmed KM, Van Geen A (2011) Arsenic migration to deep groundwater in Bangladesh influenced by adsorption and water demand. Nat Geosci 4:793–798. https://doi.org/10.1038/ngeo1283

64. Chidambaram S, Thilagavathi R, Thivya C, Karmegam U, Prasanna MV, Ramanathan A, Tirumalesh K, Sasidhar P (2017) A study on the arsenic concentration in groundwater of a coastal aquifer in south-east India: an integrated approach. Environ Dev Sustain 19:1015–1040. https://doi.org/10.1007/s10668-016-9786-7

65. Arnórsson S, Óskarsson N (2007) Molybdenum and tungsten in volcanic rocks and in surface and <100 °C ground waters in Iceland. Geochim Cosmochim Acta 71:284–304. https://doi.org/10.1016/j.gca.2006.09.030

66. Seiler RL, Stollenwerk KG, Garbarino JR (2005) Factors controlling tungsten concentrations in ground water, Carson Desert, Nevada. Appl Geochem 20:423–441. https://doi.org/10.1016/j.apgeochem.2004.09.002
67. McEwan AG, Ridge JP, McDevitt CA, Hugenholtz P (2002) The DMSO reductase family of microbial molybdenum enzymes; molecular properties and role in the dissimilatory reduction of toxic elements. Geomicrobiol J 19:3–21. https://doi.org/10.1080/014904502317246138
68. Bondu R, Cloutier V, Benzaazoua M, Rosa E, Bouzahzah H (2017) The role of sulfide minerals in the genesis of groundwater with elevated geogenic arsenic in bedrock aquifers from western Quebec, Canada. Chem Geol 474:33–44. https://doi.org/10.1016/j.chemgeo.2017.10.021
69. Mukherjee A, Fryar AE, Eastridge EM, Nally RS, Chakraborty M, Scanlon BR (2018) Controls on high and low groundwater arsenic on the opposite banks of the lower reaches of River Ganges, Bengal basin, India. Sci Total Environ 645:1371–1387. https://doi.org/10.1016/j.scitotenv.2018.06.376
70. Samanta G, Clifford DA (2005) Preservation of inorganic arsenic species in groundwater. Environ Sci Technol 39:8877–8882. https://doi.org/10.1021/es051185i
71. Rakshit S, Sallman B, Davantés A, Lefèvre G (2017) Tungstate (VI) sorption on hematite: an in situ ATR-FTIR probe on the mechanism. Chemosphere 168:685–691. https://doi.org/10.1016/j.chemosphere.2016.11.007
72. Martin DP, Seiter JM, Lafferty BJ, Bednar AJ (2017) Exploring the ability of cations to facilitate binding between inorganic oxyanions and humic acid. Chemosphere 166:192–196. https://doi.org/10.1016/j.chemosphere.2016.09.084
73. Das D, Samanta G, Mandal BK, Chowdhury TR, Chanda CR, Chowdhury PP, Basu GK, Chakraborti D (1996) Arsenic in groundwater in six districts of West Bengal, India. Environ Geochem Health 18:5–15. https://doi.org/10.1007/BF01757214
74. Bhattacharyya R, Jana J, Nath B, Sahu SJ, Chatterjee D, Jacks G (2003) Groundwater As mobilization in the Bengal Delta Plain, the use of ferralite as a possible remedial measure—a case study. Appl Geochem 18:1435–1451. https://doi.org/10.1016/S0883-2927(03)00061-1
75. Bauer M, Blodau C (2006) Mobilization of arsenic by dissolved organic matter from iron oxides, soils and sediments. Sci Total Environ 354:179–190. https://doi.org/10.1016/j.scitotenv.2005.01.027
76. Harvey CF, Swartz CH, Badruzzaman ABM, Keon-Blute N, Yu W, Ali MA, Jay J, Beckie R, Niedan V, Brabander D, Oates PM, Ashfaque KN, Islam S, Hemond HF, Ahmed MF (2002) Arsenic mobility and groundwater extraction in Bangladesh. Science 298:1602–1606. https://doi.org/10.1126/science.1076978
77. Gault AG, Islam FS, Polya DA, Charnock JM, Boothman C, Chatterjee D, Lloyd JR (2005) Microcosm depth profiles of arsenic release in a shallow aquifer, West Bengal. Mineral Mag 69:855–863. https://doi.org/10.1180/0026461056950293
78. McMahon PB, Chapelle FH, Bradley PM (2011) Evolution of redox processes in groundwater. ACS Symp Ser 1071:581–597. https://doi.org/10.1021/bk-2011-1071.ch026
79. Shen Y, Huang PC, Huang C, Sun P, Monroy GL, Wu W, Lin J, Espinosa-Marzal RM, Boppart SA, Liu WT, Nguyen TH (2018) Effect of divalent ions and a polyphosphate on composition, structure, and stiffness of simulated drinking water biofilms. Npj Biofilms Microbiomes 4:1–9. https://doi.org/10.1038/s41522-018-0058-1
80. Rango T, Vengosh A, Dwyer G, Bianchini G (2013) Mobilization of arsenic and other naturally occurring contaminants in groundwater of the main ethiopian rift aquifers. Water Res 47:5801–5818. https://doi.org/10.1016/j.watres.2013.07.002
81. Mastrocicco M, Di Giuseppe D, Vincenzi F, Colombani N, Castaldelli G (2017) Chlorate origin and fate in shallow groundwater below agricultural landscapes. Environ Pollut 231:1453–1462. https://doi.org/10.1016/j.envpol.2017.09.007
82. Shultz CD, Bailey RT, Gates TK, Heesemann BE, Morway ED (2018) Simulating selenium and nitrogen fate and transport in coupled stream-aquifer systems of irrigated regions. J Hydrol 560:512–529. https://doi.org/10.1016/j.jhydrol.2018.02.027
83. Mills TJ, Mast MA, Thomas J, Keith G (2016) Controls on selenium distribution and mobilization in an irrigated shallow groundwater system underlain by Mancos Shale, Uncompahgre

River Basin, Colorado, USA. Sci Total Environ 566–567:1621–1631. https://doi.org/10.1016/j.scitotenv.2016.06.063

84. National Water Quality Assessment Program Redox conditions in selected principal aquifers of the United States (2009). https://doi.org/10.3133/fs20093041

85. Sigrist M, Albertengo A, Brusa L, Beldoménico H, Tudino M (2013) Distribution of inorganic arsenic species in groundwater from Central-West Part of Santa Fe Province, Argentina. Appl Geochem 39:43–48. https://doi.org/10.1016/j.apgeochem.2013.09.018

86. Liu S, Guo H, Lu H, Zhang Z, Zhao W (2019) The provenance of deep groundwater and its relation to arsenic distribution in the northwestern Hetao Basin, Inner Mongolia. Environmental Geochemistry and Health. https://doi.org/10.1007/s10653-019-00433-0

87. Wang S, Mulligan CN (2006) Effect of natural organic matter on arsenic release from soils and sediments into groundwater. Environ Geochem Health 28:197–214. https://doi.org/10.1007/s10653-005-9032-y

88. Kim MJ, Nriagu J, Haack S (2000) Carbonate ions and arsenic dissolution by groundwater. Environ Sci Technol 34:3094–3100. https://doi.org/10.1021/es990949p

89. Telfeyan K, Johannesson KH, Mohajerin TJ, Palmore CD (2015) Vanadium geochemistry along groundwater flow paths in contrasting aquifers of the United States: Carrizo Sand (Texas) and Oasis Valley (Nevada) aquifers. Chem Geol 410:63–78. https://doi.org/10.1016/j.chemgeo.2015.05.024

90. Wright MT, Stollenwerk KG, Belitz K (2014) Assessing the solubility controls on vanadium in groundwater, northeastern San Joaquin Valley, CA. Appl Geochem 48:41–52. https://doi.org/10.1016/j.apgeochem.2014.06.025

91. Peel HR, Martin DP, Bednar AJ (2017) Extraction and characterization of ternary complexes between natural organic matter, cations, and oxyanions from a natural soil. Chemosphere 176:125–130. https://doi.org/10.1016/j.chemosphere.2017.02.101

92. Torres J, Tissot F, Santos P, Ferrari C, Kremer C, Kremer E (2016) Interactions of W(VI) and Mo(VI) Oxyanions with metal cations in natural waters. J Solution Chem 45:1598–1611. https://doi.org/10.1007/s10953-016-0522-6

93. Anantharaman K, Brown CT, Hug LA, Sharon I, Castelle CJ, Probst AJ, Thomas BC, Singh A, Wilkins MJ, Karaoz U, Brodie EL, Williams KH, Hubbard SS, Banfield JF (2016) Thousands of microbial genomes shed light on interconnected biogeochemical processes in an aquifer system. Nat Commun 7:1–11. https://doi.org/10.1038/ncomms13219

94. Smith RJ, Paterson JS, Launer E, Tobe SS, Morello E, Leijs R, Marri S, Mitchell JG (2016) Stygofauna enhance prokaryotic transport in groundwater ecosystems. Sci Rep 6:1–7. https://doi.org/10.1038/srep32738

95. Morita M, Uemoto H, Watanabe A (2007) Reduction of selenium oxyanions in wastewater using two bacterial strains. Eng Life Sci 7:235–240. https://doi.org/10.1002/elsc.200620188

96. Xiang R, Xu Y, Liu YQ, Lei GY, Liu JC, Huang QF (2019) Isolation distance between municipal solid waste landfills and drinking water wells for bacteria attenuation and safe drinking. Sci Rep 9:17881. https://doi.org/10.1038/s41598-019-54506-2

97. Gowrisankar G, Chelliah R, Ramakrishnan SR, Elumalai V, Dhanamadhavan S, Brindha K, Antony U, Elango L (2017) Data descriptor: chemical, microbial and antibiotic susceptibility analyses of groundwater after a major flood event in Chennai. Sci Data 4:1–13. https://doi.org/10.1038/sdata.2017.135

98. Yu R, Gan P, MacKay AA, Zhang S, Smets BF (2010) Presence, distribution, and diversity of iron-oxidizing bacteria at a landfill leachate-impacted groundwater surface water interface. FEMS Microbiol Ecol 71:260–271. https://doi.org/10.1111/j.1574-6941.2009.00797.x

99. Naseem S, McArthur JM (2018) Arsenic and other water-quality issues affecting groundwater, Indus alluvial plain, Pakistan. Hydrol Process 32:1235–1253. https://doi.org/10.1002/hyp.11489

100. Jiang Z, Li P, Wang Y, Liu H, Wei D, Yuan C, Wang H (2019) Arsenic mobilization in a high arsenic groundwater revealed by metagenomic and Geochip analyses. Sci Rep 9:1–10. https://doi.org/10.1038/s41598-019-49365-w

101. Zhao J, Al T, Chapman SW, Parker BL, Mishkin KR, Cutt D, Wilkin RT (2017) Determination of Cr(III) solids formed by reduction of Cr(VI) in a contaminated fractured bedrock aquifer: evidence for natural attenuation of Cr(VI). Chem Geol 474:1–8. https://doi.org/10.1016/j.che mgeo.2017.10.004
102. Shakya AK, Paul S, Ghosh PK (2019) Bio-attenuation of arsenic and iron coupled with nitrate remediation in multi-oxyanionic system: batch and column studies. J Hazard Mater 375:182–190. https://doi.org/10.1016/j.jhazmat.2019.04.087
103. Li M, Dopilka A, Kraetz AN, Jing H, Chan CK (2018) Layered double hydroxide/chitosan nanocomposite beads as sorbents for selenium oxoanions. Ind Eng Chem Res 57:4978–4987. https://doi.org/10.1021/acs.iecr.8b00466
104. Omorogie MO, Agunbiade FO, Alfred MO, Olaniyi OT, Adewumi TA, Bayode AA, Ofomaja AE, Naidoo EB, Okoli CP, Adebayo TA, Unuabonah EI (2018) The sequestral capture of fluoride, nitrate and phosphate by metal-doped and surfactant-modified hybrid clay materials. Chem Pap 72:409–417. https://doi.org/10.1007/s11696-017-0290-9
105. Terry PA (2004) Characterization of Cr ion exchange with hydrotalcite. Chemosphere 57:541–546. https://doi.org/10.1016/j.chemosphere.2004.08.006
106. Oladoja NA, Adelagun ROA, Ololade IA, Anthony ET, Alfred MO (2014) Synthesis of nano-sized hydrocalumite from a Gastropod shell for aqua system phosphate removal. Sep Purif Technol 124:186–194. https://doi.org/10.1016/j.seppur.2014.01.018
107. Oladoja NA, Ahmad AL (2013) Gastropod shell as a precursor for the synthesis of binary alkali-earth and transition metal oxide for Cr(VI) Abstraction from Aqua System. Sep Purif Technol 116:230–239. https://doi.org/10.1016/j.seppur.2013.05.042
108. Asiabi H, Yamini Y, Shamsayei M (2017) Highly selective and efficient removal of arsenic(V), chromium(VI) and selenium(VI) oxyanions by layered double hydroxide intercalated with zwitterionic glycine. J Hazard Mater 339:239–247. https://doi.org/10.1016/j.jhazmat.2017.06.042
109. Koilraj P, Srinivasan K (2011) High sorptive removal of borate from aqueous solution using calcined ZnAl layered double hydroxides. Ind Eng Chem Res 50:6943–6951. https://doi.org/10.1021/ie102395m
110. Mamindy-Pajany Y, Hurel C, Marmier N, Roméo M (2011) Arsenic (V) adsorption from aqueous solution onto goethite, hematite, magnetite and zero-valent iron: effects of pH, concentration and reversibility. Desalination 281:93–99. https://doi.org/10.1016/j.desal.2011.07.046
111. Zhang R, Leiviskä T, Tanskanen J, Gao B, Yue Q (2019) Utilization of ferric groundwater treatment residuals for inorganic-organic hybrid biosorbent preparation and its use for vanadium removal. Chem Eng J 361:680–689. https://doi.org/10.1016/j.cej.2018.12.122
112. Bringas E, Saiz J, Ortiz I (2015) Removal of As(V) from groundwater using functionalized magnetic adsorbent materials: effects of competing ions. Sep Purif Technol 156:699–707. https://doi.org/10.1016/j.seppur.2015.10.068
113. Holm TR (2016) Journal (American Water Works Association). Am Water Works Assoc 94:174–181
114. Brechbühl Y, Christl I, Elzinga EJ, Kretzschmar R (2012) Competitive sorption of carbonate and arsenic to hematite: combined ATR-FTIR and batch experiments. J Colloid Interface Sci 377:313–321. https://doi.org/10.1016/j.jcis.2012.03.025
115. El-Moselhy MM, Ates A, Çelebi A (2017) Synthesis and characterization of hybrid iron oxide silicates for selective removal of arsenic oxyanions from contaminated water. J Colloid Interface Sci 488:335–347. https://doi.org/10.1016/j.jcis.2016.11.003
116. Lai Keith CK, Lo IMC (2008) Removal of chromium (VI) by acid-washed zero-valent iron under various groundwater geochemistry conditions. Environ Sci Technol 42:1238–1244. https://doi.org/10.1021/es071572n
117. Qiu B, Xu C, Sun D, Yi H, Guo J, Zhang X, Qu H, Guerrero M, Wang X, Noel N, Luo Z, Guo Z, Wei S (2014) Polyaniline coated ethyl cellulose with improved hexavalent chromium removal. ACS Sustain Chem Eng 2:2070–2080. https://doi.org/10.1021/sc5003209
118. Oladoja NA, Ahmad AL, Adesina OA, Adelagun ROA (2012) Low-cost biogenic waste for phosphate capture from aqueous system. Chem Eng J 209:170–179. Doi: 10.1016/j.cej.2012.07.125

119. Oladoja NA, Ololade IA, Olaseni SE, Olatujoye VO, Jegede OS, Agunloye AO (2012) Synthesis of Nano calcium oxide from a gastropod shell and the performance evaluation for Cr (VI) removal from aqua system. Ind Eng Chem Res 51:639–648. https://doi.org/10.1021/ie201189z
120. Oladoja NA, Ololade IA, Adesina AO, Adelagun ROA, Sani YM (2013) Appraisal of gastropod shell as calcium ion source for phosphate removal and recovery in calcium phosphate minerals crystallization procedure. Chem Eng Res Des 91:810–818. https://doi.org/10.1016/j.cherd.2012.09.017
121. Zhang X, Ma J, Lu X, Huangfu X, Zou J (2015) High efficient removal of molybdenum from water by $Fe_2(SO_4)_3$: effects of pH and affecting factors in the presence of co-existing background constituents. J Hazard Mater 300:823–829. https://doi.org/10.1016/j.jhazmat.2015.08.026
122. Opiso E, Sato T, Yoneda T (2009) Adsorption and co-precipitation behavior of arsenate, chromate, selenate and boric acid with synthetic allophane-like materials. J Hazard Mater 170:79–86. https://doi.org/10.1016/j.jhazmat.2009.05.001
123. Wu X, Guo X, Zhang L (2019) Fabrication of porous zirconia microspheres as an efficient adsorbent for removal and recovery of Trace Se(IV) and Te(IV). Ind Eng Chem Res 58:342–349. https://doi.org/10.1021/acs.iecr.8b04288
124. Lu Z, Yu J, Zeng H, Liu Q (2017) Polyamine-modified magnetic graphene oxide nanocomposite for enhanced selenium removal. Sep Purif Technol 183:249–257. https://doi.org/10.1016/j.seppur.2017.04.010
125. Fu C-C, Tran HN, Chen X-H, Juang R-S (2019) Preparation of polyaminated Fe_3O_4@chitosan core-shell magnetic nanoparticles for efficient adsorption of phosphate in aqueous solutions. J Ind Eng Chem. https://doi.org/10.1016/j.jiec.2019.11.033
126. Bandara PC, Perez JVD, Nadres ET, Nannapaneni RG, Krakowiak KJ, Rodrigues DF (2019) Graphene oxide nanocomposite hydrogel beads for removal of selenium in contaminated water. ACS Appl Polym Mater 1:2668–2679. https://doi.org/10.1021/acsapm.9b00612
127. Koilraj P, Kannan S (2010) Phosphate uptake behavior of ZnAlZr ternary layered double hydroxides through surface precipitation. J Colloid Interface Sci 341:289–297. https://doi.org/10.1016/j.jcis.2009.09.059
128. Hu C, Chen Q, Chen G, Liu H, Qu J (2015) Removal of Se(IV) and Se(VI) from drinking water by coagulation. Sep Purif Technol 142:65–70. https://doi.org/10.1016/j.seppur.2014.12.028
129. Goh K-H, Lim T-T (2004) Geochemistry of inorganic arsenic and selenium in a tropical soil: effect of reaction time, pH, and competitive anions on arsenic and selenium adsorption. Chemosphere 55:849–859. https://doi.org/10.1016/j.chemosphere.2003.11.041
130. Guo X, Wu Z, He M (2009) Removal of antimony(V) and antimony(III) from drinking water by coagulation–flocculation–sedimentation (CFS). Water Res 43:4327–4335. https://doi.org/10.1016/j.watres.2009.06.033
131. Pallier V, Feuillade-Cathalifaud G, Serpaud B, Bollinger J-C (2010) Effect of organic matter on arsenic removal during coagulation/flocculation treatment. J Colloid Interface Sci 342:26–32. https://doi.org/10.1016/j.jcis.2009.09.068
132. Ahmad A, Rutten S, Eikelboom M, de Waal L, Bruning H, Bhattacharya P, van der Wal A (2020) Impact of phosphate, silicate and natural organic matter on the size of Fe(III) precipitates and arsenate co-precipitation efficiency in calcium containing water. Sep Purif Technol 235:116117. https://doi.org/10.1016/j.seppur.2019.116117
133. Lee Y, Um IH, Yoon J (2003) Arsenic(III) oxidation by iron(VI) (Ferrate) and subsequent removal of arsenic(V) by iron(III) coagulation. Environ Sci Technol 37:5750–5756. https://doi.org/10.1021/es034203+
134. Bilici Baskan M, Pala A (2010) A statistical experiment design approach for arsenic removal by coagulation process using aluminum sulfate. Desalination 254(2010):42–48. https://doi.org/10.1016/j.desal.2009.12.016
135. Oladoja NA, Aliu YD, Ofomaja AE (2011) Evaluation of snail shell as a coagulant aid in the alum precipitation of aniline blue from aqueous solution. Environ Technol 32:639–652. https://doi.org/10.1080/09593330.2010.509868

136. Oladoja NA (2016) Advances in the quest for substitute for synthetic organic polyelectrolytes as coagulant aid in water and wastewater treatment operations. Sustain Chem Pharm 3:47–58. https://doi.org/10.1016/j.scp.2016.04.001
137. Hu C, Liu H, Chen G, Qu J (2012) Effect of aluminum speciation on arsenic removal during coagulation process. Sep Purif Technol 86:35–40. https://doi.org/10.1016/j.seppur.2011.10.017
138. McDonald KJ, Reynolds B, Reddy KJ (2015) Intrinsic properties of cupric oxide nanoparticles enable effective filtration of arsenic from water. Sci Rep 5:1–10. https://doi.org/10.1038/srep11110
139. Nguyen CM, Bang S, Cho J, Kim K-W (2009) Performance and mechanism of arsenic removal from water by a nanofiltration membrane. Desalination 245:82–94. https://doi.org/10.1016/j.desal.2008.04.047
140. Saitúa H, Campderrós M, Cerutti S, Padilla AP (2005) Effect of operating conditions in removal of arsenic from water by nanofiltration membrane. Desalination 172:173–180. https://doi.org/10.1016/j.desal.2004.08.027
141. Urase T, Oh J, Yamamoto K (1998) Effect of pH on rejection of different species of arsenic by nanofiltration. Desalination 117:11–18. https://doi.org/10.1016/S0011-9164(98)00062-9
142. He Y, Tang YP, Chung TS (2016) Concurrent removal of selenium and arsenic from water using polyhedral oligomeric silsesquioxane (POSS)–polyamide thin-film nanocomposite nanofiltration membranes. Ind Eng Chem Res 55:12929–12938. https://doi.org/10.1021/acs.iecr.6b04272
143. Chen ASC, Wang L, Lytle DA, Sorg TJ (2018) Arsenic/iron removal from groundwater with elevated ammonia and natural organic matter. J Am Water Works Assoc 110:E2–E17. https://doi.org/10.5942/jawwa.2018.110.0020
144. Su C, Puls RW (2004) Nitrate reduction by zerovalent iron: effects of formate, oxalate, citrate, chloride, sulfate, borate, and phosphate. Environ Sci Technol 38:2715–2720. https://doi.org/10.1021/es034650p
145. Banerjee K, Helwick RP, Gupta S (1999) Treatment process for removal of mixed inorganic and organic arsenic species from groundwater. Environ Prog 18:280–284. https://doi.org/10.1002/ep.670180415
146. Katsoyiannis IA, Voegelin A, Zouboulis AI, Hug SJ (2015) Enhanced As(III) oxidation and removal by combined use of zero valent iron and hydrogen peroxide in aerated waters at neutral pH values. J Hazard Mater 297:1–7. https://doi.org/10.1016/j.jhazmat.2015.04.038
147. Han X, Song J, Li Y-L, Jia S-Y, Wang W-H, Huang F-G, Wu S-H (2016) As(III) removal and speciation of Fe (Oxyhydr)oxides during simultaneous oxidation of As(III) and Fe(II). Chemosphere 147:337–344. https://doi.org/10.1016/j.chemosphere.2015.12.128
148. Zhou Q, Niu W, Li Y, Li X (2020) Photoinduced fenton-simulated reduction system based on iron cycle and carbon dioxide radicals production for rapid removal of Cr(VI) from wastewater. J Cleaner Prod 120790. https://doi.org/10.1016/j.jclepro.2020.120790
149. Oladoja NA, Anthony ET, Ololade IA, Saliu TD, Bello GA (2018) Self-propagation combustion method for the synthesis of solar active Nano Ferrite for Cr(VI) reduction in aqua system. J Photochem Photobiol A 353:229–239. https://doi.org/10.1016/j.jphotochem.2017.11.026
150. Fujishima A, Honda K (1972) Electrochemical photolysis of water at a semiconductor electrode. Nature 238:37–38. https://doi.org/10.1038/238037a0
151. Fontana KB, Lenzi GG, Seára ECR, Chaves ES (2018) Comparision of photocatalysis and photolysis processes for arsenic oxidation in water. Ecotoxicol Environ Saf 151:127–131. https://doi.org/10.1016/j.ecoenv.2018.01.001
152. Mac Mahon J, Gill LW (2016) Development of a continuous flow solar oxidation process for the removal of arsenic for sustainable rural water supply. J Environ Chem Eng 4:1181–1190. https://doi.org/10.1016/j.jece.2016.01.027
153. Samad A, Furukawa M, Katsumata H, Suzuki T, Kaneco S (2016) Photocatalytic oxidation and simultaneous removal of arsenite with CuO/ZnO photocatalyst. J Photochem Photobiol A: Chem 325:97–103. https://doi.org/10.1016/j.jphotochem.2016.03.035

154. Cid LdC, Grande Mdc, Acosta EO, Ginzberg B (2012) Removal of Cr(VI) and humic acid by heterogeneous photocatalysis in a laboratory reactor and a pilot reactor. Ind Eng Chem Res 51:9468–9474. https://doi.org/10.1021/ie2010687
155. Chen X, Burda C (2008) The electronic origin of the visible-light absorption properties of C-, N- and S-doped TiO_2 nanomaterials. J Am Chem Soc 130:5018–5019. https://doi.org/10.1021/ja711023z
156. Nogueira AE, Lopes OF, Neto ABS, Ribeiro C (2017) Enhanced Cr(VI) photoreduction in aqueous solution using $Nb_2 O_5$/CuO heterostructures under UV and visible irradiation. Chem Eng J 312:220–227. https://doi.org/10.1016/j.cej.2016.11.135
157. Szabó M, Bellér G, Kalmár J, Fábián I (2017) Chapter One—The kinetics and mechanism of complex redox reactions in aqueous solution: the tools of the trade. In: van Eldik R, C.D.B.T.-A, Hubbard IC (eds) Inorg React Mech. Academic Press, pp 1–61. https://doi.org/10.1016/bs.adioch.2017.02.004
158. Richardson DE, Sharpe P (1993) Vibrational and electronic contributions to entropy changes for oxidation-reduction reactions of metal complexes. Inorg Chem 32:1809–1812. https://doi.org/10.1021/ic00061a043
159. Ma Z, Zhang M, Guo J, Liu W, Tong M (2018) Facile synthesis of ZrO_2 coated BiOCl0.5I0.5 for photocatalytic oxidation-adsorption of As(III) under visible light irradiation. Chemosphere 211:934–942. https://doi.org/10.1016/j.chemosphere.2018.08.003
160. Vaiano V, Iervolino G, Sannino D, Rizzo L, Sarno G (2016) MoOx/TiO_2 immobilized on quartz support as structured catalyst for the photocatalytic oxidation of As(III) to As(V) in aqueous solutions. Chem Eng Res Des 109:190–199. https://doi.org/10.1016/j.cherd.2016.01.029
161. Wang Z, Murugananthan M, Zhang Y (2019) Graphitic carbon nitride based photocatalysis for redox conversion of arsenic(III) and chromium(VI) in acid aqueous solution. Appl Catal B: Environ 248:349–356. https://doi.org/10.1016/j.apcatb.2019.02.041
162. Katsoyiannis IA, Zouboulis AI, Jekel M (2004) Kinetics of bacterial As(III) oxidation and subsequent As(V) removal by sorption onto biogenic manganese oxides during groundwater treatment. Ind Eng Chem Res 43:486–493. https://doi.org/10.1021/ie030525a
163. Ungureanu G, Filote C, Santos SCR, Boaventura RAR, Volf I, Botelho CMS (2016) Antimony oxyanions uptake by green marine macroalgae. J Environ Chem Eng 4:3441–3450. https://doi.org/10.1016/j.jece.2016.07.023
164. Song W, Gao B, Zhang X, Li F, Xu X, Yue Q (2019) Biological reduction of perchlorate in domesticated activated sludge considering interaction effects of temperature, pH, electron donors and acceptors. Process Saf Environ Prot 123:169–178. https://doi.org/10.1016/j.psep.2019.01.009
165. Zhang Y, Frankenberger WT (2006) Removal of selenate in river and drainage waters by *Citrobacter braakii* enhanced with zero-valent iron. J Agric Food Chem 54:152–156. https://doi.org/10.1021/jf0581240
166. Song W, Gao B, Xu X, Zhang T, Liu C, Tan X, Sun S, Yue Q (2015) Treatment of dissolved perchlorate by adsorption-microbial reduction. Chem Eng J 279:522–529. https://doi.org/10.1016/j.cej.2015.05.063
167. Dieguez-Alonso A, Anca-Couce A, Frišták V, Moreno-Jiménez E, Bacher M, Bucheli TD, Cimò G, Conte P, Hagemann N, Haller A, Hilber I, Husson O, Kammann CI, Kienzl N, Leifeld J, Rosenau T, Soja G, Schmidt HP (2019) Designing biochar properties through the blending of biomass feedstock with metals: impact on oxyanions adsorption behavior. Chemosphere 214:743–753. https://doi.org/10.1016/j.chemosphere.2018.09.091
168. Chao HP, Wang YC, Tran HN (2018) Removal of hexavalent chromium from groundwater by Mg/Al-layered double hydroxides using characteristics of in-situ synthesis. Environ Pollut 243:620–629. https://doi.org/10.1016/j.envpol.2018.08.033
169. Gal H, Weisbrod N, Dahan O, et al (2009) Perchlorate accumulation and migration in the deep vadose zone in a semiarid region. J Hydrol 378:142–149. https://doi.org/10.1016/j.jhydrol.2009.09.018

170. Kumar M, Ramanathan AL, Rahman MM, Naidu R (2016) Concentrations of inorganic arsenic in groundwater, agricultural soils and subsurface sediments from the middle Gangetic plain of Bihar, India. Sci Total Environ 573:1103–1114. https://doi.org/10.1016/j.scitotenv.2016. 08.109

171. Shahid M, Niazi NK, Dumat C, et al (2018) A meta-analysis of the distribution, sources and health risks of arsenic-contaminated groundwater in Pakistan. Environ Pollut 242:307–319. https://doi.org/10.1016/j.envpol.2018.06.083

172. Kuiper N, Rowell C, Shomar B (2015) High levels of molybdenum in Qatar's groundwater and potential impacts. J Geochemical Explor 150:16–24. https://doi.org/10.1016/j.gexplo.2014. 12.009

173. Vinson DS, Mcintosh JC, Dwyer GS, Vengosh A (2011) Arsenic and other oxyanion-forming trace elements in an alluvial basin aquifer: Evaluating sources and mobilization by isotopic tracers (Sr, B, S, O, H, Ra). Appl Geochemistry 26:1364–1376. https://doi.org/10.1016/j.apg eochem.2011.05.010

174. Park JD, Choi SJ, Choi BS, et al (2016) Arsenic levels in the groundwater of Korea and the urinary excretion among contaminated area. J Expo Sci Environ Epidemiol 26:458–463. https://doi.org/10.1038/jes.2016.16

175. Levakov I, Ronen Z, Dahan O (2019) Combined in-situ bioremediation treatment for perchlorate pollution in the vadose zone and groundwater. J Hazard Mater 369:439–447. https://doi. org/10.1016/j.jhazmat.2019.02.014

Chapter 4
Occurrence and Management of Selenium Oxyanions in Water

Chidinma G. Ugwuja, Ajibola A. Bayode, Damilare Olorunnisola, and Emmanuel I. Unuabonah

Abstract Selenium is a metalloid and an essential micronutrient needed by animals and humans at low concentrations but extremely toxic at high concentrations. It is found in the natural environment and can be present in soil, food, air, plant, and water. The chemistry of Se is linked to its different chemical forms. Se mobility and toxicity are strongly dependent on its redox state; from highly soluble oxyanions like selenate, selenite, and hydroselenite to elemental Se. However, selenate and selenite are the two predominant Se species in the water system because of their high solubility and low adsorption by sediments and soil. This provides most techniques the platform to focus on the removal of both selenite and selenate among other Se species. This chapter is focused on the occurrence and effective management of Selenate and Selenite in water.

Keywords Selenite · Selenate · Occurrence · Management · Water

4.1 Introduction

Selenium (Se) is one of the sixth main group elements called chalcogens in the periodic table. As an element, Se occurs naturally in the environment with anthropogenic sources contributing to its presence in the environment. Examples of natural sources include volatilization of water bodies and plants, terrestrial weathering, and volcanic

C. G. Ugwuja · A. A. Bayode (✉) · D. Olorunnisola · E. I. Unuabonah
Department of Chemical Sciences, Faculty of Natural Sciences, Redeemer's University, PMB 230, Ede, Osun State, Nigeria
e-mail: bayodea@run.edu.ng

African Center of Excellence for Water and Environmental Research (ACEWATER), Redeemer's University, PMB 230, Ede, Osun State, Nigeria

A. A. Bayode
Laboratório de Química Analítica Ambiental e Ecotoxicologia (LaQuAAE), Departamento de Química e Física Molecular, Instituto de Química de Sao Carlos, Universidade de Sao Paulo, Sao Carlos, Brazil

© Springer Nature Switzerland AG 2021
N. A. Oladoja and E. I. Unuabonah (eds.), *Progress and Prospects in the Management of Oxyanion Polluted Aqua Systems*, Environmental Contamination Remediation and Management, https://doi.org/10.1007/978-3-030-70757-6_4

activities [1, 2], while fossil fuels, mining, oil refining, and fossil fuel combustion are examples of anthropogenic sources.

Selenium can be found in many environmental matrices around the world, including plants, soil, food, air, and drinking water, but it is generally detected at very low concentrations in drinking water [3]. It is an essential micronutrient needed by animals and humans at low concentrations, but extremely toxic at high concentrations in water, atmosphere, and soil [4]. For instance, acute exposure to compounds of Se may lead to neurological challenges and severe respiratory problems. Investigations have pointed out that high exposure to Se, through drinking water, is the cause of low blood glutathione peroxidase activity, signifying an unexpected inverse relationship occurring between enzymatic activity and Se intake [5]. Besides, high Se intake decreases sperm movement in healthy men [6], increases the rate of some cancer development like pancreatic and skin cancers, and increases the risk of type-2 diabetes due to its negative impact on insulin signaling [7]. Health challenges associated with both Se deficiency and toxicity have been reported in different parts of the world, because of the narrow range that exists between dietary deficiency (<40 ug/day) and toxic levels (>400 ug/day) [2, 8]. So, the nutritional and toxicological properties of Se have generated strong interest and argument between the general public and the scientific community [9]. On this basis, the United States Environmental Protection Agency (USEPA) and the World Health Organization (WHO) have set the maximum acceptable concentration of Se in drinking water to be between 40 and 50 ppb, while in Europe and Japan, the permissible limit is 10 ppb [10].

The elimination of Se from the contaminated water has attracted tremendous research. The toxicity and mobility have been found to be strongly dependent on its redox state. In nature, Se is commonly observed as selenate [SeO_4^{2-}, Se(VI)], selenite [SeO_3^{2-} Se(IV)], selenide [Se^2; Se(II)], and elemental selenium as seen in Fig. 4.1 [11].

The principal aqueous forms of inorganic selenium are selenate and selenite. Selenate is typically more difficult to remove from water because it predominates in oxidizing environments as SeO_4^{2-}, while selenite is present in moderately reducing conditions as SeO_3^{2-} or $HSeO_3$. In acidic medium, selenium remains in the oxyanion form of selenite, while at the neutral and alkaline range, selenate predominates [12]. It is, therefore, the goal of this chapter, to compile several management techniques that have been used for to the removal of Se from aqua systems based on its aqueous chemistry.

4.2 Occurrence and Sources of Selenium Oxyanions in Water

The major precursor of selenate and selenite, which is selenium, is naturally occurring and it is rarely recovered in a free state. The chemistry of selenium is similar to that of sulfur and tellurium, and it occurs in the environment in different forms. The species

Fig. 4.1 Four (4) natural
oxidation states of selenium

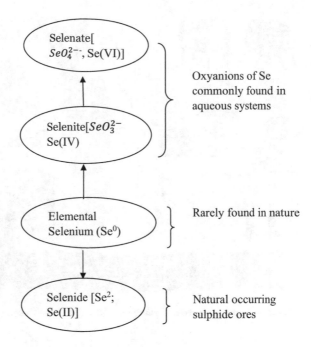

of selenium that occur in water are the oxidized water-soluble forms and they exhibit
the oxidation state of +6 (selenate) and +4 (selenite) [13, 14].

The main source of selenate and selenite in water stems from selenium, and they
gain access into water bodies through natural and anthropogenic sources. As shown
in Fig. 4.2, leachates from Se-laddened rock and waste streams, runoffs from landfills
containing e-wastes and agricultural irrigation, mineral ore deposit, runoffs due to
downpour from rainfall and from process tailings derived from the mining activities,
and metal smelting have greatly contributed to the dispersion of Se in downstream
environments [15]. The selenium in water further undergoes redox reaction, to get
converted to selenate and selenite.

The behavior, fate, and mobility of selenate and selenite in water are mostly influenced by various biological, geological, and chemical processes such as precipitation,
complexation, sorption, oxidation–reduction, and biomethylation [16–17]. These
processes are susceptible to different environmental factors like pH, redox conditions, natural organic matter, selenium concentration, competing anions, biological
and microbial activity [18–19].

The oxidation and reduction of selenium are influenced by the redox potential
and the pH value of water. Under reducing conditions and in the presence of acidic
mine effluent deposited into water, with redox potential ranging from 0 to 200 mV,
selenites are the predominant species. When the redox potential increases to values
above 200 mV, it is oxidized to selenate, and above 400 mV redox potential, selenite

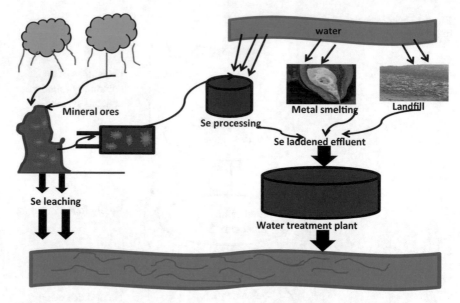

Fig. 4.2 Schematic diagram showing the pathway to which selenium gets into the water system

species dominates by 80% [19, 20]. It is safe to say that selenium mobility and bioavailability are inhibited in aqueous systems that are acidic in nature.

The mobility of selenate and selenite is related to their adsorption, either on organic matter or mineral surfaces. The adsorption is mediated by its oxidation state. Thus, selenate has a greater adsorption capacity than selenite, with a good affinity for metallic oxide and hydroxide [21]. Competition from natural organic matter and anions, such as SO_4^{2-}, PO_4^{3-}, CO_3^{2-}, and Cl^- affects the number of adsorption sites available which invariably influences the mobility and bioavailability of selenite in water [22, 23].

Generally, pH value is considered an important parameter that control the ion distribution between a solid–liquid interface during an adsorption process. It does not only affect the selenium species in solutions, but also changes the surface charge of the adsorbent [2]. $\left[SeO_4^{2-}\right]\left[SeO_3^{2-}\right]$ Selenium remains in oxyanion form of SeO_3^{2-} at acidic condition, but at neutral and alkaline range, SeO_4^{2-} is the dominant form. Se oxyanion species are water soluble and, especially, the selenate is prone to less adsorption onto the soil surface and hence more tendency to leach into the aqueous phase. However, mobility, bioavailability, and speciation of selenium species are highly dependent on the pH value and redox conditions [24].

Although, several studies have reported that Se(IV) species in aqueous solution include selenious acid (H_2SeO_3), biselenite ($HSeO_3$), and selenite SeO_3^{2-}, the biselenite ion is the predominant ion in water between pH 3.5 and 9.0. Above pH 9.0, selenite species dominates, but at solution pH below pH 3.5, selenious acid dominates [25]. For example, Zhu et al. [26] conducted an experiment on species distribution of Se oxyanions as a function of pH. As presented in Fig. 4.3, the percentage of

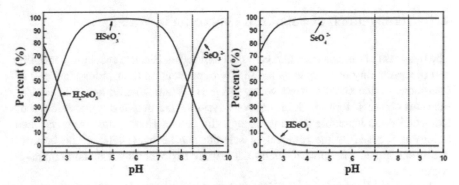

Fig. 4.3 Speciation of Se(IV) and Se(VI) as a function of pH values [26]

$HSeO_3^-$ and SeO_3^{2-} are 95% and 5% at pH 7.0, respectively, while SeO_3^{2-} dominates at pH 8.5 (63%). On the other hand, Se (VI) was present as only one specie, SeO_4^{2-}, above the pH value of 8.5. For this reason, the charge of Se oxyanions species greatly depends on pH values [26].

Furthermore, the removal of Se oxyanions has a strong correlation with variation in pH value of the medium. Several studies on the adsorption of Se oxyanions have established that the adsorption is highly favorable in acidic medium, while its efficiency decreases rapidly in basic medium [12, 25, 26]. For instance, Meher et al. [12] reported that the removal of selenite [SeO_3^{2-}] was reduced from 80.35% at pH of 4.0 to 38.22% at pH of 9.0. Similar trend was observed in the case of selenate, as the adsorption efficiency was drastically reduced from 72.34 to 37.35%, with a change in the solution pH value from 4.0 to 9.0 [12].

In another study by El-Shafey [25], it was reported that high percentage removal of SeO_3^{2-} was achieved at pH 1.5, but a decrease in efficiency was observed with increasing solution pH to 7.0 [25]. In addition, Zhang et al. [27] stated that the removal efficiency of iron-coated granular adsorbent for SeO_3^{2-} was not significantly affected within a pH range of 2.0–8.0, but a decrease in the removal efficiency was observed at pH greater than 8.0, which then decreases to zero removal efficiency at pH 12.0 [27]. Again, Chen et al. [28] established the impact of pH on oxyanion removal by taking into account the diverse species of Se in aqueous solution as a function of pH. The surface of the adsorbent Mg–FeCO$_3$ layered double hydroxide (LDH) is positively charged when pH < pHpzc (point of zero charge of Mg–FeCO$_3$ LDH); pHpzc = 8.78 [29], aiding the adsorption of the oxyanions. However, at pH > pHpzc, the LDHs surface becomes negatively charged, which repel the oxyanions and consequently cause low adsorption.

In literature, for most of the adsorbents, the adsorption capacity for Se oxyanion is drastically reduced in alkaline pH [23, 30–31]. The lowering of adsorption efficiency at the alkaline pH range may be due to the competition between the Se oxyanions and OH$^-$ for the active sites on the adsorbent, resulting in enhanced repulsion toward the anionic selenium species [12]. This results in the decrease of adsorption capacity [26].

4.3   Management of Selenium Oxyanion in Water

The treatment of Se-contaminated water is challenging because speciation, mobility, and bioavailability of Se species are highly dependent on their redox state and pH conditions of the contaminated water [32, 33]. While selenite and selenates are the most common form of Se in aqueous systems and they are water soluble, the elemental Se is insoluble and not biologically active. Hence, most removal technology is focused on the removal of selenite and selenate. Due to the bioavailability and higher solubility of selenite, it is more toxic than selenate and organo-Se compounds.

A graphical illustration of the different methods reviewed in this chapter for the management of Se oxyanions is presented in Fig. 4.4.

4.3.1   Selenium Removal by Reduction Technique

One of the most widely applied nanomaterials for the remediation and treatment of groundwater, soil, and wastewater is the nanoscale zero-valent iron (nZVI) [34]. This nano-material has being employed for the detoxification of a wide range of pollutants in water including azo dyes, perchlorate, halogenated hydrocarbons, hexavalent chromium, nitroaromatic compounds, radionuclides, antibiotics, and heavy metal ions [35–36]. Zero-valent iron can be used as a reducing agent for aqueous selenium species (selenate and selenite) in a way that the iron acts as both a catalyst and an electron donor for the reduction reaction. Basically, the ZVI surface is oxidized,

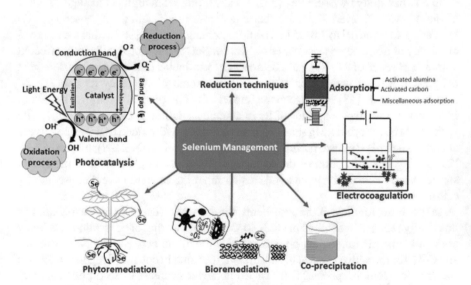

Fig. 4.4 Graphical illustration depicting the various techniques used for Se management in water

providing both ferric and ferrous iron adsorption sites for aqueous Se oxyanions. The ZVI surface is known as green rust, which is the Fe form of ZVI needed to chemically reduce both selenite and selenate to elemental Se. For example, selenate, which is more recalcitrant of the two species, is first reduced to selenite by green rust, which in turn is either adsorbed on the ferrihydrite amorphous solids created, via the redox reaction with ZVI, or further reduced to elemental Se [37].

According to a study carried out by Olegario et al. [38], the reduction of water-soluble selenate [Se(VI)] to insoluble selenide [Se(II)] was achieved using ZVI nanoparticles. From kinetic studies, it was observed that the rate of reduction of Se(VI) was four times greater with ZVI than with conventional Fe° powder. The reduction in the oxidation state of the oxyanion was confirmed from the X-ray absorption near-edge structure (XANES) spectroscopy, where it was revealed that Se(VI) was successfully reduced to Se(II) [38].

Das et al. [39] also investigated the removal of dissolved Se(VI) in simulated mining water via a reduction method using three commercial zero-valent irons (ZVIs) materials under oxic conditions. The experiment was carried out in the presence of NO_3^- and SO_4^{2-}, to create a water system impacted by mining activities. Results from the batch experiments revealed a reduction in the removal rate of Se(VI) in the presence of NO_3^- and SO_4^{2-}. However, all the three ZVI materials displayed the ability to partially remove NO_3^- by adsorption and/or reduction mechanism, and the kinetic studies, showed that the process followed a first-order reaction. The XANES analysis revealed that the Se(VI) was reduced to a mixture of Se(IV) and Se(0) linked with nZVI solids [39].

Ling et al. [34] reported the use of nanoscale zero-valent iron for the removal of selenite [Se(IV)] from water, where 5 g/L of nZVI removed 1.3 mM of selenite, by reduction, within 3 min. The process mechanism was elucidated by scanning transmission electron microscope (STEM) incorporated with energy-dispersive X-ray spectroscopy (EDX). The STEM-EDX analysis (Fig. 4.5) revealed that Se(IV) was reduced to Se(II) and Se(0), which accumulated on the iron oxide and Fe(0) interface and formed a thin (~0.5 nm) layer of reduced Se. Penetration or diffusion of Se(IV) across the Fe oxide layer is accelerated by its chemical reduction, mediated by ferrous and/or Fe(0), with the reduced products [i.e., Se(−II) and Se(0)] accumulating on the oxide/Fe(0) interface.

Additionally, the STEM mapping is also revealed that defects on the surface layer of the nZVI increased the diffusion of Se oxyanions, which further enhanced the capacity of the nZVI material for Se sequestration. This study showed that the reduction and sequestration mechanisms accounted for the rapid removal of Se(IV) and the large capacity of nZVI material for Se sequestration.

4.3.2   Adsorption

The adsorption process is one of the most promising techniques for Se removal from water. It is also a preferred method for Se management due to the process simplicity,

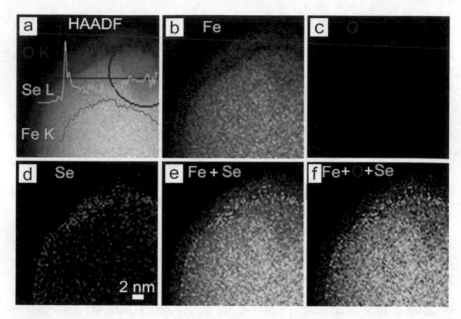

Fig. 4.5 STEM-EDX images nZVI nanoparticle collected from a reactor after 48-h reactions with 1.3 mM selenite (**a**) of Fe-Se reactions: **a** HAADF image with XEDS-STEM elemental line scans of the Fe(yellow), Se(green), O(red) over a defect area (the red circle), **b** Fe, **c** O, **d** Se, **e** Fe + Se, **f** Fe + O + Se. Signals collected from nZVI after 48-h reactions with 1.3 mM selenite [34]

economy, and efficiency. Adsorption onto various surfaces such as powdered activated carbon, activated alumina, chitosan-clay composites, activated double hydroxides, nano-TiO_2, and granular activated carbon have been investigated for the removal of Se oxyanions in aqueous systems.

Activated alumina has been recommended as one of the best available materials for the removal of Se from drinking water because of the low cost, ease of operation, and abundance. It has been listed as one of the best materials for water and wastewater treatment, due to its amphoteric property [40]. Various activated alumina have been successfully used for the removal of Se from water, and due to its amphoteric property, the performance efficiency is expected to be influenced by pH. Adsorption is strongly a pH-dependent process, as both the surface charge of the adsorbent and that of the main species of the adsorbate are influenced by the pH of the system. For instance, Yamani et al. [41] reported the use of nanocrystalline aluminum oxide (n-Al_2O_3) impregnated chitosan beads (AICB), for simultaneous adsorption of selenite and selenate, which was more effective than n-Al_2O_3 or chitosan alone. For a selenite system, the main active adsorbent is n-Al_2O_3, because chitosan has low affinity for selenite, while for the selenate system, chitosan was the main active adsorbent. However, in the case of a mixed system of selenite and selenate, the AICB was the most robust option, because it exhibited the most consistent performance. An indirect relationship between the AICB performance and the pH of the solution was observed. Thus increasing the pH of the aqueous system corresponded with decreasing adsorbent performance. Increasing background sulfate (anion) concentration was also reported to negatively influence the adsorption efficiency of AICB for both selenite and selenite [41]. This clearly demonstrates the competing effect of other anions with Se oxyanions for adsorption sites.

Meher et al. [12] showed the concurrent removal of selenite and selenate from drinking water, using mesoporous activated alumina (MAA). The adsorption capacity obtained for both selenite and selenate were 9.02 μg/g and 5.38 μg/g, respectively. While the adsorption of selenite followed a pseudo-second-order kinetic model that of selenate followed the pseudo-first-order kinetic model. The efficiency of the MAA adsorbent was also investigated at different pH values and in the presence of competitive co-ions. At near-neutral pH value and lower concentrations of competitive ions, MAA adsorbent removed the different forms of the Se from water. The reusability and regeneration study of the spent MAA adsorbent validated its suitability for column studies in a continuous flow mode. Although the adsorption capacity gradually decreased after the fourth regeneration cycle and up to the tenth regeneration cycle, the average adsorbent weight loss during each regeneration cycle was negligible [12].

Due to its porous internal structure and large specific surface area, activated carbon is commonly used in industrial purification and separation processes. Recent research interests have been directed toward the use of various types of carbonaceous waste for the production of activated carbon, due to their richness in carbon elements, low cost, and natural abundance [42]. For example, several kinds of agricultural waste, such as rice husks, walnut shells, etc., have been studied as low cost materials. One of the benefits of using activated carbon obtained from agricultural waste is that most agricultural waste possesses components such as lignin, tannins, saponins, and hemicelluloses, which add functional properties that influence the overall performance of the activated carbon [42].

Zhang et al. [27] reported the use of iron-coated granular activated carbons for the removal of selenite from aqueous solution. Ferrous chloride with sodium hypochlorite, was used in oxidizing five different types of granular activated carbon. The Darco 12 × 20 GAC, coated with 0.1 M ferrous chloride, exhibited the highest efficiency for selenite removal (97.3% removal efficiency) [27]. The high percentage removal was not considerably affected when the reaction was carried out within the pH range of 2.0–8.0. However, at pH value above 8.0, a significant decrease in the removal efficiency until pH 12.0, when zero efficiency was achieved. Hence, the adsorption of selenite onto iron-coated granular carbon is best effected at pH values <8.0, and the regeneration of the adsorbent can be carried out by raising the pH, since lower adsorption was experienced at pH >8.0 [27].

Wasewar et al. [43] investigated the use of granular activated carbon (GAC) and powdered activated carbon (PAC) for the removal of selenite from aqueous solution, and the adsorption process was observed to be highly dependent on the solution pH value. There was an increase in the percentage removal of Se(IV) by GAC and PAC at pH 2.0–3.0, while the removal percentage decreased significantly at higher pH values. However, the adsorption of Se was closely related to the distribution of its chemical forms and the surface properties of the adsorbent. Powdered activated carbon, obtained from charcoal, impregnated with iron nitrate has also been applied for the adsorption of Se [44]. The effects of different ratios of the carbon mass to the impregnating agents against the Se adsorption capacity and selectivity were investigated, and it was observed that the carbon impregnated with 10% $Fe(NO_3)_2$ and heated at 200 °C exhibited the highest adsorption capacity and selectivity toward the Se oxyanions. The highest adsorption capacity was observed at pH below 5.0 for the adsorption of Se(VI). The XPS and SEM–EDX analysis revealed that the interaction between the iron oxide and selenite enhanced the adsorption onto the Fe-loaded activated carbon [44].

The use of a carbonaceous adsorbent for the removal of selenite from water was reported by El-Shafey [25]. The carbonaceous adsorbent was obtained from peanut shells via sulfuric acid treatment. Surface functional groups, such as –OH and –COOH, were present on the surface of the acid activated carbonaceous adsorbent, which possess reduction properties for selenite. In this study, two different sorbent types were used, namely the wet and dry adsorbent. The difference between the two adsorbents is that after carbonization, the wet adsorbent was not dried in the oven but left under suction in a Gooch crucible, while the dry adsorbent was dried in an oven at 120 °C. The removal of selenite from aqueous solution was studied under varying conditions, such as temperature, pH, time, selenite concentration, and adsorbent status (wet and dry). The removal of Se(IV) by both the wet and dry adsorbent followed a pseudo-first-order kinetic model, but the removal by wet adsorbent was faster. This observation was linked to the contraction and compaction of the dry adsorbent after drying, which results in narrower pores for the diffusion of Se(IV) ions. Conversely, the adsorption capacity of the wet sorbent was also higher than that of the dry adsorbent. At low pH value, the removal of Se(IV) by the activated carbonaceous adsorbent increased, but a decrease in value was observed with increasing pH value. The high extent of adsorption at low pH values was attributed to excess protons in the system that allowed more reduction of Se(IV) to elemental selenium on the adsorbent surface. The reduction of Se(IV) to elemental Se involves redox process and carbon oxidation, as shown below [25];

$$H_2SeO_3 + 4H^+ + 4e- \rightarrow Se^0 + 3H_2O \tag{4.1}$$

$$\sim C{-}H + Se(\text{oxidized}) + H^+ \rightarrow \sim C{-}OH + Se(\text{reduced}) + H_2O \tag{4.2}$$

$$\sim C{-}H/\sim C{-}OH + Se(\text{oxidized}) + H^+ \rightarrow \sim C{=}O + Se(\text{reduced}) + H_2O \tag{4.3}$$

$$\sim C{-}H/\sim C{-}OH + Se(\text{oxidized}) + H^+ \rightarrow \sim COOH + Se(\text{reduced}) + H_2O \tag{4.4}$$

Both the activated and non-activated adsorbents exhibited increased adsorption of Se(IV) with increasing temperature. This was attributed to the fact that an increase in temperature may cause increased swelling of the adsorbent, thereby allowing more active sites to be available for the uptake of Se(IV) ions [25].

Another class of adsorbent which have been utilized for the effective removal of toxic Se oxyanions are the layered double hydroxide (LDH) and layered rare-earth hydroxides (LRHs). LDH and LRH are important materials consisting of positively charged host layer and counterions in their interlayer space. They are capable of playing versatile roles such as acting as scavengers, two-dimensional nano-reactors, and adsorbents due to their excellent intercalation and anion-exchange capability. The anion-exchange capability facilitates their use for the uptake of toxic oxyanions.

Ma et al. [2] reported the use of MoS_4^{2-} intercalated Mg/Al LDH sorbent (MoS_4-LDH) for concurrent removal of Se toxic oxyanions (selenite and selenate), Hg^{2+}, Cu^{2+}, and Cd^{2+} from water. The presence of metal ions, coexisting with the Se oxyanions, promoted the rapid uptake of both selenite and selenate. For example, the removal of Se(VI) increased with the presence of Hg^{2+} while the presence of other metal ions (i.e., Hg^{2+}, Cu^{2+}, and Cd^{2+}) facilitated the removal of Se(IV). For the individual oxyanion without the metal ions, 81.0% and 99.1% removal were achieved, respectively, while 100% removal was achieved in the presence of metal ions. The presence of metal ions extremely enhanced the Se oxyanions capture because of the reactions of the interlayer MoS_4^{2-} anions with the metal ions [2].

Zhu et al. [26] reported the use of cationic layered rare-earth hydroxide (LRH), yttrium hydroxide ($Y_2(OH)_5Cl.^3/_2$ H_2O) for the effective removal of SeO_3^{2-} and SeO_4^{2-} ion from aqueous solution. Among other inorganic adsorbents, this adsorbent presented new records, with maximum sorption capacities of 207 and 124 mg/g for SeO_3^{2-} and SeO_4^{2-}, respectively. At low concentration region and even in the presence of competitive anions like Cl^-, NO_3^-, HPO_4^{2-}, CO_3^{2-} and SO_4^{2-}, the LRH removed almost all the Se oxyanions from aqueous solution. The residual Se oxyanions were less than 10 μg/L, which was far below the limit for drinking water. The mechanism provided for the higher uptake of selenite over selenate by the LRH was that the SeO_3^{2-} ions were directly bound through strong bidentate nuclear inner-sphere complexation to the Y^{3+} center in the positively charged layer of $[Y_2(OH)_5Cl.^3/_2$ $H_2O]^+$ [26].

The adsorption mechanism of both SeO_4^{2-} and SeO_3^{2-} onto the cationic LRH adsorbent was unraveled by a combination of energy-dispersive spectroscopy (EDS), Fourier transform infrared spectrometer (FTIR), Raman, powder X-ray diffraction (PXRD), and extended X-ray absorption fine structure (EXAFS) techniques. The FTIR of the spent material showed a new peak at 860 cm^{-1}, which was attributed to v(Se–O) vibrations of SeO_4^{2-} while a broad peak between 575 and 700 cm^{-1} was assigned to Se–O vibration of a mixture of $HSeO_3^-$ and SeO_3^{2-} specie, suggesting the adsorption of SeO_4^{2-} and SeO_3^{2-} onto $Y_2(OH)_5Cl.^3/_2$ H_2O. The result from the EDS analysis showed that the number of chloride ions greatly decreased after the sorption process, indicating an anion-exchange mechanism of SeO_4^{2-} and SeO_3^{2-} with the interlayer chloride ions in $Y_2(OH)_5Cl.^3/_2$ H_2O. From the XRD analysis, the interlayer spacing of $Y_2(OH)_5Cl.^3/_2$ H_2O increased after sorption with both SeO_4^{2-} and SeO_3^{2-}, indicating that both selenate and selenite have entered into the crystallographic interlayer space. EXAFS spectra gave confirmatory evidence on the interaction mechanism, where it was concluded that the sorption of SeO_4^{2-} onto $Y_2(OH)_5Cl.^3/_2$ H_2O occurred through simple ion exchange of Cl^- in the original structure, by forming an outer-sphere complexation. In the case of SeO_3^{2-} sorption

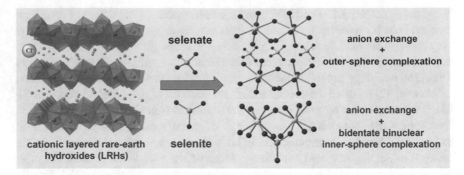

Fig. 4.6 Inner-sphere complexation of Se oxyanions with cationic layered rare-earth hydroxides (Yttrium Hydroxide) [26]

mechanism, a bidentate binuclear inner-sphere complex was formed during the anion-exchange reaction (Fig. 4.6).

Presented in Table 4.1 are selected adsorbents used for selenate and selenite adsorption with their adsorption parameters.

4.3.3 Photocatalysis

Photocatalysis is a phenomenon, where an electron–hole pair is generated on exposing semiconducting materials to light of suitable energy. Thus, the chemical reactions that occur in the presence of a semiconductor and light are collectively termed as photocatalytic reactions. The photogenerated holes (h$^+$) react with hydroxyl groups or adsorbed water on the photocatalyst surface, forming hydroxyl radicals (OH·), which are generally accepted to be responsible for initiating the oxidation pathway that occurs in the valence band. The photogenerated electrons are responsible for initiating photoreduction reactions that take place in the conduction band [53].

TiO$_2$-assisted photocatalysis was employed for the removal of aqueous phase selenite and selenate species, using ethylenediaminetetraacetic acid (EDTA) as hole (h$^+$) scavenger [54]. The binary selenite/EDTA and selenate/EDTA systems presented high selenite and selenate removal at pH 4.0 and pH 6.0, with faster removal kinetics, noted for the selenite species. The selenite and selenate removal were ascribed to their reduction by the electrons (e$^-$) generated at the conduction band. The effect of pH studies on the removal efficiency showed high selenite, selenate, and EDTA removal in the acidic pH range, with the following specific trend: pH 4.0 > pH 6.0 > pH 12.0. Aside the EDTA, a different h$^+$ scavenger (thiocyanate) was used, but the use of thiocyanate did not initiate the reduction of Se oxyanions, due to its adsorption onto the semiconductor surface (TiO$_2$). However, the addition of EDTA, as h$^+$ scavenger at pH 4.0, gave a near-complete removal of selenite and selenate. The findings from

Table 4.1 Some examples of adsorbents, including their adsorption parameters

Type of adsorbent	Type of Se oxyanion	Adsorption model	Optimum pH	Adsorption capacity (Q_{max})	Kinetic model	References
Mesoporous activated carbon	Selenate Selenite	Freundlich	6.0–8.0	5.38 µg/g 9.02 µg/g	Pseudo-first order	[12]
Cu-coated activated carbon	Selenate	Langmuir	6.0	4.48 mg/g	Pseudo-second order	[45]
Fe-coated granular activated carbon	Selenite	Langmuir	7.0	2.50 mg/g	Pseudo-second order	[27]
Cationic layered rare-earth hydroxide [Y$_2$(OH)$_5$Cl.^{3}/$_2$H2O]	Selenite, Selenate	Langmuir	7.0	207 mg/g; 124 mg/g	-	[26]
Granular activated carbon and powdered activated carbon	Selenite	Langmuir	2.0–3.0	9.45 µg/g and 12.45 µg/g respectively	Pseudo-first order	
Carbonaceous sorbent from peanut shell	Selenite	Langmuir	1.5	Nd	Pseudo-first order	[25]
Layered double hydroxide (MoS$_4$$^{2-}$ intercalated Mg/Al LDH)	Selenite, Selenate	Langmuir	6.3–8.7	294 mg/g; 85 mg/g	Pseudo-second order	[2]
Ligand functionalized organic–inorganic adsorbent	Selenite	Langmuir	2.0	111.12 mg/g	–	[46]

(continued)

Table 4.1 (continued)

Type of adsorbent	Type of Se oxyanion	Adsorption model	Optimum pH	Adsorption capacity (Q_{max})	Kinetic model	References
MgO nanosheets	Selenite Selenate	Langmuir	10.5	103.52 mg/g 10.28 mg/g	Pseudo-second order	[47]
MFe$_2$O$_4$ nanoparticles (M = Mn, Co, Cu)	Selenite Selenate	Langmuir Freundlich	6.0, 8.0, 8.1, respectively	Langmuir: 3.9, 11.6, 14.1 mg/g Freundlich: 5.27,5.55, 5.97 (mg/g)	Pseudo-second order	[48]
Rutile TiO$_2$	Selenite Selenate	–	–	2.1 μmol/g 14.8 μmol/g	–	[49]
Supported magnetite particles on zeolite	Selenite	Langmuir	3.5	23 mg/g	Pseudo-second order	[50]
Fe oxide/hydroxide nanoparticles	Selenite Selenate	Freundlich	4.0	95 mg/g 15.1 mg/g	–	[51]
Fe-coated granular activated carbon	Selenate	Freundlich	2–5	0.22 mg/g	Pseudo-second order	[52]

this study implied that the use of TiO_2-mediated photocatalysis and h^+ scavenging agent, EDTA, can successfully remove selenate and selenite from aqueous phase [54].

Tan et al. [55] reported the use of photocatalysis for the removal of Se from aqueous system. The effect of organic h^+ scavengers (e.g., formic acid, methanol, acetic acid, ethanol, salicylic acid, and sucrose) on the photocatalytic reduction of selenate and selenite over UV-illuminated TiO_2 photocatalyst was studied. Out of the six scavengers used, the photoreduction of Se ions to elemental Se was only observed in the presence of formic acid, methanol, and ethanol, and the rate of Se ion photoreduction was found to be in the order of formic acid > methanol > ethanol.

Holmes et al. [56] also reported the photocatalytic reduction of selenate to solid elemental Se (Se^0) and gaseous hydrogen selenide (H_2Se). In this study, TiO_2 was used as the catalyst for selenate removal from mine impacted water, containing high concentrations of sulfate and nitrate. Despite the presence of high concentration of sulfate, nitrate, carbonate, and trace metals in the wastewater, the process was able to reduce selenate, under different reaction conditions, from >5 00 to < 1 $\mu g\ L^{-1}$. It was reported that below the point of zero charge of TiO_2 (PZC $TiO_2 = 5.6$–6.2), faster kinetics were achieved, due to the positive zeta potential of the catalyst surface, increased electrostatic attraction, and outer-sphere adsorption of selenate. The highest pH value, at which significant Se removal was achieved, was at pH 6.04. The impact of temperature was significant, and an activation energy of 31.83 $kJ\ mol^{-1}$ was required for the reduction of selenate. Higher temperatures lead to higher selectivity toward H_2Se as well as faster selenate removal kinetics.

The removal of selenate from a simulated wet flue gas desulfurization (FGD) wastewater was reported by Tsuneori et al. [57], using TiO_2 as the photocatalysis and formic acid (HCOOH) as the hole scavenger. Excess amount of formic acid was needed to achieve efficient selenite removal from wastewater. The presence of excess anions, such as Cl^-, SO_4^{2-} and NO^{3-} inhibited the process efficiency, because of the adsorption of the anions on the catalyst surface, which hindered the selenate (SeO_4^{2-}) access to the photocatalyst surface.

Holmes et al. [58] reported the application of heterogeneous nanoscale photocatalysts, synthesized by depositing noble metal nanoparticles [gold (Au), silver (Ag), platinum (Pt), and palladium (Pd)] onto TiO_2 for photocatalytic reduction of selenate to elemental Se (Se^0) or hydrogen selenide gas (H_2Se). Se concentration of 5 mg L^{-1} was reduced in the presence of 300 mg L^{-1} formic acid. The rate of selenate removal decreased in order of increasing work function of the metals, with Ag–TiO_2 exhibiting most rapid kinetics, while Pd–TiO_2 was the slowest. The noble metal deposits on the surface of TiO_2 acted as electron sinks for the photogenerated electrons in the conduction band of TiO_2. The higher the work function of the noble metal deposit, the further the fall in the reduction potential, when there is transfer of electrons noble metal sink, and the more positive is the reduction potential of the electrons. With a more positive reduction potential, the less likely the photogenerated electrons are able to reduce the Se^{6+}/Se^0 reduction couple ($E_0 = 0.5$ V vs. SHE). For example, Au, Pt, and Pd have work functions 0.66 eV, 0.76 eV, and 1.06 eV versus SHE respectively, all of which would decrease the rate at which selenate can be reduced by drawing electrons away from the reactive conduction band of rutile TiO_2 (-0.28 V vs. SHE). Since the Se-noble metal-TiO_2 (Se–NM–TiO_2) photocatalytic system was structured in a direct Z-scheme arrangement, when Au, Ag, or Pt were used, it allowed for high selectivity toward H_2Se. In contrast, Pd acted as an electron sink, which decreased reducibility of the photogenerated electrons, ultimately causing a higher selectivity toward Se^0. Among the catalyst tested, Au–TiO_2 offered the largest H_2Se selectivity, while Pd–TiO_2 offered the highest selectivity to solid Se^0 generation.

4.3.4 Coagulation

It has been demonstrated that metal salts, such as ferric chloride ($FeCl_3$) and aluminum sulfate ($Al_2(SO_4)_3$), are effective coagulants for removing particles [59]. The actions of trivalent metal ions mainly involve two effects in particle removal: charge neutralization and sweep flocculation [60].

Hu et al. [61] reported the removal of Se oxyanions species (SeO_4^{2-} and SeO_3^{2-}) from drinking water by coagulation. Different parameters, such as coagulant type, Se oxyanion species, interfering ions, and pH value were used to evaluate the performance of Se removal by coagulation. Fe-based coagulant was much more efficient in Se removal than Al-based coagulant, and SeO_3^{2-} was more easily removed than SeO_4^{2-}. Adsorption onto the hydroxide flocs was the domineering coagulation mechanism for Se removal, while precipitation with the formation of insoluble $Al_2(SeO_3)_3$ or $Fe_2(SeO_3)_3$ played a role at low dosage, especially for SeO_3^{2-} removal with Fe/Al coagulant. High coagulant dosage and weakly acidic pH could enhance the formation of more hydroxide flocs, having more active adsorption sites and high zeta potential, which favors the removal of Se oxyanion from water [61]. Even though sulfur and Se belong to the chalcogen family, the influence of sulfate (SO_4^{2-}) on the Se removal was the lowest among the four oxyanions. The negative influence of anions on SeO_3^{2-} removal followed the order: $PO_4^{3-} > SiO_3^{2-} > CO_3^{2-} > SO_4^{2-}$.

The use of a combined sulfite/UV/Fe(III) coagulation process was employed for the efficient selenite, Se(VI) removal from sulfate-rich water [62]. In the presence of sulfate (1000 mg L^{-1}), over 99% of Se(VI) (initial conc. $= 10$ mg L^{-1}) was reduced by sulfite (5.0 mM), with a UV dose of 16 J cm^2 (within 20 min) to Se(IV). An alkaline pH (>9) was required for the reduction process, which was obtained with the addition of sulfite. Scavenging experiment with N_2O and NO_3 indicated that hydrated electrons (e_{aq}^-) were responsible for Se(VI) reduction by sulfite/UV. There was a negligible influence on Se(VI) reduction in the presence of chloride, sulfate, phosphate, and carbonate (up to 10 mM), while nitrate and humic acid inhibited Se(VI) reduction to varying extents, depending on their concentrations. Using an Fe(III) coagulant, Se(IV) was effectively removed in the co-presence of sulfite and sulfate, at an OH/Fe molar ratio of 1.8–2.8. The removal of Se(IV) by Fe(III) coagulant was not affected by the presence of chloride, nitrate, or sulfate (up to 10 mM). The magnitude of Se(VI) removal was severely affected at high levels of carbonate (10 mM) and phosphate (1 mM).

Wan et al. [63] reported the removal of Se^0 particles in simulated wastewater with amorphous ferric hydroxide $(Fe(OH)_3)$ produced from the hydrolysis of ferric chloride (the coagulant). A sharp rise in Se^0 removal (%), from ~20 to ~60 was observed when the $FeCl_3$ dosage was increased from 1 to 5 mg L^{-1}. The removal efficiency slightly increased at a dosage higher than 5 mg L^{-1}, to attain a removal efficiency of ~90% at 50 mg L^{-1}. Further increase in the $FeCl_3$ dosage to 100 mg L^{-1} did not contribute any significant improvement to the removal efficiency.

4.3.5   Electrocoagulation

Electrocoagulation, a unique method for water treatment, is based on the electrochemical dissolution of sacrificial metal electrodes into soluble or insoluble species that improves the coagulation, adsorption, or precipitation of soluble or colloidal pollutants with high removal efficiency. Electrocoagulation process involves applying an electric current to sacrificial electrodes (mostly iron and aluminum) inside a reactor tank, where the current generates a coagulating agent and gas bubbles [64].

In principle, an electrocoagulation technique uses an electrochemical cell to treat the water. In the simplest form, an electrochemical cell consists of two electrodes, the anode and the cathode, immersed in a conducting solution or an electrolyte, all connected via an electrical circuit, which includes a current source and control device [65]. The metallic cations, generated from the anode, hydrolyze to form hydroxides, poly hydroxides, and poly hydroxyl metallic compounds with a strong affinity for the dispersed particles and counterions, thus causing coagulation [66]. Hence, electrocoagulation (EC), which is also known as radiofrequency diathermy or short-wave electrolysis, is fast becoming popular as an alternative method for the treatment of water.

Recently, Hansen et al. [67] investigated the potential of an electrocoagulation technique for the treatment of Se in wastewater. In this study, the sacrificial electrode was iron, to generate the necessary ferrous ions for the process. Approximately, 90% of the Se removal was observed after 6 h of treatment, indicating a decrease in concentration from 0.30 to 0.03 mg/L. Thus, the electrocoagulation process was recommended as a feasible technique for Se removal from water [67].

Although, elemental Se [Se(0)] is particulate and less bioavailable, yet when released into surface waters, Se(0) can be oxidized to Se oxyanions [68]. Se(0) also exhibits colloidal properties that adversely affect aquatic ecosystems and membrane separation processes [61]. For this reason, Staicu et al. [61] investigated the removal potential of colloidal Se(0) by electrocoagulation process. Colloidal Se(0) is produced by a strain of *Pseudomonas fluorescens* and show limited gravitational settling. So, iron (Fe) and aluminum (Al) sacrificial electrodes were used in a batch reactor under galvanostatic conditions. An Se(0) removal of 97% was achieved using iron electrodes at 200 mA, while 96% of colloidal Se(0) was removed by aluminum electrodes at a higher current intensity (300 mA).

Additionally, the leaching tests of the electrogenerated Fe and Al sediments were evaluated following the toxicity characteristics leaching procedure (TCLP) method. The amount of Se leached from the Fe–Se sediment was below the regulatory level (1 mg/L), but the Se concentration leached from the Al-Se sediment was about 20 times above the limit. The use of Fe electrodes as a soluble anode in the electrocoagulation process provides a better option than Al electrodes [61].

4.3.6 Bioremediation

Bioremediation involves the use of microbes to consume or breakdown environmental pollutants. Environmental microorganisms can carry out a variety of transformations of Se species such as methylation, demethylation, oxidation, and reduction. However, due to the low rates at which microbial demethylation and oxidation reactions proceed, they are often not considered for Se remediation [69]. Bacteria Se reduction has been considered as one of the most efficient and economical ways of removing soluble Se oxyanions from water. Biofilms or communities of microorganisms can play a vital role in transforming Se species to less toxic chemical forms (i.e., elemental Se) [70].

Escherichia coli strain, EWB32213, isolated from sediment slurry was utilized for the reduction of selenate and selenite]. In this study, a two-stage microbial Se reduction process (i.e., from SeO_4^{2-} to SeO_3^{2-}, followed by SeO_3^{2-} to Se(0) was evaluated using monod-type kinetic expressions [71]. The rate of SeO_4^{2-} oxyanion reduction to SeO_3^{2-} in the first stage was faster, when compared with that of SeO_3^{2-} reduction to Se(0) in the second stage. Thus, the SeO_3^{2-} reduction stage was the rate-limiting step. For instance, within 1.5 days, the reduction of SeO_4^{2-} concentration that ranged between 10 and 400 mg/L was completed, whereas, after 10 days only a fraction of concentration higher than 100 mg/L of SeO_3^{2-} was reduced to Se(0). Also, during the first stage, significant amount of SeO_3^{2-} accumulated in the *E. coli* strain at high levels of initial SeO_4^{2-} [71].

In another study by Ji et al. [72], a co-culture of microorganisms was used to accelerate SeO_3^{2-} reduction to Se (0), in continuation of the first reduction stage of SeO_4^{2-} to SeO_3^{2-} [71]. In order to achieve a maximum reduction of SeO_3^{2-}, a SeO_3^{2-} reducing strain, *Pantoea vagans* EWB32213-2, isolated from sediment samples, were co-cultured with a SeO_3^{2-}-reducing strain, *Shigella fergusonii* TB42616, isolated from sludge samples. Before the co-culture study, the selected pure culture strains were used for the reduction of their specific Se oxyanions species, and the optimal variable suitable for the reduction of SeO_3^{2-} was selected, since the main objective was to achieve a maximum reduction of SeO_3^{2-} to Se(0). For example, the optimal temperature for SeO_4^{2-} reduction by *Shigella fergusonii* TB42616 was observed at 40 °C, while that of SeO_3^{2-} reduction by *Pantoea vagans* EWB32213-2 was observed at 30 °C. As a result of this, a temperature of 30 °C was selected for the co-culture study.

Therefore, for the co-culture experiment, an optimal condition for selenium reduction by a defined co-culture of *Shigella fergusonii* TB42616 and *Pantoea vagans* EWB32213-2 was conducted under a temperature of 30 °C, pH of 8.0 and a co-culture bacteria composition of 1:1. The results showed that the reduction of SeO_4^{2-} to SeO_3^{2-} was accelerated by the co-culture bacteria and less SeO_3^{2-} accumulation was observed as indicated from the low concentration of SeO_3^{2-} in the defined co-culture. Previously, in a study carried out by Ji and Wang [73], SeO_3^{2-} reduction by single culture could not be achieved after 6 days at a high initial concentration of 400 mg/L of SeO_4^{2-}. But in this study, approximately 75.53% SeO_3^{2-} accumulation was observed at a relatively high initial SeO_4^{2-} concentration of 1100 mg/L. Hence, the addition of SeO_3^{2-} reducing bacteria strain had a significant impact on the reduction of SeO_4^{2-} to SeO_3^{2-} [72].

Complete reduction of selenate (SeO_4^{2-}) to Se^0 was achieved in a membrane biofilm reactor, where methane (CH_4) was the only electron donor used [74]. Results from the scanning electron microscopy coupled with energy-dispersive spectroscopy tool (SEM–EDS) revealed that the main selenate reduction product was nano-spherical Se^0.

Biofilms, which are a collection of one or more types of microorganisms that can grow on different surfaces can both detoxify and sequester Se, playing vital roles in their fate and effects in aquatic environments. They can also play an important role in bio-transforming elements to less toxic chemical forms. However, a twist to the bioremediation of Se species, which involves the use of different tools to characterize the fate of Se oxyanions in multi-species biofilm was reported by Yang et al. [70]. Confocal laser scanning microscopy (CLSM) revealed distinct biofilm morphology at high Se oxyanion concentration. Micro-X-ray fluorescence imaging, combined with CLSM showed highly localized reduced Se species in the biofilm. The result from x-ray absorption near-edge spectroscopy (XANES) revealed that biofilm biotransformation of Se oxyanions while the extended X-ray absorption fine structure (EXAFS) confirmed the production of the Se oxyanion bioreduction to be elemental Se [70].

4.3.7 Phytoremediation

Unlike bioremediation, which uses a microorganism-based approach, phytoremediation utilizes the plants-based approach for the remediation of contaminated water. Phytoremediation, including phytoextraction and phytovolatilization, has been considered as feasible and effective approaches for the elimination of Se from a Se-contaminated environment [75, 76]. They are also relatively cheap since they are solar-driven and executed in-situ. Plants play a more direct role in remediation. For example, they may be used to accumulate metals/metalloids in their harvestable biomass (phytoextraction), and they can also convert and release certain metals/metalloids in a volatile form (phytovolatilization) [77]. Plants that can reasonably tolerate and accumulate elevated Se levels (up to 1000 mg Se/ kg) are referred to as Se accumulators. For example, true Se hyperaccumulator species, from the genera *Astragalus, Stanleya, Oonopsis,* and *Xylorhiza,* can conveniently accumulate 1000–15,000 mg Se/kg DW without any toxicity symptoms [78].

In a study by Feng and Wei [79], an arsenic (As) hyperaccumulator plant named *Pteris vittata* L. (also known as Chinese Brake fern) was found to be a Se accumulator too. To prove their assumption, they investigated the Se accumulation mechanism of this plant, focusing on antioxidant responses, up to six concentration levels of selenite (0, 1, 2, 5, 10, and 20 mg/L). The root of the Chinese brake adsorbed more Se than the fronds of the plant, with the highest Se concentration of 1.536 mg/kg in roots and 242 mg/kg in fronds. This indicated that Chinese brake fern can accumulate a large amount of Se without any obvious toxic symptoms and a noticeable decrease in its biomass. The activities of different enzymes were studied, to know their roles in the Se accumulation of Chinese Brake fern. The enzymes of catalase (CAT), ascorbate peroxidase (APX), and peroxidase (POD), contributed their antioxidative functions only under low dosages of Se, as revealed by their increased activities at Se levels ≤ 5 mg/L and decreased activities at Se levels >5 mg/L [79]. The concentration of glutathione (GSH) and enzyme activity of glutathione reductase (GR) were stimulated by levels of Se ≥ 5 mg/L. Superoxide dismutase (SOD) activity was also enhanced by 20 mg/L Se. Consequently, the results suggested that SOD, GSH, and GR were likely responsible for Se accumulation in Chinese Brake fern while enzymes of POD, APX, and CAT played limited roles in the Se accumulation of the plant [79].

An alternative phytoremediation approach, which depicts a real-life scenario for the remediation of contaminated water, is the use of constructed wetlands. Constructed wetlands consist of a complex ecosystem including microbes, plants, and sediments, which act together as a biogeochemical filter, efficiently removing dilute contaminants from very large volumes of wastewater [75]. Also, the anoxic environment and organic matter in wetlands encourage chemical and biological processes that transform contaminants into immobile or less toxic forms [80]. Thus, constructed wetlands have been used as a low-cost treatment technology to eliminate a wide range of contaminants from Se-polluted water.

For instance, an experimental wetland was constructed at Tulare Lake Drainage District (TLDD) to evaluate its potential for the removal of Se from agricultural irrigation drainage water. The wetland consisted of ten unlined cells that were continuously flooded and planted with either a non-vegetated or vegetated plants (singly or combined), including sturdy bulrush [*Schoenoplectus robustus* (Pursh) M. T. Strong], Baltic rush (*Juncus balticus Willd.*), smooth cordgrass (*Spartina alterniflora Loisel.*), rabbits foot grass [*Polypogon monspeliensis* (L.) Desf.], saltgrass [*Distichlis spicata* (L.) Greene], cattail (*Typha latifolia* L.), tule [*Schoenoplectus acutus* (Muhl. ex Bigelow) A. Love & D. Love], and widgeon grass (*Ruppia maritima* L.). The vegetated wetland cells removed Se more efficiently than the unvegetated cell. On average, the wetland cells removed 69% of the total Se from the inflow [80].

Furthermore, genetically engineered plants can be used to remove Se oxyanions just like it was used to remove organomercurial compounds with improved efficiency [81]. For instance, metal-hyperaccumulating plants and microbes with unique abilities to tolerate, accumulate, and detoxify metals and metalloids represent an important reservoir of unique genes that could be transferred to fast-growing plant species for enhanced phytoremediation [75].

4.3.8 Co-precipitation

The chemical properties of Se and sulfur are very similar. Thus, Se chemically behaves like sulfur and is most often associated with sulfur [82]. However, increasing the concentration of sulfur coexisting with Se significantly reduces the effective removal of Se. Nonetheless, co-precipitation has been utilized as a means of circumventing this problem for the effective removal of Se from a polluted solution. The advantages of this method include its simplicity, low cost, short treatment time, and its ability to preserve substituent ions for a long time in the crystal lattice, making it work as constructed barriers to retain various ions, including Se oxyanions [83].

Co-precipitation is a process that immobilizes the trace element in the mineral during crystal growth. Various salt such as barite and sulfide have been employed for this process. These salts act as the engineered barrier to retain several ions. Barite ($BaSO_4$) has been used as a sequestering phase for the removal of toxic and/or radioactive elements from polluted solutions, because of the following characteristics; (i) its high crystal stability under wide ranges of pH, pressure, and temperature conditions, (ii) it's extremely low solubility(ca. Ksp = 10–9.98 at 25 °C, 1 atm), (iii) its ability to selectively incorporate a large amount of several ions, (iv) its high density (4.5 g/cm^3) when compared with other minerals, which is an advantage for rapid sedimentation during co-precipitation process [84, 85].

Tokunaga and Yoshio [85] employed the co-precipitation process for the effective removal of selenite and selenate ions from aqueous solution using barite as a sequestering phase. The Se(IV) uptake by barite was found to be dependent on pH, sulfate concentration in the initial solution, and coexistence of calcium ion. All these can be attributed to the effects on their structural similarity and chemical affinity. Whereas, the uptake of Se(VI) was largely dependent on the concentration of sulfate in the initial concentration, which is only linked to the structural similarity. It was also reported that the mechanism involved in the effective removal of Se(IV) and Se(VI) from aqueous solution was the strong crystal stability formed between barite and water ($Ba–SeO_3–SeO_4–SO_4$), which shows various ways for Se. This strong crystal stability formed prevents the leaching out of Se regardless of the surrounding environment, once it has been incorporated into the barite lattice [85].

4.4 Future Perspectives

Due to globalization, an increase in industrial and anthropogenic activities has a very huge impact on the occurrence of selenate and selenite in water bodies. The anionic forms, which exist in water, are the most toxic to all forms of life when it exceeds the permissible limit. Therefore, more studies should be conducted on the removal of Se oxyanions from water and different techniques should be explored for their remediation in water. It is safe to apply preventive measures, as there are a lot of challenges accompanying Se removal from water. The Se released into the water can be minimized by controlling and treating mine effluent before discharging into the water bodies. Several other methods should be investigated to reduce the release of Se into the water bodies.

Most advanced techniques developed these days are very expensive, in terms of acquisition and maintenance. Therefore, more efficient and cheap materials (i.e., adsorbents and photocatalysts) should be explored. Industries should be monitored, to ensure that they follow the guidelines and limits provided for the discharge of their wastewater into the environment. Stringent penalties should be put in place for defaulters.

References

1. Reich HJ, Hondal RJ (2016) Why nature chose selenium. ACS Chem Biol 11:821–841
2. Ma L, Islam SM, Xiao C, Zhao J, Liu H, Yuan M, Sun G, Li H, Ma S, Kanatzidis MG (2017) Rapid simultaneous removal of toxic anions $[HSeO_3]^-$, $[SeO_3]^{2-}$, and $[SeO_4]^{2-}$, and metals Hg^{2+}, Cu^{2+}, and Cd^{2+} by MoS_4^{2-} intercalated layered double hydroxide. J Am Chem Soc 139:12745–12757
3. Mehdi Y, Hornick J-L, Istasse L, Dufrasne I (2013) Selenium in the environment, metabolism and involvement in body functions. Molecules 18:3292–3311
4. He Y, Xiang Y, Zhou Y, Yang Y, Zhang J, Huang H, Shang C, Luo L, Gao J, Tang L (2018) Selenium contamination, consequences and remediation techniques in water and soils: a review. Environ Res 164:288–301
5. Lee KH, Jeong D (2012) Bimodal actions of selenium essential for antioxidant and toxic pro-oxidant activities: the selenium paradox. Mol Med Rep 5:299–304
6. Hawkes WC, Turek PJ (2001) Effects of dietary selenium on sperm motility in healthy men. J Androl 22:764–772
7. Rayman MP (2012) Selenium and human health. The Lancet 379:1256–1268
8. Fairweather-Tait SJ, Bao Y, Broadley MR, Collings R, Ford D, Hesketh JE, Hurst R (2011) Selenium in human health and disease. Antioxid Redox Signal 14:1337–1383
9. Vinceti M, Crespi CM, Bonvicini F, Malagoli C, Ferrante M, Marmiroli S, Stranges S (2013) The need for a reassessment of the safe upper limit of selenium in drinking water. Sci Total Environ 443:633–642
10. WHO (2011) Guidelines for drinking-water quality. World Health Organization 216:303–304
11. Onoguchi A, Granata G, Haraguchi D, Hayashi H, Tokoro C (2019) Kinetics and mechanism of selenate and selenite removal in solution by green rust-sulfate. Roy Soc Open Sci 6:182147
12. Meher AK, Jadhav A, Labhsetwar N, Bansiwal AJAWS (2020) Simultaneous removal of selenite and selenate from drinking water using mesoporous activated alumina. Appl Water Sci 10:10

13. Di Gregorio S (2008) Selenium: a versatile trace element in life and environment. In: Trace elements as contaminants and nutrients: consequences in ecosystems and human health, pp 593–622
14. Cutter GA, Bruland KW (1984) The marine biogeochemistry of selenium: a re-evaluation 1. Limnol Oceanogr 29:1179–1192
15. Sharma S, Vance GF (2007) Dissolution chemistry of inorganic selenium in alkaline mine soils. In: Sarkar D, Datta R, Hannigan R, (eds) Developments in Environmental Science. Elsevier, pp 362–380
16. Alfthan G, Eurola M, Ekholm P, Venäläinen E-R, Root T, Korkalainen K, Hartikainen H, Salminen P, Hietaniemi V, Aspila P (2015) Effects of nationwide addition of selenium to fertilizers on foods, and animal and human health in Finland: from deficiency to optimal selenium status of the population. J Trace Elem Med Biol 31:142–147
17. Pyrzynska K (2009) Selenium speciation in enriched vegetables. Food Chem 114:1183–1191
18. Lenz M, Gmerek A, Lens PN (2006) Selenium speciation in anaerobic granular sludge. Int J Environ Anal Chem 86:615–627
19. Xing K, Zhou S, Wu X, Zhu Y, Kong J, Shao T, Tao X (2015) Concentrations and characteristics of selenium in soil samples from Dashan Region, a selenium-enriched area in China. Soil Sci Plant Nutr 61:889–897
20. Antanaitis A, Lubyte J, Antanaitis S, Staugaitis G, Viskelis P (2008) Selenium concentration dependence on soil properties. J Food Agric Environ 6:163
21. Rodríguez-Valencia C, López-Álvarez M, Cochón-Cores B, Pereiro I, Serra J, Gonzalez P (2013) Novel selenium-doped hydroxyapatite coatings for biomedical applications. J Biomed Mater Res, Part A 101:853–861
22. Goh K-H, Lim T-T (2004) Geochemistry of inorganic arsenic and selenium in a tropical soil: effect of reaction time, pH, and competitive anions on arsenic and selenium adsorption. Chemosphere 55:849–859
23. Jordan N, Ritter A, Foerstendorf H, Scheinost A, Weiß S, Heim K, Grenzer J, Mücklich A, Reuther H (2013) Adsorption mechanism of selenium (VI) onto maghemite. Geochim Cosmochim Acta 103:63–75
24. Meher AK, Jadhav A, Labhsetwar N, Bansiwal A (2020) Simultaneous removal of selenite and selenate from drinking water using mesoporous activated alumina. Appl Water Sci 10:10
25. El-Shafey E (2007) Removal of Se(IV) from aqueous solution using sulphuric acid-treated peanut shell. J Environ Manage 84:620–627
26. Zhu L, Zhang L, Li J, Zhang D, Chen L, Sheng D, Yang S, Xiao C, Wang J, Chai Z (2017) Selenium sequestration in a cationic layered rare earth hydroxide: a combined batch experiments and EXAFS investigation. Environ Sci Technol 51:8606–8615
27. Zhang N, Lin L-S, Gang D (2008) Adsorptive selenite removal from water using iron-coated GAC adsorbents. Water Res 42:3809–3816
28. Chen M-L, An M-I (2012) Selenium adsorption and speciation with Mg–FeCO₃ layered double hydroxides loaded cellulose fibre. Talanta 95:31–35
29. Yang L, Shahrivari Z, Liu PK, Sahimi M, Tsotsis TT (2005) Removal of trace levels of arsenic and selenium from aqueous solutions by calcined and uncalcined layered double hydroxides (LDH). Ind Eng Chem Res 44:6804–6815
30. Das S, Hendry MJ, Essilfie-Dughan J (2013) Adsorption of selenate onto ferrihydrite, goethite, and lepidocrocite under neutral pH conditions. Appl Geochem 28:185–193
31. Su T, Guan X, Gu G, Wang J (2008) Adsorption characteristics of As(V), Se(IV), and V(V) onto activated alumina: effects of pH, surface loading, and ionic strength. J Colloid Interface Sci 326:347–353
32. Torres J, Pintos V, Gonzatto L, Domínguez S, Kremer C, Kremer E (2011) Selenium chemical speciation in natural waters: Protonation and complexation behavior of selenite and selenate in the presence of environmentally relevant cations. Chem Geol 288:32–38
33. Scheidegger A, Grolimund D, Cui D, Devoy J, Spahiu K, Wersin P, Bonhoure I, Janousch M (2003) Reduction of selenite on iron surfaces: a micro-spectroscopic study. J de Phys IV (Proceedings) 417–420 (EDP sciences)

34. Ling L, Pan B, Zhang W (2015) Removal of selenium from water with nanoscale zero-valent iron: mechanisms of intraparticle reduction of Se(IV). Water Res 71:274–281
35. Raychoudhury T, Tufenkji N, Ghoshal S (2014) Straining of polyelectrolyte-stabilized nanoscale zero valent iron particles during transport through granular porous media. Water Res 50:80–89
36. Zhang W-X (2003) Nanoscale iron particles for environmental remediation: an overview. J Nanopart Res 5:323–332
37. Holmes AB, Gu FX (2016) Emerging nanomaterials for the application of selenium removal for wastewater treatment. Environ Sci Nano 3:982–996
38. Olegario JT, Yee N, Miller M, Sczepaniak J, Manning B (2010) Reduction of Se(VI) to Se(–II) by zerovalent iron nanoparticle suspensions. J Nanopart Res 12:2057–2068
39. Das S, Lindsay MB, Essilfie-Dughan J, Hendry MJ (2017) Dissolved selenium (VI) removal by zero-valent iron under oxic conditions: influence of sulfate and nitrate. ACS Omega 2:1513–1522
40. Su T, Guan X, Tang Y, Gu G, Wang J (2010) Predicting competitive adsorption behavior of major toxic anionic elements onto activated alumina: a speciation-based approach. J Hazard Mater 176:466–472
41. Yamani JS, Lounsbury AW, Zimmerman JB (2014) Adsorption of selenite and selenate by nanocrystalline aluminum oxide, neat and impregnated in chitosan beads. Water Res 50:373–381
42. Ma Q, Yu Y, Sindoro M, Fane AG, Wang R, Zhang H (2017) Carbon-based functional materials derived from waste for water remediation and energy storage. Adv Mater 29:1605361
43. Wasewar KL, Prasad B, Gulipalli S (2009) Removal of selenium by adsorption onto granular activated carbon (GAC) and powdered activated carbon (PAC). CLEAN—Soil Air, Water 37:872–883
44. Dobrowolski R, Otto M (2013) Preparation and evaluation of Fe-loaded activated carbon for enrichment of selenium for analytical and environmental purposes. Chemosphere 90:683–690
45. Zhao X, Zhang A, Zhang J, Wang Q, Huang X, Wu Y, Tang C (2020) Enhanced selenate removal in aqueous phase by copper-coated activated carbon. Materials 13:468
46. Awual MR, Yaita T, Suzuki S, Shiwaku H (2015) Ultimate selenium (IV) monitoring and removal from water using a new class of organic ligand based composite adsorbent. J Hazard Mater 291:111–119
47. Cui W, Li P, Wang Z, Zheng S, Zhang Y (2018) Adsorption study of selenium ions from aqueous solutions using MgO nanosheets synthesized by ultrasonic method. J Hazard Mater 341:268–276
48. Sun W, Pan W, Wang F, Xu N (2015) Removal of Se(IV) and Se(VI) by MFe_2O_4 nanoparticles from aqueous solution. Chem Eng J 273:353–362
49. Svecova L, Dossot M, Cremel S, Simonnot M-O, Sardin M, Humbert B, Den Auwer C, Michot LJ (2011) Sorption of selenium oxyanions on TiO_2 (rutile) studied by batch or column experiments and spectroscopic methods. J Hazard Mater 189:764–772
50. Verbinnen B, Block C, Lievens P, Van Brecht A, Vandecasteele C (2013) Simultaneous removal of molybdenum, antimony and selenium oxyanions from wastewater by adsorption on supported magnetite. Waste Biomass Valorization 4:635–645
51. Zelmanov G, Semiat R (2013) Selenium removal from water and its recovery using iron (Fe^{3+}) oxide/hydroxide-based nanoparticles sol (NanoFe) as an adsorbent. Sep Purif Technol 103:167–172
52. Zhang N, Gang D, Lin L-S (2010) Adsorptive removal of parts per million level selenate using iron-coated GAC adsorbents. J Environ Eng 136:1089–1095
53. Wang C-Y, Pagel R, Dohrmann JK, Bahnemann DW (2006) Antenna mechanism and deaggregation concept: novel mechanistic principles for photocatalysis. C R Chim 9:761–773
54. Labaran B, Vohra M (2014) Photocatalytic removal of selenite and selenate species: effect of EDTA and other process variables. Environ Technol 35:1091–1100
55. Tan T, Beydoun D, Amal R (2003) Effects of organic hole scavengers on the photocatalytic reduction of selenium anions. J Photochem Photobiol A: Chem 159:273–280

56. Holmes A, Giesinger K, Ye J, Milan E, Ngan A, Gu F. A non-biological option for Se removal by photocatalytic reduction of selenate in mine-impacted water containing high concentrations of sulfate and nitrate

57. Nakajima T, Yamada K, Idehara H, Takanashi H, Ohki A (2011) Removal of selenium (VI) from simulated wet flue gas desulfurization wastewater using photocatalytic reduction. J Water Environ Technol 9:13–19

58. Holmes AB, Khan D, de Oliveira Livera D, Gu F (2020) Enhanced photocatalytic selectivity of noble metallized TiO_2 nanoparticles for the reduction of selenate in water: tunable Se reduction product $H_2Se(g)$ vs. Se(s). Environ Sci Nano

59. Staicu LC, van Hullebusch ED, Oturan MA, Ackerson CJ, Lens PN (2015) Removal of colloidal biogenic selenium from wastewater. Chemosphere 125:130–138

60. Li T, Zhu Z, Wang D, Yao C, Tang H (2006) Characterization of floc size, strength and structure under various coagulation mechanisms. Powder Technol 168:104–110

61. Staicu LC, Van Hullebusch ED, Lens PN, Pilon-Smits EA, Oturan MA (2015) Electrocoagulation of colloidal biogenic selenium. Environ Sci Pollut Res 22:3127–3137

62. Wang X, Liu H, Shan C, Zhang W, Pan B (2018) A novel combined process for efficient removal of Se(VI) from sulfate-rich water: sulfite/UV/Fe(III) coagulation. Chemosphere 211:867–874

63. Wang J, Li J, Xie L, Liu Q, Zeng H (2020) Understanding the interaction mechanism between elemental selenium and ferric hydroxide in wastewater treatment. Ind Eng Chem Res 59:6662–6671

64. Malakootian M, Yousefi N, Fatehizadeh A (2011) Survey efficiency of electrocoagulation on nitrate removal from aqueous solution. Int J Environ Sci Technol 8:107–114

65. Canizares P, Carmona M, Lobato J, Martinez F, Rodrigo M (2005) Electrodissolution of aluminum electrodes in electrocoagulation processes. Ind Eng Chem Res 44:4178–4185

66. Abbas SH, Ali WH (2016) Electrocoagulation technique used to treat wastewater: a review. Am J Eng Res (AJER) 10:74–88

67. Hansen HK, Peña SF, Gutiérrez C, Lazo A, Lazo P, Ottosen LM (2019) Selenium removal from petroleum refinery wastewater using an electrocoagulation technique. J Hazard Mater 364:78–81

68. Zhang Y, Zahir ZA, Frankenberger WT (2004) Fate of colloidal-particulate elemental selenium in aquatic systems. J Environ Qual 33:559–564

69. Eswayah AS, Smith TJ, Gardiner PH (2016) Microbial transformations of selenium species of relevance to bioremediation. Appl Environ Microbiol 82:4848–4859

70. Yang SI, Lawrence JR, Swerhone GD, Pickering IJ (2011) Biotransformation of selenium and arsenic in multi-species biofilm. Environ Chem 8:543–551

71. Ji Y, Wang Y-T (2017) Selenium reduction by batch cultures of *Escherichia coli* strain EWB32213. J Environ Eng 143:04017009

72. Ji Y, Wang Y (2018) Selenium reduction by a defined co-culture in batch reactors. In: World environmental and water resources congress 2018: water, wastewater, and stormwater; urban watershed management; municipal water infrastructure; and desalination and water reuse. American Society of Civil Engineers Reston, VA, pp 246–254

73. Ji Y, Wang Y-T (2016) Selenium reduction in batch bioreactors. World Environ Water Resour Congr 2016:175–184

74. Lai C-Y, Wen L-L, Shi L-D, Zhao K-K, Wang Y-Q, Yang X, Rittmann BE, Zhou C, Tang Y, Zheng P (2016) Selenate and nitrate bioreductions using methane as the electron donor in a membrane biofilm reactor. Environ Sci Technol 50:10179–10186

75. LeDuc DL, Terry N (2005) Phytoremediation of toxic trace elements in soil and water. J Ind Microbiol Biotechnol 32:514–520

76. Wu Z, Bañuelos GS, Lin Z-Q, Liu Y, Yuan L, Yin X, Li M (2015) Biofortification and phytoremediation of selenium in China. Front Plant Sci 6:136

77. Yasin M, El Mehdawi AF, Jahn CE, Anwar A, Turner MF, Faisal M, Pilon-Smits EA (2015) Seleniferous soils as a source for production of selenium-enriched foods and potential of bacteria to enhance plant selenium uptake. Plant Soil Sci Plant Nutr 386:385–394

78. El Mehdawi A, Pilon-Smits E (2012) Ecological aspects of plant selenium hyperaccumulation. Plant Biol 14:1–10
79. Feng R, Wei C (2012) Antioxidative mechanisms on selenium accumulation in *Pteris vittata L.*, a potential selenium phytoremediation plant. Plant, Soil Environ Chem 58:105–110
80. Gao S, Tanji K, Lin Z, Terry N, Peters D (2003) Selenium removal and mass balance in a constructed flow-through wetland system. J Environ Qual 32:1557–1570
81. Ruiz ON, Hussein HS, Terry N, Daniell H (2003) Phytoremediation of organomercurial compounds via chloroplast genetic engineering. Plant Physiol 132:1344–1352
82. Moore L, Mahmoudkhani A (2011) Methods for removing selenium from aqueous systems. Proc Tailings Mine Waste 6–9
83. Tokunaga K, Yokoyama Y, Kawagucci S, Sakaguchi A, Terada Y, Takahashi Y (2013) Selenium coprecipitated with barite in marine sediments as a possible redox indicator. Chem Lett 42:1068–1069
84. Bosbach D, Böttle M, Metz V (2010) Experimental study on Ra^{2+} uptake by barite ($BaSO_4$). Kinetics of solid solution formation via $BaSO_4$ dissolution and $RaxBa_{1-x}SO_4$ (re) precipitation. Swedish Nuclear Fuel and Waste Management Co.
85. Tokunaga K, Takahashi Y (2017) Effective removal of selenite and selenate ions from aqueous solution by barite. Environ Sci Technol 51:9194–9201
86. Fernández-Martínez A, Charlet L (2009) Selenium environmental cycling and bioavailability: a structural chemist point of view. Rev Environ Sci Bio/Technol 8:81–110
87. Saha U, Fayiga A, Sonon L (2017) Selenium in the soil-plant environment: a review. Int J Appl Agric Sci 3:1
88. Shahid M, Niazi NK, Khalid S, Murtaza B, Bibi I, Rashid MI (2018) A critical review of selenium biogeochemical behavior in soil-plant system with an inference to human health. Environ Pollut 234:915–934
89. Wei X, Bhojappa S, Lin L-S, Viadero RC Jr (2012) Performance of nano-magnetite for removal of selenium from aqueous solutions. Environ Eng Sci 29:526–532
90. Tang SC, Lo IM (2013) Magnetic nanoparticles: essential factors for sustainable environmental applications. Water Res 47:2613–2632
91. Tratnyek PG, Johnson RL (2006) Nanotechnologies for environmental cleanup. Nano Today 1:44–48

Chapter 5
Advances in the Management of the Neglected Oxyanions (*Antimoniate, Borates, Carbonates, and Molybdate*) in Aqua System

Isiaka A. Lawal, Moses Gbenga Peleyeju, and Michael Klink

Abstract The need for the removal of contaminants from the aqueous environments cannot be overemphasised. This is because of the significant threats they pose to the health and consequently the well-being of both man and other living creatures. In this chapter, we consider the techniques which have been reported for the remediation of water polluted by the often-overlooked oxyanions. While methods such as chemical coagulation, electrocoagulation, reverse osmosis have been utilised for the management of few of these inorganic pollutants, adsorption technique appears to have received more attention in removing all of them. This is possibly because of the simplicity, relatively low-cost and effectiveness of the technique. In particular, ion-exchange technology was reported to offer a very versatile and effective means of removing these oxyanions from water/wastewater.

Keywords Oxyanions · Antimoniate · Borates · Carbonates · Molybdate

5.1 Introduction

Civilization, fast urban development, rapid industrialization and explosive population growth have considerably polluted the environment, directly or indirectly. These pollutions (water, air, solid and noise) constitute threat to the present and future generations. Raw industrial wastes from pharmaceuticals, leather tanning, textiles, printing, batteries, paper, rubber, oils and food processing industries amongst others contain contaminants such as polycyclic aromatic hydrocarbons, personal care

I. A. Lawal (✉) · M. G. Peleyeju · M. Klink
Faculty of Applied and Computer Science, Vaal University of Technology, Vanderbijlpark 1900, South Africa
e-mail: lawalishaq000123@yahoo.com

M. G. Peleyeju
e-mail: mgpeleyeju@gmail.com

M. Klink
e-mail: michaelk1@vut.ac.za

© Springer Nature Switzerland AG 2021   129
N. A. Oladoja and E. I. Unuabonah (eds.), *Progress and Prospects in the Management of Oxyanion Polluted Aqua Systems*, Environmental Contamination Remediation and Management, https://doi.org/10.1007/978-3-030-70757-6_5

products, dyes, inorganics compounds. These contaminants significantly impact the quality of surface and ground waters, leading to scarcity of clean water [1–3].

Specifically, heavy metals are prevalent in automobiles, batteries, electroplating, cosmetic products, metal plating and mining operations among others [4]. Environmental pollution by harmful metals and metalloids has increased tremendously due to the increasing industrial activities. These elements are mostly redox sensitive and some of their oxidation states form oxyanions in solution. Oxyanions (also referred to as oxoanions) are negatively charged polyatomic ions that contain oxygen with generic formula $A_xO_y^{z-}$. "A" is the chemical element and "O" is an oxygen atom [5, 6]. There are a lot of literature on the management/treatment of oxyanions, such as arsenate (AsO_4^{3-}), chromate (CrO_4^{2-}), dichromate ($Cr_2O_7^{2-}$), permanganate (MnO_4^-) [7, 8]. Unfortunately, reports on the management of some equally toxic oxyanions, such as borates (BO_3^{3-}), antimonate (SbO_4^{3-}), carbonates (CO_3^{2-}), molybdate (MoO_4^{2-}) and plumbate (PbO_3^{2-}) are abysmally low. Premised on the fact that the effects of these under-reported oxyanions on human and the ecosystem are no less dangerous than the more focused ones, this chapter is aimed at a critical review of the advances in the management of these genre of oxyanions in aqua system.

5.2 Boron Oxyanion

Boron, the element from which the borate oxyanion is derived, is a vital nutrient for the growth of plant. Borate occurs in aqueous solution, majorly as boric acid (H_3BO_3) and tetrahydroxyborate complex $[B(OH)_4]^-$, because of the buffering nature of the weak acid/weak base conjugate pair. Tetrahydroxyborate $[B(OH)_4]^-$ is the major anionic specie of boron at concentration lower than 0.025 M of boron solution, while polyborates are the main species at higher concentration. Also, $[B(OH)_4]^-$ is the predominant species at ~9.0 to 11.0 pH [9].

Reports have shown that low boron intake in animals and humans causes cellular dysfunctions [10]. However, if boron is taken at high dosage, it can lead to acute headache, kidney damage, nausea, diarrhoea or even death from circulatory collapse [11]. Erosion of rocks, earth's crust and soil rich in boron leads to the contamination of water bodies with boron oxyanions [12]. Boron mainly finds its way to the environment via human activities, such as industrial and agricultural productions. Industries, manufacturing insecticides, dyestuffs, fertilizers, borosilicate glasses, thin film transistor, and bleaching amongst others use boron in one way or the other [13–15]. The concentrations of boron in groundwater range from ~0.30 to 100 mg/L throughout the world, and an average of 4.5 mg/kg of seawater constituent [11]. Provisional guideline of drinking water concentration of boron by World Health Organization (WHO) is 0.5 mg/L [16]. Apart from human activities, borates get into the aquatic environments and atmosphere naturally from clay-rich sedimentary rocks (weathering), oceans, steams and geothermal process [9], though in small quantity [17]. When water-containing borate is used on crops, it affects their growth, especially when the concentration is above 5 ppm.

Generally, removing oxyanions from aqueous medium is not easy, and there is no simple method for their removal from polluted water or wastewater [18, 19]. The different methods that have been explored in the management of borate oxyanions include ion-exchange [20], coagulation [13, 21], hybrid gels [22], adsorption [23, 24], complexation with organic compounds [25], membrane processes [26], reverse osmosis [27, 28], ultrafiltration [29], electrocoagulation, adsorption/ion-exchange and liquid–liquid extraction [18].

Using coagulation–flocculation operations, organic and inorganic coagulants have been used for the removal of borate/boron from aqueous solutions. Small particles usually have surface charges that are negative, which hinders them from settling and aggregating. Coagulation occurs when these particle charges are destabilized. Thus, when positively charged coagulants are added, they balance out the charges (neutralize them). This enables the particles to aggregate and form a stable and well suspended submicron floc. Meanwhile, flocculation will increase the size of the microflocs, to form visible and denser flocs that settle out faster at the end of the process. The flocculant adsorbs to the submicron flocs and facilitates bridging of gaps between the flocs. Bringing particles closer together creates the effective range for Van Der Waals attraction force to reduce the energy barrier for flocculation and loosely packed flocs formed. Aggregation, binding and strengthening of flocs occur until visibly suspended macroflocs formed. This process is dependent on the nature of water, temperature, pH, types and dose of the coagulant used, intensity and duration of the rapid mixed [30].

Lime was reported by Farmer and Kydd [31] to remove boron in water in the range of 400 mg/L to $\geq$ 1000 mg/L. Similarly, Remy et al. [32] reported reduction in concentration of boron from 700 mg/L to less than 50 mg/L using lime powder. Organic polyelectrolyte was also used to coagulate boron to a concentration $\leq$400 mg/L [33, 34], and the solution pH had significant effect on the removal efficiency. The removal of low concentration of boron (<10 mg/L) using aluminium sulphate has also been reported [33]. Similarly, Chang and Burbank Jr [35] have reported the use of lime, aluminium sulphate and sodium aluminate for the removal of boron. Iron salts (Fe^{2+} and Fe^{3+}) have been reported to effectively remove boron from aqueous solutions at pH value that ranged between 8 and 9 [18]. However, the study showed that a large quantity of base was required for pH adjustment, which led to high salinity. High coagulant dosage is also required, even at low borate concentration (<50 mg/L), which generated large amount of waste with the attendant cost of disposal [18]. In a separate study, ferric chloride and alum were assessed as coagulants to remove boron from wastewater from saline flowback, but very high dosages were required for about 80% removal [36]. In another study, Hiraga and Sigemoto used a mixture of $CaCO_3$–$Al(OH)_3$ and $Ca(OH)_2$–$Al(OH)_3$ in varied ratios for the removal of borate. About 0.1 g of the coagulation was added to 1.0 mmol/dm^3 of H_3BO_3 solution and left to stand at room temperature for a day. A reduction in the concentration was observed and concentration lower than 0.2 mmol/dm^3 was recorded after the reaction [37]. Co-precipitation of borate with hydroxyapatite was studied, and $Ca(OH)_2$ was used as a mineralizer. The initial borate concentration was 68.1 mM, and the maximum removal efficiency, as expressed with B/Ca, was 0.40 Q/mmol-B.mmol^{-1}-Ca. [38]. The use of alumina-lime-soda for the treatment of boron-containing water was reported [39]. The amount of boron in the contaminated water was reduced by 88% after treatment, at optimized experimental conditions [39]. Other coagulant salts, which include manganese sulphate, magnesium oxide [40, 41], nickel(II), zirconium(IV), chromium(II) [42] and zinc have also been used to remove borate/boron from water [43].

Electrocoagulation (EC) is among the methods that have been reported for the removal of boron oxyanions from water. This method involves the production of coagulants by electrochemical process, and the coagulant generated then work to eliminate the pollutants by charge neutralization [44]. Electrocoagulation comprises a cell with iron or aluminium (metal anode) and uses direct electrical current. Three stages are involved in the process, namely:

(i)   formation of coagulant by electrical oxidation at the anode;
(ii) destabilization of pollutants and suspended substances by emulsion breaking, and
(iii) formation of flocs from destabilized particles.

This method is simple because it utilizes a simple equipment that can be operated easily with enough operational latitude and also requires only small amount of chemicals [44]. The use of aluminium, iron and zinc, as the anode in EC, has all been reported for the removal of boron from wastewater [45, 46]. Aluminium and iron were simultaneously used in the reactor as electrodes. The results showed that the EC process for boron removal strongly depended on the current density, initial boron concentrations and time. The process was examined under varying indices, to determine the optimal operating conditions. At 50-min retention time, initial boron concentration of 100 ppm and current density of 30 mA cm^{-2}, about 70% boron was removed by the aluminium electrode and 62% was removed by the iron electrode. As the initial boron concentration was increased to 1000 ppm, 95% boron removal for both iron and aluminium electrodes was reported [45]. Electrochemically generated zinc hydroxide was used in the removal of boron, and different operating parameters were investigated. These parameters include initial pH, current density, electrode configuration, inter-electrode distance, co-existing ions and temperature. With initial concentration of 5 mg/L, the optimum removal efficiency of 93.2% was achieved, at a current density of 0.2 A dm^{-2} and pH of 7.0, using zinc as anode and stainless steel as cathode, with the inter-electrode distance of 0.005 m [46].

Yilmaz et al. [13] compared conventional chemical coagulation, using aluminium chloride, and electrocoagulation process in the removal of borate from aqueous solution. The EC process gave higher boron removal at pH 8.0 and 7.45 g/L of aluminium dose, with 94.0% and 24.0% removal efficiencies for EC and chemical coagulation, respectively. The use of Al and Fe electrodes produced 84 and 75% of boron removal, under 60 min, using current density of 20 mA cm^{-2} [45]. When the current density was increased to 30 mA cm^{-2}, the removal efficiency increased to 90% and 85%, respectively. Electrocoagulation, using aluminium as electrode, for the removal of boron oxyanion from mining wastewaters has been reported [47, 48]. The effects of pH, time and current density were reported and the optimum removal efficiency of 70%, at pH value of 4.0, under 90 min, and current density of 18.75 mA/cm^2. Electrocoagulation of boron oxyanion was reported by Kartikaningsih et al. [49], using metallic aluminium as electrode, and as the pH value increased from 4.0 to

8.0, the boron removal efficiency also increased. An increase in the electrolytic effi-
cacy was observed as the current density increased from 1.25 to 5.0 mA cm^{-2}. At
boron concentration of 100 mg/L, using aluminium electrode, at pH value of 8.0,
within 60 min of 5.5 mA cm^{-2}, current density removal efficiency of about 70% was
recorded [50]. In a report, the EC process was used in the production of magnetic
ferrites ($CuFe_2O_4$, $NiFe_2O_4$, and $CoFe_2O_4$) for boron removal from water [51]. The
ferrites were obtained using a sacrificial iron anode in Cu, Ni and Co electrolytes. A
95% of the initial boron concentration of 10 ppm was removed using current density
of 3.75 mA cm^{-2}, at pH value of 8.0, within 60 min, [52]. In another study, metallic Ni
foam was used as electrodes for electrocoagulation of boron from wastewater, using
electrolyte with pH values of 8.0 and 9.0 and current density of 0.6–2.5 mA cm^{-2}.
At the optimal process conditions, 92% of boron was removed [53].

Chemical oxo-precipitation has recently been reported for the removal of boron
oxyanion, using hydrogen peroxide to enhance precipitation of metal perborate salt
from boric acid [54]. This method removed boron from aqueous solution at relatively neutral pH and room temperature. As presented in Fig. 5.1, boric acid produces
several forms of perborates under certain reaction conditions (nucleophilic substitution) with hydrogen peroxide. At pH value that ranged between 8.5 and 12.5, the
$B(OH)_3/H_2O_2/H_2O$ system are mostly dominated by species (C) and (F) perborates
(Fig. 5.1) [55]. Equations (1), (2) and (3) show the dissociation reaction and the two
mass balance reactions.

$$H_2O_2 \leftrightarrow HO_2^- + H^+ \quad Ka = 3:09 \times 10^{-12} M^{-1} \tag{5.1}$$

$$B_t = [B(OH)_3] + [B(OH)_4^-] + [B(OH)_3OOH^-] \\ + [B(OH)_2OOH] + [B(OH)_2(OOH)_2^-] \tag{5.2}$$

$$P_t = [H_2O_2] + [HO_2^-] + [B(OH)_3OOH^-] + [B(OH)_2OOH] + 2[B(OH)_2(OOH)_2^-] \\ + [B(OH)_3OOB(OH)_3^{2-}] + 2[B(OH)_2(OO)_2B(OH)_2^{2-}] \tag{5.3}$$

Another advantage of chemical oxo-precipitation is the ability to work with small
quantity of precipitant dosage at room temperature. Perborates produced from the
reaction of boric acid and H_2O_2 can simply precipitate at room temperature with alkaline earth metals. Barium is the most effective precipitant in the alkaline earth metals
[55]. It was reported that boron oxyanion significantly reduced using this method of
precipitation, from 1000 ppm to < 3 ppm [56, 57]. The efficiency of calcium as a
precipitant was also tested at 60% crystallization ratio, using the following conditions: molar ratios of $[H_2O_2]/[B] = 2$, $[Ca]/[B] = 0.6$, initial boron concentration
= 1000 ppm, bed height = 80 cm, effluent pH = 10.6 and hydraulic retention time
= 18 min [58]. Waste-derived mesoporous aluminosilicate, in the presence of H_2O_2
and barium ions, was used to recover boron from polluted water [59].

Adsorption is a treatment technique that provides efficient way of removing boron
oxyanion from liquid phase at low concentration [60]. Several adsorbents, which

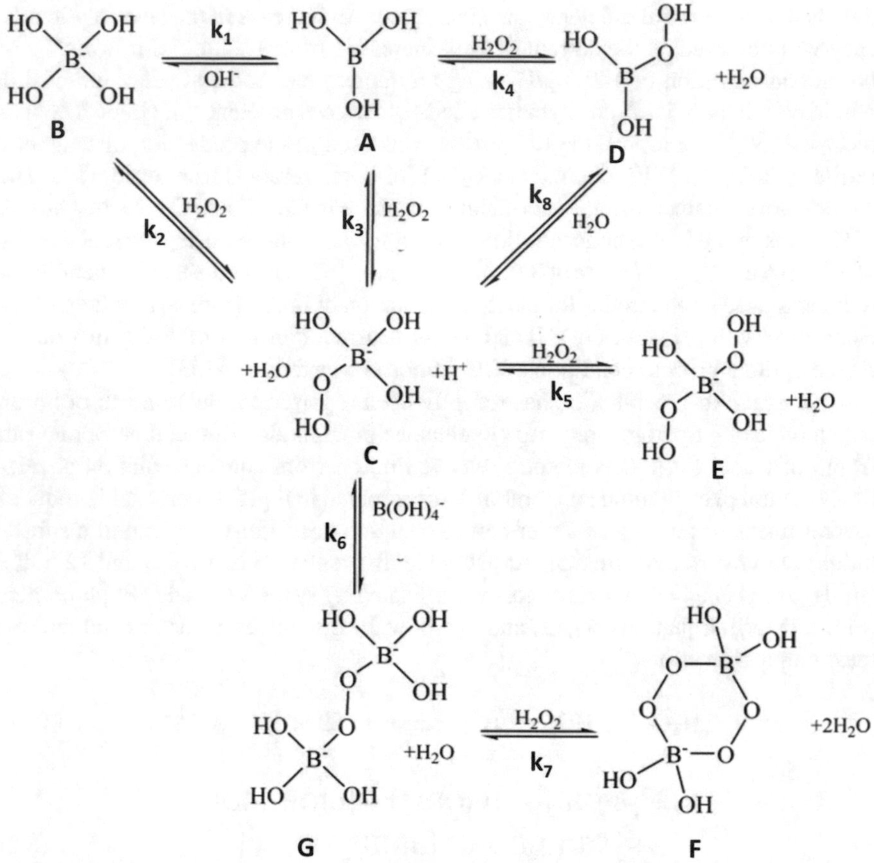

Fig. 5.1 Scheme showing the reactions between perborates, hydrogen peroxide and boric acid

include activated carbon, mesoporous silica, layered double hydroxides, fly ash, biological materials, selective resins, clays, natural minerals, oxides, nanoparticles and complexing membranes [44, 60–67] have been used in this regard.

Activated carbon (AC) is one of the most used adsorbents due to its high specific surface area, which is highly beneficial for high uptake of pollutants from water. However, limited literatures are available on the application of AC in removing boron oxyanion from wastewater. The lack of active sites for boron on the AC, which leads to low adsorption capacity, was reported to be the main reason [60]. AC from olive bagasse was used in the adsorption of boron, at pH value of 5.5, and the maximum adsorption capacity obtained was 3.5 mg/g [68]. When AC was modified with salicylic acid for the removal of boron, the adsorption equilibrium was reached within 10 min. Two grams of the adsorbent were added to 100 mL boron solutions, with concentrations of 5, 10, 25 and 50 mg/L and pH value of 4.68. The adsorption capacity of 1.777 mg/g was reported and both Freundlich and

Langmuir isotherms equations fitted the data [69]. AC modified curcumin has been reported for the removal of boron and the modified AC displayed superior removal efficiency to that of the unmodified AC [70]. The study explored both fixed bed and batch adsorption protocols. At the optimal pH of 5.5, contact time of 120 min, the curcumin modified AC gave adsorption capacity of 5.00 mg/g while the unmodified AC had 0.59 mg/g. A 99% removal was achieved in the first 5 min for the fixed-bed experiment, with 890 mg/L as inlet concentration at a flow rate of 8.0 mL/min [70]. In a related work, AC impregnated with zirconium dioxide, activated alumina and silica aerosil led to a moderate removal of boron oxyanion of initial concentration of 5 mg/L, at alkaline region, with above 90% removal [71]. Similarly, AC from Filtrasorb was reported to remove more than 90% of the 5 mg/L initial concentration of boron from water [72].

Mg/Al and Mg/Fe layered double hydroxides have been prepared and utilized for the adsorption of boric acid/borate from aqueous solution. It was reported that the removal of boric acid/borate was not dependent on pH and up to 92% of 14.0 mg/g was removed, with adsorbent dosage of 2.5 g/L [73]. Also, Jiang et al. [74] used thermally activated Mg/Al layered double hydroxide for the removal of boron and the activated material performed better. The solution pH value had no significant effect on the removal efficiency. Removal capacity greater than 90% was reported for the thermally activated hydroxide, while the one without thermal activation gave 80% removal. In another study, Ay et al. [75] reported removal efficiency greater than 95%, for both thermally activated and the unactivated Mg/Al layered double hydroxides, with nitrate interlayer anion. Cerium (IV) oxide was used for the effective uptake of low concentration of borate from water. The desorption and regeneration of the adsorbent were achieved in the study at low (i.e. 2–4) and high (i.e. 12–14) pH values, respectively [17]. Amongst all the adsorbents that have been explored for the removal of boron oxyanion, chelating resins are said to be the most efficient and selective for the removal of low concentration of boron from aqueous medium [60]. Chelating resins have macroporous polystyrene matrix joined to the hydroxyl or two adjacent phenolic hydroxyl functional groups, which are often in a cis position (vis-diols) that have high affinity for boron only [60, 63, 76, 77]. They also possess tertiary amine groups, which neutralizes the proton during complexation process [76]. The complexation reaction involves the formation of borate esters with boric acid, which dissociates quickly to release protons. Thereafter, the proton reacts with diols to produce borate complex on the resin [78] (Fig. 5.2 and Eq. 5.4).

$$H_3BO_3 + 3ROH \rightarrow B(OR)_3 + 3\ H_2O (R\ is\ alkyl\ or\ aryl) \quad\quad (5.4)$$

N-methyl-D-glucamine, the most investigated chelating resin, has 5 hydroxyl (polyols) functional groups and tertiary amine end, which help in complexation with boron [18]. Purolite S108, Diaion WA30, XSC-700, Diaion CRB 02, Dowex 2 × 8, Dowex XUS 43594.00, Purolite S110 and Amberlite IRA 743 are all modified forms of N-methyl-D-glucamine, which are available in the market [61, 64, 79].

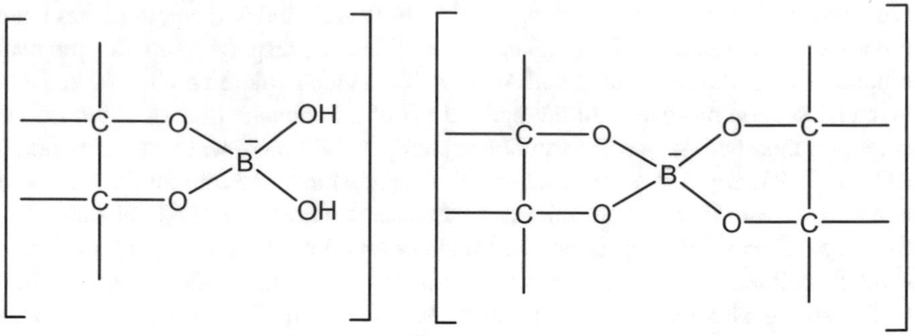

Fig. 5.2 The borate complex structure formed from the complexation reaction of chelating resin adsorbent with boron oxyanion in aqueous solution

Scientists have reported boron removal with polystyrene modified with sorbitol functional group and recorded adsorption capacity of 13.18 mg/g [80]. Fixed-bed column was loaded with two separate resins: a combination of vinyl benzyl chloride—N-methyl-D-glucamine (VBC–NMG) and iminodipropylene glycol functionalized glycidyl methacrylate—polyvinyl chloride (GMA–PVC). The breakthrough curve obtained from the former fixed bed was sharper than that of the later and Thomas model gave the best fit to the breakthrough curve. The study showed that both resins could be used industrially for the removal of boron from the aqueous system [81]. Iminopropylene glycols was supported on glycidyl methacrylate–methyl methacrylate–divinyl benzene and the removal capacity of 32 mg/g was reported [82]. The synthesised N-methyl-D-glucamine—hydroxypropyl methacrylate gave an adsorption capacity of 145.9 mg/g for boron removal. Regeneration studies showed that HCl can easily regenerate the resin and can be easily reused. Resin produced from the reaction of glycidyl metacrylate–methyl methacrylate–ethylene glycol dimethacrylate and diallylamine was used in the removal of boron [83]. Kabay et al. [84] used N-glucamine—like resins (Purolite S 108, Diaion, CRB 01 and 02) for batch adsorption removal of boron from geothermal plant wastewaters. At about 3 g of the resins, 90% of boron was removed from 1 L of wastewater. Similarly Xiao et al. [79] investigated the use of XSC-700 for the removal of boron at different temperatures, concentrations, stirring speeds and resin/brine ratios at constant diameter. Removal capacity decreased when the resin/brine ratio increased, but as the boron concentration and temperature were increased, an increase in the removal efficiency was also recorded. Freundlich isotherm gives the best correlation for the adsorption with maximum saturation capacity (K_f) of 2.9234 (mg mL^{-1})(l/mg)$^{1/n}$ [79]. In another study, the poly(VBC-co-DVB) was treated with N-methyl-d-glucamine and used for the removal of boron. Approximately, 93% of 11.0 mg/L of boron was removed using 4 g of the resin in about 20 min, and about 97% of both 4.8 and 5.4 mg/L of boron was removed with 4 g of resin in about 10 min [85]. In another study, multi-hydroxyl iminobis (propylene glycol) was supported on chitosan for the removal of boron [86]. The removal capacity was about 29.19 mg/g and the removal kinetic followed pseudo-second order. The recoverability study was conducted and about 97% of the resin was recovered and reusable, after treatment with acid [86]. In a report by Suzuki et al. [87], Lewattitt MP 500WS (a resin) was modified with chromotropic acid and used to remove boron. Boric acid/borate complexation was observed using acid–base titration and NMR and the complexation reaction was favoured in acidic medium (pH 4.5). Langmuir isotherm equation best explained the process and the removal capacity of 8.87 mg/g was observed. Other adsorbents that have been reported for boron oxyanion removal from aqueous solution are presented in Table 5.1 [88–94].

Donnan dialysis (polyethyleneimine-filled porous Celgard membranes) is another method that has been reported for the removal of borate Bryjak and Duraj [102]. High concentration of boron was used in the study (20 mg/L) and it was reported that the driving salt (NaCl) moved more rapidly towards the membranes than borate. When the receiving phase was NaCl, a 40 t0 45% removal efficiency was achieved.

Table 5.1 Removal of borate oxyanion from aqueous solution using different adsorbents

Adsorbents	Adsorption capacity (mg/g)	References
Amino modified tannin gel	24.3	[95]
Tannin gel	11.4	
Fly ash	6.9	[96]
	2.3	[97]
	0.0275 (mmol/g)	[98]
Aluminium-based water treatment residuals	0.980	[99]
Bentonite	0.51	[100]
Bentonite-Fe	0.83	
Waste calcite	1.05	
Waste calcite-Fe	1.60	
Kaolinite	0.60	
Kaolinite-Fe	0.80	
Zeolite	0.53	
Zeolite-Fe	0.76	
Rice residue	9.26	
Walnut shell residue	7.04	
Wheat residue	5.59	
Rice-Fe	9.17	
Walnut shell-Fe	7.58	
Wheat-Fe	6.06	
Magnesite tailing	65.79	[101]
Layered double hydroxides	14.0	[73]
	37.90	[15]

5.3   Antimony Oxyanion

Antimony (Sb) metalloid exists in aqueous solution as antimonate (SbO_4^{3-}) and antimonite (SbO_3^{3-}) species. Antimony (Sb) is considered toxic, with likely carcinogenic effect on humans [51, 103, 104]. Excessive Sb intake has been reported to cause many diseases in humans [16, 105]. SbO_3^{3-} is less toxic than SbO_4^{3-}, but SbO_4^{-3} is more stable. The European Union (EU) and United States Environmental Protection Agency (USEPA) have both listed antimony and its compounds as priority pollutants [56, 106] and the maximum contaminant level (MCL) given by the USEPA is $6\ \mu g\ L^{-1}$ [57]. Antimony has many industrial uses; thus, it is one of the most mined metals [107]. It is used in the production of lead-acid batteries, bullets, glass flame retardant, among others [108, 109]. Mining and extraction industries are among the major routes of antimony oxyanions to water bodies. Sb in trace amounts in water is used up by terrestrial and aquatic organisms, but in certain concentrations, it is reported to be toxic, particularly to aquatic organisms [108]. Under acidic condition, Sb(III) easily undergoes oxidation to form Sb(V) in aqueous solution [108, 110], which has led to more focus on the removal of Sb(V) than Sb(III).

Coagulation–flocculation has been used for the removal of Sb from polluted water, using Fe or Al salts, but Fe salt showed better promise than Al salt [21, 111]. Removal of Sb(V) from wastewater was studied via coagulation–flocculation–sedimentation processes using ferric chloride as coagulant. Optimum removal was achieved at pH range 4.5–5.5 [21]. The authors noted that the removal capacity of about 98% was attained at initial concentrations that ranged between 50 and $500\ \mu g\ L^{-1}$. Meanwhile, they reported a lower removal capacity, when aluminium sulphate was used as coagulant [21]. A similar observation was made by Kang et al. [111] that aluminium was not a very good coagulant for Sb(V).

Adsorption technique has been used in the treatment of water contaminated by Sb(V). It is in fact one of the best available treatment technologies for Sb removal from aqueous solution, owing to its simplicity, high efficiency and user-friendliness. Zhao et al. [112] reported nanoscale zero-valent iron stabilized with polyvinyl alcohol for the removal of Sb(V). They investigated the effect of time, concentration and pH, and the maximum adsorption capacity obtained was 1.65 mg/g, at pH below 5. Batch and column adsorption of Sb(V) ions on calcareous soils were investigated Martínez-Lladó et al. [113] and the kinetic analysis revealed that it took about seven days to achieve the maximum adsorption capacity. Thomas model described the column experiments and convective–dispersive equation explained the breakthrough curves. The desorption experiment gave about 90% recovery of the adsorbed Sb(V) ions from calcareous soils.

A study made use of orange waste loaded with Zr(IV) and Fe(III) ions for the adsorption of Sb(V) ions from water. The maximum adsorption capacity for both adsorbents was the same (i.e. 1.19 mmol/g). The presence of other anions in the polluted water had no effect on the adsorption of Sb(V) [114]. Carbon decorated with zirconium oxide (ZrO_2) was studied for the removal of antimonate, and the adsorption capacity of 57.17 mg/g was reported [115]. In another study, Sb(V) was adsorbed onto iron oxyhydroxides, and the amount of Sb(V) adsorbed depended on the pH value of the aqueous solution and the highest removal efficiency was attainable within the acidic region. In a report, where iron oxide/hydroxides were employed for the removal of Sb(V) from water, adsorption capacities of 201, 240, 280, 192, 936 μmol/g were reported for α-FeOOH, β-FeOOH, γ-FeOOH, α-Fe$_2$O$_3$ and hydrous ferric oxide, respectively [116]. Activated carbon was also used for the uptake of Sb(V) from aqueous solution and an adsorption capacity of 92 mg/g was observed [117].

The removal of Sb(V), using chelating agents, has been experimented. A study used a chelate-forming group, iminodiethanol, to modify a porous hollow-fibre membrane and thereafter used the modified membrane to adsorb Sb(V) from aqueous medium. The membrane had a thickness of 0.7 mm, porosity of 70%, an imin-odiethanol group of 1.6 mol/kg of the membrane and a water flux of 0.95 m/h at 0.1 MPa and 298 K. The breakthrough curves of antimony overlapped, irrespective of the permeation rate of the antimony solution. At antimony concentrations below 10 mg/l (pH 4.0), a linear adsorption isotherm was obtained. The removal capacity was 15 mg/g, and the adsorbed antimony was quantitatively eluted by permeation of 2 M hydrochloric acid through the pores of the membrane. [117]. Furthermore, an impurity (monophosphoric acid ester M2EHPA) from commercial grades of D2EHPA was used in the removal of Sb(V) from copper refinery electrolyte. The experiment was left to run over a period of 32 h, and 0.6 g/L of Sb in the electrolyte was reduced to 0.13 g/L [118]. Sb(V) ions have also been removed from copper electrolytes by aminophosphonic resins (Duolite C-467). High loadings of up to 39 g Sb(V) per kg of wet resin were obtained, when synthetic copper electrolyte containing 220 mg/L of Sb(V), at temperature of 30 °C and 2 g of Duolite C-467 were allowed to stand for 48 h [68]. In contrast, Schilde, Kraudelt and Uhlemann [119] reported that Sb(V) was not removed by Wofatit MK 51 chelating resin, which has methylaminoglucitol groups.

The abilities of Al-rich phases, (i.e. hydrous Al oxide, and reduced and oxidized nontronite and kaolinite) were investigated for the removal of Sb(V). It was posited that Sb(V) adsorbed in an inner-sphere mode on the surfaces of the studied substrates. The observed adsorption geometry is mostly bidentate corner-sharing, with some monodentate complexes. The kinetics of adsorption is relatively slow, and equilibrium adsorption isotherms were best fitted using the Freundlich model. The maximum adsorption capacity of 0.82 mmoles/g for kaolinite, 1.08 mmoles/g for reduced nontronite and 0.71 mmoles/g for oxidized nonotronite was reported [120]. Goethite and activated alumina have also been explored for the removal of Sb(V). The adsorption of Sb(V) on goethite was investigated in 0.01 and 0.1 M KClO$_4$ M solutions as a function of pH and Sb concentration. The results showed that Sb(V) formed inner-sphere surface complexes at the goethite surface. The maximum adsorption density for Sb(V) was 136±8 μmol/g at pH 3. The adsorption data of Sb(V) was best fitted by the modified triple-layer surface complexation model [121]. Activated alumina (commercially available) was used in the removal of Sb(V) ions from aqueous solutions. The optimum pH was 4.3, Sb(V) adsorption was temperature dependent and adsorption increased with increases in temperature. A concentration of 0.164 mM of Sb(V) solution and 0.5 mg/ml (w/v) was mixed for 1 h at varying temperature. A maximum adsorption of above 260 μmol/g was reported [122]. Also, the adsorption of Sb(V) on sand from Haro river was investigated and important parameters such as effect of electrolyte, amount of sand, contact time and concentration of Sb(V) were considered [123].

Bioremediation of metal(loid)-containing water/wastewater has also been proposed as a cost-effective technique. In the work of Zhang et al. [124], sulphate-reducing bacteria adsorbent was prepared for the removal of Sb (V). They reported 93% removal efficiency, at pH 7 and 50 mg/L initial Sb(V) concentration, for 11-day period of the batch experiment. Sb(V) adsorption on Fe-modified aerobic granules was also studied, and an adsorption capacity of 36.6 mg/g was reported [125].

Other adsorbents that have been used for removing Sb oxyanion from aqueous solution are presented in Table 5.2.

5.4 Carbon Oxyanion

The presence of calcium carbonate ($CaCO_3$) poses a major challenge in wastewater and groundwater treatment because it causes the formation of scale deposits [140, 141]. Scale formation induces water cooling blockage, reduces heat transfer and encourages corrosion of the carbon steel. Calcite (rhombohedral), aragonite (orthorhombic) and vaterite (hexagonal) are the three main anhydrous polymorphs of calcium carbonate [142]. Calcite has better stability under normal conditions, vaterite has the highest solubility and aragonite crystallizes generally at temperature above 60 °C [143]. There are reports that link carbonates concentrations (especially calcium carbonate and magnesium carbonate) and atopic eczema, Alzheimer's and cardiovascular diseases [144, 145]. Though, the link is weak, and many uncertain variables exist in the reported studies [146].

Reverse osmosis has been reported for the removal of calcium carbonate in the literature [147–149]. The precipitation of calcium carbonate from aqueous solutions using reverse osmosis cell, containing cellulose acetate membranes (@40 bar), has been reported [149]. They reported significant calcium carbonate precipitation (90%), which they attributed to the high ability of carbon dioxide to permeate through the membrane. Also, Pervov and Andrianov [147] reported reverse osmosis process for the removal of calcium carbonate using seeded crystallization. Due to calcium carbonate precipitation, there was drastic reduction of hardness and total dissolved solids concentrate and the rate of precipitation depended on the composition feed water and amount of introduced crystal. High rate of recovery was recorded, which made it possible to reuse the concentrate [147]. Calcium carbonate was recently reported to be removed by fluidized-bed reactors [150] and liming [151]. Carbonation is another conventional methods used in removing carbonates from water [152]. This method is inexpensive and can effectively remove the carbonates of calcium and magnesium. Removal of carbonates by electrolysis process has also been reported Agostinho et al. [153].

Table 5.2 Removal of antimoniate from aqueous solution using different adsorbents

Adsorbents	Adsorption capacity (mg/g)	References
Synthetic manganite	784.53 (μmol g $-$ 1)	[126]
Ferric hydroxide	99.84	[127]
	18.5	[125]
Goethite	18.3	[128]
Ferrihydrite	27.9	[129]
Akaganeite	450	[130]
Alpha-Fe_2O_3	7	[131]
Hydrous ferric oxide	114	[116]
γ-FeOOH	34.09	
β-FeOOH	29.22	
α-FeOOH	24.47	
	60.4	[125]
Fe-Cu binary oxide	104.95	[132]
Fe-Zr-D201	73.75	[133]
Fe(II)-loaded saponified orange waste	144.88	[114]
Zr(IV) and Fe(III) loaded orange waste	227.67	
Polymeric anion exchanger D201 loaded with nanohydrated ferric oxide	60.9	[134]
Calcite sands loaded with nanohydrated ferric oxide	39.9	
Iron-modified attapulgite Nano-FeO(OH) modified clinoptilolite tuff	31.79	[135]
Fe(III)-treated bacteria aerobic granules	22.6	[136]
Fe(III)-treated fungi aerobic granules	19	[137]
Zeolite-supported magnetite	19	[138]
Red soil	1.68	[139]

Wastewater treatment plants are amongst the major greenhouse gas emitters. These emissions originate from nitrogen, biological carbon, sludge management, phosphate removal and off gas. Wastewater treatment plants emit about 0.77 Gt carbon dioxide-equivalent greenhouse gases in 2010 through degradation of organics, which is approximately 1.57% of global greenhouse gases emitted [154]. Carbon dioxide (CO_2) converts to carbonic acid readily and reversibly when dissolved in water. The Paris Agreement on Climate Change aims to peg average temperature increase, globally, to less than 2 °C, but achieving this requires effective reduction of greenhouse gases, such as CO_2 in the environment [155]. Many methods have been developed for the removal/capturing of CO_2 during wastewater treatment.

Microbial electrolytic carbon capture is a method that uses wastewater as an electrolyte for microbial supported water electrolysis [156]. This process consists of anode with electroactive bacteria (microorganisms) in its chamber. The biodegradable constituents in wastewater are oxidized to give CO_2, protons and electrons. The anode accepts the electrons and moved it to the cathode via an external circuit, where water is reduced to H_2 and OH^- [157]. The anolyte that is H^+-rich can release Ca^{2+}, Mg^{2+}, etc. (metal ions) from ample of waste materials, such as coal fly ash, or from silicate minerals such as wollastonite ($CaSiO_3$). When these metal ions move to the OH^- catholyte, metal hydroxide is produced, which later reacts with CO_2 to produce bicarbonate or carbonate that is stable [158–160]. The stable carbonates can be put to further use, such as in the production of cement.

$$CH_3COOH \text{ (aq)} + 2H_2O + 6CaCSiO_3\text{(s)}$$

$$+ 4CO_2\text{(g)} \xrightarrow{V_{d,c}} 6CaCO_3\text{(s)} + 6SiO_2 + 4H_2\text{(g)} \qquad (5.5)$$

Another method of removing CO_2 from wastewater is microbial electrosynthesis, which is similar to the microbial electrolytic carbon capture. But microbial electrosynthesis uses autotrophic bacteria at the cathode for capturing and converting CO_2 into useful organic compound [161]. The organic substances present in the wastewater are oxidized to generate current at the anode by electroactive bacteria [161].

Microalgae cultivation is another method that has been reported for removing CO_2 from wastewater. This method is complementary to microbial electrolytic carbon capture and microbial electrosynthesis. The ability of microalgae to remove CO_2 simultaneously with other nutrients has attracted more attention in recent time [156, 162]. During autotrophic growth, microalgae cultivation fixes CO_2 in the process of nutrients (N and P) assimilation in wastewater [163, 164]. During autotrophic growth, microalgae fixes approximately 1.8–2.4 kg of CO_2/kg of biomass [165, 166].

Constructed wetland is another method that has been used in capturing CO_2 from wastewater. Constructed wetlands are modified system of wastewater treatment that use natural processes, including vegetation, soils and associated microbial ecosystems [167, 168]. Various classes of constructed wetlands can have a wide range of carbon capturing profile. Their CO_2 capturing depends on various conditions such as season, system and level of treatment [156]. It has been presented in an extensive review that free water surface constructed wetland has about 30% lower CO_2 than subsurface flow [169]. It was further explained that plants can assimilate CO_2 of approximately 57,000–76,000 $kgCO_2$ ha^{-1} yr^{-1} [169]. The summaries of the methods for capturing CO_2 generated from wastewater are presented in Table 5.3.

5.5 Molybdates

Molybdenum is necessary for some important metabolic processes in both animals and plants. In humans for instance, Mo is a co-factor for a few enzymes, including sulphite oxidase, xanthine oxidase/dehydrogenase and aldehyde oxidase, which

Table 5.3 Different CO_2 capturing methods in wastewater treatment [156]

Method	Mechanism	Main advantages	Major challenges
Microbial electrolytic carbon capture (MECC)	Microbial electrolysis to enable wastewater treatment and mineralization of CO_2 to carbonates.	• High-rate CO_2 capture • Sequestration of CO_2 to a stable carbonate • Effect removal of organic with low sludge	• Carbonates production value are low • Cost of running is high • Ability of remove nutrient is limited • Large-scale validation is needed
Microbial electrosynthesis (MES)	Electrons are recovered from wastewater in the anode for cathodic CO_2 reduction to organic chemicals catalysed by electrotrophs	• Potential to produce high-titre chemicals • High conversion efficiency • Self-sustainable biocatalysts • Efficient organic removal with low sludge	• Low rate of biocatalytic electron uptake • Poor selectivity of high-value products • High cost with current design • Limited nutrient removal and scalability
Microalgae cultivation	Naturally occurring microalgalcommunities are enriched to take up nitrogen and phosphorus while also assimilating CO_2 into biomass	• Effective nutrient removal with high CCU • Biomass generation for biofuels and bioproducts • Achieves organics polishing • Undergoing pilot-/full-scale testing	• Uncertain performance reliability • Large land area may be required • High cost/energy for biomass harvesting • Limited organics removal
Constructed wetlands	Engineered wetland systems that integrate vegetation, soils and microbial ecosystems to treat wastewater and capture CO_2 to plant biomass	• Simultaneous organic carbon and nutrient removal • Multifunction in addition to CCU and waste treatment • Easy maintenance and low cost • Mature process for centralized and distributed uses	• Mixed results as GHG source or sink • Nutrient removal can be limited • Large land area and limited success in cold climates

catalyse relevant reactions in the body. It is found in varying amounts in food sources like grains, legumes and meat organs such as liver and kidney. Although a very useful element in dietary amount, excess Mo in the body poses a significant risk to human health. The oxidation states of Mo range between -2 to $+6$ but exist predominantly as oxyanion in the $+6$ state in the aqueous environments. It is found naturally in minerals such as molybdenite, ferrimolybdite, powellite, jordisite and wulfenite [6, 170]. Eqs. (5.6 and 5.7) below show the formation of the oxyanion under different chemical conditions from two of its naturally occurring ores [171, 172]:

$$2MoS_2 + 9O_2 + 6H_2O \rightarrow 2MoO_4^{2-} + 4SO_4^{2-} + 12H^+$$

(molybdenite)   (molybdate anion)   (5.6)

$$PbMoO_4 + 2OH \rightarrow 2MoO_4^{2-} + Pb(OH)_2$$

(wulfenite)   (5.7)

Mo has found wide applications in metallurgy and in the manufacture of pigments, lubricants and catalysts for industrial processes [173]. Some processes such as mining, coal combustion (Mo in the air is returned to the earth by wet and dry deposition) and metallurgical activities can lead to a substantial increase in the amount of this element in water bodies. Because of its toxic effects at elevated level, the World Health Organization recommended a maximum of 0.07 mgL^{-1} Mo in drinking water. It is therefore pertinent to develop methods/technologies for remediating water containing high level of this element. In this regard, some researchers have undertaken and reported a number of studies on removal of Mo from water.

Molybdate removal has been reported [174], using hydrous Fe_2O_3 grafted polystyrene anion exchanger. Although polystyrene anion exchanger is known to exhibit considerable affinity for the oxyanion, the incorporation of Fe_2O_3 nanoparticles led to significant enhancement of the adsorptive capacity. Values as high as 213 mg/g were obtained, when the initial concentration of Mo(VI) was 300 mgL^{-1}. This improvement was attributed to the possible complexation between the molybdate and ferric oxide at acidic and near neutral pH range [175, 176]. In a report by Huang et al. [177], hybridised iron material was employed for the removal of molybdate from water. The hybrid adsorbent comprises of zero-valent iron (Fe^0), Fe_2O_3 and Fe^{2+} system. The Fe^{2+} in the system could lead to the formation iron-molybdate complex, resulting in the abatement of the oxyanion in solution. The removal efficiency was 97% at an initial molybdate concentration of 0.5 mM. The removal of the molybdate was possible by rapid reduction of the Mo(VI) to lower valences. It was postulated that each of the reactive irons in the hybrid material played a role, which was favourable for the removal of the Mo(VI) from the aqueous solution.

In the quest to advance an effective method for the removal of Mo(VI) from aqueous media, Mo(VI)-imprinted chitosan/trietholamine gel beads, using ion-imprinting technology, were prepared [178]. At pH value of 6.0 and initial molybdate concentration of 8.0 gL^{-1}, an adsorption capacity of 458 mgg^{-1} was reported for the material. Ion-imprinting technology was employed to prepare adsorbent for the sequestration of molybdate anion from water [179]. The MoO_4^{2-} surface imprinted material (IIP-quaternized dimethylaminoethyl methacrylate polymer/SiO_2) showed impressive performance for the removal of the anion and the performance was attributed to the specific recognition and the combination affinity of the template anion [180]. The adsorbent showed high adsorption capacity and good selectivity for the anion. Specifically, a binding capacity of 0.46 $mmolg^{-1}$ was obtained at the optimized experimental conditions. In a similar work, chitosan was used for the uptake of molybdate ion from standard samples and contaminated groundwater [181]. The polysaccharide exhibited significantly high adsorption capacity (265 mgg^{-1}), at low pH range. The hydroxyl groups on the material were reported to bind the oxyanion in the adsorption process [179]. The removal of molybdate anion using chemically modified magnetic chitosan has been reported [181]. The adsorption capacity of the material for the oxyanion was 8.9 $mmolg^{-1}$. In a recent study conducted by Fu et al. [182], the adsorption capacity of ion-exchange resin (D301) was evaluated for the removal of molybdate from a high acidic solution. The resin exhibited high adsorption capacity (up to 463.63 mg/g) for the oxyanion and the adsorption process was controlled by particle diffusion.

Iron oxide nanocluster was employed to serve the dual purposes of a coagulant and an adsorbent for molybdate oxyanion. The material, under optimized condition, yielded more than 90% removal of the oxyanion (with initial concentration of 0.55 mgL^{-1}) from water. Based on this report, iron oxide nanocluster was recommended for the remediation of water contaminated by molybdate anion via multiple purification processes of adsorption, flocculation and filtration [183].

From the foregoing, it is apparent that adsorption has been at the centre of molybdate uptake from water and a degree of success has been recorded in spite of the difficulty associated with the removal of this pollutant from water. One of the best adsorption technologies that has been utilized for the removal of molybdate from water is ion-exchange. Typically, an ion-exchange resin comprises porous and insoluble beads which have ionic groups evenly distributed all through them. These ions on the beads are exchanged for ions of the same or similar electrical charge in an aqueous solution when the solution is in contact with the beads. Therefore, in order to remediate water or wastewater that is contaminated by molybdate (MoO_4^{2-}), a resin containing negative ions is sought. While authors have advanced various materials and mechanisms for the adsorption of molybdate, ion-exchange technique provides for necessary manipulation/modification of the adsorbents for effective removal of the anion.

5.6 Conclusion

In this chapter, a brief overview and techniques for removal of some oxyanions from aqueous media are presented. Despite the fact that the removal of these anions has not been given so much attention like other inorganic pollutants, it is interesting to note that there have been some advances in the techniques for their removal.

Innovative optimization of certain parameters in some known water treatment technologies has yielded substantial outcomes. The many possibilities of altering the surface chemistry of a wide range of materials have positioned adsorption technique as a preferred means of cleaning water plagued by these oxyanions. Reports have shown that carefully modified adsorbents like ion-exchange resins produced desirable results.

References

1. Shannon MA, Bohn PW, Elimelech M, Georgiadis JG, Marinas BJ, Mayes AM (2010) Science and technology for water purification in the coming decades. Nanosci Technol Collect Rev Nat J World Sci 337–346
2. Fosso-Kankeu E, Mittal H, Waanders F, Ray SS (2017) Thermodynamic properties and adsorption behaviour of hydrogel nanocomposites for cadmium removal from mine effluents. J Ind Eng Chem 48:151–161
3. Song JY, Jhung SH (2017) Adsorption of pharmaceuticals and personal care products over metal-organic frameworks functionalized with hydroxyl groups: quantitative analyses of H-bonding in adsorption. Chem Eng J 322:366–374
4. Gumpu MB, Sethuraman S, Krishnan UM, Rayappan JBB (2015) A review on detection of heavy metal ions in water—an electrochemical approach. Sens Actuators B: Chem 213:515–533
5. H.I. Adegoke, F.A. Adekola, O.S. Fatoki, B.J. Ximba, Sorptive Interaction of Oxyanions with Iron Oxides: A Review, Polish Journal of Environmental Studies 22 (2013)
6. Weidner E, Ciesielczyk F (2019) Removal of hazardous oxyanions from the environment using metal-oxide-based materials. Materials 12:927
7. Salmani MH, Sahlabadi F, Eslami H, Ghaneian MT, Balaneji IR, Zad TJ (2019) Removal of Cr(VI) oxoanion from contaminated water using granular jujube stems as a porous adsorbent. Groundwater Sustain Dev 8:319–323
8. Deng S-Q, Mo X-J, Zheng S-R, Jin X, Gao Y, Cai S-L, Fan J, Zhang W-G (2019) Hydrolytically stable nanotubular cationic metal-organic framework for rapid and efficient removal of toxic oxo-anions and dyes from water. Inorg Chem 58:2899–2909
9. Parks JL, Edwards M (2005) Boron in the environment. Crit Rev Environ Sci Technol 35:81–114
10. Epa U (2008) Drinking water health advisory for boron, Health and ecological criteria division. US Environmental Protection Agency, Washington, DC
11. Koilraj P, Srinivasan K (2011) High sorptive removal of borate from aqueous solution using calcined ZnAl layered double hydroxides. Ind Eng Chem Res 50:6943–6951
12. Woods WG (1994) An introduction to boron: history, sources, uses, and chemistry. Environ Health Perspect 102:5–11
13. Yilmaz AE, Boncukcuoğlu R, Kocakerim MM (2007) A quantitative comparison between electrocoagulation and chemical coagulation for boron removal from boron-containing solution. J Hazard Mater 149:475–481
14. Miyazaki T, Takeda Y, Akane S, Itou T, Hoshiko A, En K (2010) Role of boric acid for a poly (vinyl alcohol) film as a cross-linking agent: Melting behaviors of the films with boric acid. Polymer 51:5539–5549
15. Kentjono L, Liu JC, Chang WC, Irawan C (2010) Removal of boron and iodine from optoelectronic wastewater using Mg–Al (NO$_3$) layered double hydroxide. Desalination 262:280–283
16. Who G (2011) Guidelines for drinking-water quality. World health organization 216:303–304

17. Nomura J, Ishibashi Y, Kaneda A (1986) Selective separation of borate ions in water. Google Patents
18. Xu Y, Jiang J-Q (2008) Technologies for boron removal. Ind Eng Chem Res 47:16–24
19. Waggott A (1969) An investigation of the potential problem of increasing boron concentrations in rivers and water courses. Water Res 3:749–765
20. Yılmaz AE, Boncukcuoglu R, Yılmaz MT, Kocakerim MM (2005) Adsorption of boron from boron-containing wastewaters by ion exchange in a continuous reactor. J Hazard Mater 117:221–226
21. Guo X, Wu Z, He M (2009) Removal of antimony (V) and antimony (III) from drinking water by coagulation–flocculation–sedimentation (CFS). Water Res 43:4327–4335
22. Liu H, Ye X, Li Q, Kim T, Qing B, Guo M, Ge F, Wu Z, Lee K (2009) Boron adsorption using a new boron-selective hybrid gel and the commercial resin D564. Colloids Surf, A 341:118–126
23. Liu H, Qing B, Ye X, Li Q, Lee K, Wu Z (2009) Boron adsorption by composite magnetic particles. Chem Eng J 151:235–240
24. Xi J, He M, Lin C (2011) Adsorption of antimony (III) and antimony (V) on bentonite: kinetics, thermodynamics and anion competition. Microchem J 97:85–91
25. Geffen N, Semiat R, Eisen MS, Balazs Y, Katz I, Dosoretz CG (2006) Boron removal from water by complexation to polyol compounds. J Membr Sci 286:45–51
26. Melnyk L, Goncharuk V, Butnyk I, Tsapiuk E (2005) Boron removal from natural and wastewaters using combined sorption/membrane process. Desalination 185:147–157
27. Öztürk N, Kavak D, Köse TE (2008) Boron removal from aqueous solution by reverse osmosis. Desalination 223:1–9
28. Kang M, Kawasaki M, Tamada S, Kamei T, Magara Y (2000) Effect of pH on the removal of arsenic and antimony using reverse osmosis membranes. Desalination 131:293–298
29. Blahušiak M, Schlosser Š (2009) Simulation of the adsorption—microfiltration process for boron removal from RO permeate. Desalination 241:156–166
30. Rossini M, Garrido JG, Galluzzo M (1999) Optimization of the coagulation–flocculation treatment: influence of rapid mix parameters. Water Res 33:1817–1826
31. Farmer J, Kydd J (1979) Removal of boron from solution with inorganic precipitants, Part III, Lime
32. Remy P, Muhr H, Plasari E, Ouerdiane I (2005) Removal of boron from wastewater by precipitation of a sparingly soluble salt. Environ Prog 24:105–110
33. Farmer J, Kydd J (1979) The removal of boron solution with inorganic precipitants: Part II. Aluminium Compounds, Technical Report TR-78–10
34. Sanderson B (1977) Removal of boron from solution: an examination of the co-removal of boron with polyvinyl Alcohol, Technical Report TR-77–16
35. Chang Y, Burbank N Jr (1997) The removal of boron from incinerator quench water: hydrous metallic oxides versus on ion-soecufuc resin. In: Proceedings of the 32nd industrial WASTE conference, engineering extension series
36. Chorghe D, Sari MA, Chellam S (2017) Boron removal from hydraulic fracturing wastewater by aluminum and iron coagulation: mechanisms and limitations. Water Res 126:481–487
37. Hiraga Y, Shigemoto N (2011) Behavior in removal of chromate and borate from solution employing calcination products of $Ca(OH)_2-$ and $CaCO_3- Al(OH)_3$ Mixtures. J Chem Eng Japan advpub 1111100297–1111100297
38. Sasaki K, Toshiyuki K, Ideta K, Miki H, Hirajima T, Miyawaki J, Murayama M, Dabo I (2016) Removal mechanism of high concentration borate by co-precipitation with hydroxyapatite. J Environ Chem Eng 4:1092–1101
39. Nebgen J, Shea E, Chiu S (1973) The alumina-lime soda water treatment process (1973)
40. Farmer J, Kydd J (1979) Removal of boron from solution with inorganic precipitants: part VIII. Manganese (II) compounds, Technical Report TR-79–22
41. Nicolaı M, Rosin C, Morlot M, Hartemann P, Nicolassimonnot M, Castel C, Sardin M (2006) Statement of knowledge on the means of eliminating boron in water. Borax Europe Ltd., Guildford, UK

42. J. Kydd, Removal of Boron from Solution with Inorganic Precipitants: Part XVIII. Use of Zirconium(IV), Nickel(II) and Chromium(III) Compounds, Technical Report TR-83–23, Borax Technical, Ltd: London, 1983

43. Farmer J, Kydd J (1979) Removal of boron from solution with inorganic precipitants: part VI Zinc compounds, Technical Report TR-79-13, Borax Technical, Ltd., London

44. Wolska J, Bryjak M (2013) Methods for boron removal from aqueous solutions—A review. Desalination 310:18–24

45. Sayiner G, Kandemirli F, Dimoglo A (2008) Evaluation of boron removal by electrocoagulation using iron and aluminum electrodes. Desalination 230:205–212

46. Vasudevan S, Lakshmi J, Sozhan G (2013) Electrochemically assisted coagulation for the removal of boron from water using zinc anode. Desalination 310:122–129

47. da Silva Ribeiro T, Grossi CD, Merma AG, dos Santos BF, Torem ML (2019) Removal of boron from mining wastewaters by electrocoagulation method: modelling experimental data using artificial neural networks. Miner Eng 131(2019):8–13

48. Can BZ, Boncukcuoğlu R, Yılmaz AE, Fil BA (2016) Arsenic and boron removal by electrocoagulation with aluminum electrodes. Arab J Sci Eng 41:2229–2237

49. Kartikaningsih D, Shih Y-J, Huang Y-H (2016) Boron removal from boric acid wastewater by electrocoagulation using aluminum as sacrificial anode. Sustain Environ Res 26:150–155

50. Dolati M, Aghapour AA, Khorsandi H, Karimzade S (2017) Boron removal from aqueous solutions by electrocoagulation at low concentrations. J Environ Chem Eng 5:5150–5156

51. Oorts K, Smolders E, Degryse F, Buekers J, Gascó G, Cornelis G, Mertens J (2008) Solubility and toxicity of antimony trioxide (Sb_2O_3) in soil. Environ Sci Technol 42:4378–4383

52. Widhiastuti F, Lin J-Y, Shih Y-J, Huang Y-H (2018) Electrocoagulation of boron by electrochemically co-precipitated spinel ferrites. Chem Eng J 350:893–901

53. Kartikaningsih D, Huang Y-H, Shih Y-J (2017) Electro-oxidation and characterization of nickel foam electrode for removing boron. Chemosphere 166:184–191

54. Shih Y-J, Liu C-H, Lan W-C, Huang Y-H (2014) A novel chemical oxo-precipitation (COP) process for efficient remediation of boron wastewater at room temperature. Chemosphere 111:232–237

55. Lin J-Y, Shih Y-J, Chen P-Y, Huang Y-H (2016) Precipitation recovery of boron from aqueous solution by chemical oxo-precipitation at room temperature. Appl Energy 164:1052–1058

56. USEPA (1979) Water related fate of the 129 priority pollutants, Doc. 745-R-00–007. In: USEPA (Ed), Washington, DC

57. U. EPA (2002) Edition of the drinking water standards and health advisories. US Environmental Protection Agency Washington. eDC DC

58. Vu X, Lin J-Y, Shih Y-J, Huang Y-H (2018) Reclaiming boron as calcium perborate pellets from synthetic wastewater by integrating chemical oxo-precipitation within a fluidized-bed crystallizer. ACS Sustain Chem Eng 6:4784–4792

59. Tsai C-K, Lee N-T, Huang G-H, Suzuki Y, Doong R-A (2019) Simultaneous recovery of display panel waste glass and wastewater boron by chemical oxo-precipitation with fluidized-bed heterogeneous crystallization. ACS Omega 4:14057–14066

60. Guan Z, Lv J, Bai P, Guo X (2016) Boron removal from aqueous solutions by adsorption—A review. Desalination 383:29–37

61. Hilal N, Kim GJ, Somerfield C (2011) Boron removal from saline water: a comprehensive review. Desalination 273:23–35

62. Bodzek M (2016) The removal of boron from the aquatic environment–state of the art. Desalination Water Treat 57:1107–1131

63. Wang B, Guo X, Bai P (2014) Removal technology of boron dissolved in aqueous solutions – A review. Colloids Surf, A 444:338–344

64. Nasef MM, Nallappan M, Ujang Z (2014) Polymer-based chelating adsorbents for the selective removal of boron from water and wastewater: a review. React Funct Polym 85:54–68

65. Theiss FL, Ayoko GA, Frost RL (2013) Removal of boron species by layered double hydroxides: a review. J Colloid Interface Sci 402:114–121

66. Kabay N, Güler E, Bryjak M (2010) Boron in seawater and methods for its separation—A review. Desalination 261:212–217
67. Güler E, Kaya C, Kabay N, Arda M (2015) Boron removal from seawater: state-of-the-art review. Desalination 356:85–93
68. Riveros PA, Dutrizac JE, Lastra R (2008) A study of the ion exchange removal of antimony(III) and antimony(V) from copper electrolytes. Can Metall Q 47:307–316
69. Can BZ, Ceylan Z, Kocakerim MM (2012) Adsorption of boron from aqueous solutions by activated carbon impregnated with salicylic acid: equilibrium, kinetic and thermodynamic studies. Desalin Water Treat 40:69–76
70. Halim AA, Roslan NA, Yaacub NS, Latif MT (2013) Boron removal from aqueous solution using curcumin-impregnated activated carbon. Sains Malaysiana 42:1293–1300
71. Kluczka J, Ciba J, Trojanowska J, Zolotajkin M, Turek M, Dydo P (2007) Removal of boron dissolved in water. Environ Prog 26:71–77
72. Choi W-W, Chen KY (1979) Evaluation of boron removal by adsorption on solids. Environ Sci Technol 13:189–196
73. Ferreira OP, de Moraes SG, Durán N, Cornejo L, Alves OL (2006) Evaluation of boron removal from water by hydrotalcite-like compounds. Chemosphere 62:80–88
74. Jiang J-Q, Xu Y, Quill K, Simon J, Shettle K (2007) Laboratory study of boron removal by Mg/Al double-layered hydroxides. Ind Eng Chem Res 46:4577–4583
75. Ay AN, Zümreoglu-Karan B, Temel A (2007) Boron removal by hydrotalcite-like, carbonate-free Mg–Al–NO3-LDH and a rationale on the mechanism. Microporous Mesoporous Mater 98:1–5
76. Wang N, Wei RQ, Cao FT, Liu XN (2012) Synthesis of a new acetyl-meglumine resin and its adsorption properties of boron. Kao Teng Hsueh Hsiao Hua Heush Hsueh Pao 33:2795–2800
77. Li X, Liu R, Wu S, Liu J, Cai S, Chen D (2011) Efficient removal of boron acid by N-methyl-d-glucamine functionalized silica–polyallylamine composites and its adsorption mechanism. J Colloid Interface Sci 361:232–237
78. Zeebe RE, Sanyal A, Ortiz JD, Wolf-Gladrow DA (2001) A theoretical study of the kinetics of the boric acid–borate equilibrium in seawater. Mar Chem 73:113–124
79. Xiao X, Chen B-Z, Shi X-C, Chen Y (2012) Boron removal from brine by XSC-700. J Central South Univ 19:2768–2773
80. Biçak N, Şenkal BF (1998) Sorbitol-modified poly(N-glycidyl styrene sulfonamide) for removal of boron. J Appl Polym Sci 68:2113–2119
81. Bilgin Simsek E, Beker U, Senkal BF (2014) Predicting the dynamics and performance of selective polymeric resins in a fixed bed system for boron removal. Desalination 349:39–50
82. Senkal BF, Bicak N (2003) Polymer supported iminodipropylene glycol functions for removal of boron. React Funct Polym 55:27–33
83. Bicak N, Gazi M, Senkal BF (2005) Polymer supported amino bis-(cis-propan 2,3 diol) functions for removal of trace boron from water. React Funct Polym 65:143–148
84. Kabay N, Yılmaz I, Yamac S, Samatya S, Yuksel M, Yuksel U, Arda M, Sağlam M, Iwanaga T, Hirowatari K (2004) Removal and recovery of boron from geothermal wastewater by selective ion exchange resins. I. Laboratory tests. React Funct Polym 60:163–170
85. Samatya S, Tuncel SA, Kabay N (2015) Boron removal from RO permeate of geothermal water by monodisperse poly(vinylbenzyl chloride-co-divinylbenzene) beads containing N-methyl-d-glucamine. Desalination 364:75–81
86. Gazi M, Shahmohammadi S (2012) Removal of trace boron from aqueous solution using iminobis-(propylene glycol) modified chitosan beads. React Funct Polym 72:680–686
87. Suzuki TM, Tanaka DP, Yokoyama T, Miyazaki Y, Yoshimura K (1999) Complexation and removal of trace boron from aqueous solution by an anion exchange resin loaded with chromotropic acid (disodium 2, 7-dihydroxynaphthalene-4, 5-disulfonate). J Chem Soc Dalton Trans 1639–1644
88. Yavuz E, Gursel Y, Senkal BF (2013) Modification of poly (glycidyl methacrylate) grafted onto crosslinked PVC with iminopropylene glycol group and use for removing boron from water. Desalination 310:145–150

89. Ince A, Karagoz B, Bicak N (2013) Solid tethered imino-bis-propanediol and quaternary amine functional copolymer brushes for rapid extraction of trace boron. Desalination 310:60–66

90. Santander P, Rivas BL, Urbano BF, Yılmaz İpek İ, Özkula G, Arda M, Yüksel M, Bryjak M, Kozlecki T, Kabay N (2013) Removal of boron from geothermal water by a novel boron selective resin. Desalination 310:102–108

91. Ju YH, Webb OF, Dai S, Lin JS, Barnes CE (2000) Synthesis and characterization of ordered mesoporous anion-exchange inorganic/organic hybrid resins for radionuclide separation. Ind Eng Chem Res 39:550–553

92. Liu W, Liu Y, Gu X, Wang D (2009) Monte-carlo simulation for the reliability analysis of multi-status network system based on breadth first search. Second Int Conf Inf Comput Sci 2009:280–283

93. Wang B, Lin H, Guo X, Bai P (2014) Boron removal using chelating resins with pyrocatechol functional groups. Desalination 347:138–143

94. Yu S, Xue H, Fan Y, Shi R (2013) Synthesis, characterization of salicylic-HCHO polymeric resin and its evaluation as a boron adsorbent. Chem Eng J 219:327–334

95. Morisada S, Rin T, Ogata T, Kim Y-H, Nakano Y (2011) Adsorption removal of boron in aqueous solutions by amine-modified tannin gel. Water Res 45:4028–4034

96. Polowczyk I, Ulatowska J, Koźlecki T, Bastrzyk A, Sawiński W (2013) Studies on removal of boron from aqueous solution by fly ash agglomerates. Desalination 310:93–101

97. Kluczka J, Trojanowska J, Zołotajkin M (2015) Utilization of fly ash zeolite for boron removal from aqueous solution. Desalin Water Treat 54:1839–1849

98. Yüksel S, Yürüm Y (2010) Removal of boron from aqueous solutions by adsorption using fly ash, zeolite, and demineralized lignite. Sep Sci Technol 45:105–115

99. Irawan C, Liu JC, Wu C-C (2011) Removal of boron using aluminum-based water treatment residuals (Al-WTRs). Desalination 276:322–327

100. Jalali M, Rajabi F, Ranjbar F (2016) The removal of boron from aqueous solutions using natural and chemically modified sorbents. Desalin Water Treat 57:8278–8288

101. Kıpçak İ, Özdemir M (2012) Removal of boron from aqueous solution using calcined magnesite tailing. Chem Eng J 189–190:68–74

102. Bryjak M, Duraj I (2013) Anion-exchange membranes for separation of borates by Donnan dialysis. Desalination 310:39–42

103. Wu Z, He M, Guo X, Zhou R (2010) Removal of antimony (III) and antimony (V) from drinking water by ferric chloride coagulation: Competing ion effect and the mechanism analysis. Sep Purif Technol 76:184–190

104. Gurnani N, Sharma A, Talukder G (1994) Effects of antimony on cellular systems in animals: a review. Nucleus-Calcutta-Int J Cytol 37:71–96

105. Lu H, Zhu Z, Zhang H, Zhu J, Qiu Y (2015) Simultaneous removal of arsenate and antimonate in simulated and practical water samples by adsorption onto Zn/Fe layered double hydroxide. Chem Eng J 276:365–375

106. C.C.o.t.E. Communities) (1976) Council Directive 76/Substances discharged into aquatic environment of the community. Official J L 23–29

107. Filella M, Belzile N, Chen Y-W (2002) Antimony in the environment: a review focused on natural waters: II. Relevant Solution Chem Earth-Sci Rev 59:265–285

108. McComb KA, Craw D, McQuillan AJ (2007) ATR-IR spectroscopic study of antimonate adsorption to iron oxide. Langmuir 23:12125–12130

109. Essington M, Stewart M, Vergeer K (2017) Adsorption of antimonate by Kaolinite. Soil Sci Soc Am J 81:514–525

110. Seiler H, Sigel A, Sigel H (1994) Handbook on metals in clinical and analytical chemistry. CRC Press

111. Kang M, Kamei T, Magara Y (2003) Comparing polyaluminum chloride and ferric chloride for antimony removal. Water Res 37:4171–4179

112. Zhao X, Dou X, Mohan D, Pittman CU, Ok YS, Jin X (2014) Antimonate and antimonite adsorption by a polyvinyl alcohol-stabilized granular adsorbent containing nanoscale zero-valent iron. Chem Eng J 247:250–257

113. Martínez-Lladó X, Valderrama C, Rovira M, Martí V, Giménez J, de Pablo J (2011) Sorption and mobility of Sb(V) in calcareous soils of Catalonia (NE Spain): Batch and column experiments. Geoderma 160:468–476

114. Biswas BK, Inoue J-I, Kawakita H, Ohto K, Inoue K (2009) Effective removal and recovery of antimony using metal-loaded saponified orange waste. J Hazard Mater 172:721–728

115. Luo J, Luo X, Crittenden J, Qu J, Bai Y, Peng Y, Li J (2015) Removal of Antimonite (Sb(III)) and Antimonate (Sb(V)) from Aqueous Solution Using Carbon Nanofibers That Are Decorated with Zirconium Oxide (ZrO$_2$). Environ Sci Technol 49:11115–11124

116. Guo X, Wu Z, He M, Meng X, Jin X, Qiu N, Zhang J (2014) Adsorption of antimony onto iron oxyhydroxides: adsorption behavior and surface structure. J Hazard Mater 276:339–345

117. Nishiyama S-Y, Saito K, Saito K, Sugita K, Sato K, Akiba M, Saito T, Tsuneda S, Hirata A, Tamada M, Sugo T (2003) High-speed recovery of antimony using chelating porous hollow-fiber membrane. J Membr Sci 214:275–281

118. Cupertino D, Tasker P, King M, Jackson J (1994) Removal of antimony and bismuth from copper tankhouse electrolytes, Hydrometallurgy'94. Springer. pp 591–600

119. Schilde U, Kraudelt H, Uhlemann E (1994) Separation of the oxoanions of germanium, tin, arsenic, antimony, tellurium, molybdenum and tungsten with a special chelating resin containing methylaminoglucitol groups. React Polym 22:101–106

120. Ilgen AG, Trainor TP (2012) Sb(III) and Sb(V) sorption onto Al-Rich phases: hydrous Al oxide and the clay minerals kaolinite KGa-1b and oxidized and reduced nontronite NAu-1. Environ Sci Technol 46:843–851

121. Leuz A-K, Mönch H, Johnson CA (2006) Sorption of Sb (III) and Sb (V) to goethite: influence on Sb (III) oxidation and mobilization. Environ Sci Technol 40:7277–7282

122. Xu YH, Ohki A, Macda S (2001) Adsorption and removal of antimony from aqueous solution by an activated Alumina: 1, Adsorption capacity of adsorbent and effect of process variables. Toxicol Environ Chem 80:133–144

123. Hasany SM, Chaudhary MH (1996) Sorption potential of Haro river sand for the removal of antimony from acidic aqueous solution. Appl Radiat Isot 47:467–471

124. Zhang G, Ouyang X, Li H, Fu Z, Chen J (2016) Bioremoval of antimony from contaminated waters by a mixed batch culture of sulfate-reducing bacteria. Int Biodeterior Biodegradation 115:148–155

125. Li X, Dou X, Li J (2012) Antimony(V) removal from water by iron-zirconium bimetal oxide: performance and mechanism. J Environ Sci 24:1197–1203

126. Wang X, He M, Lin C, Gao Y, Zheng L (2012) Antimony(III) oxidation and antimony(V) adsorption reactions on synthetic manganite. Geochemistry 72:41–47

127. Xu W, Liu R, Qu J, Peng R (2012) The adsorption behaviors of Fe–Mn binary oxide towards Sb (V). Acta Sci Circum 32:270–275

128. Xi J, He M, Wang K, Zhang G (2013) Adsorption of antimony(III) on goethite in the presence of competitive anions. J Geochem Explor 132:201–208

129. Wang H, Li X, Li W, Sun Y (2017) Effects of pH and complexing agents on Sb (V) adsorption onto birnessite and ferrihydrite surface, Huan jing ke xue = Huanjing kexue 38:180–187

130. Kolbe F, Weiss H, Morgenstern P, Wennrich R, Lorenz W, Schurk K, Stanjek H, Daus B (2011) Sorption of aqueous antimony and arsenic species onto akaganeite. J Colloid Interface Sci 357:460–465

131. Ambe S (1987) Adsorption kinetics of antimony(V) ions onto -ferric oxide surfaces from an aqueous solution. Langmuir 3:489–493

132. Hu XX (2016) Study on the performance and mechanism of the removal of antimony from mine wastewater by a new type of Fe-Cu binary oxide. Hunan University of Science and Technology, Xiangtan, China, pp 1–73

133. Liu XW, Chen JH, Chen JQ (2016) Removal of antimony from water by supported iron-zircon bimetal oxide polymeric anion exchange resin. Ion Exchange Adsorption. 32:244–252

134. Miao YY (2013) Removal of aqueous Sb(V) by adsorbents loaded with nano hydrated ferric oxides. Nanjing University, Nanjing, China, pp 1–51

135. Zhang H (2015) Removal of antimony(V) by adsorption of iron-modified attapulgite adsorbent. Guizhou University, Guiyang, China, pp 1–53
136. Wang L, Wan C, Lee D-J, Liu X, Zhang Y, Chen XF, Tay J-H (2014) Biosorption of antimony(V) onto Fe(III)-treated aerobic granules. Biores Technol 158:351–354
137. Wan C, Wang L, Lee D-J, Zhang Q, Li J, Liu X (2014) Fungi aerobic granules and use of Fe(III)-treated granules for biosorption of antimony(V). J Taiwan Inst Chem Eng 45:2610–2614
138. Verbinnen B, Block C, Lievens P, Van Brecht A, Vandecasteele C (2013) Vandecasteele, simultaneous removal of molybdenum, antimony and selenium oxyanions from wastewater by adsorption on supported magnetite. Waste Biomass Valorization 4:635–645
139. Li L, Zhang H, Zhou S, Xu S (2014) Pentavalent antimony adsorption behavior in two types of soils typical to South China. Acta Pedol Sin 51:278–285
140. Pytkowicz RM (1965) Rates of inorganic calcium carbonate nucleation. J Geol 73:196–199
141. Korchef A, Touaibi M (2020) Effect of pH and temperature on calcium carbonate precipitation by CO_2 removal from iron-rich water. Water Environ J n/a
142. Gopinath CS, Hegde SG, Ramaswamy AV, Mahapatra S (2002) Photoemission studies of polymorphic $CaCO_3$ materials. Mater Res Bull 37:1323–1332
143. Han YS, Hadiko G, Fuji M, Takahashi M (2006) Factors affecting the phase and morphology of $CaCO_3$ prepared by a bubbling method. J Eur Ceram Soc 26:843–847
144. Neri LC, Johansen HL (1978) Water hardness and cardiovascular mortality. Ann N Y Acad Sci 304:203–219
145. Sengupta P (2013) Potential health impacts of hard water. Int J Prev Medicine 4:866
146. Comstock GW (1979) Water hardness and cardiovascular diseases. Am J Epidemiol 110:375–400
147. Pervov A, Andrianov A (2015) Removal of calcium carbonate from reverse osmosis concentrate by seed crystallization. Pet Chem 55:373–388
148. Ayoub GM, Korban L, Al-Hindi M, Zayyat R (2018) Removal of fouling species from brackish water reverse osmosis reject stream. Environ Technol 39:804–813
149. Mitsoyannis E, Saravacos GD (1977) Precipitation of calcium carbonate on reverse osmosis membranes. Desalination 21:235–240
150. Bond R, Veerapaneni S (2013) In: Proceedings of 2013 IDA World congress on desalination and water Reuse, International Desalination Association, Tianjin
151. Gilron J (2012) Flow reversal as a tool in use of crystallizer/secondary RO treatment for high recovery desalination. In: The 3-rd Sede boqer conference on water technologies
152. Ahn MK, Chilakala R, Han C, Thenepalli T (2018) Removal of hardness from water samples by a carbonation process with a closed pressure reactor. Water 10:54
153. Agostinho L, Nascimento L, Cavalcanti B (2012) Water hardness removal for industrial use: application of the electrolysis process. Open Access Sci Rep 1:460–465
154. C.C. IPCC (2014) Mitigation of climate change, Contribution of working group III to the fifth assessment report of the intergovernmental panel on climate change
155. den Elzen M, Höhne N, Jiang K, Cantzler J, Drost P, Fransen T, Fekete H, Kuramochi T, Lee D, Levin K (2017) Methodology of national and global studies-the emissions gap report 2017 Chapter 3 Appendix A, The Emissions Gap Report 2017: A UN Environment Synthesis Report (2017)
156. Lu L, Guest JS, Peters CA, Zhu X, Rau GH, Ren ZJ (2018) Wastewater treatment for carbon capture and utilization. Nat Stain 1:750–758
157. Wang H, Ren ZJ (2013) A comprehensive review of microbial electrochemical systems as a platform technology. Biotechnol Adv 31:1796–1807
158. Salek SS, Kleerebezem R, Jonkers HM, Witkamp G-J, van Loosdrecht MC (2013) Mineral CO_2 sequestration by environmental biotechnological processes. Trends Biotechnol 31:139–146
159. Lu L, Huang Z, Rau GH, Ren ZJ (2015) Microbial electrolytic carbon capture for carbon negative and energy positive wastewater treatment. Environ Sci Technol 49:8193–8201

160. Zhu X, Logan BE (2014) Microbial electrolysis desalination and chemical-production cell for CO_2 sequestration. Biores Technol 159:24–29
161. Rabaey K, Rozendal RA (2010) Microbial electrosynthesis—revisiting the electrical route for microbial production. Nat Rev Microbiol 8:706–716
162. Shoener B, Bradley I, Cusick R, Guest JS (2014) Energy positive domestic wastewater treatment: the roles of anaerobic and phototrophic technologies. Environ Sci Processes Impacts 16:1204–1222
163. Barry A, Wolfe A, English C, Ruddick C, Lambert D (2016) National algal biofuels technology review, USDOE office of energy efficiency and renewable energy (EERE). Bioenergy
164. Moody JW, McGinty CM, Quinn JC (2014) Global evaluation of biofuel potential from microalgae. Proc Natl Acad Sci 111:8691–8696
165. Li Y, Leow S, Fedders AC, Sharma BK, Guest JS, Strathmann TJ (2017) Quantitative multiphase model for hydrothermal liquefaction of algal biomass. Green Chem 19:1163–1174
166. Leow S, Witter JR, Vardon DR, Sharma BK, Guest JS, Strathmann TJ (2015) Prediction of microalgae hydrothermal liquefaction products from feedstock biochemical composition. Green Chem 17:3584–3599
167. Wu H, Fan J, Zhang J, Ngo HH, Guo W, Liang S, Hu Z, Liu H (2015) Strategies and techniques to enhance constructed wetland performance for sustainable wastewater treatment. Environ Sci Pollut Res 22:14637–14650
168. Wu S, Kuschk P, Brix H, Vymazal J, Dong R (2014) Development of constructed wetlands in performance intensifications for wastewater treatment: a nitrogen and organic matter targeted review. Water Res 57:40–55
169. Mander Ü, Lõhmus K, Teiter S, Mauring T, Nurk K, Augustin J (2008) Gaseous fluxes in the nitrogen and carbon budgets of subsurface flow constructed wetlands. Sci Total Environ 404:343–353
170. Giussani A (2011) Molybdenum in the environment and its relevance for animal and human health
171. Smedley PL, Kinniburgh DG (2017) Molybdenum in natural waters: A review of occurrence, distributions and controls. Appl Geochem 84:387–432
172. Keng X, Wang H-B, Wang S-D (2019) Direct leaching of molybdenum and lead from lean wulfenite raw ore. Trans Nonferrous Metals Soc China 29:2638–2645
173. Atia AA, Donia AM, Awed HA (2008) Synthesis of magnetic chelating resins functionalized with tetraethylenepentamine for adsorption of molybdate anions from aqueous solutions. J Hazard Mater 155:100–108
174. Li J, Chen D, Liao X, Pan B (2019) Selective adsorption of molybdate from water by polystyrene anion exchanger-supporting nanocomposite of hydrous ferric oxides. Sci Total Environ 691:64–70
175. Brinza L, Vu HP, Neamtu M, Benning LG (2019) Experimental and simulation results of the adsorption of Mo and V onto ferrihydrite. Sci Rep 9:1–12
176. Gustafsson JP, Tiberg C (2015) Molybdenum binding to soil constituents in acid soils: an XAS and modelling study. Chem Geol 417:279–288
177. Huang YH, Tang C, Zeng H (2012) Removing molybdate from water using a hybridized zero-valent iron/magnetite/Fe (II) treatment system. Chem Eng J 200:257–263
178. Zhang L, Xue J, Zhou X, Fei X, Wang Y, Zhou Y, Zhong L, Han X (2014) Adsorption of molybdate on molybdate-imprinted chitosan/triethanolamine gel beads. Carbohyd Polym 114:514–520
179. Bertoni FA, González JC, García SI, Sala LF, Bellú SE (2018) Application of chitosan in removal of molybdate ions from contaminated water and groundwater. Carbohyd Polym 180:55–62
180. Gao B, Li X, Chen T, Fang L (2014) Preparation of molybdate anion surface-imprinted material for selective removal of molybdate anion from water medium. Ind Eng Chem Res 53:4469–4479
181. Elwakeel KZ, Atia AA, Donia AM (2009) Removal of Mo (VI) as oxoanions from aqueous solutions using chemically modified magnetic chitosan resins. Hydrometallurgy 97:21–28

182. Fu Y-F, Xiao Q-G, Gao Y-Y, Ning P-G, Xu H-B, Zhang Y (2018) Direct extraction of Mo (VI) from acidic leach solution of molybdenite ore by ion exchange resin: Batch and column adsorption studies. Trans Nonferrous Metals Soc China 28:1660–1669

183. Ma W, Sha X, Gao L, Cheng Z, Meng F, Cai J, Tan D, Wang R (2015) Effect of iron oxide nanocluster on enhanced removal of molybdate from surface water and pilot scale test. Colloids Surf, A 478:45–53

Chapter 6
Trends in the Management of Arsenic Contamination in Potable Water

Eric T. Anthony and Nurudeen A. Oladoja

Abstract The continuous incidence of arsenic contamination in potable water has been linked to the abundance of naturally occurring arsenic minerals in the earth crust and the indiscriminate use of arsenic-rich resource in industrial and agricultural operations. In potable water, the presence of oxyanions of arsenite and arsenate is of great concern. In an aqueous system, the stoichiometric balance between arsenite and arsenate is altered by the change in the pH value and redox potential of the system, the organic matter concentration, the amount of dissolved and undissolved minerals and salts, and competing ions. Many strategies employed in removing arsenic favor the removal of arsenic as arsenate, rather than arsenite. This is due to the relatively lower toxic value of arsenate, the limited mobility, and the higher affinity for the common substrate. Thus, many of the technologies used to remove arsenic are preceded by the oxidation of arsenite to arsenate, and the removal of the resulting arsenate occurred via trapping co-precipitation or surface adsorption. The materials employed are characterized as low cost and ecofriendly, exhibiting improved textural capacity, stability, and ease of recovery.

Keywords Potable water · Arsenate · Arsenite · Occurrence · Removal and management · Treatment strategies

6.1 Introduction

Arsenic (As), a pro-oxidant metallic element, has consistently presented imperious challenges in aqueous solution, due to its mobility, solubility, and stability. It is considered a threat in all its states of matter. It is lethal in the gaseous state [1], impose chronic and acute toxicity in aqueous solution [2], and threaten food security

E. T. Anthony
Department of Chemistry, University of Fort Hare, Alice, South Africa

N. A. Oladoja (✉)
Hydrochemistry Research Laboratory, Department of Chemical Sciences, Adekunle Ajasin University, Akungba Akoko, Nigeria
e-mail: nurudeen.oladoja@aaua.edu.ng

© Springer Nature Switzerland AG 2021   155
N. A. Oladoja and E. I. Unuabonah (eds.), *Progress and Prospects in the Management of Oxyanion Polluted Aqua Systems*, Environmental Contamination Remediation and Management, https://doi.org/10.1007/978-3-030-70757-6_6

by leaching into soil and tissue, thus, impeding plants, eggs, and tissue development
[3, 4]. Arsenic toxicity includes hearing loss [5], genotoxicity [6], respiratory disease
[7], cytotoxicity and mutagenicity [8], cardiovascular disease [9], and cancer [10].
In potable water, the incidence of arsenic is of unusual interest due to its economic
importance. At concentrations above 10 μg/L, inorganic arsenic (H_3AsO_4, $H_2AsO_4^-$,
$HAsO_4^{2-}$, and AsO_4^{3-}) is an outright nuisance in potable water, with no nutritional
benefit. Though the organic and inorganic forms of arsenic can be consumed via
inhalation (as aerosols), the extreme threat is posed by foods (water and biota)
ingestion [11]. Moderate exposure to inorganic arsenic, via ingestion, can result
in sex-specific effects during a vital window of child development [12].

Arsenic is abundant in the earth's crust, with an average estimation of 2–5 mg/kg.
It is inherent in rocks, sediments, and ash. In lieu, the natural sources of arsenic
contamination are enormous. Such that, natural occurrences like volcanic incidences,
seismic events, and aquifer depletion and recharge have disturbed the balance of
arsenic bound minerals, thus, elevating its average prevalence in the environment
[13]. The vast distribution of arsenic in drinking water in the Southwest region of
Ireland is presented in Fig. 6.1. The map showed that several sources of drinking
water in the region exceeded the acceptable arsenic limit [14]. The weathering of
arsenopyrite (FeAsS), realgar (AsS), and orpiment (As_2S_3) is a typical incidence
for the release of arsenic. The occupational release of arsenic into the environment
is associated with metal mining and smelting, petroleum exploration combustion,
incineration services, pesticide manufacture and application, and wood preservatives. Common wood preservatives and pesticides that have been used extensively in
Canada include arsenic pentoxide, chromate copper arsenate, and calcium arsenate,
respectively [15].

The arsenic contaminant is classified, based on its chemical form in the environment, as organic and inorganic. Both forms are found within the living-host and get
distributed in the ecosystem. For example, arsenosugars are contained in seaweeds,
which can metabolize to dimethyl arsenic, thio-dimethylarsinoylethanol, and thio-
dimethylarsinoylacetate [16]. In human urine, arsenic is detected as arsenobetaine
((CH_3)$_3$$AsCH_2COOH$), arsenite ($As(OH)_3$), arsenate ($AsO(OH)_3$), methylarsenic
acid ($CH_3AsO(OH)_2$), and dimethylarsenic acid ((CH_3)$_2$$AsO(OH)$) [17]. Inorganic
arsenic has been found in adults [18] and infants [19] toe nails and human placenta
[20]. Arsenite was the prevalent form of arsenic detected in the roots, stem, and
fruits of strawberry plants [21]. The concentration of total arsenic detected in rice
cultivated in Brazil was 86.5 ng/g, and it was distributed as arsenate (16%), arsenite
(63%), dimethyl arsenic acid (24%), and monomethylarsonic acid (12%) [22]. An
in vitro digestion studies revealed that the transformation of inorganic arsenic occurs
via arsenobetaine, and 100% content of arsenic in gastropod is bioaccessible to
humans as arsenite and arsenocholine [23]. After coal burning in Nitra Valley in
central Slovakia, the mean value of arsenic detected around the power station was
26, 11.6, and 9.4 μg/g, in soil, house dust, and composite, respectively [24]. This is
an indication that a singular source of arsenic can result in multiple pollutions, and in
different forms.

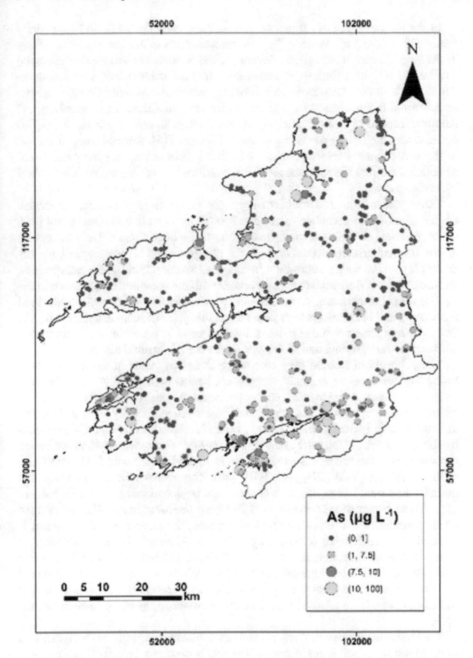

Fig. 6.1 Spatial distribution of arsenic in drinking water of Southwest Ireland [14]. Reprinted with permission from Elsevier, 2020

In the atmosphere, both forms of arsenic have been detected, and its toxicity was related to the particle size [25]. The aerosol forms of arsenic are released due to the combination of natural incidences (volcanic emission and marine aerosols) and human activities (industrial emission). The predominant forms of arsenic in the atmosphere are arsenate, arsenic trioxide, and arsine. Volatile organic arsenic, arsenic trioxide, and arsine are associated with microbial actions, metal smelting, and landfills, respectively. In the smelting of arsenious gold, arsenic trioxides (As_2O_3) accompanies the release of other gaseous substances [15]. Aerosol-trapped arsenic in the atmosphere is returned to the soil after precipitation, and the precipitated arsenic is sunk into the soil, washed, or immobilized according to the solution/soil hydrochemistry.

The exposure of man to arsenic can be from the intake of plants and vegetables. Higher plants can accumulate more than 70% of the arsenic pollutant in soil [26], which is stored in the root, stem, leave, and fruits of the plant. The root uptake of arsenate and arsenite occurs through the phosphate and silicon carriers, respectively [27]. In the aquifer, accommodated within volcaniclastic sedimentary rocks, the occurrence of dimethylarsinate correlates with the occurrence of arsenite in the aquifer [28]. Furthermore, some amount of arsenic can be released during food processing, and the rate of release depends on the type of food, i.e. species-specific [29]. The inorganic arsenic content of foods is more susceptible to release during boiling, steaming, frying, and soaking, than the organic arsenic content.

The geocycle of arsenic may commence from any point in the environment. Using the marine as an example, the process begins with the uptake of inorganic arsenic (mostly arsenate) by the microscopic marine algae. The inorganic arsenic is a known constituent of rocks, sediments, and soluble solutes in many water bodies. The ingested arsenic is used as nourishment, which is retained as lipid-soluble (arseno-lipids) compounds, which can be substituted for nitrate during scarcity. Detoxification process secrets the excess ingested arsenic as methylarsonic acid ($CH_3AsO(OH)_2$) or dimethylarsinic acid ($(CH_3)_2AsO(OH)$) [30]. The microscopic marine algae that initially consumed inorganic arsenate are a vital organism that helps in the balance of food web, because they serve as food for higher marine animals. When consumed by fish, mollusks, and other marine invertebrates, the arsenic metabolizes and is retained as arsenobetaine ($(CH_3)_3AsCH_2COOH$). Alternatively, the arsenobetaine stored in marine invertebrates can be metabolized and released as methylarsonic acid ($CH_3AsO(OH)_2$ and inorganic arsenic [31]. When these animals are consumed by man, the arsenic is absorbed by the human body (resulting in impaired health) and can eventually be released back into the environment as waste to complete the cycle.

The increasing incidence of arsenic contamination in freshwater has been ascribed to its proximity to soil and sediments. The soil is obtained from rock degradation; thus, soils generated from the weathering of rocks-harbouring arsenic, are rich in arsenic. Methylated arsenic is also released when plant and animal decay to form sediments. Thus, the concentration of arsenic in soil and sediment can vary, depending on the origin of the contamination. A typical example is the concentration of soil arsenic in industrial and non-industrial-based environment. The concentrations of arsenic in the soil around Ottawa, Canada, ranged between 1.7 and 9.9 mg/kg, due to minimal industrial activities [32]; however, at a depth of 15 cm, an alarming concentration (7860 mg/kg) was detected in Zloty Stok area of Poland, as a result of mining activities [33]. The polluted sediments and soil are eventually washed into freshwater.

Similar to soil and sediment, the concentration of arsenic in freshwater may be region-specific. Some regions of the world have experienced moderate arsenic contamination while disturbing concentrations have been detected in Bangladesh freshwater supplies [34]. This has manifested in the elevated arsenic concentration in potable water around this region. The concentrations of arsenic in Santa Fe province of Argentina are presented in Fig. 6.2. Many of these regions have relied on groundwater with arsenic exceeding 50 μg/L [35]. Even in many non-arid regions of the world, arsenic in drinking water is still a significant concern. In a study conducted by Huq et al. (2020), the strong oxidizing condition and moderate alkalinity were responsible for the elevated arsenic level (>10 μg/L) in 59 out of 64 wells used as potable water in Bangladesh [36]. In the USA, as a result of redox reactions occurring within wells, the concentration of arsenic in the potable water increased over time, and arsenic concentration up to 80 μg/L was detected [37]. In Nicaragua, over 55,700 persons have been exposed to arsenic contamination via potable water, with

Fig. 6.2 Map showing the elevated concentration of As pollution in the freshwater of Santa Fe province of Argentina [35]. Reprinted with permission from Elsevier, 2020

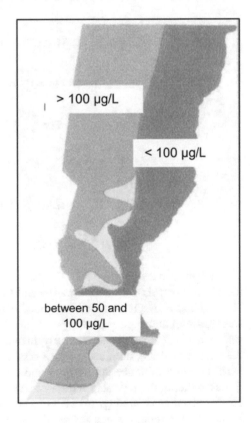

concentrations up to 1320 µg/L with a mean concentration of 48.3 µg/L detected in the wells from this region [38]. In Ireland, the elevated level of arsenic has been linked with arsenic harbouring rocks (e.g. rhyolite, limestone, sandstone, and shale).

6.2   The Dynamics of Arsenic in a Potable Water System

The partially filled 4p orbitals of As $[1s^2 2s^2 2p^6 3s^2 3p^6 3d^{10} 4s^2 4p^3]$ allow As to participate in many redox reaction, forming oxyanions in aqueous solution. Thus, arsenic exists in four oxidation states, namely; arsenate ((+5) (HAs O_4^{2-}), arsenite (+3) $(HAsO_3^{2-})$, arsenic (As^0) (0), and arsine (AsH$_3$) (−3), with oxidation state +5 and +3 being the most stable. The stability and subsequent transformation of arsenic in potable water depend on the water biogeochemical condition (i.e. pH value, redox potential, organic matter concentration, un/dissolved minerals, and competing species). In potable water with pH value range (6.5–8.5), the stable form of arsenic (i.e. (H$_3$AsO$_3$) (+3) and (H$_3$AsO$_4$) (+5)) is mobile and soluble, and can form oxide, carbonate, hydroxide and phosphate, depending on the pH value (Fig. 6.3). The dissociation reactions of H$_3$AsO$_3$ (Eq. 6.1–6.3) and H$_3$AsO$_4$ (Eq. 4.4–6.6) are depicted below:

$$H_3AsO_3 \leftrightarrow H_2AsO_3^- + H^+ \quad pKa = 2.24 \qquad (6.1)$$

$$H_2AsO_3^- \leftrightarrow HAsO_3^{2-} + H^+ \quad pKa = 6.69 \qquad (6.2)$$

$$HAsO_3^{2-} \leftrightarrow AsO_3^{3-} + H^+ \quad pKa = 11.5 \qquad (6.3)$$

$$H_3AsO_4 \leftrightarrow H_2AsO_4^- + H^+ \quad pKa = 9.2 \qquad (6.4)$$

$$H_2AsO_4^- \leftrightarrow HAsO_4^{2-} + H^+ \quad pKa = 12.1 \qquad (6.5)$$

$$HAsO_4^{2-} \leftrightarrow AsO_4^{3-} + H^+ \quad pKa = 13.4 \qquad (6.6)$$

In addition to the solution pH value, solution redox condition is another important factor that controls the transformation and stability of arsenic in aqueous solution. H$_3$AsO$_4$ and H$_2$AsO$_3$ dominate in oxidizing and reducing conditions, respectively. In oxidizing condition, H$_3$AsO$_4$ predominate as H$_2$AsO$_4^-$ and AsO$_4^{-2}$, at low (pH <6.9) and high pH values, respectively. In reducing condition, at pH value below 9.2, the neutral H$_2$AsO$_3$ dominate, which dissociates according to Eq. 6.1–6.3. These factors do not influence the stability and transformation of arsenic independently. For example, in the Northern region of La Pampa, Argentina, the source water was characterized as oxidizing, Na-HCO$_3$ and Na-SO$_4$-Cl type, and with elevated pH in the alkaline region which favoured arsenate formation (97%). The arsenate was

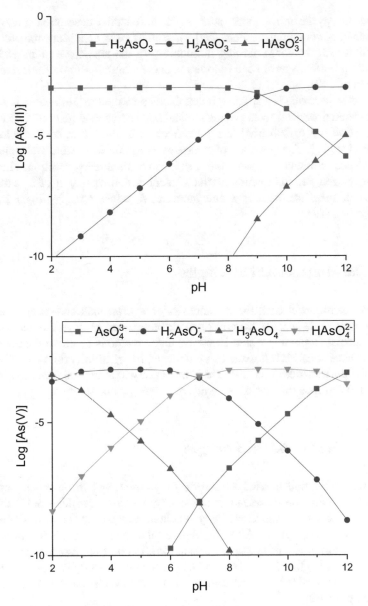

Fig. 6.3 pH-dependent speciation diagram of arsenite and arsenate (*Simulated using Visual MINTEQ™ software*)

released from pyrite or iron oxides surface, due to reductive desorption or dissolution from minerals containing arsenic [39]. In similar oxic condition, minerals show high affinity for arsenic, retarding its mobility in the aqueous system. Halloysite and chlorite minerals have been reported to show a higher affinity for arsenite than illite mineral [40].

Hydrogen peroxide is a strong oxidant that is prevalent in freshwater. It is generated in the environment from photochemical and radioactive reactions. Thus, high concentration (up to 200 μM) has been detected in rainwater, surface water, and groundwater [41]. The presence of hydrogen peroxide in potable water increases the incidence of arsenic exposure because it can initiate redox reaction. The waterborne hydrogen peroxide causes surface leaching and oxidation of arsenite from mineral, and the dissolution of arsenic increases with increasing hydrogen peroxide concentration [42].

6.3   The Management Strategies

Different arsenic species coexist in potable water, with arsenate and arsenite being the dominating mobile species. Several methods have been employed to manage these arsenic species in potable water. However, many methods favoured the removal of arsenic since arsenate is less toxic, less mobile, with high affinity for many substrates, rather than the more toxic arsenite. Therefore, during treatment, arsenite is preferably oxidized to arsenate for ease of removal from the aqueous solution.

6.3.1   Oxidation-Based Strategies

In practice, oxidation is used as a form of pretreatment in the management of arsenic. The oxidation of arsenic is induced by electron transfer from a chemical species or photons. Oxidation is highly beneficial to the overall removal of arsenic. For example, Moore et al. (2008) highlighted that after the oxidation of arsenite, the removal efficiency of arsenate, in nanofiltration and reverse osmosis systems, increased from negligible to 50% and 90%, respectively [43]. Two procedures that have been employed for the oxidation of arsenic include chemical and light-assisted oxidation processes.

6.3.1.1   Chemical Oxidation

In chemical oxidation, the oxidant is either introduced directly into the reaction medium or generated in situ for the oxidation of arsenite. In a simple oxidation process arsenopyrite (FeAsS), a mineral harbouring arsenite is oxidized in the presence of hydrogen peroxide (Eq. 6.7). The increasing concentration of hydrogen peroxide was found beneficial to the removal of arsenite, as the removal efficiency increased from 28.1 to 59.7% when the concentration of hydrogen peroxide increased from 0.25 to 3.0 mol/L, respectively [44]. However, this is not practicable because of cost and secondary pollution issues.

$$FeAsS + 2H_2O_2 \rightarrow FeS + H_3AsO_4 + H^+ \tag{6.7}$$

Biochar, a pyrogenic carbonaceous material with solid-phase free radicals [45], is a good candidate for the oxidation of arsenite and subsequent adsorption of the resulting arsenate. Biochar was used for the oxidation of arsenite in aqueous solution [46]. The solution pH and dissolved oxygen were essential in the oxidation process. In the absence of biochar, dissolved oxygen can initiate the oxidation of arsenite at pH 9.5 with 21% efficiency in 24 h. This is due to the higher reduction potential of $O_2/\cdot HO_2$ ($E_{pH=9.5} = -0.127$ V) than As(V)/As(III) ($E_{pH=9.5} = -0.397$ V). When coupled with biochar, the oxidation of arsenite was completed (99%) under 4 and 12 h in oxic and anoxic conditions, respectively.

Biochar/bimetallic composite was prepared by the incorporation of Fe–Mn oxides for the oxidation and subsequent removal of arsenate from simulated potable water. The oxidation of arsenite was rapid, within 30 min, at neutral pH, which coincided with the increasing concentration of arsenate. The oxidation was attributed to the reactive oxygen species and Mn(IV) on the surface of the biochar composite adsorbent [47]. The biochar composite also exhibited an improved adsorption capacity (8.8 mg/g).

6.3.1.2 Light-Assisted Oxidation

In the light-assisted oxidation system, the oxidation phase may be homogeneous or heterogeneous. In the heterogeneous system, a solid photocatalyst absorbs photon of light to undergo a series of redox reaction. The process is initiated when the photocatalyst absorbs photon of energy that is equal or greater than its energy bandgap; thus, an electron (e^-) is ejected from the conduction band into the valence band, generating a vacant reactive hole (h^+) at the conduction band [48].

In the presence of terbuthylazine, a refractory organic contaminant, hydrogen peroxide, coupled with UV-radiation of wavelength 253.7 nm, was employed for the oxidation of arsenite in groundwater. When applied separately, the UV-radiation accounted for 40%, at a dosage of 2000 mJcm^{-2}, while hydrogen peroxide accounted for only 8.0%, at hydrogen peroxide concentration of 15 mg/L in 315 min reaction time. When coupled, the oxidation efficiency increased significantly, even at lower hydrogen peroxide dosage and UV irradiation. Thus, at 5 mg/L of hydrogen peroxide, the conversion of arsenite to arsenate was 54 and 85% at UV dosage of 1200 and 2000 mJcm^{-2}, respectively [49]. This is an indication that the coupling of oxidants can help improve the oxidation of arsenite to arsenate. However, the use of excess oxidant may also be detrimental to the oxidation reaction. Therefore, finding a balance between the concentration of oxidant and arsenite to be converted is an important factor to consider during the oxidation process. The stoichiometric balance was found to be in the ratio 3:2 for $As^{3+}:Fe^{6+}$ [50].

Ag@AgCl was utilized as a photocatalyst in a visible-light-driven heterogeneous photocatalysis reaction for the oxidation of arsenite. The oxidation process was influenced by the initial arsenite solution pH value. The cooperative effect of h^+ and $\cdot O_2^-$ (superoxide) reactive species was responsible for the oxidation of arsenite. However, in acidic solution, the generated $\cdot O_2^-$ reactive radical was consumed by the presence of the abundant H^+ in acidic media (pH 3) to form hydrogen peroxide, a less potent oxidant, when compared with $\cdot O_2^-$. Thus, it decreased the oxidation rate to 39.6%, from 75.55 and 83.13% in neutral (pH 7) and alkaline media (pH 10), respectively [51].

TiO_2 is regarded as a model photocatalyst; however, it suffers severe setbacks, due to the recombination of reactive species and broad bandgap (3.2–3.35 eV), which makes it inactive in the visible region, thus limiting its application in the larger part of the electromagnetic spectrum. To obviate this challenge, element(s) are incorporated into the lattice of the TiO_2 or coupled with other semiconductors [52]. Fe_2O_3 is abundant in the aquifer; thus, it is considered the right candidate for the adsorption of arsenic. When coupled with TiO_2 in the presence of polyaniline, a suitable electron donor and photogenerated hole (h+) carriers, the resulting composite was useful in the oxidation of arsenite to arsenate and subsequent adsorption of the resulting arsenate onto the surface of the composite [53]. The superoxide free radical and photogenerated holes were responsible for the oxidation of arsenite to arsenate. However, the presence of common competitive anions (phosphate, carbonate, chloride, and sulphate) significantly retarded subsequent adsorption of arsenate after the oxidation process. The composite exhibited good adsorption capacity (7–9 mg/g) in a broad pH range (2–10).

6.3.2 Coagulation–Flocculation-Based Strategies

In this process, positively charged molecules are introduced into the reacting medium to destabilize and neutralize the negatively charged arsenate, to form a stable floc. In practice, the removal of arsenic from aqueous system using common coagulants (ferric chloride and alum) is favourable in the absence of phosphate and dissolved organic matter [54]. The overall success of the coagulation process can be accounted for on four fronts: coagulant dose; energy input; flocs size and stability, and sludge generated. Both the chemical coagulation and electrocoagulation have been employed in the removal of arsenic species from solution.

6.3.2.1   Chemical Coagulation

The use of conventional coagulant agents ($AlCl_3$ and $FeCl_3$) has been investigated in the removal of organic and inorganic arsenic from drinking water. Monomethylarsonic acid ($CH_3AsO(OH)_2$ and dimethylarsonic acid ($(CH_3)_2As(OH)_2$) are common organic forms of arsenic that has been detected in drinking water. In a comparative study, the use of $FeCl_3$ was more effective in the removal of methylated arsenic from drinking water than polyaluminum chloride and $AlCl_3$ [55]. The degree of arsenic methylation was the defining factor in the coagulation process. Thus, the trend of the removal efficiency was arsenate > mononmethylarsenate > dimethylarsenate. At 0.6 mmol/L of polyaluminum chloride, $AlCl_3$, and $FeCl_3$, the removal of dimethylarsenate was negligible, 20%, and 50%, respectively. At $FeCl_3$ coagulant dosage of 0.2 mmol/L, the removal of efficiency was approximately 20, 95, and 98% for dimethylarsenate, mononmethylarsenate, and arsenate, respectively. The natural occurrence of dissolved organic carbon in freshwater created a significant challenge during the removal of arsenic from aqueous system. The removal of arsenic from freshwater was impeded with an increasing concentration of dissolved organic carbon [56]. The threshold for dissolved organic carbon, phosphate, and carbonate was 9.22 mg/L, 16 mg/L, and 300 mg/L, respectively, using $FeCl_3$ as coagulant [57].

To avoid the additional cost incurred during the separation of generated flocs from treated water, it is vital to control the size and stability of flocs generated during the coagulation process. For instance, a microfilter (0.22 μm pore size) was required to remove settled flocs in the range of 0.5 and 20 μm from treated water after the coagulation process, using ferric sulphate as a coagulant. The addition of coagulant aid ($CaCO_3$) improved the minimum size of the flocs generated; thus, with a microfilter of pore size 2.5 μm, the removal efficiency of arsenic was 99% [58]. In another study, cactus mucilage suspension was used as a coagulant aid with Fe(III) for the removal of arsenic from aqueous solution [59]. The addition of mucilage suspension increased the rate of coagulation and flocculation of the precipitate.

6.3.2.2   Electrocoagulation

In electrocoagulation, the polluted water is electrified to disturb the surface charge of the contaminant, as the feedwater flows on the metal plates along the electrodes. This causes the anode to lose an electron to the cathode, thus, generating oxidized metal ion (M^{n+}) (Eqs. 6.8 and 6.9), and the oxidation of arsenite to arsenate is initiated by M^{n+}. The in situ generated M^{n+} is also oxidized by aerated oxygen to form hydrous metal oxides, which is used for the coagulation of arsenate. At this point, the distance between the electrode increases until the coagulation process ceases [60].

$$\text{Anode} \quad Fe_{(s)} \rightarrow Fe^{2+}_{(aq)} + 2e^- \tag{6.8}$$

$$Al_{(s)} \rightarrow Al^{3+}_{(aq)} + 3e^- \tag{6.9}$$

$$2H_2O \rightarrow O_2 + O_2 + 4H^+ + 4e \tag{6.10}$$

$$2OH^- \rightarrow O_2 + 2H^+ + 2e^- \tag{6.11}$$

$$\text{Cathode} \quad O_2 + 4H^+ + 4e^- \rightarrow 2H_2O \tag{6.12}$$

$$2H_2O + 2e^- \rightarrow H_2 + 2OH^- \tag{6.13}$$

$$2H^+ + 2e^- \rightarrow H_2 \tag{6.14}$$

In practice, a salty medium is required for electrocoagulation, which is one of the challenges with the management of arsenic in potable water. The arsenic-contaminated freshwater is low in conductivity and contamination level [61]. In the removal of trace amounts of arsenic contamination from water, electrodes (iron and aluminium), post-treatment, pH, and water matrix were varied to have an insight into the optimized condition [62]. The choice of electrodes is vital in the removal of arsenite since iron species generated in situ during electrocoagulation have the potential of oxidizing arsenite to arsenate, which is easier to remove as adsorbed solute or complex from solution.

Arsenite is the more toxic form of arsenic and tends to predominate, due to its less affinity for the substrate. Therefore, oxidation of arsenite to arsenate is advantageous water purification systems. During electrocoagulation, the overall performance of the process is partly dependent on the oxidation of arsenite to arsenate and subsequent sorption/precipitation of arsenate in the solution. The balance between the ratio of coagulant dosage to arsenite enhances the removal. This factor was explained by the charge dosage rate. At low charge dosage rate (1 Coulombs/L/min), dissolved Fe(II) in the solution was minimal with a higher concentration of dissolved oxygen, causing enhanced oxidation of arsenite to arsenate. Higher charge dosage rate (24 Coulombs/L/min) resulted in an opposing effect [60].

Constituents of common freshwater tend to alter the mechanism of electrocoagulation. The in situ formation of hydrous metal oxides is an essential part of electrocoagulation. In a batch system, which utilized iron electrodes for the removal of arsenic, during electrocoagulation, the presence of dissolved SiO_2 caused the overall reaction step to change from precipitation as $FeAsO_{4(s)}$ to adsorption. In this case, the arsenic was adsorbed onto lepidocrocite in the presence of dissolved SiO_2, but formed a new crystal phase when SiO_2 was absent in the solution [63].

The speciation of arsenic in aqueous solution also responds to electrocoagulation differently. This is partly because of the high affinity of arsenate to conventional coagulants. In a pilot study that involved the removal of arsenite and arsenate using an iron electrode, at initial concentration of 100 µg/L and pH 7, 15 and 90 min were required to achieve <1 µg/L residual arsenate and arsenite, respectively [63].

In practice, the flow rate of the feedwater is a vital parameter that is considered during electrocoagulation. The concentration of residual arsenic in feedwater is expected to decrease over-time during treatment. Thus, at the early stage of electrocoagulation, the initial concentration of arsenic is higher, with a limited amount of in situ generated hydrous metal oxides available for adsorption or precipitation. Therefore, at higher flow rate, arsenic in feedwater can elude treatment because of reduced contact time and limited available sites of reaction with in situ generated hydrous metal oxides. In a pilot study, the residual arsenic at feed flow rate of 0.875, 1.75, and 3.5 L/min was approximately 20, 35, and 50 g/L in 1 min, respectively [64].

6.3.3   Membrane Technology

Membrane filtration is a robust treatment method, employed in the removal of soluble and insoluble arsenic from potable water. It is embedded in a full-scale drinking water plant or as decentralized onsite treatment [65]. In membrane filtration, contaminants, usually solute, (retentate) are excluded or trapped as the feed, under external force (gravity, vacuum, hydraulic pressure, centrifugal, and electrostatic force) passed through a permeable membrane filter. The permeable membrane may be fabricated from pure or blend of polymers, ceramics, and metallic materials. The pore of the membrane and the flow of the feed are the most significant consideration during filtration. Thus, the technology is classified based on the mean pore size of the membrane and the flow of the feed. In pore size classifications, pore sizes between the range of 10 to 100, 0.1 to 10.0, 0.001 to 0.1, ~0.001 μm, less than 0.001 μm, are traditional filtration, microfiltration, ultrafiltration, nanofiltration, and reverse osmosis, respectively. The mechanism of arsenic separation could be sieving/size exclusion (steric exclusion) and non-sieving/charge exclusion (Donnan and dielectric exclusion, and adsorptive exclusion). The sieving mechanism is involved when contaminants and the membrane filter are neutrally charged. During non-sieving mechanism, charged pollutants are rejected due to adsorption within the pore or interaction with the charged membrane [66].

6.3.3.1   Microfiltration

Generally, microfiltration technology is used when a particle size greater than 0.1 μm is to be retained. Thus, microfiltration is not used singly; rather, it is combined with other treatment methods, as hybrid technology, which can transform dissolved arsenic to its insoluble form [67–69]. In a comparative study, on the performance of micro and nanofilters on the removal of arsenic from aqueous solution, the removal efficiency of 57 and 81% were associated with nanofiltration, for arsenite and arsenate removal, respectively. Microfilter membrane performed poorly with 37 and 40%, for arsenite and arsenate, respectively. The large pore size (0.4 μm) of the microfilter allowed the passage of arsenite and arsenate [70]. However, with the addition of nano-zerovalent iron, using microfiltration technology, the percentage exclusion of arsenite and arsenate improved to 84% and 90%, respectively. The success of the process depends on the quality of raw water [71]. In fact, Brandhuber et al. 1998 concluded that microfiltration may be recommended, only if arsenic is present in the particulate form [72]. A similar study showed that the efficiency of microfiltration in the retention of arsenate was only comparable with ultrafiltration after pretreatment [73].

6.3.3.2 Ultrafiltration

In comparison with microfiltration, ultrafiltration can solely be employed for the removal of arsenic in potable water. The cost of operation can be further minimized depending on the quality of the raw water. Ultrafiltration offers an increased rejection of arsenic due to the increased compactness of the membrane. Polyphenylsulphone is a traditional polymer that has been employed as a membrane filter for water treatment purposes. However, it suffers many shortcomings; susceptibility to fouling, hydrophobicity, water permeability, and pore size that exceeds the size of arsenic ions. To obviate these shortcomings, ceramic material (ZrO_2) and cellulose were incorporated [74]. The cellulose offers the negative surface functionalities (-COOH and -NH), while the ZrO_2 imbed porosity, membrane stability at pH 4.3–9.1, antifouling, hydrophilicity, and clean water permeability and improved membrane regeneration. The unmodified polyphenylsulphone membrane retained 18.89% of dissolved arsenate, but after the modification, 87.24% of the arsenate was retained, at 1 ppm initial arsenate concentration [74]. The membrane surface charge (carboxylic and sulphonate groups) and water quality were the most important factors that affected the retention of arsenate. Increasing the initial arsenate concentration, ionic strength and the presence of divalent competing cations were detrimental to the retention of arsenate [75]. Other studies have also revealed the significance of surface charge. In a pilot ultrafiltration study, aerating the membrane system caused the oxidation of arsenite to arsenate, and the resulting arsenate was rejected due to the electrostatic repulsion between the surface of the membrane (amino-functionalized coffee cellulose) and arsenate [76].

Carbon quantum dot has received significant attention in drinking water treatment due to its abundance and non-toxicity. Na^+/carbon quantum dot membrane was synthesized via citric acid pyrolysis, followed by alkalization and polymerization with piperazine. The resulting membrane exhibited high (99.5%) rejection of arsenate due to the small pore and narrow pore size distribution and repulsion of arsenate at pH 8.0 [77].

When a membrane with a fixed charge is employed, Donna exclusion can be relied on as one of the mechanisms for the removal of arsenate (H_3AsO_4) from natural water because of the speciation. Arsenate exists as neutral specie (H_3AsO_4) at extreme acidity (pH 1.0–2.2), and deprotonation starts at pH 2.2, 7.08, and 11.5 to negative charge oxyanions, $H_2AsO_4^-$, $HAsO_4^{2-}$, AsO_3^{3-}, respectively. Under the similar condition, the retention of arsenite is governed by a different mechanism (size exclusion), because arsenite exists as neutral oxyanions (H_3AsO_3) between pH range 1.0–9.22, which dissociate to form $H_2AsO_3^-$, and $HAsO_3^{2-}$, between pH 9.22 to 12.30, respectively [78]. Hydrous tri-metal oxide (iron/nickel/manganese), in ratio 3:2:1, was fused into polysulphone, and the resulting membrane was used for the sieving of arsenite from potable water [79]. The loading of the tri-metal oxides caused an increase in specific surface area, density, porosity, and permeability. At 0.5 bar pressure, the synthesized membrane achieved the removal of arsenite below <0.01 mg/L, for 3665 mL of permeate collected. The neutral arsenite (H_3AsO_3) was formed via inner-sphere complex with the membrane, by the substitution of - OH group on the membrane with arsenite ion (Eq. 6.15), as the permeate is excluded from the feed.

$$Membrane - OH + H_3AsO_3 \leftrightarrow Membrane - H_2AsO_3 + H_2O \qquad (6.15)$$

6.3.3.3   Nanofiltration

Nanofiltration is a technology that has found diverse applications in water treatment. It is best used when arsenic is present in trace amount because it is specifically sensitive to pressure, concentration, temperature, pH and interfering ions, and susceptible to fouling. The applied pressure is vital because it is the force used to drive permeate through the membrane, but it does not translate to increased arsenic rejection. Temperature is also a factor to be considered in nanofiltration technology, because increasing temperature causes the entropy of the system to increase, thus an increase in the rate of arsenic diffusion. Therefore, more arsenic ions are transported through as permeate.

In a comparative study, two commercially available membranes, N30F and NF-90, with a lower molecular weight cut-off of ~400 Da and ~200 Da, respectively, were employed for the removal of arsenate [80]. The rejection efficiency of NF-90 and N30F for arsenate was 94% and 78%, respectively. The more compact NF-90 retained more arsenate than the N30F nano-membrane. Increasing pressure from 2 to 12 bar, and temperature from 15 to 40 °C, enhanced the permeate flux but did not improve the rejection of arsenate. In a comprehensive study, negatively charged GE-HL nanofilter was employed for the removal of arsenate [81]. The nanofilter reached optimum rejection (93%), at arsenate initial concentration of 122.54 µg/L in 40 min, with a stability of 6 h. The effect of initial concentration was insignificant at an initial concentration below ~70 µg/L, as arsenate was completely retained on the nano-membrane. Above the threshold, increasing arsenate initial concentration caused a reduction in the rejection of arsenate in the permeate. The Donna exclusion, which prevailed at neutral pH, was beneficial to the rejection of arsenate. Thus, the rejection efficiency increased from 38.1 to 96.1%, with increasing pH value from pH 3.0 to 6.7, and a further reduction by 17.1% at pH 9.9. The effect of monovalent ions (Na^+ and Cl^-) was insignificant; however, divalent ions (SO_4^{2-} and Ca^{2+}) caused significant reduction in the sieving of arsenate from the aqueous system.

The rejection of arsenic ion, due to the difference in pore size, is not sufficient to describe the mechanism of the removal of arsenic ion from potable water. In a comparative study, using four different nano-membrane for the removal of arsenate in potable water concentration, the nano-membrane (NF90) with the largest pore size (0.463 nm), though characterized by lower water permeability, was most effective (91.1%) for the sieving of arsenate from potable water. The nano-membrane filter (M#1) of pore size 0.430 nm had rejection efficiency of 86.2%. The superior rejection of NF90 was attributed to the surface charge of the membrane. The zeta potential of NF90 and M#1 nano-membrane was −57.8 and −54.7, respectively.

The sorption capabilities of potential membranous materials have been explored to improve the overall performance, which includes selectivity, rejection, and fouling. Therefore, adsorbent that shows high adsorption capacity towards arsenic oxyanions has been immobilized on membranous materials to improve its tendencies for the removal of arsenic oxyanions from potable water. Oxides of iron have found tremendous application in the removal of arsenic, due to their affinity towards arsenic oxyanions. Fe_3O_4 microsphere was fused to polyethersulphone membrane via the phase inversion technique. The rejection of arsenate by adsorption occurred as arsenate reacts with Fe_3O_4 microsphere, within the pore of the membrane to form a spherical complex [82]. Zinc oxide was incorporated into the matrix of cellulose-acetate, and the nano-membrane obtained was operated at near-neutral pH for the sieving of neutral arsenite, at an initial concentration of 1000 mg/L [83]. The incorporation of ZnO into cellulose-acetate framework increased the rejection of arsenite from 40.32 to 61.44% and improved the water flux by 20% and the membrane tensile strength. The bare cellulose-acetate ruptured at a pressure of 5 bar, whereas the ZnO/cellulose-acetate membrane was still operational above this pressure.

In nanofiltration, potable water essential minerals (Ca^{2+}, Mg^{2+}) are also prone to be sieved; thus, re-mineralization of permeate may be required. Graphene oxide/polyamide nanocomposite membrane was used for the removal of arsenate from potable water [84]. The membrane showed high rejection (85%) of arsenate from potable water, at low operating pressure (5 kgf/cm^2). Increasing the pressure to 15 kgf/cm^2 resulted in 98.5% rejection of arsenate. In addition to the arsenic rejection, 90% and 60% of Ca^{2+} and Mg^{2+} in the initial feed were detected in the permeate. This is an indication that the nano-membrane filter also captured high proportion of the essential minerals.

6.3.4   Adsorption

Adsorption is a robust method used for the transfer of arsenic oxyanions from contaminated potable water onto the surface of a solid phase. The mechanism of adsorption may proceed via ion-exchange, surface complexation, or electrostatic attraction. Many process conditions can influence the adsorption of arsenic oxyanions onto a solid surface. In a particular system, the affinity of the predominant arsenic oxyanions towards the solid surface is mostly governed by the adsorbent physicochemical properties. Some of these characteristics include specific surface area, pore volume, amorphous, particle size, and surface functionality. For example, the adsorption capacity of graphene oxide/iron nanoparticles composite [85] and graphene oxide/magnetite composite [86] were 306.1 and 431.41, and 85 and 38 mg/g, for arsenite and arsenate, respectively. The difference in adsorption capacity was mainly attributed to the difference in the adsorbent. Selected adsorbents that have been used for the management of arsenic, at specified aqua system conditions, are presented in Table 6.1.

Some adsorbent that has been employed for the removal of arsenic oxyanions includes naturally occurring minerals, zeolite, alumina, iron oxide, and soil. The removal efficiency of naturally occurring iron oxides minerals was tested for the removal of arsenate in drinking water [105]. XRD analysis showed that these minerals were multiphase. For example, magnetite contained traces of haematite, haematite-contained minor magnetite, while goethite was near pure. The phase of laterite was significantly goethite, talc, quartz, and gibbsite. The adsorption study showed that the removal of arsenate was a function of the specific surface area. Arsenate ions are strongly bonded to the surface of the iron mineral adsorbent via inner-sphere complexes or precipitation. The order of the specific surface area was laterite, goethite, magnetite, and haematite are 81.2 >12.1 >6.58 >3.77 m^2/g, respectively. The multicomponent nature of the laterite improved the specific surface area, thus the adsorption capacity. The removal of arsenite using nano-zerovalent iron was spontaneous, with increasing adsorbent dosage. The adsorption site on the nano-zerovalent iron was identified as the amorphous Fe(II)/(III) and magnetite. The presence of silica and phosphate completely blocked the adsorption of arsenite [106].

Among the traditional adsorbents, aluminium oxide/hydroxides are common adsorbents that have been employed for the removal of arsenic oxyanions. Amorphous aluminium hydroxides are unstable at pH <4.3 and >8.5, and 6% soluble at pH values that ranged between the values of 4.3–8.5. The adsorption may occur either below or above the isoelectric point pH (i.e. 8.5) [107]. The isoelectric point is the pH at which a molecule carries no net electrical charge or is electrically neutral in the statistical mean. In this study, below the isoelectric point, the mechanism of adsorption was controlled by the coulombic force of attraction, and above the isoelectric point, the adsorption mechanism occurred via chemical adsorption, with increasing initial arsenate concentration. To achieve arsenate residual concentration of < 10 μg/L, which was attained with the use of Fe_2O_3, more dosage (approximately 6) of Al_2O_3 was required. The low adsorption capacity was attributed to the low specific surface area (0.55 m^2/g) of Al_2O_3, as compared with 5.05 m^2/g for Fe_2O_3 [108]. Since the adsorption efficiency of non-porous adsorbent is controlled by the specific surface area, traditional adsorbents exhibit lower adsorption capacity in the removal of arsenic oxyanions. The uptake of arsenite in neutral pH, using zirconia nanoparticles, followed similarly characteristics; low adsorption capacity (1.85 mmol/g) and high equilibrium time, 48 h [109].

Table 6.1 Performance efficiencies of selected adsorbents for the management of arsenic contamination in drinking water

Adsorbent	Arsenic initial concentration (mg/L)	Adsorbent dosage (g/L)	Adsorption capacity (mg/g)		pH	Equilibrium time (mins)	References
			Arsenite	Arsenate			
Fe$_3$O$_4$-graphene oxide	550	0.1	85	38	7	1440	[86]
CeO$_2$-graphene oxide	200	0.5	185	212	7	240	[87]
Reduced graphene oxide-Fe$_3$O$_4$-TiO$_2$	10	0.2	147.05	–	7	120	[88]
Nano-zerovalent iron/activated carbon	2	1.0	18.2	–	6.5	4320	[89]
Graphene oxide-iron	550	0.25	306	431	7	1500	[85]
Mesoporous activated carbon/iron/calcium	1.00	0.05	3.38	2.80	7	180	[90]
Chitosan-magnetic graphene oxide	10.0	5	45	–	7.3	240	[91]
Activated silica-based catalytic media	0.06	20	0.318	0.237	5–8.5	1440	[92]
Clay Iron oxide/pillared clay Manganese oxide/pillared	2	0.1	0.004 0.017 0.025	0.004 0.025 0.026	6	120	[93]
Cellulose nanocrystal/Iron oxide	50	0.05	13.87	–	7	480	[94]
Activated charcoal/zirconium magnesium	30	0.4	132.28	95.6	6–10	3600	[95]

(continued)

Table 6.1 (continued)

Adsorbent	Arsenic initial concentration (mg/L)	Adsorbent dosage (g/L)	Adsorption capacity (mg/g)		pH	Equilibrium time (mins)	References
			Arsenite	Arsenate			
Starch/Iron-magnesium binary oxide	30	0.4	161.29	–	7	3600	[96]
Starch/FeMnOx-reduced graphene oxide	7.0	0.2	78.74	55.56	7	3600	[97]
Chitosan/Iron	5.0	1.0	–	120.77	7	3600	[98]
Eggshell	7.0	1.0	11.69	8.43	7	120	[99]
Java plum			4.63	4.62			
Water chestnut shell			9.61	7.73			
Corn cob			4.33	5.71			
Tea waste			7.36	4.92			
Pomegranate peel			5.57	4.50			
Chitosan/Clay/magnetite	336	5	–	5.9	5	600	[100]
Fe_2O_4	10	0.2	49.3	44.1	3–7	1600	[101]
$MnFe_2O_4$			93.8	90.4			
$CoFe_2O_4$			100.3	73.8			
Perlite	20	0.05	–	0.0025	3–6	400	[102]
γ-Fe_2O_3				4.64			
α-MnO_2				7.09			
ZnO_xS_{1-x}	2	0.05	299.4	299.4	7	15	[103]
Rice husk/iron oxide	10	2.5		82	6	60	[104]

Though the above-mentioned adsorbents have shown high potential in the removal of arsenic oxyanions from drinking water, the application of these conventional adsorbents has been limited, mostly because of the powdered nature. The finely powdered nature of these conventional adsorbents easily results in agglomeration [110], therefore, limiting accessible active sites and specific surface area during the adsorption process. It also retards the recovery of used adsorbents, thereby increasing cost and enhanced leaching, which may result in secondary contamination of treated drinking water. While some of these shortcomings may be addressed by adjusting experimental design, and process variables during adsorbent synthesizes, finely powdered adsorbents are inapplicable in fixed-bed water treatment systems. Therefore, to increase the application of conventional adsorbents in the removal of arsenic oxyanions, granular adsorbent has been employed [111]. Granular TiO_2 was used for the removal of arsenate and arsenite from the USA and Bangladesh groundwater. The removal efficiency varied at different region. In Bangladesh groundwater, the removal of arsenite and arsenate was similar (40 mg/g); however, in the US groundwater, the removal of arsenate was higher (41.4 mg/g), than the removal of arsenite (32.4 mg/g). The adsorption of arsenate and arsenite was controlled by the surface charge of TiO_2 within the solution pH value range. In basic solution, the removal of arsenite was higher, while in the acidic solution, the removal of arsenate was higher [112].

In response to the numerous shortcomings in adsorption technology, materials are engineered to improve the overall process. As mentioned in the previous section, powdered adsorbents can be processed to granulated adsorbent to overcoming some reaction shortcomings. Similarly, powdered adsorbents may be loaded on a solid support or fused to mechanical stable materials to improve the stability, reactive sites, specific surface area, recovery, and applications. Iron (elemental or oxides) is a good candidate for the removal of arsenic oxyanions from potable water. However, it suffers some setbacks, due to the rapid agglomeration and oxidation. Graphene oxide/iron nanoparticle composite was synthesized via the sol-gel method, for the removal of arsenate and arsenite from water. The composite achieved a high removal capacity (>98%) for both oxyanions at neutral pH. Bare graphene oxide adsorption of arsenic oxyanions was negligible. The contribution of graphene oxide to the overall surface adsorption of arsenate and arsenite at 5 mg/L initial concentration was associated with the effective dispersion of the composite, rather than the active surface adsorption [85]. The occurrence of arsenic in potable water at low concentration is of particular interest because it present long-time chronic toxicity [113]. Nano-zerovalent iron/reduced graphite oxide composite was used for the removal of arsenite and arsenate from drinking water at low concentration. At neutral pH, the removal of arsenate and arsenite was ~16 mg/g and ~17 mg/g, respectively. The difference in adsorption capacity was attributed to the mechanism of adsorption [114].

The performance of graphene oxide/magnetite and reduced graphene oxide/magnetite composite was compared for the removal of arsenic oxyanions from water. Graphene oxide/magnetite showed more affinity for arsenite and arsenate than reduced graphene oxide/magnetite. The affinity was a function of time, magnetite content, surface charge, and pH value. The removal of arsenite and arsenate followed surface complexation and electrostatic attraction mechanism, respectively [86].

Activated carbon and nano-zerovalent iron are typical adsorbents that have been used for heavy metal water remediation. These separate adsorbents were fused by dispersing nano-zerovalent iron into the activated carbon matrix and employed for the removal of arsenite and arsenate from water [89]. Arsenite showed more affinity (18.2 mg/g) to the resulting adsorbent than arsenate (12.0 mg/g). The presence of phosphate and silicate in the water was detrimental to the removal of the oxyanions at neutral pH. Among the competing ions, phosphate and silicate were the most important ions that adversely affected the removal of arsenic oxyanions in potable water. Activated charcoal was used as support for $ZrO(OH)_2$ and MnO_2 composite for the removal arsenite in the presence of competing anions. The metal oxides (i.e. MnO_2 and $ZrO(OH)_2$) and the activated charcoal in the composite functioned as the oxidizing agent, and adsorbent for the oxidized arsenate, respectively [95]. At 30 mg/L concentration of coexisting ions (i.e. Cl^-, F^-, NO_3^-, SiO_3^{2-}, PO_4^{3-}, SO_4^{2-}, HCO_3^-), the impacts on the removal of arsenate were negligible, but the presence of phosphate impacted negatively on the removal of arsenite due to the similarities in the chemistry of the two species.

The hydroxyl group, generated from the hydration of metal oxides in an aqueous system, plays a significant role in the removal of arsenic oxyanions from water. The hydroxyl group acts as acid sites for Lewis and Bronsted acid sites interaction [115], and these reactive sites may be improved by increasing the number of acid sites on the metal oxides adsorbents. The adsorption capacity of TiO_2 was compared with $Ti(OH)_4$ and $Ti(OH)_4$ bearing sulphate functional group ($Ti(OH)_4/SO_4^{2-}$), modified by wetness impregnation protocol, for the removal of arsenite from water at neutral pH. The impregnation of the sulphate group on titanium hydroxide increased the Lewis acid sites [116]. The arsenite chemically adsorbed at the Lewis acid site; thus, the adsorption efficiency was 60, 64 and 80%, for TiO_2, $Ti(OH)_4$ and ($Ti(OH)_4/SO_4^{2-}$), respectively.

The quantity or mole of the participating elements in a composite also influences the adsorption capacity of an adsorbent. Goethite was modified by varying the mole of Mn in goethite, and Birnessite was used as the Mn source [117]. The incorporation of excess Mn in goethite altered the texture, electrical and chemical properties of the adsorption, and slightly improved the removal efficiency of arsenate at neutral pH, from 96.6 to 99%, at 0.5 mg/L initial concentration. Further studies revealed that the role of Mn and Fe in bimetallic composite included the oxidation of arsenite to arsenate by Mn and subsequent adsorption of the resulting arsenate on the surface of Fe, respectively [96]. Thus, Fe/Mn ratio between 1.0 and 2.0 was found to be the ideal proportion in Fe/Mn bimetallic adsorbent.

Bismuth and aluminium oxides usually suffer from low adsorption capacity and slow reaction time in the removal of arsenic oxyanions. Thus, bismuth was fused into aluminium oxide via calcination to alleviate these shortcomings. The removal efficiency of the resulting adsorbent increased from 38.4 to 55.7% for bismuth oxide and aluminium oxide, respectively, to 89.2% for bismuth/aluminium oxide composite,

[118]. The impregnation of bismuth into aluminium oxide caused a reduction in the specific surface but an increase in the total pore volume and average particle size.

6.4 Conclusion

Access to clean and safe potable is vital in achieving the United Nation Sustainable Development goals. While much energy has been expended on achieving the set objectives, arsenic oxyanions in potable water seem to be one of the numerous challenges that require more attention. Arsenic enters the water via occupational and natural source. Elevated concentration, above the threshold (10 μm/L), has been detected in the feedwater and treated water. The existing strategies for the removal of arsenic oxyanions from drinking water have been promising, and if properly optimized, can expunge arsenic oxyanions from potable water. One of the most encouraging practices is the oxidation of the more recalcitrant arsenite to arsenate, before removal, by adsorption, coagulation, or filtration. The engineering of novel reactive materials is highly promising for the effective arsenic oxyanions removal from potable water. The promising engineered suitable materials are characterized by improved specific surface area, pore size and volume, amorphosity, surface functionality, stability, non-toxicity, magnetism, and low cost.

References

1. Gwaltney-Brant SM (2013) Heavy metals. In: Haschek WM, Rousseaux CG, Wallig MABT-H, RH, TP, Third E, (eds) Haschek and rousseaux's handbook of toxicologic pathology. Academic Press, Boston, pp 1315–1347. https://doi.org/10.1016/B978-0-12-415759-0.000 41-8
2. Kelepertsis A, Alexakis D, Skordas K (2006) Arsenic, antimony and other toxic elements in the drinking water of eastern thessaly in greece and its possible effects on human health. Environ Geol 50(1):76–84. https://doi.org/10.1007/s00254-006-0188-2
3. Muehe EM, Wang T, Kerl CF, Planer-Friedrich B, Fendorf S (2019) Rice production threatened by coupled stresses of climate and soil arsenic. Nat Commun 10(1):4985. https://doi.org/10.1038/s41467-019-12946-4
4. Zhang XY, Zhou MY, Li LL, Jiang YJ, Zou XT (2017) Effects of arsenic supplementation in feed on laying performance, arsenic retention of eggs and organs, biochemical indices and endocrine hormones. Br Poult Sci 58(1):63–68. https://doi.org/10.1080/00071668.2016.121 6945
5. He T, Ohgami N, Li X, Yajima I, Negishi-Oshino R, Kato Y, Ohgami K, Xu H, Ahsan N, Akhand AA, Kato M (2019) Hearing loss in humans drinking tube well water with high levels of iron in arsenic-polluted area. Sci Rep 9(1):9028. https://doi.org/10.1038/s41598-019-455 24-1
6. Banerjee M, Banerjee N, Bhattacharjee P, Mondal D, Lythgoe PR, Martínez M, Pan J, Polya DA, Giri AK (2013) High arsenic in rice is associated with elevated genotoxic effects in humans. Sci Rep 3(1):2195. https://doi.org/10.1038/srep02195

7. Burgess JL, Kurzius-Spencer M, Poplin GS, Littau SR, Kopplin MJ, Stürup S, Boitano S, Clark Lantz R (2014) Environmental arsenic exposure, selenium and sputum alpha-1 antitrypsin. J Eposure Sci Environ Epidemiol 24(2):150–155. https://doi.org/10.1038/jes.2013.35

8. Kalman DA, Dills RL, Steinmaus C, Yunus M, Khan AF, Prodhan MM, Yuan Y, Smith AH (2014) Occurrence of trivalent monomethyl arsenic and other urinary arsenic species in a highly exposed juvenile population in Bangladesh. J Eposure Sci Environ Epidemiol 24(2):113–120. https://doi.org/10.1038/jes.2013.14

9. Burgess JL, Kurzius-Spencer M, O'Rourke MK, Littau SR, Roberge J, Meza-Montenegro MM, Gutiérrez-Millán LE, Harris RB (2013) Environmental arsenic exposure and serum matrix metalloproteinase-9. J Eposure Sci Environ Epidemiol 23(2):163–169. https://doi.org/10.1038/jes.2012.107

10. Zhong M, Huang Z, Wang L, Lin Z, Cao Z, Li X, Zhang F, Wang H, Li Y, Ma X (2018) Malignant transformation of human bronchial epithelial cells induced by arsenic through STAT3/MiR-301a/SMAD4 Loop. Sci Rep 8(1):13291. https://doi.org/10.1038/s41598-018-31516-0

11. Hysong TA, Burgess JL, Cebrián Garcia ME, O'Rourke MK (2003) House dust and inorganic urinary arsenic in two arizona mining towns. J Expo Anal Environ Epidemiol 13(3):211–218. https://doi.org/10.1038/sj.jea.7500272

12. Hoen AG, Madan JC, Li Z, Coker M, Lundgren SN, Morrison HG, Palys T, Jackson BP, Sogin ML, Cottingham KL, Karagas MR (2018) Sex-specific associations of infants' gut microbiome with arsenic exposure in a US population. Sci Rep 8(1):12627. https://doi.org/10.1038/s41598-018-30581-9

13. Rosen MR, Binda G, Archer C, Pozzi A, Michetti AM, Noble PJ (2018) Mechanisms of earthquake-induced chemical and fluid transport to carbonate groundwater springs after earthquakes. Water Resour Res 54(8):5225–5244. https://doi.org/10.1029/2017WR022097

14. McGrory ER, Brown C, Bargary N, Williams NH, Mannix A, Zhang C, Henry T, Daly E, Nicholas S, Petrunic BM, Lee M, Morrison L (2017) Arsenic contamination of drinking water in Ireland: a spatial analysis of occurrence and potential risk. Sci Total Environ 579:1863–1875. https://doi.org/10.1016/j.scitotenv.2016.11.171

15. Wang S, Mulligan CN (2006) Occurrence of arsenic contamination in canada: sources, behavior and distribution. Sci Total Environ 366(2–3):701–721. https://doi.org/10.1016/j.scitotenv.2005.09.005

16. Taylor VF, Li Z, Sayarath V, Palys TJ, Morse KR, Scholz-Bright RA, Karagas MR (2017) Distinct arsenic metabolites following seaweed consumption in humans. Sci Rep 7(1):3920. https://doi.org/10.1038/s41598-017-03883-7

17. Rivera-Núñez Z, Meliker JR, Meeker JD, Slotnick MJ, Nriagu JO (2012) Urinary arsenic species, toenail arsenic, and arsenic intake estimates in a michigan population with low levels of arsenic in drinking water. J Eposure Sci Environ Epidemiol 22(2):182–190. https://doi.org/10.1038/jes.2011.27

18. Yu ZM, Dummer TJB, Adams A, Murimboh JD, Parker L (2014) Relationship between drinking water and toenail arsenic concentrations among a cohort of nova scotians. J Eposure Sci Environ Epidemiol 24(2):135–144. https://doi.org/10.1038/jes.2013.88

19. Davis MA, Li Z, Gilbert-Diamond D, Mackenzie TA, Cottingham KL, Jackson BP, Lee JS, Baker ER, Marsit CJ, Karagas MR (2014) Infant toenails as a biomarker of in utero arsenic exposure. J Eposure Sci Environ Epidemiol 24(5):467–473. https://doi.org/10.1038/jes.2014.38

20. Punshon T, Davis MA, Marsit CJ, Theiler SK, Baker ER, Jackson BP, Conway DC, Karagas MR (2015) Placental arsenic concentrations in relation to both maternal and infant biomarkers of exposure in a US cohort. J Eposure Sci Environ Epidemiol 25(6):599–603. https://doi.org/10.1038/jes.2015.16

21. González de las Torres AI, Giráldez I, Martínez F, Palencia P, Corns WT, Sánchez-Rodas, D (2020) Arsenic accumulation and speciation in strawberry plants exposed to inorganic arsenic enriched irrigation. Food Chem 315:126215. https://doi.org/10.1016/j.foodchem.2020.126215

22. Freire BM, Santos da VS, de Neves PCF.; Oliveira Souza Reis JM.; de Souza SS, Barbosa F, Batista BL (2020) Elemental chemical composition and as speciation in rice varieties selected for biofortification. Anal Methods (2020). https://doi.org/10.1039/d0ay00294a
23. Lyu R, Gao Z, Li D, Yang Z, Zhang T (2020) Bioaccessibility of arsenic from gastropod along the Xiangjiang river: assessing human health risks using an in vitro digestion model. Ecotoxicol Environ Saf 193:110334. https://doi.org/10.1016/j.ecoenv.2020.110334
24. Keegan T, Hong B, Thornton I, Farago M, Jakubis P, Jakubis M, Pesch B, Ranft U, Nieuwen-huijsen MJ (2002) Assessment of environmental arsenic levels in prievidza district. J Eposure Sci Environ Epidemiol 12(3):179–185. https://doi.org/10.1038/sj.jea.7500216
25. Brewer R, Belzer W (2001) Assessment of metal concentrations in atmospheric particles from Burnaby Lake, British Columbia, CAN. Atmos Environ 35(30):5223–5233. https://doi.org/10.1016/S1352-2310(01)00343-0
26. Goswami C, Majumder A, Misra AK, Bandyopadhyay K (2014) Arsenic uptake by lemna minor in hydroponic system. Int J Phytorem 16(12):1221–1227. https://doi.org/10.1080/15226514.2013.821452
27. Allevato E, Stazi SR, Marabottini R, D'Annibale A (2019) Mechanisms of arsenic assimilation by plants and countermeasures to attenuate its accumulation in crops other than rice. Ecotoxicol Environ Saf 185:109701. https://doi.org/10.1016/j.ecoenv.2019.109701
28. Maguffin SC, Kirk MF, Daigle AR, Hinkle SR, Jin Q (2015) Substantial contribution of biomethylation to aquifer arsenic cycling. Nat Geosci 8(4):290–293. https://doi.org/10.1038/ngeo2383
29. Cheyns K, Waegeneers N, Van de Wiele T, Ruttens A (2017) Arsenic release from foodstuffs upon food preparation. J Agric Food Chem 65(11):2443–2453. https://doi.org/10.1021/acs.jafc.6b05721
30. Knowles FC, Benson AA (1983) The biochemistry of arsenic. Trends Biochem Sci 8(5):178–180. https://doi.org/10.1016/0968-0004(83)90168-8
31. Mukhopadhyay R, Rosen BP, Phung LT, Silver S (2002) Microbial arsenic: from geocycles to genes and enzymes. FEMS Microbiol Rev 26(3):311–325. https://doi.org/10.1111/j.1574-6976.2002.tb00617.x
32. Rasmussen PE, Subramanian KS, Jessiman BJ (2001) A multi-element profile of house dust in relation to exterior dust and soils in the City of Ottawa, Canada. Sci Total Environ 267(1–3):125–140. https://doi.org/10.1016/S0048-9697(00)00775-0
33. Karczewska A, Krysiak A, Mokrzycka D, Jezierski P, Szopka K (2013) Arsenic distribution in soils of a former as mining area and processing. Pol J Environ Stud 22(1):175–181
34. Neumann RB, Ashfaque KN, Badruzzaman ABM, Ashraf Ali M, Shoemaker JK, Harvey CF (2010) Anthropogenic influences on groundwater arsenic concentrations in Bangladesh. Nat Geosci 3(1):46–52. https://doi.org/10.1038/ngeo685
35. Litter MI, Ingallinella AM, Olmos V, Savio M, Difeo G, Botto L, Farfán Torres EM, Taylor S, Frangie S, Herkovits J, Schalamuk I, González MJ, Berardozzi E, García Einschlag FS, Bhattacharya P, Ahmad A (2019) Arsenic in Argentina: occurrence, human health, legislation and determination. Sci Total Environ 676:756–766. https://doi.org/10.1016/j.scitotenv.2019.04.262
36. Huq ME, Fahad S, Shao Z, Sarven MS, Khan IA, Alam M, Saeed M, Ullah H, Adnan M, Saud S, Cheng Q, Ali S, Wahid F, Zamin M, Raza MA, Saeed B, Riaz M, Khan WU (2020) Arsenic in a groundwater environment in Bangladesh: occurrence and mobilization. J Environ Manage 262:110318. https://doi.org/10.1016/j.jenvman.2020.110318
37. Erickson ML, Malenda HF, Berquist EC, Ayotte JD (2019) Arsenic concentrations after drinking water well installation: time-varying effects on arsenic mobilization. Sci Total Environ 678:681–691. https://doi.org/10.1016/j.scitotenv.2019.04.362
38. Delgado Quezada V, Altamirano Espinoza M, Bundschuh J (2020) Arsenic in geoenvironments of nicaragua: exposure, health effects, mitigation and future needs. Sci Total Environ 716:136527. https://doi.org/10.1016/j.scitotenv.2020.136527
39. Mariño EE, Ávila GT, Bhattacharya P, Schulz CJ (2020) The occurrence of arsenic and other trace elements in groundwaters of the southwestern chaco-pampean plain, Argentina. J S Am Earth Sci 100:102547. https://doi.org/10.1016/j.jsames.2020.102547

40. Lin Z, Puls RW (2000) Adsorption, desorption and oxidation of arsenic affected by clay minerals and aging process. Environ Geol 39(7):753–759. https://doi.org/10.1007/s00254 0050490
41. Zuo Y, Hoigné J (1993) Evidence for photochemical formation of H_2O_2 and oxidation of SO_2 in authentic fog water. Science 260(5104):71–73. https://doi.org/10.1126/science.260. 5104.71
42. Ma Y, Qin Y, Zheng B, Zhang L, Zhao Y (2016) Arsenic release from the abiotic oxidation of arsenopyrite under the impact of waterborne H_2O_2: A SEM and XPS study. Environ Sci Pollut Res 23(2):1381–1390. https://doi.org/10.1007/s11356-015-5166-3
43. Moore KW, Huck PM, Siverns S (2008) Arsenic removal using oxidative media and nanofiltration. J Am Water Works Association 100(12):74–83. https://doi.org/10.1002/j.1551-8833. 2008.tb09800.x
44. Wang J, Wang W, Bai Y, Fu Y, Xie F, Xie G (2020) Study on pre-oxidation of a high-arsenic and high-sulfur refractory gold concentrate with potassium permanganate and hydrogen peroxide. Trans Indian Inst Met 73(3):577–586. https://doi.org/10.1007/s12666-020-01863-6
45. Fang G, Gao J, Liu C, Dionysiou DD, Wang Y, Zhou D (2014) Key role of persistent free radicals in hydrogen peroxide activation by biochar: implications to organic contaminant degradation. Environ Sci Technol 48(3):1902–1910. https://doi.org/10.1021/es4048126
46. Zhong D, Jiang Y, Zhao Z, Wang L, Chen J, Ren S, Liu Z, Zhang Y, Tsang DCW, Crittenden JC (2019) PH dependence of arsenic oxidation by rice-husk-derived biochar: roles of redox-active moieties. Environ Sci Technol 53(15):9034–9044. https://doi.org/10.1021/acs.est.9b0 0756
47. Lin L, Song Z, Huang Y, Khan ZH, Qiu W (2019) Removal and oxidation of arsenic from aqueous solution by biochar impregnated with Fe-Mn oxides. Water Air Soil Pollut 230(5):105. https://doi.org/10.1007/s11270-019-4146-5
48. Fujishima A, Honda K (1972) Electrochemical photolysis of water at a semiconductor electrode. Nature 238(5358):37–38. https://doi.org/10.1038/238037a0
49. Sorlini S, Gialdini F, Stefan M (2014) UV/H_2O_2 oxidation of arsenic and terbuthylazine in drinking water. Environ Monit Assess 186(2):1311–1316. https://doi.org/10.1007/s10661-013-3481-z
50. Lee Y, Um IH, Yoon J (2003) Arsenic(III) oxidation by iron(VI) (Ferrate) and subsequent removal of arsenic(V) by Iron(III) coagulation. Environ Sci Technol 37(24):5750–5756. https://doi.org/10.1021/es034203+
51. Qin Y, Cui Y, Tian Z, Wu Y, Li Y (2017) Synthesis of AG@AgCl core-shell structure nanowires and its photocatalytic oxidation of arsenic (III) under visible light. Nanoscale Res Lett 12(1):247. https://doi.org/10.1186/s11671-017-2017-9
52. Fujishima A, Zhang X, Tryk DA (2008) TiO_2 photocatalysis and related surface phenomena. Surf Sci Rep 63(12):515–582. https://doi.org/10.1016/j.surfrep.2008.10.001
53. Wang Y, Zhang P, Zhang TC, Xiang G, Wang X, Pehkonen S, Yuan S (2020) A magnetic γ-Fe_2O_3@PANI@TiO_2 core-shell nanocomposite for arsenic removal via a coupled visible-light-induced photocatalytic oxidation-adsorption process. Nanoscale Adv. https://doi.org/10. 1039/D0NA00171F
54. Ge J, Guha B, Lippincott L, Cach S, Wei J, Su TL, Meng X (2020) Challenges of arsenic removal from municipal wastewater by coagulation with ferric chloride and alum. Sci Total Environ 725:138351. https://doi.org/10.1016/j.scitotenv.2020.138351
55. Hu C, Chen Q, Liu H, Qu J (2015) Coagulation of methylated arsenic from drinking water: influence of methyl substitution. J Hazard Mater 293:97–104. https://doi.org/10.1016/j.jha zmat.2015.03.055
56. Watson MA, Tubić A, Agbaba J, Nikić J, Maletić S, Molnar Jazić J, Dalmacija B (2016) Response surface methodology investigation into the interactions between arsenic and humic acid in water during the coagulation process. J Hazard Mater 312:150–158. https://doi.org/ 10.1016/j.jhazmat.2016.03.002
57. Zhang G, Li X, Wu S, Gu P (2012) Effect of source water quality on arsenic (V) removal from drinking water by coagulation/microfiltration. Environ Earth Sci 66(4):1269–1277. https:// doi.org/10.1007/s12665-012-1549-7

58. Song S, Lopez-Valdivieso A, Hernandez-Campos DJ, Peng C, Monroy-Fernandez MG, Razo-Soto I (2006) Arsenic removal from high-arsenic water by enhanced coagulation with ferric ions and coarse calcite. Water Res 40(2):364–372. https://doi.org/10.1016/j.watres.2005. 09.046

59. Fox DI, Stebbins DM, Alcantar NA (2016) Combining ferric salt and cactus mucilage for arsenic removal from water. Environ Sci Technol 50(5):2507–2513. https://doi.org/10.1021/ acs.est.5b04145

60. Li L, Van Genuchten CM, Addy SEA, Yao J, Gao N, Gadgil AJ (2012) Modeling As(III) oxidation and removal with iron electrocoagulation in groundwater. Environ Sci Technol 46(21):12038–12045. https://doi.org/10.1021/es302456b

61. Dubrawski KL, Mohseni M (2013) Standardizing electrocoagulation reactor design: iron electrodes for NOM removal. Chemosphere 91(1):55–60. https://doi.org/10.1016/j.chemos phere.2012.11.075

62. Heffron J, Marhefke M, Mayer BK (2016) Removal of trace metal contaminants from potable water by electrocoagulation. Sci Rep 6(1):28478. https://doi.org/10.1038/srep28478

63. Wan W, Pepping TJ, Banerji T, Chaudhari S, Giammar DE (2011) Effects of water chemistry on arsenic removal from drinking water by electrocoagulation. Water Res 45(1):384–392. https://doi.org/10.1016/j.watres.2010.08.016

64. García-Lara AM, Montero-Ocampo C, Martínez-Villafañe F (2009) An empirical model for treatment of arsenic contaminated underground water by electrocoagulation process employing a bipolar cell configuration with continuous flow. Water Sci Technol 60(8):2153–2160. https://doi.org/10.2166/wst.2009.641

65. Gerrity D, Pecson B, Trussell R, Trussell R (2013) Potable reuse treatment trains throughout the world. J Water Supply: Res Technol AQUA 62:321–338. https://doi.org/10.2166/aqua. 2013.041

66. Park M, Snyder S.A, Attenuation of contaminants of emerging concerns by nanofiltration membrane: rejection mechanism and application in water reuse. In: contaminants of emerging concern in water and wastewater: advanced treatment processes. In: Hernández-Maldonado AJ, Blaney LBT-C, ECW W (eds) Butterworth-Heinemann pp 177–206. https://doi.org/10. 1016/B978-0-12-813561-7.00006-7

67. Mólgora CC, Domínguez AM, Avila EM, Drogui P, Buelna G (2013) Removal of arsenic from drinking water: a comparative study between electrocoagulation-microfiltration and chemical coagulation-microfiltration processes. Sep Purif Technol 118, 645–651. https://doi. org/10.1016/j.seppur.2013.08.011

68. Han B, Runnells T, Zimbron J, Wickramasinghe R (2002) Arsenic removal from drinking water by flocculation and microfiltration. Desalination, 145 (1):293–298. https://doi.org/10. 1016/S0011-9164(02)00425-3

69. Zhang T, Sun DD (2013) Removal of arsenic from water using multifunctional micro-/nano-structured MnO$_2$ spheres and microfiltration. Chem Eng J 225:271–279. https://doi.org/10. 1016/j.cej.2013.04.001

70. Nguyen VT, Vigneswaran S, Ngo HH, Shon HK, Kandasamy J (2009) Arsenic removal by a membrane hybrid filtration system. Desalination 236(1):363–369. https://doi.org/10.1016/ j.desal.2007.10.088

71. Wickramasinghe SR, Han B, Zimbron J, Shen Z, Karim MN (2004) Arsenic removal by coagulation and filtration: comparison of groundwaters from the united states and Bangladesh. Desalination 169(3):231–244. https://doi.org/10.1016/j.desal.2004.03.013

72. Brandhuber P, Amy G (1998) Alternative methods for membrane filtration of arsenic from drinking water. Desalination 117(1):1–10. https://doi.org/10.1016/S0011-9164(98)00061-7

73. Ahmad A, Rutten S, de Waal L, Vollaard P, van Genuchten C, Bruning H, Cornelissen E, van der Wal A (2020) Mechanisms of arsenate removal and membrane fouling in ferric based coprecipitation–low pressure membrane filtration systems. Sep Purif Technol 241:116644. https://doi.org/10.1016/j.seppur.2020.116644

74. Kumar M, Isloor AM, Somasekhara Rao T, Ismail AF, Farnood R, Nambissan PMG (2020) Removal of toxic arsenic from aqueous media using polyphenylsulfone/cellulose acetate

hollow fiber membranes containing zirconium oxide. Chem Eng J 393:124367. https://doi. org/10.1016/j.cej.2020.124367

75. Brandhuber P, Amy G (2001) Arsenic removal by a charged ultrafiltration membrane—Influences of membrane operating conditions and water quality on arsenic rejection. Desalination 140(1):1–14. https://doi.org/10.1016/S0011-9164(01)00350-2

76. Hao L, Wang N, Wang C, Li G (2018) Arsenic removal from water and river water by the combined adsorption—UF membrane process. Chemosphere 202:768–776. https://doi.org/ 10.1016/j.chemosphere.2018.03.159

77. He Y, Zhao DL, Chung, T-S (2018) Na+functionalized carbon quantum dot incorporated thin-film nanocomposite membranes for selenium and arsenic removal. J Membrane Sci 564:483–491. https://doi.org/10.1016/j.memsci.2018.07.031

78. Ahmed Baig J, Gul Kazi T, Qadir Shah A, Abbas Kandhro G, Imran Afridi H, Balal Arain M, Khan Jamali M, Jalbani N (2010) Speciation and evaluation of arsenic in surface water and groundwater samples: a multivariate case study. Ecotoxicol Environ Saf 73(5):914–923. https://doi.org/10.1016/j.ecoenv.2010.01.002

79. Nasir AM, Goh PS, Ismail AF (2019) Highly adsorptive polysulfone/hydrous iron-nickel-manganese (PSF/HINM) nanocomposite hollow fiber membrane for synergistic arsenic removal. Sep Purif Technol 213:162–175. https://doi.org/10.1016/j.seppur.2018.12.040

80. Figoli A, Cassano A, Criscuoli A, Mozumder MSI, Uddin MT, Islam MA, Drioli E (2010) Influence of operating parameters on the arsenic removal by nanofiltration. Water Res 44(1):97–104. https://doi.org/10.1016/j.watres.2009.09.007

81. Wang X, Liu W, Li D, Ma W (2009) Arsenic (V) removal from groundwater by GE-HL nanofiltration membrane: effects of arsenic concentration, PH, and co-existing ions. Front Environ Sci Eng China 3(4):428. https://doi.org/10.1007/s11783-009-0146-9

82. Zhang X, Fang X, Li J, Pan S, Sun X, Shen J, Han W, Wang L, Zhao S (2018) Developing new adsorptive membrane by modification of support layer with iron oxide microspheres for arsenic removal. J Colloid Interface Sci 514:760–768. https://doi.org/10.1016/j.jcis.2018. 01.002

83. Potla Durthi C, Rajulapati SB, Palliparambi AA, Kola AK, Sonawane SH (2018) Studies on removal of arsenic using cellulose acetate-zinc oxide nanoparticle mixed matrix membrane. Int Nano Lett 8(3):201–211. https://doi.org/10.1007/s40089-018-0245-3

84. Pal M, Mondal MK, Paine TK, Pal P (2018) Purifying arsenic and fluoride-contaminated water by a novel graphene-based nanocomposite membrane of enhanced selectivity and sustained flux. Environ Sci Pollut Res 25(17):16579–16589. https://doi.org/10.1007/s11356-018-1829-1

85. Das TK, Sakthivel TS, Jeyaranjan A, Seal S, Bezbaruah AN (2020) Ultra-high arsenic adsorption by graphene oxide iron nanohybrid: removal mechanisms and potential applications. Chemosphere 253:126702. https://doi.org/10.1016/j.chemosphere.2020.126702

86. Yoon Y, Park WK, Hwang T-M, Yoon DH, Yang WS, Kang J-W (2016) Comparative evaluation of magnetite–graphene oxide and magnetite-reduced graphene oxide composite for As(III) and As(V) removal. J Hazard Mater 304:196–204. https://doi.org/10.1016/j.jhazmat. 2015.10.053

87. Sakthivel TS, Das S, Pratt CJ, Seal S (2017) One-pot synthesis of a ceria-graphene oxide composite for the efficient removal of arsenic species. Nanoscale 9(10):3367–3374. https:// doi.org/10.1039/C6NR07608D

88. Benjwal P, Kumar M, Chamoli P, Kar KK (2015) Enhanced photocatalytic degradation of methylene blue and adsorption of arsenic(Iii) by reduced graphene oxide (RGO)–metal oxide (TiO_2/Fe_3O_4) based nanocomposites. RSC Adv 5(89):73249–73260. https://doi.org/10.1039/ C5RA13689J

89. Zhu H, Jia Y, Wu X, Wang H (2009) Removal of arsenic from water by supported nano zero-valent iron on activated carbon. J Hazardous Mater 172(2):1591–1596. https://doi.org/ 10.1016/j.jhazmat.2009.08.031

90. Gong X-J, Li Y-S, Dong Y-Q, Li W-G (2020) Arsenic adsorption by innovative iron/calcium in-situ-impregnated mesoporous activated carbons from low-temperature water and effects

of the presence of humic acids. Chemosphere 250:126275. https://doi.org/10.1016/j.chemos phere.2020.126275

91. Sherlala AIA, Raman AAA, Bello MM, Buthiyappan A (2019) Adsorption of arsenic using chitosan magnetic graphene oxide nanocomposite. J Environ Manage 246:547–556. https:// doi.org/10.1016/j.jenvman.2019.05.117

92. Aremu JO, Lay M, Glasgow G (2019) Kinetic and isotherm studies on adsorption of arsenic using silica based catalytic media. J Water Process Eng 32:100939. https://doi.org/10.1016/j. jwpe.2019.100939

93. Mishra T, Mahato DK (2016) A comparative study on enhanced arsenic(V) and arsenic(III) removal by iron oxide and manganese oxide pillared clays from ground water. J Environ Chem Eng 4(1):1224–1230. https://doi.org/10.1016/j.jece.2016.01.022

94. Dong F, Xu X, Shaghaleh H, Guo J, Guo L, Qian Y, Liu H, Wang S (2020) Factors influencing the morphology and adsorption performance of cellulose nanocrystal/iron oxide nanorod composites for the removal of arsenic during water treatment. Int J Biol Macromolecules 156:1418–1424. https://doi.org/10.1016/j.ijbiomac.2019.11.182

95. Yin Y, Zhou T, Luo H, Geng J, Yu W, Jiang Z (2019) Adsorption of arsenic by activated charcoal coated zirconium-manganese nanocomposite: performance and mechanism. Colloids Surf Physicochemical Eng Aspects 575:318–328. https://doi.org/10.1016/j.colsurfa.2019.04.093

96. Xu F, Chen H, Dai Y, Wu S, Tang X (2019) Arsenic adsorption and removal by a new starch stabilized ferromanganese binary oxide in water. J Environ Manage 245:160–167. https://doi. org/10.1016/j.jenvman.2019.05.071

97. Lou Z, Cao Z, Xu J, Zhou X, Zhu J, Liu X, Ali Baig S, Zhou J, Xu X (2017) Enhanced removal of As(III)/(V) from water by simultaneously supported and stabilized Fe-Mn binary oxide nanohybrids. Chem Eng J 322:710–721. https://doi.org/10.1016/j.cej.2017.04.079

98. Lobo C, Castellari J, Colman Lerner J, Bertola N, Zaritzky N (2020) Functional iron chitosan microspheres synthesized by ionotropic gelation for the removal of arsenic (V) from water. Int J Bio Macromolecules, 164:1575–1583. https://doi.org/10.1016/j.ijbiomac.2020.07.253

99. Shakoor MB, Niazi NK, Bibi I, Shahid M, Saqib ZA, Nawaz MF, Shaheen SM, Wang H, Tsang DCW, Bundschuh J, Ok YS, Rinklebe J (2019) Exploring the arsenic removal potential of various biosorbents from water. Environ Int 123:567–579. https://doi.org/10.1016/j.envint. 2018.12.049

100. Cho D-W, Jeon BH, Chon C-M, Kim Y, Schwartz FW, Lee E-S, Song H (2012) A novel chitosan/clay/magnetite composite for adsorption of Cu(II) and As(V). Chem Eng J 200– 202:654–662. https://doi.org/10.1016/j.cej.2012.06.126

101. Zhang S, Niu H, Cai Y, Zhao X, Shi Y (2010) Arsenite and arsenate adsorption on coprecipitated bimetal oxide magnetic nanomaterials: $MnFe_2O_4$ and $CoFe_2O_4$. Chem Eng J 158(3):599–607. https://doi.org/10.1016/j.cej.2010.02.013

102. Nguyen Thanh D, Singh M, Ulbrich P, Strnadova N, Štěpánek, F (2011) Perlite Incorporating γ-Fe_2O_3 and α-MnO_2 nanomaterials: preparation and evaluation of a new adsorbent for As(V) removal. Sep Purif Technol 82:93–101. https://doi.org/10.1016/j.seppur.2011.08.030

103. Uppal H, Chawla S, Joshi AG, Haranath D, Vijayan N, Singh N (2019) Facile chemical synthesis and novel application of zinc oxysulfide nanomaterial for instant and superior adsorption of arsenic from water. J Cleaner Prod 208:458–469. https://doi.org/10.1016/j.jcl epro.2018.10.023

104. Pillai P, Kakadiya N, Timaniya Z, Dharaskar S, Sillanpaa M (2020) Removal of arsenic using iron oxide amended with rice husk nanoparticles from aqueous solution. Mater Today: Proc 28:830–835. https://doi.org/10.1016/j.matpr.2019.12.307

105. Aredes S, Klein B, Pawlik M (2012) The removal of arsenic from water using natural iron oxide minerals. J Cleaner Prod 29–30:208–213. https://doi.org/10.1016/j.jclepro.2012.01.029

106. Kanel SR, Manning B, Charlet L, Choi H (2005) Removal of arsenic(III) from groundwater by nanoscale zero-valent iron. Environ Sci Technol 39(5):1291–1298. https://doi.org/10.1021/ es048991u

107. Anderson MA, Ferguson JF, Gavis J (1976) Arsenate adsorption on amorphous aluminum hydroxide. J Colloid Interface Sci 54(3):391–399. https://doi.org/10.1016/0021-9797(76)903 18-0

108. Jeong Y, Fan M, Singh S, Chuang C-L, Saha B, Hans van Leeuwen J (2007) Evaluation of iron oxide and aluminum oxide as potential arsenic(V) adsorbents. chemical engineering and processing. Process Intensification 46(10):1030–1039. https://doi.org/10.1016/j.cep.2007.05.004

109. Zheng Y-M, Yu L, Wu D, Paul Chen J (2012) Removal of arsenite from aqueous solution by a zirconia nanoparticle. Chem Eng. J. 188:15–22. https://doi.org/10.1016/j.cej.2011.12.054

110. Oladoja, NA, Bello GA, Helmreich B, Obisesan SV, Ogunniyi JA, Anthony ET, Saliu TD (2019) Defluoridation efficiency of a green composite reactive material derived from lateritic soil and gastropod shell. Sustain Chem Pharm 12. https://doi.org/10.1016/j.scp.2019.100131

111. Driehaus W, Jekel M, Hildebrandt U (1998) Granular ferric hydroxide—a new adsorbent for the removal of arsenic from natural water. J Water Supply: Res Technol-Aqua 47(1):30–35. https://doi.org/10.2166/aqua.1998.0005

112. Dutta PK, Ray AK, Sharma VK, Millero FJ (2004) Adsorption of arsenate and arsenite on titanium dioxide suspensions. J Colloid Interface Sci. 278 (2):270–275. https://doi.org/10.1016/j.jcis.2004.06.015

113. Kapaj S, Peterson H, Liber K, Bhattacharya P (2006) Human health effects from chronic arsenic poisoning–a review. J Environ Sci Health A 41(10):2399–2428. https://doi.org/10.1080/10934520600873571

114. Wang C, Luo H , Zhang Z. Wu Y, Zhang J, Chen S (2014) Removal of As(III) and As(V) from aqueous solutions using nanoscale zero valent iron-reduced graphite oxide modified composites. Journal of Hazardous Materials 268:124–131. https://doi.org/10.1016/j.jhazmat.2014.01.009

115. Oladoja NA, Aboluwoye CO, Ololade IA, Adebayo OL, Olaseni SE, Adelagun ROA (2012) Intercalation of gastropod shell derived calcium oxide in clay and application in phosphate removal from aqua medium. Ind Eng Chem Res 51(45):14637–14645. https://doi.org/10.1021/ie301520v

116. Lee SH, Jang YH, Nguyen DD, Chang, SW, Kim SC, Lee S M, Kim SS. Adsorption properties of arsenic on sulfated TiO_2 adsorbents. J Ind Eng Chem 80:444–449. https://doi.org/10.1016/j.jiec.2019.08.024

117. Alvarez-Cruz JL, Garrido-Hoyos SE. Effect of the mole ratio of Mn/Fe composites on arsenic (V) Adsorption. Sci Environ 668:47–55. https://doi.org/10.1016/j.scitotenv.2019.02.234

118. Zhu N, Qiao J, Ye Y, Yan T (2018) Synthesis of mesoporous bismuth-impregnated aluminum oxide for arsenic removal: adsorption mechanism study and application to a lab-scale column. J Environ Manage 211:73–82. https://doi.org/10.1016/j.jenvman.2018.01.049

Chapter 7
Prospects of Photocatalysis in the Management of Nitrate Contamination in Potable Water

Zeeshan Ajmal, Yassine Naciri, Abdelghani Hsini, Bianca M. Bresolin, Abdul Qadeer, Muhammad Nauman, Muhammad Arif, Muhammad Kashif Irshad, Khursheid Ahmed Khan, Ridha Djellabi, Claudia L. Bianchi, Mohamed Laabd, Abdallah Albourine, and Renjie Dong

Abstract Nitrate is one of the most widespread toxic inorganic compounds in groundwater due to its high water solubility. High level of nitrate in potable water may poses serious risks to the environment and to human health. Heterogeneous photocatalysis has been widely used for water remediation and disinfection, however,

Zeeshan Ajmal, Yassine Naciri and Abdelghani Hsini—These authors contributed equally to the formation of this book chapter.

Z. Ajmal · M. Arif · R. Dong
MoA Key Laboratory for Clean Production and Utilization of Renewable Energy, MoST National Center for International Research of BioEnergy Science and Technology, College of Engineering, China Agricultural University, 100083 Beijing, People's Republic of China

Y. Naciri · A. Hsini · M. Laabd · A. Albourine
Laboratory of Materials and Environment, Faculty of Sciences, Ibn Zohr University, City Dakhla, B.P. 8106, Agadir, Morocco

B. M. Bresolin
Department of Separation Science, Lappeenranta University of Technology, Sammonkatu 12, 50130 Mikkeli, Finland

R. Djellabi (✉) · C. L. Bianchi
Department of Chemistry, University of Milan, via C. Golgi 19, Milan 20133, Italy
e-mail: ridha.djellabi@unimi.it

A. Qadeer
National Engineering Laboratory for Lake Pollution Control and Ecological Restoration, Chinese Research Academy of Environmental Science, Beijing 10012, China

M. Nauman
State Key Laboratory of Chemical Resource Engineering, Beijing University of Chemical Technology, Beijing 100029, People's Republic of China

M. K. Irshad
Department of Environmental Science and Engineering, Government College University, Faisalabad, Pakistan

K. A. Khan
Ghazi University City Campus, Dera Ghazi Khan 32200, Pakistan

© Springer Nature Switzerland AG 2021   185
N. A. Oladoja and E. I. Unuabonah (eds.), *Progress and Prospects in the Management of Oxyanion Polluted Aqua Systems*, Environmental Contamination Remediation and Management, https://doi.org/10.1007/978-3-030-70757-6_7

less research studies, comparatively, have reported photocatalytic nitrate reduction because of the complexity of the mechanism of reaction. Mainly, nitrate photoreduction takes place directly via reaction with photo-generated electrons in the conduction band of the photocatalyst or by photo-produced reducing species under light irradiation. As a result, nitrate can be transformed into unpreferred by-products such as nitrite and ammonium, while the reduction into dinitrogen gas is much recommended due to its high importance. On the other hand, the issue of the re-oxidation of ammonium into nitrate has also been reported. The efficiency and selectivity of a photocatalytic system to reduce nitrate into dinitrogen depend on the operating parameters controlling the reaction, and more importantly, the selectivity strongly depends on the type of the photocatalytic nanomaterial. For this reason, a pool of studies have been performed in order to enhance the selectivity of nitrate reduction into dinitrogen by developing different kinds of nanomaterials. In this chapter, we examine: (i) the conventional technologies for nitrate removal/reduction, (ii) the effect of operating conditions on the photocatalytic nitrate reduction process, as well as (iii) the influence of the type of photoactive nanomaterial on the selectivity and the performance toward nitrate reduction.

Keywords Nitrate reduction · Photocatalysis · Nanomaterials · Drinking water · Selective reduction

7.1 Introduction

Due to the huge increase in worldwide population together with large agricultural and industrial activities, the discharge of big amounts of varying types of pollutants into the environment became a serious global issue [1–4]. In the year 2017, WHO and UNICEF reported that 663 million people lacked access to clean drinking in the least developed countries. Nowadays, the scientific community and engineers expend efforts to develop several water purification technologies for the remediation of different types of wastewaters. It is important to point out that the cost, efficiency, and safety are the main parameters to choose one or more technologies to treat wastewater contaminated by specific pollutants. Technologies that result in secondary waste formation or require the use of huge quantities of chemicals or energy are usually costly and complicated which should be avoided, while those simple continuous low-cost green technologies are recommended for use, especially, in least developed countries [1, 5–15]. Heterogenous photocatalysis is one of the advanced oxidation processes (AOPs), which is based on the irradiation of a photocatalyst, such as TiO_2, by light having an energy higher or equal to the band gap of the photocatalyst. This leads to the formation of photo-induced electrons/positive holes charges [16, 17]. This photoactive system has been applied widely for environmental remediation and energy production, while scientists are still developing new photoactive materials with the extension of their use for real-world applications [18–26].

Nitrogen is known as an important compound for the growth of plants as it is the key element of amino acids and proteins. Plants can absorb nitrogen as NH_4^+ or NO_3^- species that are formed by mineralization and nitrification, respectively. However, nitrogen species can reach surface waters and groundwater due to the excessive use of nitrogenous fertilizers [27, 28]. The water pollution by nitrate or nitrite can cause serious environmental and health problems. For example, a high dose of nitrate in human body causes methemoglobinemia, which could be very carcinogenic [29]. On the other hand, nitrate or nitrite can be transformed into nitroarenes in water, and nitroarenes can provoke damage in respiratory, cardiovascular, and central nervous systems [30, 31] and also high nitrate in potable water can cause cancer. Additionally, nitrate is a very widespread pollutant in waters due to its high water solubility. High level of nitrate also can stimulate heavy algal growth resulting in eutrophication in groundwaters, and serious risks to environment and human health. For this reason, World Health Organization (WHO) recommended 10 mg/L of nitrate as a limit concentration in drinking water [32]. Many conventional denitrification techniques have been developed to produce potable waters of suitable quality, such as adsorption [33], ion exchange [34], chemical reduction [35], membrane filtration [36], electrochemical [37], and biological denitrification [38]. Recently, many studies have been reported on the photocatalytic reduction of nitrate in water [39–43]. In spite of the huge amount of research papers reporting the removal of organic and inorganic pollutants by photocatalysis, the photocatalytic reduction of nitrate in water is still comparatively less studied as it is a challenge to find a photoactive material with good selectivity and efficiency, which can reduce water-soluble NO_3^- to N_2 gas. This chapter examines the current state of studies related to the photocatalytic reduction of nitrate in water. First, the conventional technologies for nitrate were summarized and their advantages and drawbacks were discussed. Secondly, the photocatalytic mechanistic pathways for nitrate reduction, along with the effects of operating parameters on the efficiency were examined. Thirdly, some novel nanomaterials for selective and enhanced nitrate photocatalytic reduction were discussed.

7.2 Nitrate Pollution in Water

7.2.1 Physicochemical and Photochemistry of Nitrate and Nitride

Nitrate (NO_3^-) and nitride (NO_2^-) contamination in drinking water have been recognized as a major challenge due to their negative impact on the environment, the ecosystems and human health. Inorganic nitrate and nitrite ions are naturally occurring species part of the nitrogen cycle, primarily produced by the fixation of nitrogen and oxygen via chemical, biological, but they can also bound to sodium or potassium metal cations when not totally solubilized [44]. Between the two inorganic species, nitrate ions show higher stability since they are chemically unreactive in water [45]. On the other hand, they can undergo reduction reaction to nitrite ions or vice versa. Nitrate is the higher oxidized form of nitrogen available in nature, with an oxidation state of nitrogen +5, and it accounts for the majority of the total available nitrogen in surface waters. Its formation is primarily linked to conversion of ammonium ion, from fertilizer and manure, through two-step oxidation process, first to nitrite and then to nitrate.

Nitrate and nitrite typically exist in an oxidation state of -1 in compounds. Chemically, nitrate is the conjugate base of the stronger acid, nitric acid, when completely dissociated in aqueous solutions. Nitrite is the conjugate base of nitrous acid, a weaker acid. Nitrite readily decomposes, yielding water and dinitrogen trioxide, or nitric acid, nitric oxide, and water. When photocatalytic reduction is applied as removal technology for such contaminants, the consideration of direct photolytic reactions and the photochemical features of nitrides and nitrates must be taken into consideration for the determination of the overall mechanism. Specifically, both NO_3^- and NO_2^- have been found to undergo photolysis, but ammonium cationic species were photo-inert; thus, the photolysis is considered negligible in water solutions [46]. In water, nitrate photolysis occurs at ultraviolet (UV) range, between 270 and 330 nm, with the maximum absorption at 300 nm. However, due to low quantum yield, a low conversion efficiency to nitrite is generally expected [47]. Therefore, nitrate is expected to be one of the major final products of the photolytic process in aqueous solution. On the one hand, the formation of nitrate radicals by photolysis is a critical concern in the troposphere under natural irradiation, due to the presence of unwanted compounds such as nitric acid or NO_x [48, 49]. On the other hand, nitrite undergoes a complex photolytic mechanism at a wavelength of betwen 200 and 400 nm [50]. The photolytic degradation recorded in previous literature demonstrates a low transformation to other nitrogenous species. The process leads to the formation of radical species that can induce oxidation of organics and other species present in solution with a negligible impact on the overall nitrite content [39].

7.2.2   Sources of Nitrate in Water

Nitrogen is a fundamental element for living things in an ecosystem, thus, it is regarded as one of the primary nutrients critical for their survival. Although, it is abundant in the atmosphere, in the form of dinitrogen gas (N_2), the different types in the natural environment include: organic (e.g., amino and nucleic acids) and inorganic cation, ammonium, being the most reduced form of nitrogen; nitrogen oxyanions such as nitrite (NO_2^-) and nitrate (NO_3^-), as the most oxidized compound. Since the last century, humans have drastically enhanced the natural rate at which nitrogen is deposited into air, land, and water ecosystems. Among the different forms of nitrogen, nitrate as a pollutant in drinking water supplies is considered a serious global issue [51].

In nature, the preferred way to transfer nitrogen from the environment into living ecosystems is represented by its fixation. However, the balance of the natural nitrogen cycle has been heavily tilted by human activities. Among the major sources of contamination, the excess application of inorganic nitrogenous fertilizers and manures, the combustion of fossil fuels, the replacement of natural vegetation with nitrogen-fixing crops, such as soybeans, wastewater treatment plants, anthropogenic inputs such as oxidation of nitrogenous waste products in human and animal excreta, must be taken into consideration [52].

In particular, synthetic fertilizers represent the major anthropogenic source of nitrogen in the environment, since approximately half of the applied nitrogen-based chemicals can drain-off from agricultural fields into surface and groundwater, leading to an exponential increase of nitrate concentrations in freshwater [27]. Such human activities, by doubling the amount of fixed nitrogen over that of the natural level, are major causes of the increase of greenhouse effect, reduction in the protective ozone layer, enhancement of smog levels, acid rain, and the contamination of the available drinking water.

7.2.3 Nitrate Toxicity

As previously mentioned, fixed nitrogen, released in particular during fertilizer application, in the form of nitrite ions, remarkably exacerbate water pollution issues. Ecological concerns arises from the role of nitrite compounds in over-enriching aquatic ecosystems, producing several environmental concerns such as acidification, eutrophication, since depletion of oxygen leads to "dead zones" in lakes and water bodies [53, 54]. Globally, fixed nitrogen in the world has dramatically increased, with critical consequences for ecosystems and public health. Nitrate discharges from anthropogenic sources may thus be a serious ecological risk, especially for certain aquatic animals. Freshwater invertebrates, fishes, and amphibians can be drastically affected by long-term exposures to nitrates even at low concentrations [55].

In the case of marine invertebrates and fishes, higher resistivity has been recorded [56]. In early 2004, a previous review have deeply investigated the health effects related to nitrate exposure from drinking water, especially in with respect to long-term effects [57]. An International Agency for Research on Cancer (IARC) Working Group have further reviewed human, animal, and mechanistic studies of cancer through mid-2006, concluding an endogenous nitrosation and probably carcinogenic effect due to the consumption of such nitrogen forms [58]. Specifically, it has been reported that nitrate in drinking water is absorbed in the upper gastrointestinal tract and is distributed in the human body. This may induce endogenous formation of N-nitroso compounds, which is involved in a wide range of physiological effects; including regulation of blood pressure and blood flow [59], the inhibition of platelet adhesion and aggregation [60], the maintenance of blood vessel tonus [61], modulation of mitochondrial function [48, 49, 62], etc. Moreover, these physiological processes can also lead to methemoglobinemia, especially because of high concentration of nitrate in water, food, and medications [57, 63]. Similar concerns have been raised about nitrate intake during pregnancy. Among the risk factor investigated for a range of pregnancy outcomes are spontaneous abortion, fetal deaths, prematurity, intrauterine growth retardation, low birth weight, congenital malformations, and neonatal deaths [64, 65].

The availability of birth defect surveillance systems around the world has led to an intense investigation on the relationship between nitrate concentration in drinking water and congenital malformations. In recent years, epidemiologic studies have been devoted to the ecologic investigation of stomach cancer mortality. Some results have also positively correlated brain, esophagus, stomach, kidney, and ovary cancer incidences with the consumption of nitrate contaminated water. Numerous animal and human studies have shown that high dose of nitrates ingestion competitively inhibits the function of thyroid gland and this results in associated thyroid disease [66–69]. Association between nitrate in drinking water and other non-cancer health effects that include type 1 childhood diabetes (T1D), blood pressure, and acute respiratory tract infections in children [57] have been established.

7.2.4 Permissible Level of Nitrates and Nitrites in Drinking Water

Managing the nitrogen cycle in water have been identified by the US National Academy of Engineers as one of the Grand Challenges for the twenty-first-century society and numerous efforts have been devoted to restore balance to the nitrogen cycle with better fertilization technologies and by capturing and recycling waste [70, 71]. The *Global Burden of Disease project* edited by the World Health Organization (WHO) attempted to quantify and compare the level of illness at both global and regional levels. The original document for the development of WHO guidelines for drinking water quality has been updated and revised in different editions. Currently, the maximum contaminant level (MCL) for nitrate in public drinking water supplies has been set to 50 mg/L as NO_3 and 0.50 mg/l for NO_2. In particular, the value has been set in order to protect infants against methemoglobinemia [27]. However, it can be said that many other health effects have not been specifically considered. For example, it should be considered that ingestion is not the only source of nitrite ions in human physiological systems but endogenous production can occur as a result of the activity of transient species such as nitric oxide [61]. Similar values have been set for nitrate by other institutions (e.g., European Union, EPA (USA), and Ministry of Environmental Protection of China). On the contrary, for nitrite, EPA and China government set a value of 1 mg/L (as N) [27, 72, 73].

Table 7.1 Advantages and disadvantages of different conventional techniques for nitrate removal

Treatment types	Advantages	Disadvantages
Ion exchange process	Simple, recoverable, medium operational cost, approx. 90% efficiency achieved	Secondary brine wastes, high cost for treatment of high concentration of nitrate
Reverse osmosis	Multiple contaminant removal, less control, and no need for extensive post-treatment	Membrane fouling and scaling, lower water recovery, operational complexity, energy demands
Adsorption	Medium operational cost, removal efficiency varies with different adsorbents, post-treatment	Need to dispose of adsorbents. Low surface area for some adsorbents
Electrodialysis	Multiple contaminant removal, higher water recovery, needs less control, and no need for extensive post-treatment	Energy demands, expensive treatment method, waste disposal
Chemical denitrification	No waste disposal is required Post-treatment is required due to the production of by-products Faster elimination of nitrate	Potential for nitrate peaking, high chemical use (salt), strongly dependent on the solution pH
Biological denitrification	Total elimination of nitrates Low probability of formation of toxic by-products Specific for nitrates	Possible sensitivity to environmental conditions Weak at low temperatures Bacteriological risks

7.3   Conventional Techniques for Nitrate Removal/Reduction

During the past three decades, several conventional techniques for nitrate removal/reduction from wastewater have been established. These techniques could be further classified into two categories: physicochemical and biological conventional techniques. Among them, several treatment strategies have been developed for nitrate removal from contaminated wastewater include reverse osmosis, resin-ion exchange, chemical reduction, electrodialysis, and adsorption microbial-based treatment methods. [74–78]. The merits and demerits of the aforementioned conventional methods for the removal of nitrates from drinking water are presented in Tables 7.1 and 7.2 [34, 79–88].

7.4   Photocatalysis for Nitrate Reduction/Removal in Water

As already reported in Sect. 7.3, we see that the conventional nitrate removal/reduction techniques encounter a wide range of limitations. Due to this, they are considered to be uneconomical, less effective as they further produce secondary

Table 7.2 Expected reactions in nitrates photoreduction [104–107]

	Equation no	Equations
Reactions NO_3^- reduction	1	$TiO_2 + h\upsilon \rightarrow TiO_2 + e_{CB}^- + h_{VB}^+$
	2	$H_2O + h_{VB}^+ \rightarrow H^+ + OH^\bullet$
	3	$(O_2)_{ads} + e_{CB}^- \rightarrow O_2^{\bullet-}$
	4	$2H^+ + 2e^- \rightarrow H_2$
	5	$R^\bullet + H^+ \rightarrow RH^{\bullet+} \rightarrow Intermediates/final\ products$
	6	$O_2^{\bullet-} + H^+ \rightarrow HO_2^\bullet$
	7	$HO_2^\bullet + e_{CB}^- \rightarrow HO_2^-$
	8	$NO_3^- + h\upsilon \leftrightarrow NO_2^\bullet + O_\bullet^-$
	9	$NO_3^- + h\upsilon \leftrightarrow ONOO^-$
	10	$ONOO^- + h\upsilon \leftrightarrow NO^\bullet + O_2^{\bullet-}$
	11	$NO^\bullet + OH^\bullet \leftrightarrow HNO_2$
	12	$NO_3^- + 10H^+ + 8e^- \leftrightarrow NH_4^+ + 3H_2O$
	13	$2NO_3^- + 10H^+ + 8e^- \rightarrow N_2O + 5H_2O$
	14	$NO_3^- + 2H^+ + 2e^- \rightarrow NO_2^- + H_2O$
	15	$NO_3^- + H_2O + 2e^- \rightarrow NO_2^- + 2OH$
	16	$2NO_3^- + 12H^+ + 10e^- \rightarrow N_2 + 6H_2O$
	17	$NO_3^- + e^- \rightarrow NO_3^{2-}$

toxic by-products. The main drawback associated with physicochemical processes in the removal of nitrate from wastewater is the formation of brine, which needs further treatment, rather than converting it into valuable product. On the other hand, biological denitrification has also some limitations since the produced metabolic substances and germs have the capability to contaminate drinking water. Moreover, it has been reported that high concentrated industrial waste effluent, higher nitrite, and ammonium concentration inhibit the nitrification process [83, 89]. Similarly, the previously reported nitrogen and ammonia nitrogen treatment strategies (i.e., ammonia stripping, constructed wetland, adsorption) also have certain restriction for their field application, such as high foaming in ammonia stripping process and leaching of adsorbed material in constructed wetland as well as saturation of adsorbent in adsorption process [90–92]. In order to address these problems, it is crucial to develop the most up-to-date effective nitrate reduction technology rather than conventional methods.

Photocatalytic oxidation is a relatively new technique, and it has the potential to be used as effective treatment method, due to the generation of highly reactive oxidative •OH radical for pollutant degradation. Several studies have reported that photocatalytic treatments have dual advantages of pollutant removal via degradation and the reduction of pollutant into useful by-product [93, 94]. Photocatalytic denitrification is regarded as a reliable method, since the initial report on photocatalytic nitrate reduction by Bems et al., in 1999 [95]. A wide range of research efforts has been devoted toward photocatalytic nitrate reduction by evaluating photo-oxidation

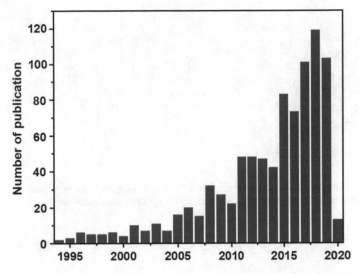

Fig. 7.1 Number of publications from 1994 until January 2020 on nitrate photocatalytic reduction

processes. The data presented in Fig. 7.1 showed the quick rise in the photocatalytic nitrate reduction research since 1990. In this context, photocatalytic nitrate reduction further shows more economical solution in water purification studies over other nitrogen removal treatments.

A large volume of research work has verified the importance of photocatalytic nitrate reduction through the use of series of various highly efficient photocatalysts (e.g., TiO_2, Fe_2O_3, ZnO, CdS, GaP, ZnS) [95–101]. Of the various photocatalysts, titanium dioxide (TiO_2) is a highly efficient and widely investigated photocatalyst, due to its promising characteristics, such as photostability, non-toxicity, and cost-effectiveness [102].

When particles are exposed to incident light, whose energy is higher than that of the semiconductor band gap, the excitement of electrons toward conduction band (e_{CB}^-) starts and the production of positive holes (h_{CB}^+), for the oxidation and reduction reaction (Fig. 7.2a) [103]. The band energy position of photocatalyst and the redox potential have very close relationship, because the reduction reaction starts when the conduction band potential is more negative than the desired species reduction potential ($E_{CB} > E_{RED}$) [103]. On the other hand, oxidation reaction requires more positive valence band than desired species oxidation potential ($E_{VB} > E_{OXD}$). In an aqueous solution, the nitrite, nitrate, and ammonium have standard potential (pH = 7) set within the bandgap of TiO_2 bandgap (Fig. 7.2b). Thus, TiO_2 photocatalyst can theoretically reduce nitrate to dinitrogen (N_2) with a potential of at least 0.75 V at pH = 7. However, the production of dinitrogen (N_2) via nitrate reduction is actually a ten-electron reduction process, and therefore it is extremely implausible to be occurring. Thus, these potentials are only used to estimate the accomplishment of reaction. The detailed studies regarding photoreduction of NO_3^- and NO_2^- are presented in

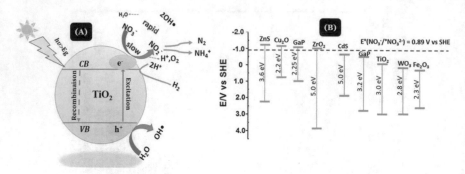

Fig. 7.2 **a** Nitrates photocatalytic reduction scheme on TiO_2, **b** band edge position of several semiconductors, using the standard hydrogen electrode (SHE) as a reference

literature, and the main reactions expected in nitrates photoreduction are given in Table 7.2.

7.4.1 Direct Inorganic Nitrogen Species Photocatalysis

7.4.1.1 Nitrate Photolysis

The NO_3^- absorption spectrum is always dominated by a weak n $\rightarrow \pi^*$ band around 302 nm ($\varepsilon = 7.2$ M^{-1} cm^{-1}) and a much stronger $\pi \rightarrow \pi^*$ band at 200 nm ($\varepsilon = 9900$ M^{-1} cm^{-1}) [108]. Indeed, NO_3^- photolysis mechanism is complex, and it reacts photochemically by several pathways in aqueous solution, with the overall stoichiometry of the best known reaction being [109]:

$$\left[NO_3^- \xrightarrow{h\upsilon} NO_3^- \right]* \tag{7.18}$$

It is well understood that stoichiometry must be maintained during whole pH range in the absence of •OH scavengers, when irradiated with $\lambda > 200$ nm [109, 110].

When irradiated at $\lambda > 280$ nm, then two primary photolytic pathways occurred [47, 109].

$$\left[NO_3^- \xrightarrow{h\upsilon} NO_3^- \right]* \tag{7.19}$$

$$\left[NO_3^- \right]* \rightarrow NO_2^- + O(^3P) \tag{7.20}$$

$$\left[NO_3^- \right]* \rightarrow NO_2^\bullet + O^{\bullet-} \xrightarrow{H_2O} NO_2 \bullet + \bullet OH + OH^- \tag{7.21}$$

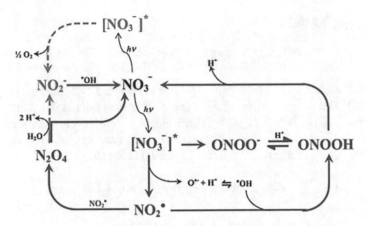

Fig. 7.3 Mechanism and main processes of nitrate photolysis in water. Highlights the pathway leading to NO_3^- as final product (solid line), highlights the pathway yielding NO_2^- as result of the photoreduction (dashed line). Adapted with permission from ref. [39], copyright 2017 Elsevier)

During irradiation at $\lambda < 280$ nm, the peroxynitrite anion ($ONOO^-$) is produced via isomerization of $[NO_3^-]$ as a result of a third primary reaction pathway [109–111].

$$[NO_3^-]^* \rightarrow ONOO^- \leftrightharpoons HOONO \quad pK_a = 6.5 \tag{7.22}$$

On the other hand, in second pathway to $ONOO^-$, the already produced radicals in Eq. 7.21 can recombine within the solvent cage by producing peroxynitrous acid (HOONO) [109, 110, 112, 113]:

$$\bullet OH + NO_2\bullet \rightarrow HOONO \quad k_{17} = 1.3 \times 10^9 M^{-1}s^{-1} \tag{7.23}$$

At pH < 7, HOONO isomerizes rapidly to NO_3^-

$$HOONO \rightarrow NO_3^- + H^+ \tag{7.24}$$

The main mechanism of nitrate photolysis in aqueous system are presented in Fig. 7.3. It clearly indicated that the conversion of nitrate, via photocatalysis, exhibited a low conversion to nitrite [47]; thus, the production of nitrate is considered to be the major final products. However, nitrate radical generation during photolysis has become the subject of major controversy in tropospheric chemistry under natural irradiation environment.

7.4.1.2 Nitrite Photolysis

Usually, nitrite photolysis is performed very well within the λ region between (200 nm and 400 nm), ($\varepsilon = 22.5$ M^{-1} cm^{-1}), where the transition n $\rightarrow \pi^*$ is possible (reaction (Eq. 7.25)).

$$NO_2^- \xrightarrow{h\upsilon} [NO_2^-]^* \tag{7.25}$$

The UV excitation of nitrite anions produced HO• and NO• radicals (Eq. 7.26) [114, 115]. The recombination of HO• and NO• may results in the production of HNO_2 (Eq. 7.27) or be scavenged by nitrite anions, at a diffusion-controlled rate, to yield OH^- and NO_2• (Eq. 7.28). NO• and NO_2• may react to give N_2O_3 (Eq. 7.29), which hydrolyzes to regenerate nitrite (Eq. 7.30) [115, 116].

$$(NO_2-)* + H_2O \rightarrow HO_\bullet + NO_\bullet + OH^- \tag{7.26}$$

$$HO_\bullet + NO_\bullet \rightarrow HNO_2 \tag{7.27}$$

$$HO_\bullet + NO_2^- \rightarrow NO_{2\bullet} + OH^- \tag{7.28}$$

$$NO_\bullet + NO_{2\bullet} \rightarrow N_2O_3 \tag{7.29}$$

$$N_2O_3 + H_2O \rightarrow 2HNO_2 \tag{7.30}$$

Usually, NO• is competitively oxidized by oxygen to produce nitrate, which lead to the depletion of oxygen in an open-air saturated solution (Eq. 7.31) [115].

$$NO^\bullet + NO_2^\bullet + O_2 + H_2O \rightarrow 2HNO_3 \tag{7.31}$$

The HO• radical is considered to be the most reactive species during nitrite photolysis (Fig. 7.4). These radicals are considered to be the more excited oxidant and the most aggressive during photocatalysis.

7.4.2 Influence of Operational Parameters

Nitrate photocatalytic reduction is greatly affected by several parameters, such as pH, irradiation time, nitrate concentration, the effect of metallization of TiO_2, the influence of metalized catalyst preparation method as well as the effect of sacrificial agent (hole scavenger). These parameters will be considered as more important as they affect nitrate photocatalytic route during its reduction in aqueous system.

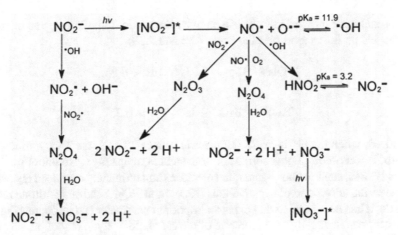

Fig. 7.4 Primary photo processes and subsequent reactions during NO_2^- photolysis. Adapted with permission from ref. [109], copyright 1999 Elsevier)

7.4.2.1   Irradiation Duration and Solution PH Value

The variations in pH during the reaction have a significant impact toward photocatalytic performance. Due to multiple interactions, the role of pH during nitrate photocatalytic reduction is very complex task to explain. Gao et al. [117] investigated the impact of pH over photocatalytic nitrate reduction using Ni-Cu/TiO$_2$ as a catalyst. They observed that up to 120 min of reaction, only traces of NH_4^+ and NO_2^- were produced, but the nitrite and ammonium concentrations increased simultaneously with increase in pH, which goes from ≈ 3 to ≈ 6 (Fig. 7.5a) [117].

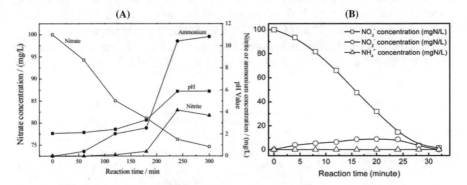

Fig. 7.5   a Concentrations of the nitrate, produced nitrite, ammonium, and pH value of the solution as a function of the reaction time over Ni–Cu/TiO$_2$ (4 wt%, Ni:Cu = 3:1) catalyst (Adapted with permission from ref. [117], copyright 2004 Elsevier). **b** Concentration curve of NO_3^-, formed NO_2^-, and NH_4^+ plotted as a function of irradiation time (Adapted with permission from ref. [118], copyright 2005, Elsevier)

The authors explained this phenomenon by the adsorption of nitrite and nitrate at various pH on the support (see the Eqs. 7.32 and 7.33).

$$TiOH + H^+ \rightarrow TiOH_2^+ \quad pH < 6.25, \tag{7.32}$$

$$TiOH \rightarrow TiO^- + H^+ \quad pH > 6.25. \tag{7.33}$$

Indeed, when the pH is <6.25 (isoelectric point of TiO_2), the TiO_2 surface will adsorb H^+ ions due to higher surface proton exchange capacity (0.46 mmol/g) [118]. These H^+ ions capture photo-generated electrons and ultimately produced H_2, which promotes the adsorption of negative ions (NO_3^- and NO_2^-) and aided nitrate reduction. So, it has been verified that pH is an important parameter and its decreasing and increasing value directly affect the rate of nitrate photocatalytic reduction.

Zhang et al. [118] investigated the impact of the irradiation time for photocatalytic nitrate reduction in the presence of Ag/TiO_2 as photocatalyst. It was shown that for the 1% Ag/TiO_2 photocatalytic system, the conversion of nitrate increased linearly. Moreover, during illumination, the nitrites appeared progressively under irradiation at 365 nm, but completely converted after 30 min of irradiation with good selectivity toward nitrogen (Fig. 7.5b).

7.4.2.2 Effect of Inorganic Species on the Nitrate Reduction by Photocatalysis

Zhang et al. [118] evaluated the possible impact of salts during photocatalytic reduction of nitrate in the presence of Ag/TiO_2 catalyst, using different concentrations of anions such as Na_2SO_4, Na_2CO_3, and $NaHCO_3$ (Fig. 7.6). They found that sulfate, carbonate, and bicarbonate had detrimental effects on the nitrate reduction. Their findings showed that SO_4^{2-} and CO_3^{2-} generated more negative impact than HCO_3^- (Fig. 7.6a–c). The distinctive interruptive impact of salts on the nitrate reduction was due to the undistributed aptitude adsorption competition of other ions with nitrate

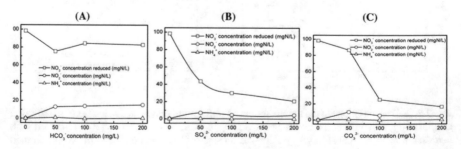

Fig. 7.6 Concentration curve of reduced NO_3^-, formed NO_2^-, and NH_4^+ plotted as a function of **a** HCO_3^- **b** SO_4^{2-} **c** CO_3^{2-} concentration. (Adapted with permission from ref. [118], copyright 2005) Elsevier)

over the catalyst active surface sites. Usually, in formic acid system, the TiO_2 surface sites is positively charged, so it favors the adsorption of anionic species (e.g., SO_4^{2-} or CO_3^{2-}) over HCO_3^- [118].

7.4.2.3   Effect of Metallic Constituents of Photocatalyst

Usually, the performance of a photocatalyst could be improved by expanding its range of photon absorption toward the visible light region. However, the band gap (BG) is characteristic of the semiconductor (BG = 3.02 eV for TiO_2, corresponding to $\lambda \leq 400$ nm, with maximum absorption at 375 nm), therefore, the possibility of improving the absorption of light energy is to metallize the semiconductor.

It has been reported that TiO_2 surface sites modification via metal oxides incorporation enhanced its light harvesting and photocatalytic activity. The modification of TiO_2 surface site with metal oxides, such as Pd, Pt, Au, and Cu, improved the solar radiation/charge separation and facilitated the charge transfer between metal and semiconductor [119–124]. Ranjit et al. [125–127] studied the conversion of nitrate to ammonia, by using noble metal oxides doped TiO_2 catalyst and reported that nitrate conversion totally relied on the nature of metal oxides for TiO_2 surface modification. A maximum metal oxide content favored higher catalytic activity, but beyond this amount the performance decreased [125–127]. It is indicated that these metals nanoparticles can act as an electron donor to promote the electron transfer from metal to the semiconductor. In addition to this function, these metals can suppress the recombination of surface electrons and holes. It is important to mention that the doping ions should have well matched combination with the photocatalyst surface, to enhance charge transfer efficiency (electron/vacancy); otherwise, these ions provides a recombination centers for photocatalyst particles [120].

During nitrate photoreduction process, the mechanism of metal oxides doped TiO_2 mainly depends on the detailed evaluation of the role of noble metals. Usually, after exposing to UV light, a series of electron and holes over TiO_2 surface sites (Fig. 7.7). As it is well understood that these metals have low Fermi levels than TiO_2 and the photoexcited electrons could be easily deposited over metal oxides surface sites from semiconductor conduction band, while photo-generated valence band holes stay on TiO_2 surface sites. Thus, transferred electrons can directly participate in nitrate reduction process. Again, there is the reduction of recombination of electron/hole by increasing the lifetime of the charge separation state. So, the accumulated electron over metal oxides surface sites would directly be accessible for nitrate reduction. Moreover, its more obvious that hole scavenger is the major contributor to enhancing process efficiency recombination of holes and electron. A similar scenario have been reported by other authors [43, 127, 128].

An array of metal-loaded TiO_2 catalysts that have been used for nitrate reduction in photocatalysis are summarized in Table 7.3.

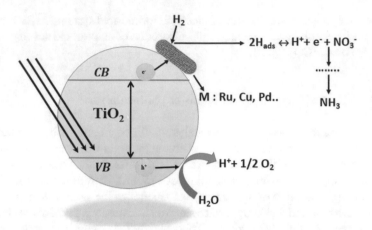

Fig. 7.7 Schematic representation of possible mechanism behind the UV-assisted nitrate reduction over metal-loaded TiO_2

7.4.2.4 Effect of Hole Scavengers

The hole scavengers are important during photocatalytic process. During photoexcitation of semiconductor material, the higher recombination phenomenon of electron and hole pair is regarded as main disadvantage of TiO_2 photocatalyst. In order to overcome this setback, it is important to block or control the movement of electron within the semiconductor through scavenging or trapping generated holes and excited electrons.

The presence of hole scavengers (or sacrificial electron donors) enhances electrons lifetime that is required for specific reactions. Thus, with the aim of improving photocatalytic reduction of NO_3^-, a series of sacrificial electron donors, (i.e., EDTA, sodium oxalate, ethanol, sucrose, formic acid, methanol, and humic acid) have been employed [95, 117–119, 126, 129–131]. The mineralization of the hole scavenger produced CO_2 and H_2 or partial oxidization into CO_2 or low molecular weight organics or acids [118, 119]. Different kinds of hole scavengers have been investigated and exhibited diverse impacts over different systems. Thus, it is important to select the best suited scavenger for better activity of photocatalytic process. Sá et al. [119] investigated nitrate photoreduction using different hole scavengers over TiO_2 surface sites (P25 or Hombicat-supported by Fe, Cu, Ag) and their findings showed that among the scavengers, formic acid proved better, with high conversion up to 95% and nitrogen selectivity up to 67% in presence of 1 Cu/P25.

The nature and impact of hole scavenger for selective nitrate transformation into nitrogen, nitrite, and ammonium over Pd–Cu/TiO_2 surface have been investigated [132]. It was revealed that the process efficiency were closely related to the nature of hole scavenger which is also associated with the nature of the reactive species produced over catalyst surface sites (Fig. 7.8) [132]. Formic acid proved to be the

Table 7.3 Photocatalytic reduction of nitrate in an aqueous suspension of metal-loaded P-25 TiO_2 in the presence of oxalic acid

Metal loaded	pH	Time (h)	NO_3^-/μmol	NO_2^-/μmol	NH_3/NH_4^+/μmol	N_2 selectivityc/%	Ref
Cu	1	12	28	0	20	0	[123]
Cu	5	12	24	5	8	0	[123]
Cu	10	12	36	13	0	0	[123]
Cu	–	3	22	0.04	20	–	[122]
Cu	8	2	28	16	3	21	[121]
Ag	11	12	38	13	0	0	[123]
Ag	–	3	43	ND	23	–	[122]
Ag	4	2	~0	20	16	20	[121]
Ag	8	2	6	41	~0	20	[121]
Ag	12 >	2	36	12	~0	29	[121]
Co	–	3	<2.2	< 0.02	1	–	[122]
Pt	11	12	48	0	0	0	[123]
Pt	–	3	3.2	0.04	<1	–	[122]
Pt	8	2	50	~0	~0	~0	[121]
Pd	11	12	47	0	0	0	[123]
Pd	–	3	1.3	0.04	<1	–	[122]
Pd	8	2	48	~0	2	6	[121]
Au	11	12	42	6	0	0	[123]
Au	–	6	22	0.03	11		[122]
Au	8	2	48	~0	~0	20	[121]
Ni	–	6	10	0.03	10	–	[122]

most efficient and the no conversion occurred in the presence of oxalic and humic acids.

After irradiation, the general redox reactions between nitrite and nitrate with formic acid are given below [118]:

$$2NO_3^- + 5HCOO^- + 7H^+ \rightarrow N_2 + 5CO_2 + 6H_2O \tag{7.34}$$

$$2NO_3^- + 12H^+ + 10e \rightarrow N_2 + 6H_2O \tag{7.35}$$

$$HCOO^- + h^+ \rightarrow H^+ + CO_2 \cdot^- \tag{7.36}$$

$$2NO_3^- + 12H^+ + 10CO_2 \cdot^- \rightarrow N_2 + 6H_2O + 10CO_2 \tag{7.37}$$

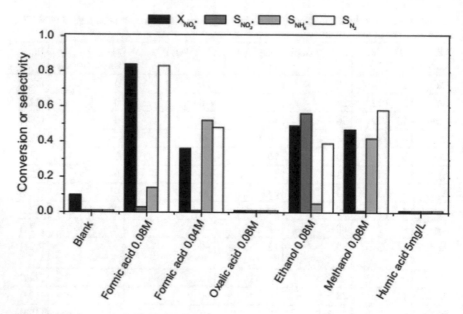

Fig. 7.8 Nitrate conversions ($X_{NO_2^-}$) and nitrite, ammonium and nitrogen selectivities ($S_{NO_2^-}$, $S_{NH_4^+}$, S_{N_2}) over the Pd–Cu/TiO2 catalyst after 4 h of reaction under UV–Vis irradiation using different hole scavengers. (Adapted with permission from ref [132], copyright 2014) Elsevier)

$$2NO_2^- + 3HCOO^- + 5H^+ \rightarrow N_2 + 3CO_2 + 4H_2O \tag{7.38}$$

$$2NO_2^- + 8H^+ + 6e \rightarrow N_2 + 4H_2O \tag{7.39}$$

$$2NO_2^- + 8H^+ + 6CO_2\bullet^- \rightarrow N_2 + 4H_2O + 6CO_2. \tag{7.40}$$

The holes and electron after photoexcitation was consumed by formic acid and nitrate, respectively. Nitrates can be reduced directly or through intermediaries of $CO_2\bullet^-$ which is being formed during the oxidation of formic acid.

Ni-Cu/TiO$_2$ catalysts were investigated for nitrates photoreduction by Wenliang et al. [117]. During photocatalytic nitrate reduction, oxalic acid and sodium oxalate were used as the hole scavenger. High conversion of nitrates in the presence of oxalic acid, as a sacrificial agent, was obtained, with a high yield of ammonium, Fig. 7.9. The overall nitrate reduction process in the presence of oxalic acid is given below (under UV illumination):

$$NO_3^- + C_2O_4^{2-} + 2H^+ \leftrightarrow NO_2^- + 2CO_2 + H_2O \tag{7.41}$$

$$NO_3^- + 4C_2O_4^{2-} + 10H^+ \leftrightarrow NH_4^+ + 8CO_2 + 3H_2O \tag{7.42}$$

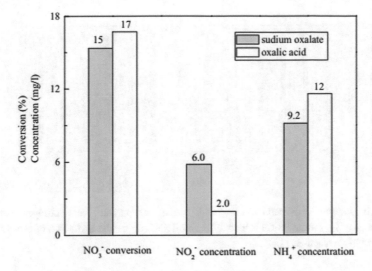

Fig. 7.9 Effect of hole scavengers (sodium oxalate and oxalic acid) in the photocatalytic reduction of nitrate ions over Ni–Cu/TiO$_2$ (4 wt%, Ni:Cu = 3:1) catalyst. (Adapted with permission from ref [117], copyright 2004, Elsevier)

Similarly, in another research performed by Yuexiang et al. [133], these authors further clarified the significant impact of oxalic acid as an effective hole scavenger. They observed that oxalic acid alone, added to the TiO$_2$ catalyst, promoted the catalytic reduction of nitrate. However, after conversion, ammonium was produced as by-product, which is undesirable (Eq. 7.43).

$$NO_3^- + 4C_2O_4^{2-} + 10H^+ \rightarrow NH_4^+ + 8CO_2 + 3H_2O \tag{7.43}$$

A mechanism involving the radical CO$_2\bullet^-$ anion has also been proposed, which is formed from partial oxidation of oxalic acid by h$^+$ (Eq. 7.44):

$$C_2O_4^{2-} + h^+ \rightarrow CO_2 + CO_2\bullet^- \tag{7.44}$$

7.4.3   Photocatalysts for Nitrate Reduction

According to literature survey, the unmodified semiconductors usually exhibit very minimum photocatalytic performance. Conversely, the surface modification of semiconductors, by doping with other elements, changing the crystalline phase, coupling with another semiconductor, and surface and junction defects, enhanced light harvesting and the nitrate catalytic reduction into N$_2$. Silva et al. [134] used TiO$_2$/carbon nanotube hybrid material (CNT), loaded with 1%Pd-1% Cu (%. wt.), for

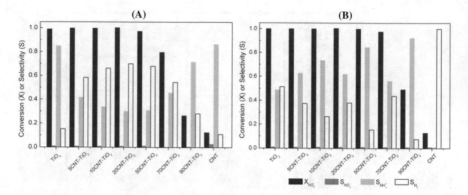

Fig. 7.10 Nitrate conversion $X_{NO_3^-}$ and nitrite, ammonium, and nitrogen selectivity ($S_{NO_2^-}$, $S_{NH_4^+}$, S_{N_2}) using Pd-Cu loaded TiO$_2$ and CNT-TiO$_2$ catalysts after 4 h of catalytic **a** and photocatalytic **b** reactions [134]

effective catalytic activity toward nitrate reduction. The composite material showed the best rate of nitrate conversion within (60 min). The progressive loss of the activity of NO$_3^-$ conversion was observed with higher CNT content. An increase in N$_2$ selectivity with increasing CNT content up to 20 wt% was observed (Fig. 7.10a). The CNT-TiO$_2$ catalysts was more efficient for the catalytic conversion of NO$_3^-$ to ammonia (Fig. 7.10b) [134].

In the presence of H$_2$ and CO$_2$, the possible photocatalytic reduction of nitrate, over Pd–Cu/CNT-TiO$_2$, was explained by the model presented in Fig. 7.11. It was

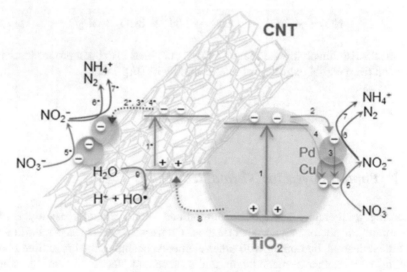

Fig. 7.11 Schematic representation of the photocatalytic reduction of nitrate over Pd–Cu/CNT-TiO$_2$ catalysts in the presence of H$_2$ and CO$_2$ under near UV to visible light irradiation [134]

posited that the simultaneous excitement of both TiO_2 and CNT occurred because the irradiation source emitted in the near UV to visible range. The transfer of electron over Pd and Cu nanoparticles present on TiO_2 and also over CNTs started after charge separation (steps **2–4**). This is because photo-generated electrons are the major contributors for the enhancement of the reduction of adsorbed nitrate and nitrite over Cu and Pd, respectively (Fig. 7.11, steps **5–7**). Conversely, in the presence of an electron donor, the H^+ could be oxidized to $HO^•$ after the migration of positively charged holes from TiO_2 to CNT surface sites (steps **8 and 9**). Moreover, hydroxyl radicals may indirectly re-oxidize by-products to NO_3^-, while $CO_2^{•-}$ which can be generated from the reduction of CO_2 by available electrons, may play a role as reducing mediator. The presence of excessive amount of H^+ ions in the reaction medium, using CNT-TiO_2 catalyst, brought about the higher selectivity toward NH_4^+ production after irradiation step **9** as shown in Fig. 7.11.

Kato et al. [135] investigated the use of Cu and Ag doped TiO_2 (CuAg/TiO_2) for highly selective reduction of nitrate into ammonia after UV irradiation. The optimum experimental conditions were examined, such as Cu and Ag molar ratio effect over TiO_2, the loaded CuAg amount, the amount of catalyst loaded and photocatalyst reusability potential. Their findings showed that higher NH_3 selectivity (85%), and higher NO_3^- conversion (96%) over (Cu0.9Ag/TiO_2) catalyst rather than simple TiO_2, where the maximum NO_3^- conversion and NH_3 selectivity were 44.9% and 33.6%, respectively. The (1 wt% Cu0.9Ag/TiO_2) catalyst, 30 mg of catalyst, CH_3OH as hole scavenger and pH = 6 are the optimum conditions for the highly effective reaction. The 1 wt% Cu0.9Ag/TiO_2 catalyst showed good stability after five cycles of photocatalytic activity. The reaction mechanism during the photoreduction of NO_3^- to NH_3 was explained by the model presented in Fig. 7.12[135].

After UV light irradiation of CuAg/TiO_2, the production of holes and electrons started. Thus, the electrons moved from conduction band to the valence band and then to the CuAg alloy surface sites, where electrons participated in NO_3^- reduction into NO_2^-, and then NH_3. The production of H_2 was also observed during CH_3OH oxidation into HCOOH, which led to the further reduction of NO_3^- [135]. The overall reaction mechanisms are presented below:

$$\text{Photocatalyst } h\nu, \text{cat.} \rightarrow h^+ + e^- \tag{7.45}$$

$$NO_3^- + 2H^+ + 2e^- \rightarrow NO_2^- + H_2O \tag{7.46}$$

$$NO_3^- + 10H^+ + 8e^- \rightarrow NH_4^+ + 3H_2O \tag{7.47}$$

$$NO_3^- + 6H^+ + 5e^- \rightarrow 1/2N_2^+ 3H_2O \tag{7.48}$$

It is widely accepted that the O vacancies in semiconductors generally introduce localized impurity states within the band gap [136]. Photoexcited carriers are assumed to be trapped at the defect states and thereby lose their mobility, which significantly

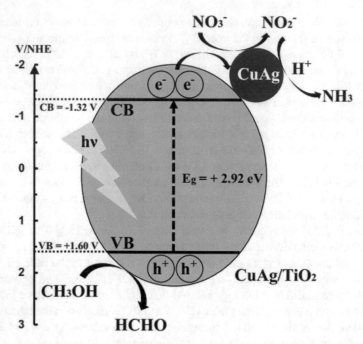

Fig. 7.12 Scheme reaction mechanism of CuAg/TiO$_2$ in NO$_3^-$ reduction to NH$_3$ [135]

influences photocatalytic activity. These defects have a significant influence on the formation, separation, migration of photo-generated electrons and holes, as well as light-absorption range. For instance, Hirakawa and coworkers [137] investigated the photocatalytic reductions of nitrate (NO$_3^-$) over simple TiO$_2$ (Fig. 7.13a). Further investigations revealed that NO$_3^-$ could be strongly coordinated with two defect-induced Ti^{3+} sites, through a bidentate structure and photoexcited electrons, which can easily reduce it into NO$_2^-$ and NO species. These intermediates induced the selective production of NH$_3$, after adsorption over surface defects sites (Fig. 7.13b). The N$_2$ molecules from N$_2$O was produced as a result of the reaction of NO$_2^-$ intermediate with other intermediates (Fig. 7.13c). They reported that the photocatalyst exhibited about 97% selectivity for the nitrate-to-ammonia transformation, due to smaller number of Lewis acid sites and large number of surface defects. Thus, such kind of nitrate reduction pathway to NH$_3$ provided a valuable approach [137].

TiO$_2$ composite is regarded as one of the most effective for nitrate reduction. For instance, Hirayama and coworkers [138] evaluated the potential photocatalytic behavior of the combined Pt/TiO$_2$ with SnPd/Al$_2$O$_3$ for effective NO$_3^-$ photocatalytic reduction in real groundwater under irradiation at $\lambda > 300$ nm, using glucose as a hole scavenger. Pt/TiO$_2$ was found to be ineffective for NO$_3^-$ decomposition and production of H$_2$ (Eq. 7.49). On the other hand, when Sn–Pd/Al$_2$O$_3$ was used, the conversion of nitrate was more significant without nitrite formation and with the desired N$_2$ selectively formation (75% selectivity at 23% conversion of NO$_3^-$).

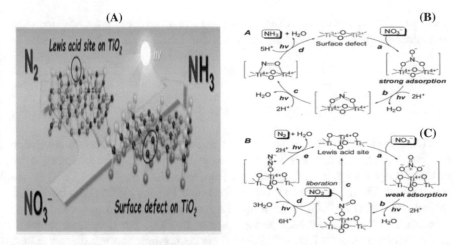

Fig. 7.13 a Schematic representation of photocatalytic nitrate reduction to ammonia or nitrogen. Proposed mechanisms for photocatalytic reduction of NO_3^- on **b** surface defect and **c** Lewis acid site of TiO_2. Reproduced with permission from Ref. [137] Copyright 2017 the American Chemical Society

According to cross blank testing, the conversion of NO_3^- was negligible even when a bimetallic catalyst $Pt/TiO_2 - SnPd/Al_2O_3$ was used. Moreover, after UV irradiation, neither NO_3^- reduction nor H_2 production was observed in the absence of glucose.

Through the coupling of two systems Pt/TiO_2 and $SnPd/Al_2O_3$ (Fig. 7.14), photocatalytically produced H_2 acted as a reactant for nitrate reduction (Eq 7.50) [138]. The

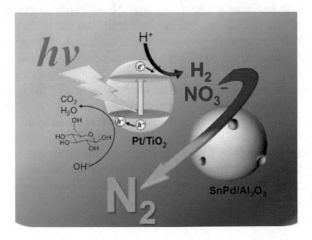

Fig. 7.14 Schematic representation of photocatalytic nitrate reduction to of NO_3^- in real groundwater in the over the $Pt/TiO_2 - SnPd/Al_2O_3$ system. Reproduced with permission from Ref. [138]. Copyright 2014 the American Chemical Society)

photocatalytic separated H_2 production and its further use for nitrate reduction were also accomplished with a substrate $Pt/TiO_2 + Sn–Pd/Al_2O_3$ in real groundwater.

$$2H^+ + 2e^- \rightarrow H_2 \tag{7.49}$$

$$NO_3^- + 5/2H_2 \rightarrow 1/2N_2 + 2H_2O + OH^- \tag{7.50}$$

The authors observed poor performance in the groundwater system, relative to the pure nitrate solutions. Furthermore, the catalytic performance was not greatly influenced by a wide variety of cations (e.g., Na^+, K^+, Mg^{2+}, and Ca^{2+}) in the groundwater. On the other hand, sulfate ions over Pt sites on Pt/TiO_2 deactivated the oxidation sites on TiO_2 by reacting with surface hydroxyl groups, which led to a decrease in the photocatalytic performance of Pt/TiO_2. The decreased photocatalytic performance of $Pt/TiO_2 - SnPd/Al_2O_3$ in the groundwater was closely related to the presence of silicate, sulfate, and other organic compounds. Because, silicate and sulfate ions deactivated Pt_e^- and $TiO_2\_h^+$ surface overs Pt/TiO_2 surface via adsorption, a decrease in photocatalytic activity and H_2 evolution ensued.

Another investigation was carried out by Challagulla et al. [139] regarding competitive nature of hydrogen generation and nitrate reduction by using noble metals doped TiO_2. The performance of the noble metal doped TiO_2 was compared with undoped TiO_2. Under identical reaction conditions, the photoreductions of Pt/TiO_2, Pd/TiO_2, and Au/TiO_2 were far lower than that of undoped pure catalyst. Comparatively, Ag/TiO_2 showed better performance than the noble metal doped catalysts, but the overall activity was lower than that of undoped TiO_2 (Fig. 7.15a). The H_2 production potential of the reaction was also determined by the authors on the TiO_2 doped by noble metals. The Pd and Pt incorporation highly favored the production of H_2 while that of Au and Ag moderately favored H_2 production. The H_2 generation increased by 400 fold after Pd and Pt doping and the photoreduction rate of nitrate decreased by about 15 fold with respect to the undoped TiO_2 (Fig. 7.15b). Previous studies clearly indicate that a noble metal doped TiO_2 has a considerable potential to generate H_2 rather than pristine undoped TiO_2 and the produced H_2 is ineffective for NO_3^- reduction. Therefore, authors confirmed that the conducting band (CB) is responsible for nitrate reduction. However, below CB minima, there exist some sub-bands after Pt doping which results in the draining of photoexcited electron to the sub-band position below the nitrate reduction potential and this and this results in a decrease in nitrate reduction process (Fig. 7.15c). This indicated that hydrogen reduction potential is actively located under Pt sub-bands and its might be more helpful for proficient hydrogen production onto the surface of Pt/TiO_2 [139]. The Pt and Pd have ability to reducing hydrogen affinity as well as catalytic properties [121, 127].

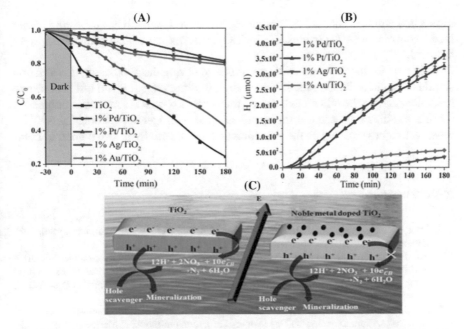

Fig. 7.15 **a** Photoreduction of nitrate over pristine and noble metals doped TiO₂ performed in 50:50 methanol-water mixture, **b** the photoreduction of nitrate over pristine and noble metal doped TiO₂ performed in 50:50 methanol-water mixture, **c** schematic representation of photoreduction of nitrate over pristine and noble metal doped TiO₂ Reproduced with permission from Ref. [139] Copyright 2017 the American Chemical Society)

7.5 Conclusions

About 75% of the world's ammonia production goes to fertilizer production for agricultural purposes. Ultimately, high levels nitrate wastewater is produced that pollutes water. Heterogenous photocatalysis has been applied successfully for the reduction of nitrate in water. To carry out the photocatalytic reduction of nitrate, operating conditions such as pH are very important. Several photocatalysts have been developed for enhanced reduction of nitrate and also to control the by-product formation. The effect of the presence of hole scavengers (or sacrificial electron donors) on the photocatalytic reduction of nitrate was summarized and discussed. It was revealed that the process efficiency was closely related to the nature of hole scavenger. In terms of photocatalytic materials, the photocatalytic efficiency and the selectivity are strongly depending on the type of photocatalyst. Therefore, we discussed the development of novel photocatalysts for nitrate reduction. For example, TiO₂/carbon nanotube hybrid material (CNT), loaded with 1%Pd-1% Cu (%. wt.), is very effective for the photocatalytic activity toward nitrate reduction. An increase in N₂ selectivity with increasing CNT content up to 20 wt% was reported. The CNT-TiO₂ catalysts were more efficient for the catalytic conversion of NO₃⁻ to ammonia. Cu and Ag doped TiO₂ (CuAg/TiO₂) were found to be highly selective for the reduction of nitrate into ammonia under UV irradiation; higher NH_3 selectivity (85%) and higher NO_3^- conversion (96%) over $(Cu_{0.9}Ag/TiO_2)$ catalyst rather than simple TiO_2 were reported.

Solutions for the recovery of N-gases formed during the photocatalytic reduction should be studied. Reactors that can ensure high mass transfer and efficient nitrate reduction, and on the other hand, a good recovery of N-gases is much recommended for investigation. Furthermore, the combination of heterogeneous photocatalysis with other techniques to enhance the reduction and the selectivity is also a good option.

References

1. Pooi CK, Ng HY (2018) Review of low-cost point-of-use water treatment systems for developing communities. npj Clean Water 1:11
2. Hsini A, Naciri Y, Laabd M, El Ouardi M, Ajmal Z, Lakhmiri R, Boukherroub R, Albourine A (2020) Synthesis and characterization of arginine-doped polyaniline/walnut shell hybrid composite with superior clean-up ability for chromium (VI) from aqueous media: Equilibrium, reusability and process optimization. J Mol Liq 316:113832
3. Abdellaoui Y, Abou Oualid H, Hsini A, El Ibrahimi B, Laabd M, El Ouardi M, Giácoman-Vallejos G, Gamero-Melo P (2021) Synthesis of zirconium-modified Merlinoite from fly ash for enhanced removal of phosphate in aqueous medium: Experimental studies supported by Monte Carlo/SA simulations. Chem Eng J 404:126600
4. Benafqir M, Hsini A, Laabd M, Laktif T, Ait Addi A, Albourine A, El Alem N (2019) Application of density functional theory computation (DFT) and process capability study for performance evaluation of orthophosphate removal process using polyaniline@hematite-titaniferous sand composite (PANI@HTS) as a substrate. Sep Purif Technol 236:116286
5. Ajmal Z, Muhmood A, Usman M, Kizito S, Lu J, Dong R, Wu S (2018) Phosphate removal from aqueous solution using iron oxides: Adsorption, desorption and regeneration characteristics. J Colloid Interface Sci 528:145–155
6. El Jaouhari A, Chennah A, Ben Jaddi S, Ait Ahsaine H, Anfar Z, Tahiri Alaoui Y, Naciri Y, Benlhachemi A, Bazzaoui M (2019) Electrosynthesis of zinc phosphate-polypyrrole coatings for improved corrosion resistance of steel. Surf Interfaces 15:224–231
7. Essekri A, Hsini A, Naciri Y, Laabd M, Ajmal Z, El Ouardi M, Ait Addi A, Albourine A (2020) Novel citric acid-functionalized brown algae with a high removal efficiency of crystal violet dye from colored wastewaters: insights into equilibrium, adsorption mechanism, and reusability. Int J Phytoremediation 0:1–11
8. Hsini A, Essekri A, Aarab N, Laabd M, Addi AA, Lakhmiri R, Albourine A (2020) Elaboration of novel polyaniline@Almond shell biocomposite for effective removal of hexavalent chromium ions and Orange G dye from aqueous solutions. Environ Sci Pollut Res, pp 1–14
9. Djellabi R, Zhang L, Yang B, Haider MR, Zhao X (2019) Sustainable self-floating lignocellulosic biomass-TiO2@Aerogel for outdoor solar photocatalytic Cr(VI) reduction. Sep Purif Technol 229:115830
10. Bhojwani S, Topolski K, Mukherjee R, Sengupta D, El-Halwagi MM (2019) Technology review and data analysis for cost assessment of water treatment systems. Sci Total Environ 651:2749–2761

11. Bolisetty S, Peydayesh M, Mezzenga R (2019) Sustainable technologies for water purification from heavy metals: review and analysis. Chem Soc Rev 48:463–487

12. Aarab N, Hsini A, Essekri A, Laabd M, Lakhmiri R, Albourine A (2020) Removal of an emerging pharmaceutical pollutant (metronidazole) using PPY-PANi copolymer: Kinetics, equilibrium and DFT identification of adsorption mechanism. Groundw Sustain Dev 11:100416

13. Aarab N, Hsini A, Laabd M, Essekri A, Laktif T, Haki MA, Lakhmiri R, Albourine A (2020) Theoretical study of the adsorption of sodium salicylate and metronidazole on the PANi. Mater Today Proc 22:100–103

14. Ba Mohammed B, Hsini A, Abdellaoui Y, Abou Oualid H, Laabd M, El Ouardi M, Ait Addi A, Yamni K, Tijani N (2020) Fe-ZSM-5 zeolite for efficient removal of basic Fuchsin dye from aqueous solutions: Synthesis, characterization and adsorption process optimization using BBD-RSM modeling. J Environ Chem Eng. 8:104419

15. Hsini A, Naciri Y, Benafqir M, Ajmal Z, Aarab N, Laabd M, Navío JA, Puga F, Boukherroub R, Bakiz B, Albourine A (2021) Facile synthesis and characterization of a novel 1,2,4,5-benzene tetracarboxylic acid doped polyaniline@zinc phosphate nanocomposite for highly efficient removal of hazardous hexavalent chromium ions from water. J Colloid Interface Sci. 585:560–573

16. Djellabi R, Fouzi Ghorab M, Smara A, Bianchi CL, Cerrato G, Zhao X, Yang B (2020) Titania–montmorillonite for the photocatalytic removal of contaminants from water: adsorb & shuttle process. In: Green Materials for Wastewater Treat. Chap. 13, pp 291–319

17. Naciri Y, Hsini A, Ajmal Z, Navío JA, Bakiz B, Albourine A, Ezahri M, Benlhachemi A (2020) Recent progress on the enhancement of photocatalytic properties of BiPO$_4$ using π–conjugated materials. Adv Colloid Interface Sci 280:102160

18. Naciri Y, Chennah A, Jaramillo-Páez C, Navío JA, Bakiz B, Taoufyq A, Ezahri M, Villain S, Guinneton F, Benlhachemi A (2019) Preparation, characterization and photocatalytic degradation of Rhodamine B dye over a novel Zn$_3$(PO$_4$)$_2$/BiPO$_4$ catalyst. J Environ Chem Eng 7:103075

19. Djellabi R, Yang B, Xiao K, Gong Y, Cao D, Sharif HMA, Zhao X, Zhu C, Zhang J (2019) Unravelling the mechanistic role of Ti–O–C bonding bridge at titania/lignocellulosic biomass interface for Cr(VI) photoreduction under visible light. J Colloid Interface Sci 553:409–417

20. Naciri Y, Ait Ahsaine H, Chennah A, Amedlous A, Taoufyq A, Bakiz B, Ezahri M, Villain S, Benlhachemi A (2018) Facile synthesis, characterization and photocatalytic performance of Zn$_3$(PO$_4$)$_2$ platelets toward photodegradation of Rhodamine B dye. J Environ Chem Eng 6:1840–1847

21. Naciri Y, Bouddouch A, Bakiz B, Taoufyq A, Ezahri M, Benlhachemi A (2019) Photocatalytic degradation of sulfadiazine by Zn$_3$(PO$_4$)$_2$/BiPO$_4$ composites upon UV light irradiation. Mater Today Proc 3:8–11

22. Bianchi CL, Cerrato G, Bresolin BM, Djellabi R, Rtimi S (2020) Digitally printed AgNPs doped TiO$_2$ on commercial porcelain-Grès tiles: synergistic effects and continuous photocatalytic antibacterial activity. Surfaces 3:11–25

23. Naciri Y, Hsini A, Ajmal Z, Bouddouch A, Bakiz B, Navío JA, Albourine A, Valmalette JC, Ezahri M, Benlhachemi A (2020) Influence of Sr-doping on structural, optical and photocatalytic properties of synthesized Ca$_3$(PO$_4$)$_2$. J Colloid Interface Sci 572:269–280

24. Barebita H, Naciri Y, Ferraa S, Nimour A, Guedira T (2020) Investigation of structural and photocatalytic behavior of Bi$_{13}$B$_{1-2x}$V$_x$P$_x$O$_{20.95}$ + 2x ($0 \leq x \leq 0.5$). Solid State Sci 108:106389

25. Ahsaine HA, Slassi A, Naciri Y, Chennah A, Jaramillo-Páez C, Anfar Z, Zbair M, Benlhachemi A, Navío JA (2018) Photo/electrocatalytic properties of nanocrystalline ZnO and La–Doped ZnO: combined DFT fundamental semiconducting properties and experimental study. ChemistrySelect 3:7778–7791

26. Shaim A, Amaterz E, Naciri Y, Taoufyq A, Bakiz B, Ezahri M, Benlhachemi A, Ouammou A, Chahine A (2019) Synthesis, characterization and photocatalytic activity of titanophosphate glasses. Mediterr J Chem 8:66–73

27. Ward MH, Jones RR, Brender JD, de Kok TM, Weyer PJ, Nolan BT, Villanueva CM, van Breda SG (2018) Drinking water nitrate and human health: an updated review. Int J Environ Res Public Health 15:1–31
28. Wick K, Heumesser C, Schmid E (2012) Groundwater nitrate contamination: Factors and indicators. J Environ Manage 111:178–186
29. Wakida FT, Lerner DN (2005) Non-agricultural sources of groundwater nitrate: a review and case study. Water Res 39:3–16
30. Ju K-S, Parales RE (2010) Nitroaromatic compounds, from synthesis to biodegradation. Microbiol Mol Biol Rev 74:250–272
31. Majumdar D, Gupta N (2000) Nitrate pollution of groundwater and associated human health disorders. Indian J Environ Health 42:28–39
32. Water S (2006) W.H. Organization, Guidelines for drinking-water quality, 1. http://apps.who.int/iris/handle/10665/43428
33. Bhatnagar A, Sillanpää M (2011) A review of emerging adsorbents for nitrate removal from water. Chem Eng J 168:493–504
34. Samatya S, Kabay N, Yüksel Ü, Arda M, Yüksel M (2006) Removal of nitrate from aqueous solution by nitrate selective ion exchange resins. React Funct Polym 66:1206–1214
35. Fanning JC (2000) The chemical reduction of nitrate in aqueous solution. Coord Chem Rev 199:159–179
36. Mautner A, Kobkeatthawin T, Bismarck A (2017) Efficient continuous removal of nitrates from water with cationic cellulose nanopaper membranes. Resour Technol 3:22–28
37. Koparal AS, Öütveren ÜB (2002) Removal of nitrate from water by electroreduction and electrocoagulation. J Hazard Mater 89:83–94
38. Jensen VB, Darby JL, Seidel C, Gorman C (2014) Nitrate in potable water supplies: alternative management strategies. Crit Rev Environ Sci Technol 44:2203–2286
39. Tugaoen HON, Garcia-Segura S, Hristovski K, Westerhoff P (2017) Challenges in photocatalytic reduction of nitrate as a water treatment technology. Sci Total Environ 599–600:1524–1551
40. Yue M, Wang R, Cheng N, Cong R, Gao W, Yang T (2016) $ZnCr_2S_4$: Highly effective photocatalyst converting nitrate into N2 without over-reduction under both UV and pure visible light. Sci Rep 6:1–11
41. Yang T, Doudrick K, Westerhoff P (2013) Photocatalytic reduction of nitrate using titanium dioxide for regeneration of ion exchange brine. Water Res 47:1299–1307
42. Adachi M, Kudo A (2012) Effect of surface modification with layered double hydroxide on reduction of nitrate to nitrogen over $BaLa_4Ti_4O_{15}$ photocatalyst. Chem Lett 41:1007–1008
43. Anderson JA (2011) Photocatalytic nitrate reduction over Au/TiO_2. Catal Today 175:316–321
44. Moorcroft MJ, Davis J, Compton RG (2001) Detection and determination of nitrate and nitrite: a review. Talanta 54:785–803
45. Fewtrell L (2004) Drinking-water nitrate, methemoglobinemia, and global burden of disease: a discussion. Environ Health Perspect 112:1371–1374
46. Vione D, Maurino V, Minero C, Pelizzetti E, Analitica C, Torino U, Giuria VP (2005) Reactions induced in natural waters by irradiation of nitrate and nitrite ions. Environ Chem, pp 221–253
47. Warneck P, Wurzinger C (1988) Product quantum yields for the 305-nm photodecomposition of NO3- in aqueous solution. J Phys Chem 92:6278–6283
48. Gankanda A, Grassian VH (2014) Nitrate photochemistry on laboratory proxies of mineral dust aerosol: wavelength dependence and action spectra. J Phys Chem C 118:29117–29125
49. Scharko NK, Berke AE, Ra JD (2014) Release of nitrous acid and nitrogen dioxide from nitrate photolysis in acidic aqueous solutions. Environ Sci Technol 48:11991–12001
50. Fischer M, Warneck P (1996) Photodecomposition of Nitrite and Undissociated Nitrous Acid in Aqueous Solution. J Phys Chem 3654:18749–18756
51. Rezvani F, Sarrafzadeh M, Ebrahimi S (2019) Nitrate removal from drinking water with a focus on biological methods: a review. Environ Sci Pollut Res 26:1124–1141

52. Chindler DAWS, Chlesinger WIHS, Ilman DAGT (1997) Human alteration of the global nitrogen cycle: sources and consequences. Ecol Appl 7:737–750
53. Erisman JW, Galloway JN, Seitzinger S, Bleeker A, Dise NB, Petrescu AMR, Leach AM, De Vries W, Erisman JW (2013) Consequences of human modification of the global nitrogen cycle. Philos Trans R Soc B, p 368
54. Ormerod SJ, Durance I (2009) Restoration and recovery from acidification in upland Welsh streams over 25 years. J Appl Ecol 46:164–17
55. Kincheloe JW, Wedemeyer GA, Koch DL (1979) Tolerance of developing salmonid eggs and fry to nitrate exposure. Bull Environ Contam Toxicol 23:575–578
56. Camargo JA, Alonso A, Salamanca A (2005) Nitrate toxicity to aquatic animals: a review with new data for freshwater invertebrates. Chemosphere 58:1255–1267
57. Ward MH, Theo M, Levallois P, Brender J, Gulis G, Nolan BT, Vanderslice J (2005) Working Group Report: Drinking-water nitrate and health-recent findings and research needs. Environ Health Perspect 113 (11):1607–1614
58. IARC Monographs on the evaluation of carcinogenic risks to humans: ingested nitrate and nitrite, and cyanobacterial peptide toxins (2010)
59. For E, Of I, Oxide N, Flow B (1992) Evidence for involvement of nitric oxide in the regulation of hypothalamic portal blood flow. Neuroscience 51:769–772
60. Radomski MW, Palmer RMJ, Moncada S (1987) The anti-aggregating properties of vascular endothelium: interactions between prostacyclin and nitric oxide. J Pharmacol 92:639–646
61. Palmer RMJ, Ferrige AG, Moncada S (1987) Nitric oxide release accounts for the biological activity of endothelium-delived relaxing factor. Nature 327:524–526
62. Larsen FJ, Schiffer TA, Weitzberg E, Lundberg JO (2012) Regulation of mitochondrial function and energetics by reactive nitrogen oxides. Free Radic Biol Med 53:1919–1928
63. Sanchez-echaniz J, Benito-ferna J (2015) Methemoglobinemia and consumption of vegetables in infants. Pediatrics 107:1024–1028
64. Aschengra A, Zierler S, Cohen A (2010) Quality of community drinking water and the occurrence of spontaneous abortion. Arch. Environ. Heal. An Int. J. 44:283–290
65. Section ET, Agency EP (1996) Health implications of nitrate and nitrite in drinking water: an update on methemoglobinemia occurrence and reproductive and developmental toxicity. Regul Toxicol Pharmacol 23:35–43
66. Van Maanen JMS, Welle IJ, Hageman G, Dallinga JW, Mertens PLJM, Kleinjans JCS (1996) Nitrate contamination of drinking water: relationship with hprtvariant frequency in lymphocyte DNA and urinary excretion of N-nitrosamines. Environ Health Perspect 104:522–528
67. De Groef B, Decallonne BR, Van Der Geyten S, Darras VM, Bouillon R (2006) Perchlorate versus other environmental sodium/iodide symporter inhibitors: potential thyroid-related health effects. Eur J Endocrinol 155:17–25
68. Aschebrook-kilfoy B, Heltshe SL, Nuckols JR, Sabra MM, Shuldiner AR, Mitchell BD, Airola M, Holford TR, Zhang Y, Ward MH (2012) Modeled nitrate levels in well water supplies and prevalence of abnormal thyroid conditions among the Old Order Amish in Pennsylvania. Environ Heal, pp 1–11
69. Elena S (2008) Possible effects of environmental nitrates and toxic organochlorines on human thyroid in highly polluted areas in slovakia. Thyroid 18:353–362
70. Afzal BM (2006) Drinking water and women's health. J Midwifery Women's Heal 51:12–18
71. Cech E (2012) Great problems of grand challenges : problematizing engineering' s understandings of its role in society. Int J Eng Soc Justice Peace. 1:85–94
72. EPA, National Primary Drinking Water Regulations (2017)
73. Official Journal of the European Communities, DIRECTIVE 2000/60/EC OF THE EUROPEAN PARLIAMENT AND OF THE COUNCILof 23 October 2000 establishing a framework for Community action in the field of water policy (2000)
74. Li E, Wang R, Jin X, Lu S, Qiu Z, Zhang X (2018) Investigation into the nitrate removal efficiency and microbial communities in a sequencing batch reactor treating reverse osmosis concentrate produced by a coking wastewater treatment plant. Environ Technol 39:2203–2214

75. Ratnayake SY, Ratnayake AK, Schild D, Maczka E, Jartych E, Luetzenkirchen J, Kosmulski M, Weerasooriya R (2017) Chemical reduction of nitrate by zerovalent iron nanoparticles adsorbed radiation-grafted copolymer matrix. Nukleonika 62:269–275
76. Riveros F, guajardo N, Valenzuela MB, Bernardes AM, Ferreira JZ, Cifuentes G (2019) Removal of nitrates from copper-containing aqueous acidic leach solutions by electrodialysis. Miner Process Extr Metall Trans Inst Min Metall 0:1–9
77. PRY, Senthil Kumar SRP (2018) Abstract, efficient removal of nitrate and phosphate using graphene nanocomposites, Springer International Publishing. https://doi.org/10.1007/978-3-319-75484-0
78. Xu Q, Liu X, Yang G, Wang D, Wang Q, Liu Y, Li X, Yang Q (2019) Free nitrous acid-based nitrifying sludge treatment in a two-sludge system obtains high polyhydroxyalkanoates accumulation and satisfied biological nutrients removal. Bioresour Technol 284:16–24
79. Choi J, Batchelor B (2008) Nitrate reduction by fluoride green rust modified with copper. Chemosphere 70:1108–1116
80. Feleke Z, Sakakibara Y (2002) A bio-electrochemical reactor coupled with adsorber for the removal of nitrate and inhibitory pesticide. Water Res 36:3092–3102
81. McAdam EJ, Judd SJ (2006) A review of membrane bioreactor potential for nitrate removal from drinking water. Desalination 196:135–148
82. Cyplik P, Grajek W, Marecik R, Króliczak P, Dembczyński R (2007) Application of a membrane bioreactor to denitrification of brine. Desalination 207:134–143
83. Lee S, Maken S, Jang JH, Park K, Park JW (2006) Development of physicochemical nitrogen removal process for high strength industrial wastewater. Water Res 40:975–980
84. Prüsse U, Hähnlein M, Daum J, Vorlop KD (2000) Improving the catalytic nitrate reduction. Catal Today 55:79–90
85. Kumar M, Chakraborty S (2006) Chemical denitrification of water by zero-valent magnesium powder. J Hazard Mater 135:112–121
86. Qadeer A, Liu S, Liu M, Liu X, Ajmal Z, Huang Y, Jing Y, Khalil SK, Zhao D, Weining D, Wei XY, Liu Y (2019) Historically linked residues profile of OCPs and PCBs in surface sediments of typical urban river networks, Shanghai: ecotoxicological state and sources. J Clean Prod 231:1070–1078
87. Ajmal Z, Usman M, Anastopoulos I, Qadeer A, Zhu R, Wakeel A, Dong R (2020) Use of nano-/micro-magnetite for abatement of cadmium and lead contamination. J Environ Manage 264:110477
88. Kizito S, Lv T, Wu S, Ajmal Z, Luo H, Dong R (2017) Treatment of anaerobic digested effluent in biochar-packed vertical flow constructed wetland columns: role of media and tidal operation. Sci Total Environ 592:197–205
89. Anthonisen AC, Loehr RC, Prakasam TBS, Srinath EG (1976) Inhibition of nitrification by ammonia and nitrous acid. J Water Pollut Control Fed 48:835–852
90. Wu S, Lv T, Lu Q, Ajmal Z, Dong R (2017) Treatment of anaerobic digestate supernatant in microbial fuel cell coupled constructed wetlands: evaluation of nitrogen removal, electricity generation, and bacterial community response. Sci Total Environ 580:339–346
91. Kizito S, Luo H, Wu S, Ajmal Z, Lv T, Dong R (2017) Phosphate recovery from liquid fraction of anaerobic digestate using four slow pyrolyzed biochars: Dynamics of adsorption, desorption and regeneration. J Environ Manage 201:260–267
92. Ajmal Z, Muhmood A, Dong R, Wu S (2020) Probing the efficiency of magnetically modified biomass-derived biochar for effective phosphate removal. J Environ Manage 253:109730
93. Compagnoni M, Ramis G, Freyria FS, Armandi M, Bonelli B, Rossetti I (2017) Innovative photoreactors for unconventional photocatalytic processes: the photoreduction of CO_2 and the photo-oxidation of ammonia. Rend Lincei 28:151–158
94. Bahadori E, Tripodi A, Ramis G, Rossetti I (2019) Semi-batch photocatalytic reduction of nitrates: role of process conditions and co-catalysts. Chem Cat Chem 11:4642–4652
95. Bems B, Jentoft FC, Schlögl R (1999) Photoinduced decomposition of nitrate in drinking water in the presence of titania and humic acids. Appl Catal B Environ 20:155–163

96. Ketir W, Bouguelia A, Trari M (2009) NO_3^- removal with a new delafossite CuCrO2 photocatalyst. Desalination 244:144–152
97. Chen Y, Crittenden J, Hackney S, Sutter L, Hand DW (2005) Preparation of a novel TiO2-based p-n junction nanotube photocatalyst. Am Chem Soc 39:1201–1208
98. Zhu N, Tang J, Tang C, Duan P, Yao L, Wu Y, Dionysiou DD (2018) Combined CdS nanoparticles-assisted photocatalysis and periphytic biological processes for nitrate removal. Chem Eng J 353:237–245
99. Wang Y, Li B, Li G, Ma X, Zhang H, Huang Y, Wang G, Wang J, Song Y (2017) A modified Z-scheme Er_3^+ : $YAlO_3$ @ (PdS/BiPO4)/(Au/rGO)/CdS photocatalyst for enhanced solar-light photocatalytic conversion of nitrite. Chem Eng J 322:556–570
100. Adamu H, McCue AJ, Taylor RSF, Manyar HG, Anderson JA (2017) Simultaneous photocatalytic removal of nitrate and oxalic acid over Cu_2O/TiO_2 and Cu_2O/TiO_2.AC composites. Appl Catal B Environ 217:181–191
101. Huo WC, Dong X, Li JY, Liu M, Liu XY, Zhang YX, Dong F (2019) Synthesis of Bi_2WO_6 with gradient oxygen vacancies for highly photocatalytic NO oxidation and mechanism study. Chem Eng J 361:129–138
102. Kobwittaya K, Sirivithayapakorn S (2014) Photocatalytic reduction of nitrate over TiO_2 and Ag-modified TiO_2. J Saudi Chem Soc 18:291–298
103. Liu G, You S, Ma M, Huang H, Ren N (2016) Removal of nitrate by photocatalytic denitrification using nonlinear optical material. Environ Sci Technol 50:11218–11225
104. Kim MS, Chung SH, Yoo CJ, Lee MS, Cho IH, Lee DW, Lee KY (2013) Catalytic reduction of nitrate in water over Pd-Cu/TiO_2 catalyst: Effect of the strong metal-support interaction (SMSI) on the catalytic activity. Appl Catal B Environ 142–143:354–361
105. Doudrick K, Yang T, Hristovski K, Westerhoff P (2013) Photocatalytic nitrate reduction in water: managing the hole scavenger and reaction by-product selectivity. Appl Catal B Environ 136–137:40–47
106. Wang J, Song M, Chen B, Wang L, Zhu R (2017) Effects of pH and H_2O_2 on ammonia, nitrite, and nitrate transformations during UV-254nm irradiation: implications to nitrogen removal and analysis. Chemosphere 184:1003–1011
107. Frank AJ, Graetzel M (1982) Sensitized photoreduction of nitrate in homogeneous and micellar solutions. Inorg Chem 21:3834–3837
108. Goldstein S, Rabani J (2007) Mechanism of nitrite formation by nitrate photolysis in aqueous solutions: the role of peroxynitrite, nitrogen dioxide, and hydroxyl radical. J Am Chem Soc 129:10597–10601
109. Mack J, Bolton JR (1999) Photochemistry of nitrite and nitrate in aqueous solution: a review. J Photochem Photobiol A Chem 128:1–13
110. Wagner I, Strehlow H, Busse G (1980) Flash photolysis of nitrate ions in aqueous solution. Zeitschrift Für Phys Chemie. 123:1–33
111. Løgager T, Sehested K (1993) Formation and decay of peroxynitric acid: a pulse radiolysis study. J Phys Chem 97:10047–10052
112. Mark G, Korth HG, Schuchmann HP, Von Sonntag C (1996) The photochemistry of aqueous nitrate ion revisited. J Photochem Photobiol A Chem 101:89–103
113. M.C. Gonzalez, A.M. Braun, Vacuum-UV photolysis of aqueous solutions of nitrate: Effect of organic matter: I. Phenol, J. Photochem. Photobiol. A Chem. 93 (1996) 7–19
114. Treinin A, Hayon E (1970) Absorption spectra and reaction kinetics of NO_2, N_2O_3, and N_2O_4 in aqueous solution. J Am Chem Soc 92:5821–5828
115. Bilski P, Chignell CF, Szychlinski J, Borkowski A, Oleksy E, Reszka K (1992) Photooxidation of organic and inorganic substrates during uv photolysis of nitrite anion in aqueous solution. J Am Chem Soc 114:549–556
116. Calfa JP, Phelan KG, Bonner FT (1982) Cyclic azide as an aqueous solution intermediate: evidence pro and con. Inorg Chem 21:521–524
117. Gao W, Jin R, Chen J, Guan X, Zeng H, Zhang F, Guan N (2004) Titania-supported bimetallic catalysts for photocatalytic reduction of nitrate. Catal Today 90:331–336

118. Zhang F, Jin R, Chen J, Shao C, Gao W, Li L, Guan N (2005) High photocatalytic activity and selectivity for nitrogen in nitrate reduction on Ag/TiO$_2$ catalyst with fine silver clusters. J Catal 232:424–431

119. Sá J, Agüera CA, Gross S, Anderson JA (2009) Photocatalytic nitrate reduction over metal modified TiO2. Appl. Catal. B Environ. 85:192–200

120. Luiz DDB, Andersen SLF, Berger C, José HJ, Moreira RDFPM (2012) Photocatalytic reduction of nitrate ions in water over metal-modified TiO$_2$. J Photochem Photobiol A Chem 246:36–44

121. Gekko H, Hashimoto K, Kominami H (2012) Photocatalytic reduction of nitrite to dinitrogen in aqueous suspensions of metal-loaded titanium (iv) oxide in the presence of a hole scavenger: An ensemble effect of silver and palladium co-catalysts. Phys Chem Chem Phys 14:7965–7970

122. Kominami H, Furusho A, Murakami SY, Inoue H, Kera Y, Ohtani B (2001) Effective photocatalytic reduction of nitrate to ammonia in an aqueous suspension of metal-loaded titanium(IV) oxide particles in the presence of oxalic acid. Catal Lett 76:31–34

123. Kominami H, Nakaseko T, Shimada Y, Furusho A, Inoue H, Murakami SY, Kera Y, Ohtani B (2005) Selective photocatalytic reduction of nitrate to nitrogen molecules in an aqueous suspension of metal-loaded titanium(IV) oxide particles. Chem Commun 3:2933–2935

124. Malato S, Fernández-Ibáñez P, Maldonado MI, Blanco J, Gernjak W (2009) Decontamination and disinfection of water by solar photocatalysis: Recent overview and trends. Catal Today 147:1–59

125. Ranjit KT, Viswanathan B (1997) Photocatalytic reduction of nitrite and nitrate ions over doped TiO$_2$ catalysts. J. Photochem. Photobiol. A Chem. 107:215–220

126. Ranjit KT, Viswanathan B (1997) Photocatalytic reduction of nitrite and nitrate ions to ammonia on M/TiO2 catalysts. J Photochem Photobiol A Chem 108:73–78

127. Ranjit KT, Varadarajan TK, Viswanathan B (1995) Photocatalytic reduction of nitrite and nitrate ions to ammonia on Ru/TiO$_2$ catalysts. J Photochem Photobiol A Chem 89:67–68

128. Sowmya A, Meenakshi S (2015) Photocatalytic reduction of nitrate over Ag-TiO$_2$ in the presence of oxalic acid. J Water Process Eng 8:e23–e30

129. Hirayama J, Kondo H, Miura YK, Abe R, Kamiya Y (2012) Highly effective photocatalytic system comprising semiconductor photocatalyst and supported bimetallic non-photocatalyst for selective reduction of nitrate to nitrogen in water. Catal Commun 20:99–102

130. Mori T, Suzuki J, Fujimoto K, Watanabe M, Hasegawa Y (1999) Reductive decomposition of nitrate ion to nitrogen in water on a unique hollandite photocatalyst. Appl Catal B Environ 23:283–289

131. Penpolcharoen M, Amal R, Brungs M (2001) Degradation of sucrose and nitrate over titania coated nano-hematite photocatalysts. J Nanoparticle Res 3:289–302

132. Soares OSGP, Pereira MFR, Órfão JJM, Faria JL, Silva CG (2014) Photocatalytic nitrate reduction over Pd-Cu/TiO$_2$. Chem Eng J 251:123–130

133. Li Y, Wasgestian F (1998) Photocatalytic reduction of nitrate ions on TiO$_2$ by oxalic acid. J Photochem Photobiol A Chem 112:255–259

134. Silva CG, Pereira MFR, Órfão JJM, Faria JL, Soares OSGP (2018) Catalytic and photocatalytic nitrate reduction over Pd-Cu loaded over hybrid materials of multi-walled carbon nanotubes and TiO$_2$. Front Chem 6:1–10

135. Kato R, Furukawa M, Tateishi I, Katsumata H, Kaneco S (2019) Novel photocatalytic NH3 synthesis by NO$_3$—reduction over CuAg/TiO$_2$. ChemEngineering. 3:49

136. Jing T, Dai Y, Wei W, Ma X, Huang B (2014) Near-infrared photocatalytic activity induced by intrinsic defects in Bi$_2$MO$_6$ (M = W, Mo). Phys Chem Chem Phys 16:18596–18604

137. Hirakawa H, Hashimoto M, Shiraishi Y, Hirai T (2017) Selective Nitrate-to-Ammonia Transformation on Surface Defects of Titanium Dioxide Photocatalysts. ACS Catal. 7:3713–3720

138. Hirayama J, Kamiya Y (2014) Combining the photocatalyst Pt/TiO_2 and the nonphotocatalyst $SnPd/Al_2O_3$ for effective photocatalytic purification of groundwater polluted with nitrate. ACS Catal 4:2207–2215
139. Challagulla S, Tarafder K, Ganesan R, Roy S (2017) All that glitters is not gold: a probe into photocatalytic nitrate reduction mechanism over noble metal doped and undoped TiO_2. J Phys Chem C 121:27406–27416

Chapter 8
Advances in the Microbial Fuel Cell Technology for the Management of Oxyanions in Water

Jafar Ali, Aroosa Khan, Hassan Waseem, Ridha Djellabi, Pervez Anwar, Lei Wang, and Gang Pan

Abstract The importance of the water-energy-food nexus is critical to sustainable development with the growing population on this planet. An upsurge in urbanization and industrialization is the major cause of water contamination with the oxyanions. Non-biodegradable nature, high toxicity, and high solubility demand the effective means to remove oxyanions from water by transforming oxyanions into harmless and/or immobilized forms. Microbial fuel cell (MFC) technology has emerged as a promising way to treat toxic wastewater with the generation of bioenergy. Utilization of various oxyanions as the electron acceptor in MFC cathode can provide the sustainable waste management of oxyanions. This chapter has summarized the contemporary advancements and modifications applied to MFC for the treatment/management of various oxyanions from aqueous solutions.

Keywords Wastewater treatment · Sustainable · Environment · Microbial fuel cell · Waste management · Oxyanions

J. Ali (✉) · H. Waseem · P. Anwar
Department of Biochemistry and Molecular Biology, University of Sialkot, Punjab 51310, Pakistan
e-mail: jafar.ali@uskt.edu.pk

J. Ali · L. Wang · G. Pan (✉)
Key Laboratory of Environmental Nanotechnology and Health Effects, Research Center for Eco-Environmental Sciences, Chinese Academy of Sciences, Beijing, China
e-mail: gpan@rcees.ac.cn

J. Ali
University of Chinese Academy of Sciences, Beijing 100049, People's Republic of China

A. Khan
Department of Microbiology, Quaid-I-Azam University, Islamabad, Pakistan

R. Djellabi
Department of Chemistry, Università Degli Studi Di Milano, Via Golgi 19, Milan, Italy

G. Pan
Centre of Integrated Water-Energy-Food Studies, School of Animal, Rural and Environmental Sciences, Nottingham Trent University, Brackenhurst Campus, Southwell NG25 0QF, UK

© Springer Nature Switzerland AG 2021
N. A. Oladoja and E. I. Unuabonah (eds.), *Progress and Prospects in the Management of Oxyanion Polluted Aqua Systems*, Environmental Contamination Remediation and Management, https://doi.org/10.1007/978-3-030-70757-6_8

8.1 Introduction

The rapid industrial development has significantly amplified environmental contamination by toxic metals and metalloids. The majority of these hazardous elements are redox-sensitive and can form oxyanions in aqueous solutions. Oxyanions are negatively charged polyatomic ions having oxygen $(A_xO_y^{z-})$. Common oxyanions of Mo, W, As, B, V, Cr, and Se are present in water and wastewater streams [1]. Oxyanions are described by their non-biodegradable nature, toxicity, and high solubility. Thus, oxyanions can accumulate in environmental streams and ultimately enter the food chain [2]. While considering their harmful impact in wastewater, oxyanions are regarded as major toxins in sewage or wastewater [3]. The primary sources of harmful oxyanions are alkaline wastes produced by various high-temperature processes, including thermal treatment of waste, combustion of fossil fuels, and smelting of ferrous and non-ferrous metals. However, oxyanions are also released by manufacturing industries, including tannery, electroplating, microelectronics, battery, fertilizers, and metal finishing [4–6].

The isotopic composition of oxyanions is often used as essential tools for studying biogeochemical cycles of sulfur and nitrogen. The high concentration of oxyanions creates an environment that enhances the dissolution of weak-acid oxyanion species, such as arsenic, antimony, or vanadium that further contaminate groundwater sources. The increasing concentration of oxyanions in groundwater sources is one of the emerging environmental problems. Oxyanions have the potential of being transmitted into living organisms through skin absorption, inhalation, and ingestion. This results in irreversible effects in metabolic reactions [7]. Excessive intake of arsenic can cause various diseases, which include vascular diseases, skin lacerations, and cancers in humans [8, 9]. Recent studies posited that trivalent methylated arsenicals induce carcinogenicity of environmental arsenic. These arsenicals are intermediate metabolites produced during the methylation step of inorganic arsenic [10, 11].

Extremely toxic contaminants pose severe threats to the environment and public health, even at low concentrations; therefore, their removal from water and wastewater is of importance to environmental and public health. Given the detrimental properties of oxyanions, their removal from water and wastewater is crucial for environmental and public health reasons [5]. The most common strategies used for the removal of oxyanions from wastewater and water sources include adsorption, ion exchange, reverse osmosis, filtration, chemical precipitation, electrodialysis, solvent extraction, evaporation, and biological approaches [12–16]. All of these methods have limitations in their efficient applications, especially where it relates to cost-effectiveness and high energy demands. Moreover, these conventional strategies do not degrade oxyanions. Instead, they convert them into less toxic secondary waste that requires further treatment or disposal technique which probably increases the cost of the treatment process. Hence, it is necessary to develop a new sustainable mix of technologies that meets the potential energy requirements for the remediation of water and wastewater contaminated with oxyanion. Biological treatments using microbial fuel cell (MFC) technology is presented as an alternative and sustainable approach because of its potential to transform these contaminants into inert products.

Microbial fuel cell technology (MFC) is an emerging environmental technology, which simultaneously treats waste and generates bioenergy. MFC has the additional advantage of a coupled mechanism of wastewater treatment and electricity production [17, 18]. The technology applies the metabolic activities of microorganisms for the conversion of biochemical energy in substrates to electrical energy which pave the way for the development of electro-microbiology [19, 20]. The simple mechanism for MFCs is that oxidation of organic substrates takes place at the anode, with the liberation of electrons, which are then transported through an external circuit (resistance) to the cathode, and the ultimate consumption of electrons will determine the nature of the by-products in the cathode chamber. There is a terminal electron acceptor (TEA) in the cathode chamber, which performs a critical role in controlling the performance of MFCs [21]. Oxyanions can also act as a TEA to generate electricity in bioelectrochemical systems, thus providing sustainable waste management of oxyanions along with bioenergy production [22–24].

This chapter is aimed at addressing the contemporary advancements and modifications applied to MFCs for wastewater treatment along with energy generation, with a focus on the management of oxyanions in aqueous solution.

8.2   Overview and Working Principles of Microbial Fuel Cells Technology

Bioelectrochemical cells/microbial fuel cells are the galvanic cells used to extract bioenergy from organic wastewater through electrochemical reaction. MFC reactors mediate exogenic reaction, which holds negative free reaction energy (Gibbs free energy) with impulsive energy release [25, 26]. The oxidation-reduction reactions, with the help of electroactive microbes, generate electricity [27]. The major structural components of MFCs are cathode, anode, and a separating membrane [28, 29]. Carbon materials, including rods, glassy carbon, graphite plates, carbon felt, carbon foam, cloth, paper, stainless steel mesh, platinum black, and reticulated transparent carbon are generally used as the anode material. These carbon materials must be non-corrosive, conductive and allow the microbes to form a biofilm.

On the contrary, the cathode can be aerobic and anaerobic, depending on the nature and function of the cathodic material. The anaerobic cathode comprises of the catalysts materials such as manganese oxide, platinum black, and polyaniline that facilitates better oxygen reduction reaction (ORR) [30]. The anaerobic cathode may contain a ferricyanide solution or microorganism for electron transfer. Electroactive microbes can also serve as biocatalysts at the cathode surface when they are allowed to form electroactive biofilm (biocathode). This facilitates the electron transfer process and catalyzes the cathodic reactions [31]. Although membrane separators are now modified according to the structural configuration of the MFCs, yet at the initial stage of the technology, salt bridges were used as membrane separators [32, 33].

8.2.1 Electron Transfer Mechanisms in MFCs

Various terminal electron acceptors readily accept electrons and produce different products, which are easily diffused out of the cell [34, 35]. Terminal electron acceptor (TEA) includes oxygen, different oxyanions, and metal oxide contaminated solutions. These mediators must be low cost, non-toxic, non-biodegradable, and high solubility in the anolyte. They must be able to cross the cell membrane easily and possess a high electrode reaction rate. Electron transfer in the MFCs takes place in the following three ways:

(a) The outer membrane cytochrome (C-type) of microorganisms (i.e., *Geobacter* and *Shewanella* species) directly transfers electrons, produced from nicotinamide adenine dinucleotide (NADH), to the electrode surface,

(b) Certain species of bacteria, such as *Pseudomonas and Shewanella,* secrete a particular chemical species (e.g., flavins, thionin, neutral red, and pyocyanin), which acts as mediator molecules and performs electron transfer through these mediators [36].

(c) Some microorganisms (i.e., *Geobacter sulfurreducens* and *Shewanella oneidensis*) species use conductive appendages (i.e., cellular outgrowth), which are 20 nm long named as "nanowires," for the transfer of electrons outside the cell [37].

Besides, the additional by-products of fermentation reaction could also serve as electron donors. These by-products may include acetic acid, formic acid, and lactic acid and formate, produced through the pyruvate oxidation [38].

The conventional and basic MFCs are used to study a specific parameter, whereas highly efficient systems are designed with advanced architectures to maximize power production [39, 40]. Researchers have significantly improved the efficiency of MFCs through the modification of the structural design, electrode materials, catalysts, and microbes [41–44]. The most common configuration is the two-chamber system. Many variations have been proposed in this basic model, with an effort to improve the power density [45–47] (Fig. 8.1).

8.3 Removal of Oxyanions from Wastewater Using MFC Technology

Oxyanions are the types of contaminants in wastewater which cannot be degraded biologically, and as such, require unique treatment protocols with high molecular selectivity for their removal or transformation. Electrochemically mediated

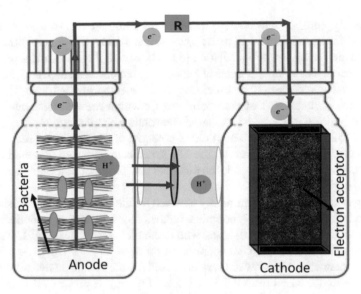

Fig. 8.1 Schematic diagram of electron transfer in microbial fuel cell

capture of heavy metals/oxyanions can significantly address the selectivity issue. As described earlier, oxyanions can be utilized as electron acceptors in MFCs due to high redox potential, therefore, reducing the overall cost of the treatment process and enhance sustainability. Also, biocathode (electroactive microbes serving as catalyst) is another sustainable approach for the bio-reduction of oxyanions in bioelectrochemical systems. Production of no secondary pollutants, easy operation, and low cost is the prerequisites for environmental remediation technologies [48]. The following section presents the array of the research reports on the use of MFCs technology for the removal of some prominent oxyanions.

8.3.1 Chromium Oxyanions

Chromium (Cr) is an element of group 6 in the Periodic Table that has broad industrial applications in the field of electroplating, tanning, dyes synthesis, and processing of wood materials [49]. The effluents from these industrial processes release Cr oxyanion into the environmental matrix with their oxidation number varying from 0 to +6. Among all oxidation states, Cr (III) is more stable and less toxic than Cr (VI), which is more water soluble, mobile and toxic. Cr(VI) is designated as a class-I carcinogen by International Research Agency (IRA) on cancer, and the permissible limit of 50 mg/L in water is as declared by Environmental Protection Agency (EPA) [50]. Cr (VI) species negatively impact the natural environment, human health, and other forms of life. Thus, the reduction/transformation of Cr(VI) is mandatory before the discharge of industrial effluents.

Generally, chromium in wastewater is removed through any of the conventional protocols that include chemical reduction, bioremediation, membrane filtration, ion exchange, and coagulation/flocculation [51–53]. But all these treatment procedures have several shortcomings, in terms of high energy requirement, high cost, and generation of secondary waste. The efficiency of the method applied also depends on the speciation, mobility, and distribution of Cr within the environmental system. Thus, the removal of Cr-based oxyanions demands a broader and more sustainable approach [52, 54]. Due to the high redox potential of Cr(VI), it can serve as TEA in the cathode of MFC. Therefore, the MFC mediated transformation/reduction of Cr(VI) holds great promise due to the potential generation of renewable energy, low cost, and high sustainability.

Wang et al., carried out preliminary research for the Cr (VI) reduction using MFC operated in fed-batch mode. A complete reduction of 100 mg/L was noticed during 150 h (with an initial pH of 2.0) along with maximum power density of 150 mW/m^2. The Coulombic efficiency (CE) decreased from 59% to 25% when the pH of the Cr (VI) contaminated wastewater was increased from 2 to 6. Moreover, open circuit potential also decreased from 0.9 V to 0.52 V [55, 56]. High Cr (VI) removal efficiency (99%) was achieved under acidic pH condition, along with the generation of power density of 767 mW/m^2 [57]. Another study showed that the reduction of chromate oxyanions (CrO_4^{2-} & $Cr_2O_7^{2-}$) using different anaerobic microorganisms (i.e., *Escherichia coli, Shewanella oneidensis, Pseudomonas dechromaticans, Desulfovibrio vulgaris, Enterobacter cloacae,* and *Aeromonas dechromatica,*) as bio-anodes as shown in Fig. 8.2 [58].

Although, 95% reduction efficiency and energy output of 89 ± 3 mW/m^2 were obtained, but the biocathode mediated reduction of Cr(VI) faced the drawback of cell toxicity. Although, MFCs technology for Cr(VI) reduction is advantageous over the conventional methods, it is still lagging in real-life applications. This is because of the high concentrations of Cr(VI) present in acidic wastewaters released by the mining and electroplating industries [59]. Recent researches have targeted the incorporation of various modifications in the MFCs design, electrode, and substrate materials, and the addition of some other ions, such as Fe(II), Fe(III), nanoparticles, and graphite cathodes, that resulted in enhanced reduction kinetics and removal efficiency of Cr oxyanions.

Electrical repulsion between a negatively charged cathode and chromate/dichromate functional groups can hamper the reduction kinetics. Slow reaction kinetics in MFC cathode can be improved by introducing electron-shuttle mediators. Iron-based nanomaterials are among the most commonly used materials for electrode modifications to improve Cr(VI) reduction kinetics via reducing the charge transfer resistance. For example, it was observed that iron species accelerated the reduction of Cr(VI) by 1.6 times in an MFC system. And cathodic Coulombic efficiency was also increased from 65% to 81%. The increased electron transfer rate and the lowered reduction potential of the experimental MFC was attributed to the Fe (III), which facilitated the reduction of Cr(VI) [60] In a recent study, the cathode in a dual-chamber MFC was modified with FeS@rGO nanocomposite for Cr reduction and renewable energy generation. This system exhibited 100% Cr(VI) reduction and

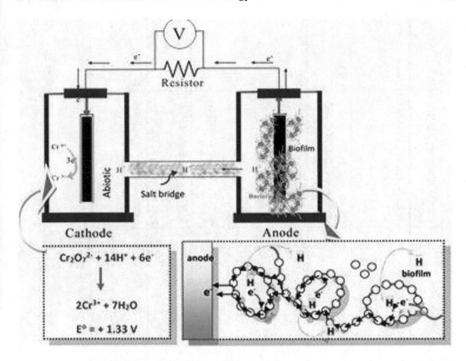

Fig. 8.2 Chromium reduction in the MFC source [58]

reduction rate was 4.6 times higher than the MFC without the modified FeS@rGO nanocomposite. The generated power density (154 mW/m^2) was 328 times, relative to the unmodified cathode [50].

Similarly, the natural pyrrhotite-coated cathode was used to enhance Cr(VI) reduction performance of an MFC. The modified electrode achieved the maximum power density of 45.4 mW/m^2, which was 1.3 times higher than that of control. In addition, Cr(VI) removal efficiency was improved from 46% to 97.5% [61]. The reduction of Cr(VI) was investigated by replacing the conventional cathode graphite with carbon cloth cathode, and 100% Cr(VI) reduction was achieved [62]. In another study, NaX zeolite-modified graphite felt was utilized as biocathode, in comparison with HNO$_3$ pretreated graphite felt. The NaX zeolite proved to be an excellent bioelectrode material, resulting in 8.2 times higher Cr(VI) reduction, relative to the unmodified electrode [63]. Though substantial reduction rates of Cr oxyanions have been obtained in MFC, however, its practical applications are still hampered by low mass transfer and charge transfer resistance. Introducing electron shuttles is expected to mitigate this by accelerating the removal of oxyanions, but the approach is not cost effective. It has been suggested that some in situ electrode modification should be adopted for bioelectrochemical reduction of Cr for long-term applications [64]. Some studies on the Cr(VI) reduction using the MFC are summarized in Table 8.1.

Table 8.1 Summary of the chromium (VI) oxyanions reduction by MFC

MFC setup used for chromium oxyanions reduction	Energy outputs (mW/m^2)	Oxyanions removal efficiencies (%)/reduction rate	Modifications applied to MFC	References
MFC cathodic modified graphite felt with iron sulfide enfolded with reduced graphene oxide (FeS@rGO) nanocomposites	154 mW/m^2	100 (1.43 mg/L/h)	Graphite cathode modified with FeS@rGO nanocomposites	[50]
Primary research of MFC on the treatment of Cr(VI) in wastewater	767.01 (2.08 mA/m^2)	99.85	Development of dual-chambered MFC with an organic electron donor	[57]
Two Cr(VI)-containing effluents were synthetically prepared in the laboratory (4 and 8 mg l − 1) to check Cr reduction	89 ± 3 mW/m^2 for 4 mg L^{-1} of Cr 69.5 ± 2.1 mW/m^2 for 8 mg L^{-1} of Cr	95 86	Using different anaerobic microorganisms as bio-anode. (E. Coli, P. dechromaticans, D. vulgaris, etc.)	[58]
Reduction of Cr(VI) via indirect electron transfer of Fe(III) mediation	0.22 mW/m2	65.6 with Fe(III)	Fe(III) as an electron-shuttle mediator	[60]
MFCs equipped with a pyrrhotite-coated cathode	45.4	97.5	MFC graphite cathode modified with pyrrhotite-coated cathode	[61]
Cr(VI) bio-electrochemically reduced to non-toxic Cr(III) in the dual-chamber MFCs with different cathodes	1221.94	100	Graphite cathode modified with carbon cloth	[62]

(continued)

Table 8.1 (continued)

MFC setup used for chromium oxyanions reduction	Energy outputs (mW/m^2)	Oxyanions removal efficiencies (%)/reduction rate	Modifications applied to MFC	References
Two kinds of NaX zeolite-modified graphite felts used as biocathode electrodes in hexavalent chromium (Cr(VI))-reducing microbial fuel cells (MFCs)	28.90 ± 3.18 mW/m2	10.39 ± 0.28 mg/L h (8.2 times higher)	NaX zeolite-modified graphite felt operated as a biocathode	[63]

8.3.2   Nitrogenous Oxyanions

Nitrogenous oxyanions, especially nitrate (NO_3^-) pollution has become a serious issue during the last few years, due to increased discharge from the fertilizer industry. Excessive concentration of nitrate can harm human health, environment, and animal health. Nitrates can also be converted into nitrites (NO_2^-), which is the more toxic form, with the help of microorganisms [64]. The permissible limit of nitrate is 10 mg/L, according to USPEA. So, the removal of nitrate is essential from wastewater and groundwater. Different physical and chemical methods, including ion exchange, catalytic reduction, and biological denitrification, have been used to remove this oxyanion from wastewater and groundwater [65–67]. For bacterial nitrification, both autotrophic and heterotrophic modes can be used for nitrate removal [68].

Bioelectrochemical systems are innovative systems for wastewater treatment and bioelectricity generation and nitrates reduction to nitrites. [69–71]. Biocathode mediated reduction offers a complete conversion of nitrate to N_2, under anaerobic conditions in MFCs [72]. The stimulation of denitrification activity of a bioelectrochemical reactor with the help of an external MFC has been reported [73]. The denitrification activity was improved when concentration of denitrifying bacteria reached up to 3.5 10^7, with the power density of 502.5 mW/m^2 and output voltage range from 500 to 700 mV [73]. Various approaches, such as electrode/architectural modifications and introduction of mediators, like Fe(III), have been adopted to increase the efficiency of nitrate removal by MFC [74, 75]. The removal of nitrate in the anode and cathode chamber of MFC, using the denitrifying bacteria on biocathode have been investigated. A power density of 2.80 ± 0.05 W/m^3 was achieved when nitrate loading of 14 ± 0.3 g N/(m^3 NC•d) was used [76]. Further studies were recommended to investigate the prominent sulfate formation in the presence of short hydraulic retention time (HRT) and low temperature.

The performance of a novel stake up-flow baffled reactor MFC was investigated for the removal of nitrates, where the optimum biocathode denitrification frequency was 148.3/1.4 g N m^{-3} NCV d^{-1} (Fig. 8.3). However, the power density of the MFC reactor in a closed-circuit system was improved by fourfolds, when compared with the open circuit system [77]. An autotrophic denitrifying biocathode-based MFC, coupled with electricity generation, was also utilized for nitrate/perchlorate reduction. Substrate removal efficiency for nitrate was 87.05%, at the stabilized power output of 3.10 A/m^3 [78]. It is proposed that, nitrate removal efficiency can be improved using the bioelectrochemical reactors. Therefore, the influence of adsorption and bio-reduction must be explored further for the management of oxyanions. These studies have proven that the MFCs/electro-microbiological procedure lowered the carbon requirements, efficiently removed nitrates, and produced sufficient energy output. Briefly, nitrogenous oxyanion removal/reduction through MFC is a sustainable and effective approach at the laboratory scale. Large-scale applications of this innovative method have not been thoroughly studied, which demands for long-term stability of biofilm and biocathode, HRT and temperature.

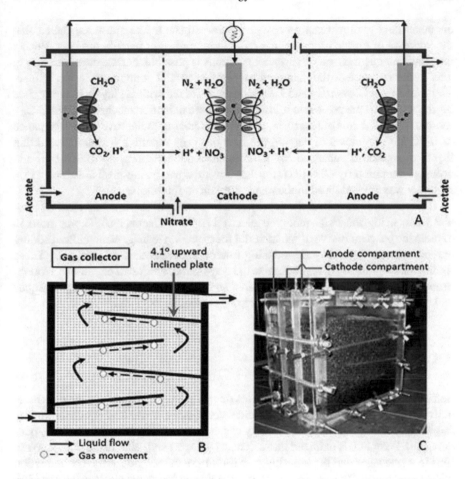

Fig. 8.3 **a** Bioelectrochemical systems for **b** internal structural design of up-flow baffled reactor; **c** assembled MFC *Source* [77]

8.3.3   Perchlorate Oxyanions

Perchlorate (ClO_4^-) is widely used in the production of match sticks, fireworks, highway safety flares, and rocket propellants. Due to its weak adsorption on mineral surfaces, high solubility, and low reactivity, it has been detected in groundwater and drinking waters. Perchlorates have negative impacts on the thyroid gland and human health; hence, it is recognized as an emerging contaminant [79]. An autotrophic denitrifying microbial community was grown on the cathode of MFC and used to remove a mixture of oxyanions (i.e., nitrate and perchlorate). The removal efficiency of ClO_4^- and NO_3^- was 53.14% and 87%, respectively, when the molar ratio of NO_3/ClO_4 was 1:1 in the influents. The substrate removal was coupled with bioelectricity generation (3 mA/m³). The perchlorate reduction and influence of nitrate concentration on membrane biofilm reactors have been investigated. The study suggested that the presence of nitrate promoted the reduction/removal of perchlorate ions. Hence, the simultaneous removal of various oxyanions is possible. Nitrate concentration in wastewater determines the nature and type of microbial community in biocathode. Thus, the *Methyloversatilis* and *Zoogloea* were are the critical players in membrane bioreactors and were posited to follow independent reduction mechanisms for ClO_4^- removal [80]. The role of resazurin, as a redox mediator, to improve the performance of MFC in the removal of perchlorate ions has been reported. It was reported that 9 μM of resazurin enhanced the perchlorate reduction ratio, up to 101.6%, and voltage generation by 40% [81]. Furthermore, the catalyzing mechanism of redox mediator was studied through the respiratory chain inhibitors.

A recent study described the perchlorate ions removal in microbial electrochemical cells, using the Fe/C modified electrodes. Dual-chamber MFC was used to explore the best cathode catalytic material for oxyanion management without adding external power or perchlorate-reducing microbial enrichment cultures [82]. These findings are significant to revolutionize the way for the removal of oxyanions. Collaborative interactions are needed to develop some hybrid approaches using microbes and catalysts for enhanced perchlorate ion reduction.

8.3.4   Selenium Oxyanions

Selenium (Se) is one of the trace nutrients that play an essential role in the metabolic activities of human and animal life cycles, yet highly toxic at higher concentration. High accumulation in water bodies may cause acute and chronic toxicities to organisms [83]. Selenium is oxidized into selenate (SeO_4^{2-}) and selenite (SeO_3^{2-}) oxyanions in wastewater and sewage sludge. MFCs have been used for the bio-reduction of Se oxyanions, with a wider variety of organic and inorganic electron donors [84]. The selenite removal, along with electricity generation, was examined in a single-chamber MFC, using glucose and acetate as the carbon sources. Using acetate as the carbon source, 88% Se removal was achieved, while 99% was achieved with glucose as the carbon source [85]. The power output of MFC was not much affected when the concentration of glucose/substrate was up to 125 mg/L, and Coulombic efficiency was also improved from 25% to 38% when Se concentration was further increased to 150 mg/L. However, the electricity production was inhibited after 200 mg/L concentrations of selenium in the MFC. In another study, Se bioremediation was investigated in a microbial fuel cell inoculated with *Shewanella oneidensis* MR-1. In this setup, 92% of 100 mg/L of Se, was removed after 100 h and Coulombic efficiency was affected even at very high concentration of 200 mg/L [86]. New approaches must be developed for further and real-life bioremediation of this oxyanion by MFC.

8.3.5   Arsenic Oxyanions

Arsenic (As) is a highly toxic metalloid present in water as an oxyanion and over 250 million people around the world consume arsenic-contaminated water. The arsenic toxicity manifests in the form of neurological damage and even death [87]. MFC/bioelectrochemical systems are gaining attention as a reliable and sustainable technology for the treatment of arsenic oxyanions. Conventional strategies used for the mitigation of arsenic have become obsolete due to low efficiency. For example, conventional zerovalent iron technology has low arsenic removal efficiency due to the low corrosion rate [88]. Thus, new innovative/sustainable methods must be developed for the management of arsenic oxyanions [89]. A hybrid system was designed by combining the conventional arsenic removal approach with MFC [90]. In another study, substrate competition between iron-reducing bacteria and MFC anode was considered as a responsible factor for lowering the concentration of Fe and As. The, As content was reduced by 53% and power density of 12.0 mW/m^2 after 50 days [91]. Impulsive anaerobic oxidation of arsenic was achieved in a study, where various arsenic-resistant and arsenic-oxidizing bacteria, along with electrochemically active bacteria interacted for the oxidation of arsenic and bioelectricity production [92]. It has been reported that iron minerals have good arsenic sorbent properties that ultimately facilitate the removal from the aqueous solution. Therefore, iron-based electrode materials could be much efficient in arsenic-based oxyanions remediation. Two different studies applied the first idea, where the pH of the water was neutralized from ~ 3.7 to ~ 7.2, with 80% removal of Fe and As in an air-cathode MFCs (Fig. 8.4). However, in the second study, MFC–zerovalent iron hybrid process was constructed to remove arsenite (As(III)) from aqueous solutions. As(III) concentration was reduced below the detection limit after 2 h [90, 93].

8.4   Conclusion and Future Recommendations

In summary, microbial fuel cells (MFCs) have emerged as a promising method for the management of oxyanions. Simultaneous wastewater (oxyanions) treatment and bioelectricity generation could be achieved sustainably. Although significant results have been reported, but the MFC technology still needs more attention for large-scale implementation. There are some areas where more work needs to be done, such as electrode surface area, system architecture, and cost of materials. Exploration of more electroactive bacteria will also help in the development of electro-microbiology for the management of oxyanions. Moreover, the integration of MFC technology with existing conventional treatment methods may also improve the effectiveness of MFC for wastewater treatment and oxyanion removal.

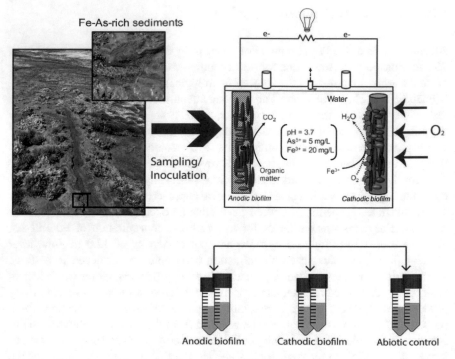

Fig. 8.4 Abstract model for As and Fe removal in SC-MFC *Source* [85]

References

1. Yin YB, Guo S, Heck KN, Clark CA, Coonrod CL, Wong MS (2018) Treating water by degrading oxyanions using metallic nanostructures. ACS Sustain Chem Eng 6(9):11160–11175
2. Khan K, Lu Y, Saeed MA et al (2018) Prevalent fecal contamination in drinking water resources and potential health risks in Swat, Pakistan. J Environ Sci (China) 72:1–12
3. Jadhav SV, Bringas E, Yadav GD, Rathod VK, Ortiz I, Marathe KV (2015) Arsenic and fluoride contaminated groundwaters: a review of current technologies for contaminants removal. J Environ Manage 162:306–325
4. Kailasam V, Rosenberg E (2012) Oxyanion removal and recovery using silica polyamine composites. Hydrometallurgy 129–130:97–104
5. Bhande R, Ghosh PK (2018) Oxyanions removal by biological processes: a review. Water Qual Manage, pp 37–54
6. Cavka A, Alriksson B, Ahnlund M, Jönsson LJ (2011) Effect of sulfur oxyanions on lignocellulose-derived fermentation inhibitors. Biotechnol Bioeng 108(11):2592–2599
7. Adegoke H, Adekola F, Fatoki OS, Ximba BJ (2013) Sorptive interaction of oxyanions with iron oxides: a review. Polit J Environ Stud 22(1):7–24
8. Zhao D, SenGupta AK, Stewart L (1998) Selective removal of Cr(VI) oxyanions with a new anion exchanger. Indus Eng Chem Res 37(11):4383–4387
9. Hajji S, Montes-Hernandez G, Sarret G, Tordo A, Morin G, Ona-Nguema G, Bureau S, Turki T, Mzoughi N (2019) Arsenite and chromate sequestration onto ferrihydrite, siderite and goethite nanostructured minerals: Isotherms from flow-through reactor experiments and XAS, Elsevier. J Hazard Mater 362:358–367

10. Do MH, Ngo HH, Guo WS et al (2018) Challenges in the application of microbial fuel cells to wastewater treatment and energy production: a mini review. Sci Total Environ 639:910–920
11. Van Halem D, Bakker SA, Amy GL, Van Dijk JC (2009) Arsenic in drinking water: a worldwide water quality concern for water supply companies. Drinking Water Eng Sci 2:29–34
12. Weidner E, Ciesielczyk F (2019) Materials Removal of hazardous oxyanions from the environment using metal-oxide-based materials, Materials (Basel) 12(6):927
13. Wang JJ, Zhou B, Li M et al (2018) An overview of carbothermal synthesis of metal-biochar composites for the removal of oxyanion contaminants from aqueous solution. Elsevier 129:674–687
14. Ungureanu G, Filote C et al (2016) Antimony oxyanions uptake by green marine macroalgae. J Environ Chem Eng 4(3):3441–3450
15. Gupta V, Fakhri A, Bharti A et al (2017) Optimization by response surface methodology for vanadium (V) removal from aqueous solutions using PdO-MWCNTs nanocomposites, Elsevier. J Mol Liq 234:117–123
16. Kołodyńska D, Budnyak T, Tertykh V (2017) Sol-gel derived organic-inorganic hybrid ceramic materials for heavy metal removal. In: Mishra A (ed) Sol-gel based nanoceramic materials: preparation, Properties and Applications. Springer, Cham, pp 253–274
17. Aghababaie M, Farhadian M, Biria D (2015) Effective factors on the performance of microbial fuel cells in wastewater treatment-a review. Environ Technol Rev 4(1):74–89
18. Ali J, Wang L, Waseem H et al (2019) Bioelectrochemical recovery of silver from wastewater with sustainable power generation and its reuse for biofouling mitigation. J Cleaner Prod 235:1425–1437
19. Liu T (2020) Practical applications of microbial fuel cell technology in winery wastewater treatment, University of British Columbia
20. Ali J, Sohail A, Wang L, Haider MR, Mulk S, Pan G (2018) Electro-microbiology as a promising approach towards renewable energy and environmental sustainability. Energies 11(7):1822
21. Zhang L, Shen J, Wang L, Ding L, Xu K, Ren H (2014) Stable operation of microbial fuel cells at low temperatures (5–10 °C) with light exposure and its anodic microbial analysis. Bioprocess Biosyst Eng 37(5):819–827
22. Raheem A, Singh Sikarwar V, He J et al (2018) Opportunities and challenges in sustainable treatment and resource reuse of sewage sludge: a review. Chem Eng J 337:616–641
23. Ali J, Sohail A, Wang L, Haider MR, Mulk S, Pan G (2018) Electro-microbiology as a promising approach towards renewable energy and environmental sustainability, Energies 11(7):1822
24. Singh A, Yakhmi MJV (2014) Microbial fuel cells to recover heavy metals. Environ Chem Lett 12:483–494
25. Scott K (2016) An introduction to microbial fuel cells. In: Microbial electrochemical and fuel cells: fundamentals and applications, pp 3–27
26. Capodaglio AG (2016) A multi-perspective review of microbial fuel- cells for wastewater treatment: Bio-electro—chemical, microbiologic and modeling aspects. AIP Conf Proc 1758:030032
27. Gude VG (2016) Microbial fuel cells for wastewater treatment and energy generation. In: Microbial electrochemical and fuel cells: fundamentals and applications, Elsevier, pp 247–285
28. Kim B, An J, Fapyane D et al (2015) Bioelectronic platforms for optimal bio-anode of bio-electrochemical systems: from nano-to macro scopes. Biores Technol 195:2–13
29. Chiao M, Lam KB, Lin L (2003) "Micromachined microbial fuel cells," The sixteenth annual international conference on micro electro mechanical systems, IEEE, Kyoto, Japan, pp 383–386
30. Varanasi J, Nayak A, Sohn Y, Pradhan D et al (2016) Improvement of power generation of microbial fuel cell by integrating tungsten oxide electrocatalyst with pure or mixed culture biocatalysts. Electrochim Acta 199:154–163
31. Shen X, Zhang J, Liu D, Hu Z, Liu H (2018) Enhance performance of microbial fuel cell coupled surface flow constructed wetland by using submerged plants and enclosed anodes. Chem Eng J 351:1–1178
32. He L, Du P, Chen Y et al (2017) Advances in microbial fuel cells for wastewater treatment. Renew Sustain Energy Rev 71:388–403

33. Pandit S, Das D (2018) Principles of microbial fuel cell for the power generation. In: Microbial fuel cell
34. Hindatu Y, Annuar MSM, Gumel AM (2017) Mini-review: anode modification for improved performance of microbial fuel cell. Renew Sustain Energy Rev 73:236–248
35. Wang Y, Ding S, Gong M et al (2016) Diffusion characteristics of agarose hydrogel used in diffusive gradients in thin films for measurements of cations and anions. Anal Chim Acta 94:47–56
36. Tkach O, Sangeetha T, Maria S, Wang A (2017) Performance of low temperature Microbial Fuel Cells (MFCs) catalyzed by mixed bacterial consortia. J Environ Sci (China) 52:284–292
37. Neto S, Reginatto V et al (2018) Microbial fuel cells and wastewater treatment. Electrochem Water Wastewater Treat, pp 305–331
38. Pedro J (2017) Microbial fuel cell for wastewater treatment, Instituto superior tecnico
39. Nikhil GN, Chaitanya DNSK, Srikanth S, Swamy YV, Mohan SV (2018) Applied resistance for power generation and energy distribution in microbial fuel cells with rationale for maximum power point. Chem Eng J 335:267–274
40. Behera BK, Varma A (2016) Microbial fuel cell (MFC). In: Microbial resources for sustainable energy. Springer International Publishing, Cham, pp 181–221
41. Martínez-Conesa EJ, Ortiz-Martínez VM, Salar-García MJ et al (2017) A Box-behnken design-based model for predicting power performance in microbial fuel cells using wastewater. Chem Eng Commun 204(1):97–104
42. Oon Y, Ong S, Ho L, Wong Y et al (2016) Synergistic effect of up-flow constructed wetland and microbial fuel cell for simultaneous wastewater treatment and energy recovery. Biores Technol 203:190–197
43. Ortiz-Martínez V, Gajda I, Salar-García M, Greenman J, Hernández-Fernández F, Ieropoulos I (2016) Study of the effects of ionic liquid-modified cathodes and ceramic separators on MFC performance. Chem Eng J 291:317–324
44. Kim K, Yang W, Ye Y, LaBarge N et al (2016) Performance of anaerobic fluidized membrane bioreactors using effluents of microbial fuel cells treating domestic wastewater. Biores Technol 208:58–63
45. Krieg T, Mayer F, Sell D, Holtmann D (2019) Insights into the applicability of microbial fuel cells in wastewater treatment plants for a sustainable generation of electricity. Environ Technol (United Kingdom) 40(9):1101–1109
46. Logan BE, Hamelers B, Rozendal R et al (2006) Critical review microbial fuel cells: methodology and technology. ACS Publ 40(17):5181–5192
47. Parkash A (2016) Microbial fuel cells: a source of bioenergy. Journal of Microbial & Biochemical Technology 8(3):247–255
48. Guan C, Hu A et al (2019) Stratified chemical and microbial characteristics between anode and cathode after long-term operation of plant microbial fuel cells for remediation of metal. Sci Total Environ 670:585–594
49. Qiu B, Xu C, Sun D et al (2014) Polyaniline coating on carbon fiber fabrics for improved hexavalent chromium removal. RSC Adv 4:29855–29865
50. Ali J, Wang L, Waseem H, Djellabi R, Oladoja NA, Pan G (2020) FeS @ rGO nanocomposites as electrocatalysts for enhanced chromium removal and clean energy generation by microbial fuel cell. Chem Eng J 384:123335
51. Shah JH, Fiaz M, Athar M et al (2019) Facile synthesis of N/B-double-doped Mn_2O_3 and WO_3 nanoparticles for dye degradation under visible light. Environ Technol (United Kingdom)
52. Djellabi R, Yang B, Sharif H, Zhang J et al (2019) Sustainable and easy recoverable magnetic TiO_2-Lignocellulosic Biomass@ Fe_3O_4 for solar photocatalytic water remediation. J Cleaner Prod 233:841–847
53. Wang L, Ali J, Zhang C, Mailhot G, Pan G (2020) Simultaneously enhanced photocatalytic and antibacterial activities of TiO_2/Ag composite nanofibers for wastewater purification. J Environ Chem Eng 8(1):102104
54. Wang L, Zhang C, Cheng R, Ali J, Wang Z, Mailhot G, Pan G (2018) Microcystis aeruginosa synergistically facilitate the photocatalytic degradation of tetracycline hydrochloride and Cr (VI) on PAN/TiO_2/Ag nanofiber mats. Catalysts 8(12):628

55. Nancharaiah Y, Mohan S et al (2015) Metals removal and recovery in bioelectrochemical systems: a review. Biores Technol 195:102–114
56. Habibul N, Hu Y, Wang YK, Chen W, Yu HQ, Sheng GP (2016) Bioelectrochemical chromium(VI) removal in plant-microbial fuel cells. Environ Sci Technol 50(7):3882–3889
57. Gangadharan P, Nambi IM (2015) Hexavalent chromium reduction and energy recovery by using dual-chambered microbial fuel cell. Water Sci Technol 71(3):353–359
58. Sophia AC, Saikant S (2016) Reduction of chromium(VI) with energy recovery using microbial fuel cell technology. J Water Proc Eng 11:39–45
59. Dhungana TP, Yadav PN (2008) Determination of chromium in tannery effluent and study of adsorption of cr(vi) on sawdust and charcoal from sugarcane bagasses. J Nepal Chem Soc 23:93–101
60. Wang Q, Huang L, Pan Y, Quan X, Li G (2017) Impact of Fe (III) as an effective electron-shuttle mediator for enhanced Cr (VI) reduction in microbial fuel cells: reduction of diffusional resistances and cathode overpotentials. J Hazard Mater 321:896–906
61. Shi J, Zhao W, Liu C, Jiang T, Ding H (2017) Enhanced performance for treatment of Cr (VI)—containing wastewater by microbial fuel cells with natural pyrrhotite-coated cathode. Water 9(12):979
62. Li M, Zhou S, Xu Y et al (2018) Simultaneous Cr (VI) reduction and bioelectricity generation in a dual chamber microbial fuel cell. Chem Eng J 334:1621–1629
63. Wu X, Tong F, Yong X, Zhou J, Zhang L, Jia H (2016) Effect of NaX zeolite-modified graphite felts on hexavalent chromium removal in biocathode microbial fuel cells. J Hazard Mater 308:303–311
64. Kim C, Lee CR, Song YE, Heo J, Choi SM, Lim DH, Cho J, Park C, Jang M, Kim JR (2017) Hexavalent chromium as a cathodic electron acceptor in a bipolar membrane microbial fuel cell with the simultaneous treatment of electroplating wastewater. Chem Eng J 328:703–707
65. Vogl A, Bischof F et al (2016) Increase life time and performance of microbial fuel cells by limiting excess oxygen to the cathodes. Biochem Eng J 106:139–146
66. Kumar M, Chakraborty S (2006) Chemical denitrification of water by zero-valent magnesium powder. J Hazard Mater 135(1–3):112–121
67. Bergquist AM, Bertoch M, Gildert G, Strathmann TJ, Werth CJ (2017) Catalytic denitrification in a trickle bed reactor: ion exchange waste brine treatment. J Am Water Works Assoc 109(5):129–151
68. Ding X, Wei D, Guo W, Wang B, Meng Z, Feng R, Du B, Wei Q (2019) Biological denitrification in an anoxic sequencing batch biofilm reactor: Performance evaluation, nitrous oxide emission and microbial community. Biores Technol 285:121359
69. Xing Wei, Li Jinlong, Li Desheng, Jincui Hu, Deng Shihai, Cui Yuwei, Yao Hong (2018) Stable-isotope probing reveals the activity and function of autotrophic and heterotrophic denitrifiers in nitrate removal from organic-limited wastewater. Environ Sci Technol 52(14):7867–7875
70. Al-Mamun A, Lefebvre O, Baawain MS, Ng HY (2016) A sandwiched denitrifying biocathode in a microbial fuel cell for electricity generation and waste minimization. Springer 13(4):1055–1064
71. Liu R, Gao C, Zhao Y et al (2012) Biological treatment of steroidal drug industrial effluent and electricity generation in the microbial fuel cells. Biores Technol 123:86–91
72. Clauwaert P, Rabaey K, Aelterman P, Schamphelaire LD, Pham TH, Boeckx P, Boon N, Verstraete W (2007) Biological denitrification in microbial fuel cells. Environ Sci Technol 41:3354–3360
73. Zhang B, Liu Y, Tong S, Zheng M, Zhao Y, Tian C, Feng C (2014) Enhancement of bacterial denitrification for nitrate removal in groundwater with electrical stimulation from microbial fuel cells. J Power Sources 268:423–429
74. Ding LJ, An XL, Li S, Zhang GL, Zhu YG (2014) Nitrogen loss through anaerobic ammonium oxidation coupled to iron reduction from paddy soils in a chronosequence. Environ Sci Technol 48(18):10641–10647
75. Shin J, Song Y, An B, Seo S et al (2014) Energy recovery of ethanolamine in wastewater using an air-cathode microbial fuel cell. Int Biodeterior Biodegradation 95:117–121

76. Zhong L, Zhang S, Wei Y, Bao R (2017) Power recovery coupled with sulfide and nitrate removal in separate chambers using a microbial fuel cell. Biochem Eng J 124:6–12
77. Al-mamun A, Said M, Egger F, Al-muhtaseb H (2017) Optimization of a baffled-reactor microbial fuel cell using autotrophic denitrifying bio-cathode for removing nitrogen and recovering electrical energy. Biochem Eng J 120:93–102
78. Jiang C, Yang Q, Wang D et al (2017) Simultaneous perchlorate and nitrate removal coupled with electricity generation in autotrophic denitrifying biocathode microbial fuel cell. Chem Eng J 308:783–790
79. Sotres A, Tey L, Bonmatí A et al (2016) Microbial community dynamics in continuous microbial fuel cells fed with synthetic wastewater and pig slurry. Bioelectrochemistry 111:70–82
80. Li H, Zhou L, Lin H, Zhang W, Xia S (2019) Nitrate effects on perchlorate reduction in a H_2/CO_2-based biofilm. Sci Total Environ 694:133564
81. Lian J, Tian X, Guo J, Guo Y, Song Y, Yue L, Liang X et al (2016) Effects of resazurin on perchlorate reduction and bioelectricity generation in microbial fuel cells and its catalyzing mechanism. Biochem Eng J 114:164–172
82. Yang Q, Zhang F, Zhan J, Gao C, Liu M (2019) Perchlorate removal in microbial electrochemical systems with iron/carbon electrodes. Frontiers Chem 7:19
83. Lounsbury A, Wang R, Plata D et al (2019) Preferential adsorption of selenium oxyanions onto 1 1 0 and 0 1 2 nano-hematite facets. J Colloid Interface Sci 537:465–474
84. Lai C, Wen L, Shi L, Zhao K, Wang Y, Yang X (2016) Selenate and nitrate bioreductions using methane as the electron donor in a membrane bio fi lm reactor. Environ Sci Technol 50(18):10179–10186
85. Catal T, Bermek ÆH, Liu ÆH (2009) Removal of selenite from wastewater using microbial fuel cells removal of selenite from wastewater using microbial fuel cells. Biotechnol Lett 31:1211–1216
86. Zhang Y, Wu Z, Wang Q, Yang L, Li M, Lin ZQ, Bañuelos G (2013) Removal of selenite from wastewater using microbial fuel cells inoculated with Shewanella oneidensis MR-1. Selenium in the Environment and Human Health, p 210
87. Su X, Kushima A, Halliday C et al (2018) Electrochemically-mediated selective capture of heavy metal chromium and arsenic oxyanions from water. Nat Commun 9:4701
88. Litter MI, Ingallinella AM et al (2019) Arsenic in Argentina: technologies for arsenic removal from groundwater sources, investment costs and waste management practices. Sci Total Environ 690:778–789
89. Waseem H, Williams MR, Stedtfeld RD, Stedtfeld TM, Shanker R, Hashsham SA (2017) Organ-on-chip systems: an emerging platform for toxicity screening of chemicals, pharmaceuticals, and nanomaterials. In Nanotoxicology, pp 203–231
90. Xue A, Shen Z, Zhao B, Zhao H (2013) Arsenic removal from aqueous solution by a microbial fuel cell – zerovalent iron hybrid process. J Hazard Mater 261:621–627
91. Gustave W, Yuan ZF, Sekar R, Chang HC, Zhang J, Wells M, Chen Z et al (2018) Arsenic mitigation in paddy soils by using microbial fuel cells. Environ Pollut 238:647–655
92. Li Y, Zhang B, Cheng M, Li Y, Hao L (2016) Spontaneous arsenic (III) oxidation with bioelectricity generation in single-chamber microbial fuel cells. J Hazard Mater 306:8–12
93. Leiva E, Leiva-aravena E, Rodríguez C, Serrano J, Vargas I (2018) Arsenic removal mediated by acidic pH neutralization and iron precipitation in microbial fuel cells. Sci Total Environ 645:471–481

Chapter 9
Managing Oxyanions in Aquasystems—Calling Microbes to Action

Aemere Ogunlaja, Grace S. Peter, and Florence A. Sowo

Abstract Oxyanions are pollutants that pose health risks to humans, impact organisms negatively and can cause environmental hazard like eutrophication. Their removal from aqua systems is therefore expedient. Although conventional chemical and physical treatments exist and have been well-exploited, the biotreatment option is the way forward as they are eco-friendly, cheaper and less technical. There are available scientific studies on the use of biological agents for the removal of oxyanions from water which ranges from the use of plants through to organisms. However, a only a few of these studies focus on the use of microorganisms for oxyanion removal in water. This chapter, therefore, focuses on the use of microorganisms for the removal of oxyanions from water. It provides a collection of reports on the laboratory and field applications of microorganism removal of oxyanions in water either singly or as a consortium and highlights their successes and weaknesses. This chapter also gives a rare insight into genes responsible for arsenic-resistant bacteria that allows them to effectively accumulate arsenate and arsenite from aqua systems. We present future perspectives that will aid further research in this area of study.

Keywords Biotreatment · Microorganisms · Aqua system · Health hazard

9.1 Introduction

Microbes are ubiquitous, and their roles cannot be overemphasized. Their metabolic activities are known to contribute to the stability in the ecosystem through processes like respiration, fermentation, degradation and transformation. With the upsurge in

A. Ogunlaja (✉) · G. S. Peter
Department of Biological Sciences, Redeemer's University, PMB 230, Ede, Osun State, Nigeria
e-mail: ogunlajaa@run.edu.ng

African Centre of Excellence for Water and Environmental Research, Redeemer's University, PMB 230, Ede, Osun State, Nigeria

F. A. Sowo
Department of Microbiology, University of Ibadan, Ibadan, Oyo State, Nigeria

© Springer Nature Switzerland AG 2021
N. A. Oladoja and E. I. Unuabonah (eds.), *Progress and Prospects in the Management of Oxyanion Polluted Aqua Systems*, Environmental Contamination Remediation and Management, https://doi.org/10.1007/978-3-030-70757-6_9

industrialization, which comes with pollution challenges in the environment, treatment and management of pollutants became expedient. The conventional methods (chemical and physical methods) recorded relative successes and their use have been in practice dating back to decades ago. However, the negative effects of their by-products [1, 2], and expensive nature of such treatments have brought biological treatment to limelight. Fortunately, harnessing microbial processes in managing pollutants is environmentally friendly and cost-effective and has become the focus in recent times.

Toxic oxyanions are mostly that of metals or metalloids. Other oxyanions like phosphates (PO_4^{3-}), nitrates (NO_3^-), sulphates (SO_4^{2-}), carbonate (CO_3^{2-}) and silicate (SiO_4^{2-}) are important in cell metabolism of organisms including microbes; hence, they are found naturally in organisms as well as in the environment. However, their presence in the environment at certain levels constitutes environmental challenges like eutrophication (as in the case of excess phosphate in water bodies) which requires treatment to avoid bloom in the water. However, metabolic activities like respiration by bacteria which require electron acceptors trigger the reduction of oxyanions. Routes of cellular transport for the transfer of materials are also used by these oxyanions as they mimic organic molecules because of their structural similarities. In addition, biologically active molecules on cell membranes and walls have an affinity to bind certain ions or molecules from matrix [3]. These mechanisms in microorganisms can be exploited in the management of harmful oxyanion.

An earlier report had shown the competitive nature of oxyanions from group VI element (molybdate, tungstate and chromate) in sulphate respiratory bacteria (*Desulfovibro* sp.) resulting in formation of Adenine monophosphate (AMP) instead of Adenine triphosphate (ATP). The ability to compete with sulphate is due to the similarities in the stereochemistry of selenate, molybdate and Chromate [4], and hence, there is anionic mimicry (Fig. 9.1). Although this procedure could be exploited to reduce oxyanions in the environment, this mechanism will be counterproductive as it can hamper the normal cellular activities (sulphate reduction) of the bacterial

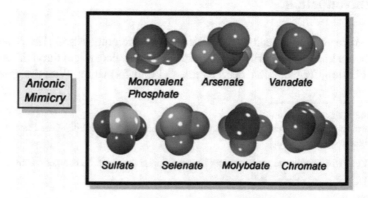

Fig. 9.1 Similar structures of monovalent phosphorus compared with arsenate and vanadate; sulphate with selenate, molybdate and chromate. Adapted from Bridges and Zalups [5]

cells, hence causing reduction in cell density. Harnessing the ability of microbes to incorporate/transform specific toxic oxyanions without obstructing normal cellular activities of the microbe therefore proves effective for pollutant clean-up.

Generally, during reduction of oxyanions by bacteria, electron donors are required but most oxyanion contaminated waters are devoid of them. Hence, donor substrate which can be oxidized by the bacteria must be added for oxyanion reduction to be effective. Conventionally, oxyanion removal by autotrophic bacteria is augmented with hydrogen gas (electron donor), which is preferred to some other organic and inorganic counterparts because H_2 is not toxic among other qualities [6]. The removal of oxyanions of Selenium, Arsenic, Chromium, Vanadium, etc., using various mechanisms by microbes will be discussed in this chapter.

9.2 Removal of Contaminants by Microbes in Aqua Systems

9.2.1 Removal of Selenate and Selenite by Microbes in Aqua Systems

Selenium, though an essential element to organisms, can also be toxic at certain concentrations. Whereas, its lack causes white muscle disease in animals, excess of it causes health problems like liver, muscle and heart diseases such as the Kashin–Beck disease [7]. This is because selenium replaces sulphur during metabolic activities [8] due to the close stereochemistry and properties between selenium and sulphur. Also, selenium usually occurs alongside with sulphur in nature and are both released into the environment during combustion of coal and oil. The most toxic form of selenium is their anionic form which occur as selenium (IV) and (VI). Pertechnetate or perrhenate is a surrogate of selenium which is of environmental concern and a focus for removal in recent times because it is a product of nuclear fission [9]. Selenium also occurs in its oxyanion forms in wastewaters [10]. Selenite (SeO_3^{2-}) and selenate (SeO_4^{2-}) account for the majority of the total Se concentration in waters. The selenate (SeO_4^{2-}) form is predominantly found in alkaline environment with some selenite (SeO_3^{2-}), selenite is stable under a more reducing condition [11], selenite is relatively more toxic than selenate [12]. Removal of selenium oxyanions by microbes is reported in literatures, and biological treatment is one of the most promising methods of selenium removal [13].

Oxidation/reduction process in removal of selenium oxyanions is rarely used in their remediation [13] as a strong chemical reducing agent is needed but a handful of report indicates the effective use of selenium reducing bacteria. In sulphate-rich slurry under anoxic condition, Oremland et al. [14] established the removal of seleno-oxyanions as unrelated to sulphate reduction. They showed that there was increased rate of reducing selenate (SeO_4^{2-}) to selenite (SeO_3^{2-}) in the presence of lactate or acetate (100 or 95% reduction respectively, after 5 days) and H_2/N_2 (99.7% after a period of 7 days). Reduction of these oxyanions was inhibited in the presence of oxygen, nitrate and manganese (IV) oxide. In San Joaquin, California, known for its enormous agricultural activities, *T. selenatis* successfully reduced selenium oxyanions to elemental selenium in the nitrate-rich wastewaters [15, 16]. Thereafter in same environment, Oremland et al. [17] investigated the removal of selenate by *Sulfurospirillum barnesii, Bacillus arsenicoselenatis and Bacillus selenitireducens* in cultured experiments, while alternating arsenate, nitrate, fumarate or thiosulphate as the electron acceptor. *S. barnesii* and *B. arsenicoselenatis* reduced selenate to elemental selenium but *B. selenitireducens* was able to reduce selenite to elemental selenium. *B. selenitireducens* further reduced elemental selenium to selenide but *S. barnesii, B. arsenicoselenatis and* Selenihalanaerobacter were ineffective in reducing elemental selenium to selenide [18]. Their findings led to the installation of an anaerobic reactor treatment plant in San Joaquin valley, California. Other bacteria reported to effectively reduce selenite to selenide include *Rhodospirillum rubrum* [19] and *Veillonella atypical, Geobacter sulfurreducens*, and *Shewanella oneidensis* [20].

Maltman and Yurkov [21] identified some shortfall in the use of some aforementioned microorganisms in the removal of selenium oxyanions which include their relatively low rate of oxyanion removal from aqua systems [16] and the inability to remove the resultant end products from the environment [22]. In this regard, they (Maltman and Yurkov [21]) investigated the use of *Erythromicrobium ramosum, Erythromonas ursincola*, the Pseudoalteromonas relative, AV-Te-18, and Shewanella relative, ER-V-8 for selenium oxyanion removal. They demonstrated that for over 48 h and under aerobic condition, *E. ramosum (E5)* removed 98 µg/ml of selenite and *E. ursincola* removed 100 µg/ml of selenite while ER-V-8 and AV-Te-18 individually, removed 103 µg/ml of selenite. However, under the same period and in anaerobic condition, ER-V-8 and AV-Te-18 removed 46 and 25 µg/ml of selenite in 24 and 48 h, respectively. These bacteria showed better performance for Se oxyanion removal than those earlier proposed by Macy et al. [15], Cantafio et al. [16], Staicu et al. [22], Hunter and Kuykendall [23], and Luek et al. [24]. The advantage of these bacteria over the earlier reported ones is their ability to mop up the end-product (elemental selenium) through cellular accumulation and higher capacity to remove higher levels of selenium oxyanions in the environment within shorter periods of time.

The remediation of selenium oxyanions also involves using microbial biomass as biosorbent. As the living cells undergo their biological processes, their active molecules have affinity to bind certain ion or molecules in the environment or pollutants attach to the functional groups of dead/live cell [13, 25] resulting in adsorption of matter. This forms the basis for removal of pollutants by biosorption, and it has been used in removal of oxyanions like selenium in wastewater. The use of bacteria as biosorbent can be easily managed, requiring less time and can be engineered genetically. Hence, its common use in selenium treatment as reported in the literature. Fujita et al. [26] demonstrated that *Bacillus* sp. (SF 1) reduced selenate to elemental selenium under anaerobic condition in a flow reactor within short retention time but there was selenite accumulation. Selenate and selenite were however converted into elemental selenium after a longer retention time. In a mine wastewater contaminated with selenate (SeO_4^{2-}) and sulphate (SO_4^{2-}), there was comparison between the effectiveness of biotrickling and up flow anaerobic sludge blanket (UASB) setup in selenate and selenite removal. Result showed that the removal of SeO_4^{2-} using UASB was unaffected by SO_4^{2-} but biotrickling filter biofilm improved selenate removal by >70% [13, 27]. Using Pseudomonadaceas and Enterobacteriaceae as biosorbent also reduced selenium concentrations from between 1.7 and 30 mg Se/L to between 160 and 1000 µg Se/L in three types of wastewater from a mine. Pseudomonadaceas and Enterobacteriaceae reduced the concentration of Se from 1.7–30 mg Se/L to between 160 and 1000 µg Se/L in three types of wastewater from a mine [28]. Pieniz et al. [29] used *Enterococcus faecalis* and *Enterococcus faecium* for the removal of selenium at initial temperature of 25 °C and pH of 7.0. 9.9 mg/L and 59.7 mg/L of selenite were further removed by *E. Faecalis* and *E. faecium*, respectively, in 24 h.

Fungi and yeast are also of great interest in their use in biotechnology as they are known to proliferate at a high rate and can be genetically manipulated easily. In addition, they (fungi and yeast) have high genetic diversity, can survive various environmental conditions and can be morphologically manipulated. Furthermore, growing them is easy and their waste biomass is readily available as industrial by-products [30]. Though their use is well-documented for treatment of other pollutants, their use as biosorbents for selenium treatment are not well exploited [31]; hence, there are only a few report on their use for removal of selenium oxyanion from the environment. Organic and inorganic forms of selenium bind with fungi and selenium can be bioaccumulated intracellularly by forming ionic bonding with cellular proteins, polysaccharide and phospholipids or extracellularly by active transport. *Phanerochaete chrysosporium* was used to remove selenite from simulated wastewater. In a 41-day experiment using a continuous flow bioreactor, there was the removal of total soluble Se at ~70% [32]. The fungal caused intracellular production of elemental selenium nanoparticles which led to its morphological changes.

Experimental studies have shown that Fusarium sp., *Aspergillus niger*, *Rhizopus arrhizus*, Mucor SK and *Trichoderma reesei* reduced selenite to its elemental form [33] while *Saccharomyces warrum* reduced selenite in 30–180 µg/mL sodium selenite. In the later study, *S. warrum* accumulated 0.6–2.2 mg/g of selenium from malt wort and 0.3–0.9 mg/g from sparge water, respectively [34]. The biosorption capacity (127 mg/g) of *Ganoderma lucidum* biomass on selenium was also reported by [35], and they revealed that the carboxyl, carbonyl, hydroxyl and amino groups played significant role in the process.

Like other cells, algae have proteins, lipid and polysaccharide on their cell walls; these molecules have sulfate, carboxyl, hydroxyl, and amino functional groups which confer on it, high binding capacity to metalloid and metals [36]. The suitability of algae as biosorbent used for pollutant removal technology is also attributed to its cost-effectiveness, abundance and sustainability [37] even though their adsorbent capacities can be improved by pre-treatment. The capacity of Spirogyra biomass with potential for selenium removal was improved by either autoclaving, heating, or chemical pre-treatment with sodium hydroxide and acetic acid [38]. It was reported that the pre-treatment by autoclaving caused the best effect on biosorption capacity [38]. Pre-treatment of Gracilaria biomass (by-product of agar production) with ferric solution and conversion to biochar showed its capacity to absorb Se (IV) and (VI) at low capacity (2.7 mg/g) [13]. Nevertheless a 98% removal of Se (IV) from mock solution by modified Gracilaria biochar has been reported [39].

9.2.2 Removal of Arsenate and Arsenite from Aqua Systems by Microbes

Arsenic is naturally found in sediments but can be released during geological activities like digging of shallow wells. Arsenic is suspected to be leached out of sediments through microbial metabolism [40]. Oxyanions of arsenic are also immobilized by natural organic matter (NOM) through mechanisms like complexation [41]. Weathering of rocks and other anthropogenic activities like the use or production of chemicals containing arsenic can also contribute to arsenic levels in the environment. High concentrations of arsenic were found in domestic drinking water from shallow-well in the American Midwest and Bengalese [40, 41]. Arsenate (AsO_4^{3-}) and Arsenite (AsO_3^{3-}) are the most common toxic oxyanions of Arsenic found in aqua systems [42]. Arsenate exists readily in aqua systems due to its relatively high solubility although they could form phosphate precipitate and other types of precipitates in the environment in the presence of insoluble iron compounds and calcium [40]. Arsenate is stable in aerobic aqua systems like surface water and shallow groundwater while arsenite can be found in deep groundwater usually under moderately reducing anaerobic condition [43]. Arsenate toxicity is due to its structural similarity with phosphate (Fig. 9.1) while Arsenite in water can mimic non-ionized glycerol and hence, can be transferred by glyceroporin membrane channel proteins to cells [44, 45]. Its affinity to protein thiol groups, protein-DNA and its DNA-DNA cross-linking also make it more toxic than Arsenate [46–48]. Arsenite is reported as 25–50 times more toxic than Arsenate [49, 50]. Hence, the removal of Arsenic oxyanions with the aid of microbes is addressed in this section.

An option used in treatment plants for Arsenic removal is the oxidation of Arsenite to its less toxic form, Arsenate [51]. Although chemical oxidation is conventionally used by adding reagents, they constitute environmental hazards because of toxic by-products from the process. This drawback is overcomed when biological pre-treatment technique is used for the oxidation process [52]. The oxidation of arsenite can be achieved by a wide range of bacteria like heterotrophic arsenite oxidizers through detoxification reactions or by the chemolithoautotrophic growth strategy [40, 53, 54]. Another option for Arsenic removal is the reduction of Arsenate. Bacteria are known to cause the reduction of Arsenate and their uses in Arsenic removal are well-documented [55, 56].

Most times, the treatments of oxyanions in aqua systems are achieved by a combination of biological/physicochemical adsorption processes like biogenic iron and manganese oxides used in oxidation and adsorption [57] for Arsenic removal. Other successful combined methods for Arsenic removal are biofilters coupled with activated granular Fe-hydroxides, metallic Fe adsorbents and alumina [58, 59] and biological/iron/manganese oxidation systems [60]. In a study, Corsini et al. investigated the use of twenty (20) indigenous bacterial isolates in removing arsenate and arsenite from groundwater in Lombardia, Italy [61]. The isolates majorly from the genera *Achromobacter*, *Rhodococcus* and *Pseudomonas* with the chromosomally encoded arsenical resistance (ars) operon genes were able to reduce 75 mg/L arsenate to arsenite. Among three arsenite oxidizing isolates, *Rhodococcus* sp., *Achromobacter* sp. and *Aliihoeflea* sp., the strains of *Achromobacter* sp. (1L) and *Aliihoeflea* sp. (2WW) had aioA genes for arsenite oxidase. The combination of oxidation by 2WW strain with goethite adsorbent caused the removal of Arsenite (from 200 to 8 μg/l) in groundwater which accounted for 95% removal of arsenite. In another work, Akhter et al. [62] reported the presence of genes responsible for arsenite oxidation and arsenic resistance in *Pseudomonas stutzeri* TS44 indicating its potential for arsenite removal from the environment. Later, the use of functionalized melanin (Fe and Cu impregnated) extracted from *P. stutzeri* for arsenite and arsenate removal in aqueous system was studied [63]. Results indicated that within pH of 4.0–6.0 there was >99% removal of both oxyanions in 50 min and 80 min for Fe-melanin and Cu-melanin, respectively. 99% adsorption efficiency was recorded after four reuse cycles of the adsorbent by desorption of Fe and Cu bound to the oxyanions and re-functionalizing with Fe and Cu, which suggests their sustainable use for arsenic removal.

Individual or consortiums of microbes have also been tested for their potential to remove arsenic. Findings show that *Momordica charantia* oxidized arsenite in 45 min, *Inonotus hispidus* biosorbed both arsenate and arsenite in 30 min and *Staphylococus xylosus* transformed both Arsenite and Arsenate in 30/150 min, respectively [50, 64–66]. In 1 mg/L arsenite and arsenate contaminated water, Teclu et al. [46] reported 70% and 87% removal of arsenite and arsenate, respectively, after 14 days of using a mixed culture of sulphate-reducing bacteria (SRBs) in a batch mode after a 14 days. Under similar conditions, with 5 mg/L each of the oxyanions, there were 61% and 81% removal of arsenite and arsenate, respectively, due to increased concentration of the oxyanions which impacted the bioremoval efficiency. In the same study, 50 ml of SRB cell pellets were also examined for their capacity to sequester As (III) and As (V) from the 1 mg/L and 5 mg/L contaminated waters. A 6.6% and 10.5% of 1 mg/L As (III) and As (V), respectively, were sequestered in the contaminated water while 6.4% and 10.0% of 5 mg/L As (III) and As (V), respectively, were sequestered in the contaminated water after 24 h contact. As (III) removal was lower than that of As (V) and efficiencies of removal for the actively growing SRB were better than the SRB cell pellets for both oxyanions in this study. The sulphide from metabolic activity of SRB is known to react with the dissolved arsenic oxyanions to form an arsenic sulphide precipitate, which can lead to a decrease in the availability of dissolved arsenic. However, due to the formation of soluble thioarsenite as by-product during this process, the procedure is not stable and it is also not satisfactory [67].

Sun et al. [68] in their study aimed at overcoming this hindrance by proposing a sulphur-reducing process using a community of sulphur-reducing bacteria (SRB) other than sulphate reduction in the aforementioned treatment system. Their procedure indicated the use of reduced numbers of electron donors, and the rate of sulphide production was relatively higher [68–70]. Their findings also showed that the produced sulphides precipitated arsenite causing >99% removal, thioarsenite formation was eliminated and the procedure efficiently removed arsenite at acidic pH natural to arsenic-contaminated waters. Their findings indicated that treatment of arsenite using sulphur reducing bacteria is a better alternative than sulphate-reducing bacteria. Two strains of *Bacillus cereus* (P1C1Ib, P2Ic) and one of *Lysinibacillus boronitolerans* (P2IIB) species out of 38 strains isolated from mine waste that impacted soils in Paracatu, Brazil, showed potential for arsenite removal [71]. These three strains were resistant to 3000 mg/L of arsenite which is among the highest concentrations of arsenite tolerated by arsenic resistant bacteria earlier reported by Escalante et al. [72], Shakya et al. [73], Majumder et al. [74], Huang et al. [75]. Even though their use in arsenic treatment is not yet reported, they could be exploited.

The ability for organisms to exist in the presence of pollutants is linked to their genes that induces the expression of resistance proteins. This knowledge is applied in genetic engineering. Several genes are related to arsenic resistance and metabolism. The *asoA* and *asoB* genes in *A. Faecalis* [40], *arsC* gene in more than 50 organisms ranging from bacteria, yeasts and protists [76], *arsB* and *acr3* genes in arsenic-resistant bacteria [77] are well documented. Among such resistant proteins expressed by the arsenic-resistant genes is ArsR. The ArsR gene has specific affinity for As (III) but not for other oxyanions like phosphate and sulfate. Kostal et al. [51] studied the expression of ArsR gene by bacterial cells in the accumulation of arsenate and arsenite. They genetically modified *E. coli* such that the strain over expressed ArsR gene which resulted in selective accumulation of arsenite. Although they recorded relative success, high expressions of ArsR led to reduction in cell density. This set back was rectified by the addition of fusion partner (elastin-like polypeptide) in the protein. The modified ArsR (ELP153AR) gene showed better performance as their over expression did not affect cell density but caused 5 times and 60 times higher accumulations levels of Arsenate and Arsenite respectively. Other reports on Arsenic removal using microbes include work by Kim et al. [78] and Ahmad et al. [79].

9.2.3　Removal of Chromate from Aqua Systems by Microbes

The oxyanions of chromium are chromate and chromite. Chromate is soluble and toxic while chromite is insoluble at ≥ 7.0 pH. Chromite is less toxic because it cannot easily be transported through cell membrane [80, 81]. However, they are micronutrients to organisms [82]. Chromate at high concentration inhibits the growth of microbes. It causes mutagenic and carcinogenic effects in humans [83] but at low levels in the environment, some microbes can reduce chromate to chromite [84]. Species of microbes that are resistant to chromate live in chromate-contaminated environment with levels as high as 500 mg/L [85] and 250 mg/L [86]. Species like *Pseudomonas fluorescens* LB300 and yeast grew in 270 mg/L and 500 mg/L of Cr

(IV) [84, 87], respectively, while *P. Ambigua* showed tolerance to 2000 mg/L of Cr (IV). However, for the synthesis of RNA, DNA and proteins were inhibited. It caused 87% and 13% removal of soluble and insoluble fractions of chromate, respectively [88]. Other chromate-tolerant bacteria are *Bacillus firmus* (49,400 mg/L), from soil contaminated with electroplating effluent [89] and CMBL Crl3 strain (45,000 mg/L) in wastewater from leather tanning industry [90]. A major medium of cellular transport for chromate from the environment is the sulphate transport pathway [91, 92], although sulphate intake was reported to be unaffected by the presence of Cr (IV) [84, 93, 94]. There is also the outward translocation of Cr (IV) through cell membrane [95]. Other genes associated with transport of chromate are MtrC and OmcA [96–98] while genes associated with chromate reduction include ChrR, NemA and NfsA [99, 100].

Microbes tolerant to chromate at high concentrations are potential candidates for chromate removal and such studies are well-exploited and reported but not all resistant or tolerant bacteria are efficient in chromate removal [101]. Thus, the selection of microbes useful for chromate removal does not only depend on their tolerance to high levels of chromate but testing for their chromate reduction capability. Studies showed that even though two species of bacteria had similar Cr (IV) reducing capacity (40 μg/L) and tolerance levels (80 μg/L), the removal of Cr (IV) was 10 μg/L and 30 μg/L (*Bacillus* sp. and Arthrotobacter sp.) respectively, after a 46 h period [102]. Also, *Brevibacterium* AKR2 strain tolerated 1000 and 1500 mg/L of chromate showing 91% and 86% chromate removal respectively, in 24 h [103].

Chromate reduction to chromite could be achieved aerobically by using chromate reductase, anaerobically by using Cr (IV) as electron acceptor or reduction or by chemical reactions with sugars, amino acids, organic acids, glutathione, nucleotides or vitamins [104]. Chromite is then immobilized as species of hydroxide [105, 106]. Several experimental reports indicate the use of individual microbes for the removal of chromate [107–109]. Nonetheless, they prove challenging practically. Instead, the use of bacterial consortiums is more feasible in nature. For instance, Ma et al. [110] monitored the use of a consortium of bacteria for chromate removal and established that the removal was by extracellular enzyme and not by adsorption. However, the diversity and richness of the bacteria decreased with reduction of chromate. Although, the success of using free cells for chromate removal is usually challenging because of chromate toxicity and the cellular damage they impose on cells, yet the use of immobilized cells has proven effective in biological treatment of chromate in wastewater [81]. Immobilized *Microbacterium* sp. removed 100 μM Cr (VI) within four days [111] and immobilized *Desulfovibrio vulgaris* caused reduction of 0.5 mM Cr(VI) to 0.1 mM in 22 h [112]. Other immobilized bacteria cells used in chromate removal include *Desulfovibrio desulfuricans* [113], *Streptomyces griseus* [114], *Pannonibacter phragmitetus* LSSE-09 [115], *Pseudomonas* S4 [116], etc. The reuse of the immobilized cells of *Pseudomonas* S4 [116] and *S. griseus* [114] was proven sustainable as the reduction of chromate with recycled cells recorded similar removal efficiency. A combination of adsorption, biosorption and bioaccumulation processes using substrate and microorganisms has proved efficient in oxyanion removal. Gao et al. [117] investigated chromate removal using immobilized bacteria on maifanite/ZnAl-LDHs compared to natural maifanite. Furthermore, constructed rapid infiltration systems (CRIS) has been developed as a technology with the potential to be exploited for oxyanion removal from aqueous systems.

9.2.4   Removal of Phosphate from Aqua Systems by Microorganisms

Phosphorus (P) normally originates from human and animal wastes, food processing effluents, commercial fertilizers, industrial wastewater, agricultural land runoffs and household detergents, it is a major nutrient contaminant in water. The forms of oxyanions of phosphorus include orthophosphate, polyphosphates and organic phosphorus which enter into water bodies through mining, industrial and agricultural activities and sewage discharges. Excessive concentrations of phosphorus in water usually lead to eutrophication [118, 119]. This consequently causes the deterioration of water quality due to the growth of plants such as algae that disrupt the ecological balance of the waters affected [120] as low levels of oxygen can cause death of some aquatic lives [121, 122]. Algae bloom is potentially risky to human health because of the consumption of aquatic food contaminated with algal toxins or the direct exposure to waterborne toxins [123].

Polyphosphates and organic phosphate are converted to orthophosphate by hydrolysis/or microbial mobilization [124, 125]. Orthophosphate is soluble, and it is directly assimilated by most plants, including algae, and they also have the ability to strongly adsorb onto inorganic particles, matters/sediments in water. Several strains of bacterial and fungal species have been described and investigated in detail for their phosphate-solubilizing capabilities [125, 126]. Such isolates include species of *Pseudomonas* and *Bacillus* [127], fungi, actinomycetes and even algae. Other bacteria that have been reported as phosphorus solubilizer include *Rhodococcus, Arthrobacter*, Serratia, Chryseobacterium, Gordonia, *Phyllobacterium, Delftia* sp. [128, 129], *Azotobacter* [130], *Xanthomonas* [131], *Enterobacter, Pantoea,* and *Klebsiella* [132], *Vibrio proteolyticus, Xanthobacter agilis* [133]. Many different strains of these bacteria have been identified as phosphate solubilising bacteria (PSB), including *Pantoea agglomerans* (P5), *Microbacterium laevaniformans* (P7) and *Pseudomonas putida* (P13) strains which are highly efficient insoluble phosphate solubilizers. It was reported by Bass et al. [134] that a consortia of four bacteria synergistically solubilize phosphorus at a much faster rate than any single strain. Solubilized phosphorus can then be finally removed by chemical or physical methods.

9.2.5 Removal of Perchlorate and Chlorate from Aqua Systems by Microbes

The oxyanions of chlorine, perchlorate (ClO_4^-) and chlorate (ClO_3^-) are highly soluble, strong oxidants that are deposited in the environment through both anthropogenic and natural processes [135–137]. Perchlorate is a common chemical which has diverse range of industrial uses ranging from pyrotechnics to lubricating oils [138]. However, it is predominantly used as an energetic booster or an oxidant in solid rocket fuels by the munitions industry [139, 140]. It is toxic to humans because it inhibits the uptake of iodine by the thyroid gland [141], thereby disrupting the production of thyroid hormones, potentially leading to hypothyroidism in both infants and young children [142, 143]. Humans usually gets exposed to it through oral ingestion as contamination is widespread in soil, fertilizers, and groundwater, allowing it to readily move into the food chain [144]. It also has a negative effect on ecosystems because it leads to loss of environmental quality. According to Smith et al. [145], short-term exposure has been shown to affect the nervous, respiratory, immune, and reproductive systems.

One of the most promising, effective, and economically friendly method utilised for the removal of perchlorate is the use of bacteria in biotechnological systems, because they are capable of reducing and eliminating the oxyanion. Xu and Logan [144] reported that the use of bacteria in removing perchlorates usually leads to the complete degradation of perchlorate ions into Cl^- and O_2. Perchlorate and chlorate are reduced to chlorite (ClO_2^-) by the enzyme perchlorate reductase (pcr; EC 1.97.1.-), and further broken down into O_2 and Cl^- by the enzyme chlorite dismutase (EC 1.13.11.49) [146, 147]. The biological degradation pathway is as follows: ClO_4^-(perchlorate) $\rightarrow$ ClO_3^-(chlorate) $\rightarrow$ ClO_2^-(chlorite) $\rightarrow$ Cl^- (chloride) $+ O_2$.

Studies on the kinetics of microbial reduction and gene regulation of ClO_4^- have been reported [148, 149]. In both reports, the diversity of perchlorate-reducing microorganisms (PRMs) is presently confined to bacteria and Achaea alone. Majority of them, both cultured and uncultured, belong to the bacteria domain. They are, generally called perchlorate-reducing bacteria (PRB) and among bacteria, proteobacteria represented the most prominent species (α, β, γ, δ and ε subgroups, dominated by β), and a minor fraction by Firmicute. Achaea and protozoa community in ClO_4^- reducing bioreactors is scarcely reported. An archaea reducing ClO_4^- from a marine environment (Archaeoglobus fulgidus) from the phylum Euryarchaeota, has been reported [150]. Perchlorate-reducing bacteria are phylogenetically diverse; these include alphaproteobacteria, betaproteobacteria, gammaproteobacteria and deltaproteobacteria classes, with betaproteobacteria being the most commonly detected class [151]. In a study carried out by Acevedo-Barrios et al. [152], bacterial strains belonging to the betaproteobacteria class showed biological capacity to reduce concentrations of $KClO_4$ between 10 and 25%, they also reported that the genera nesiotobacter and salinivibrio showed the highest percentage (25%) of perchlorate reduction, while the genera vibrio, bacillus, and staphylococcus presented the lowest proportion of $KClO_4$ reduction, with 14%, 12% and 10%, respectively. Recent studies have shown that the amount of perchlorate reduced may be inversely proportional to increased salinity in an aquatic system [153].

According to Youngblut et al. [154] and Coleman et al. [155], dissimilatory (per)chlorate reducing bacteria (DPRB) use a highly conserved perchlorate reductase, PcrABC, to reduce perchlorate to chlorate and subsequently to chlorite. The chlorite is rapidly removed by another highly conserved enzyme, chlorite dismutase (Cld), to produce molecular oxygen (O_2) and innocuous chloride (Cl^-). The oxygen that is produced is then respired by the same organism generally through the use of a high-affinity cytochrome cbb3-oxidase. A report by Brundrett et al. [156] has shown that due to their unique ability to generate molecular oxygen under anaerobic conditions, they are capable of stimulating oxygenase-dependent anaerobic metabolisms. This was first shown in anoxic co-cultures of DPRB with obligate aerobic hydrocarbon utilizing *Pseudomonas* spp. [157, 158]. In these studies, degradation of benzene and naphthalene was demonstrated under anoxic conditions when the cultures were amended with chlorite. The chlorite was directly dissimulated into O_2 and Cl^- by the active DPRB, and the biogenic O_2 was subsequently available for the aerobic *Pseudomonas* to use as a co-substrate and an electron acceptor for the hydrocarbon metabolism in an oxygenase-dependent manner. The production of oxygen by these organisms makes them suitable for the bioremediation of a broad diversity of recalcitrant xenobiotic compounds under anoxic or oxygen limiting conditions, both insitu and in bioreactors. The optimal temperature range for perchlorate reduction is 28–37 °C [159, 160].

9.2.6   Removal of Nitrate and Nitrites from Aqua Systems by Microbes

Nitrogen is an essential nutrient available and utilized in various forms by living organisms. It is largely inaccessible as nitrogen gas (N_2) and must undergo various transformations involving reduction and oxidation. Other inorganic forms of nitrogen include ammonia and nitrate while organic forms include amino and nucleic acids. Nitrite (NO_2^-) is an oxyanion of nitrogen with symmetrical structure made up of one nitrogen atom bonded to two oxygen atoms. It is also an intermediate in the nitrogen cycle wherein NH_4^+ is converted into NO_2^- by microbes during nitrification. Nitrates and nitrites exist as highly soluble inorganic salts in the environment and are soluble in water. Nitrate is produced from nitric acid and is known as the conjugate base of nitric acid. It is also a metabolic product of nitrite oxidation during the latter stages of the nitrogen cycle. Both nitrate and nitrite are intentionally introduced to the environment majorly for agricultural purposes and are thus present in food crops and the water we drink [161].

Ingested nitrates and nitrites have been implicated in the endogenous synthesis of *N*-nitroso compounds (NOCs) such as nitrosamines, a carcinogenic compound associated with stomach and colorectal cancer (CRC). Long-term dietary nitrate derived from animals are associated with rectal cancer while colorectal cancer (CRC) is associated with nitrate exposure in drinking water [162]. Nitrate and nitrite are also contributing factors to methemoglobinemia (blue blood syndrome) which is an uncommon condition of the blood where the oxygen transporting ability of haemoglobin is impaired. This is brought about by the attachment of nitrites to the oxygen atoms in the blood resulting in a deficiency of oxygen thus changing haemoglobin to methemoglobin. Nitrites have also been implicated in stomach cancer. The primary and major source of nitrate contamination in different water bodies especially ground and surface water is associated with agricultural activities. Other means of contamination include runoffs from landscapes exposed to nitrogen-based fertilizer usage, animal manure and also sewage. This results in eutrophication, distrophication or hypertrophication of the water body resulting in hypoxia; the depletion of oxygen and formation of dead zones incapable of supporting life. Although, it is not as lethal as nitrite, nitrate levels measuring above 30 parts per million (ppm) are capable of inhibiting growth, ability to reproduce, and impairs the immune system of aquatic species. In animals, nitrate poisoning results in an increased heart rate, respiration, colour change of blood and tissues from red to blue or brown. Plants also suffer from an elevated nitrate level resulting in asphyxiation and plant death.

Several processes have been established for the removal of nitrite in water. The oxygen-limited autotrophic nitrification and denitrification (OLAND) process uses normal nitrifiers, dominated by ammonium oxidizers using nitrite as electron acceptors, ammonia is oxidized to nitrogen gas. This is followed by the action of hydroxylamine oxidoreductase (HAO) or its related enzyme which is responsible for the loss of nitrogen [163]. The downside of the OLAND process includes the need for aeration when converting ammonia into nitrites and an additional external carbon source such as methanol over an extended period of time [164]. The completely autotrophic nitrogen removal over nitrite (CANON) process is aimed at complete autotrophic nitrogen removal over nitrite in a single reactor. Using this process, 85% of ammonia is mainly converted into nitrogen gas and 15% is recovered as nitrate with a negligible production of nitrite at 0.1%. Unlike the OLAND process, CANON does not require an external carbon source [164, 165]. The SHARON–ANAMMOX process is complicated as it uses a double-stage partial nitration/anaerobic ammonium oxidation technique and requires the dilution of ammonia in high concentrations [164, 166].

Given the foregoing and the setback in the above illustration for nitrate conversion to nitrogen gas, the total removal or conversion of excess nitrate and nitrite to gaseous nitrogen has been the focus for several scientists. Several organisms categorized as ammonia oxidizers and nitrite oxidizers are required in a two-step sequence in converting ammonia to nitrite, and nitrite to nitrate called nitrification. Nitrification occurs at the aerobic/anaerobic interface, but nitrifying bacteria have a high affinity for oxygen. Some nitrite-oxidizing bacteria includes: *Nitrospira* (δ-proteobacteria), *Nitrobacter* (α-proteobacteria), *Nitrococcus* (γ-proteobacteria) and *Nitrospina*. Ammonia-oxidizing organisms use ammonia or nitrite as electron donors and CO_2 as their sole carbon source. In the presence of iron oxide (ferrihydrite or magnetite), *S. oneidensis* MR-1 was capable of removing nitrite with gaseous nitrogen as the end product. Both ferrihydrite and magnetite compounds remained stable with gaseous nitrogen yield of 65.56% and 23.13%, respectively [167].

Nitrification has been adopted in the removal and reduction of nitrogen in recycled water using either autotrophic or heterotrophic organisms. A heightened ammonium and nitrite removal were reported following autotrophic microbial activities in the presence of heterotrophic strains. Several *Janthinobacterium* species were capable of nitrite removal at temperatures as low as 25 and 15 °C. In an experimental setup of heterotrophic ammonium and nitrite (HAN) removal complex made up of *Janthinobacterium* and *Dyadobacter* species applied in an trout (*Oncorhynchus mykiss*) aquaculture system, *Dyadobacter* species removed un-ionized ammonia below the recommended level for trout culture by converting ammonia to nitrite which was also removed by *Janthinobacterium* sp. to a level below the range limit (1000 $\mu g\,l^{-1}$) in the aquaculture system. The pH was the only affected variable in the designed system owning to the acid production during nitrification. Following this experiment, trout stresses were alleviated and the aquaculture productivity increased in comparison with the control. It was however established that *Janthinobacterium* sp. can grow and degrade nitrite within a range of 2–15 °C at a pH of 6.8, a condition capable of antibacterial and fungal roles [168]. Another study established the capacity of nitrate/nitrite-dependent anaerobic methane oxidation (n-DAMO) archaea coupled with anaerobic ammonium oxidation (Anammox) in a membrane biofilm reactor (MBfR) for the removal of nitrogen from landfill leachate. The n-DAMO archaea utilizes methane in reverse methanogenesis to provide electrons to reduce nitrate to nitrite [169], and the presence of anaerobic ammonium oxidation bacteria (AnAOB) removes the electrons needed by n-DAMO for its denitrification process. To avoid this, the growth of n-DAMO archaea was influenced in an MBfR using ammonium and nitrate as a start-up, loaded at a consistent low level to enable accumulation of n-DAMO archaea biomass for nitrate removal [170]. The activities of AnAOB (converting nitrites to nitrates) were stimulated following the addition of nitrite into the MBfR already fed with ammonium and nitrate. This made nitrates available for the denitirification process of n-DAMO archaea. The nitrate removal rate of n-DAMO archaea reached about 160.0 mg NO_3^--N L^{-1} d^{-1} within 200 days of the study. The n-DAMO and Anammox process are said to be efficient in the removal of nitrogen from landfill leacheate [171].

Nitrosomonas, *Nitrospira* and *Candidatus brocadia* are described as ammonium oxidation bacteria (AOB), nitrite oxidation bacteria (NOB) and anammox bacteria (AnAOB), respectively [172]. To evaluate nitrification ability in recycled water or seawater, AOB and NOB were subjected to different saline conditions. A moving bed bioreactor (MBBR) with immobilized microbial granules (IMG) was tested on recycled synthetic aquaculture wastewater for the nitrification at 2.5 mg/L NH_3-N daily. Although increase in salinity from near zero to 35.0 g/L, decreased the microbial activity of NOB by 86.32%. At high salinity of 35.0 g/L NaCl, the IMG effectively converted ammonia into nitrate up to 92%. *Nitrosomonas* sp. and *Nitrospira* sp. were established as the dominant genera for AOB and NOB at different salinity levels [173]. Nitrifying biofilms developed in brackish water were compared to nitrifying biofilms developed in freshwater to acknowledge the most robust when subjected to salinity changes. After 60 days, the brackish water biofilm had half the nitrification capacity of the freshwater biofilm, less diverse microbial community, lower proportion of nitrifiers, and a significantly different nitrifying community composition. *Nitrosomonas* and *Nitrosospira* species were the main ammonia oxidizers in the brackish water biofilms. *Nitrotoga* was the dominant nitrite oxidizer in both treatments while *Nitrosomonas* sp. was dominant in the freshwater biofilm. However, the low concentrations of ammonia and nitrite, and the rapid increase of nitrate concentration, indicate the complete nitrification in both reactors within the 60 days. This study concludes that biofilms develop nitrification in brackish water in comparable time as in freshwater, and brackish start-up can be a strategy for bioreactors with varying salinity [174].

9.2.7 Removal of Bromate and Its Related Oxyanions from Aqua Systems by Microbes

Bromate is a conjugate of bromic acid, slightly soluble in water and denser than water. It is commonly formed by the reaction of bromide and ozone: $Br^- + O_3 \rightarrow BrO_3^-$. In water containing bromide, photoactivation can cause liquid or gaseous bromine to generate bromate. Following ozonization of water, several bromated ions are formed via multistage oxidation of bromides by molecular ozone or hydroxyl radicals [175, 176]. Bromine exists in several forms: Bromide ion (Br^-), Hypobromite (BrO^-), Bromite ion (BrO_2^-), bromate ion (BrO_3^-) and perbromate ion (BrO_4^-). Bromate formation is a multistep oxidization process by the transfer of oxygen from ozone (O_3) [175, 177]. Following the formation of bromite, the activity of ozone in water triggers the production of intermediary radicals such as hydroxyl radical ($\cdot OH$), $BrO_2\cdot$, and $O_3\cdot^-$. Bromide ion may also react with the hydroxyl radical ($\cdot OH$) creating the bromine radical ($Br\cdot$) which is oxidized by ozone to produce the intermediate radical, $BrO\cdot$ that is hydrolysed to Bromite ion (BrO_2^-). The last stage of oxidation by ozone yields bromate ion BrO_3^- [178] (Fig. 9.2).

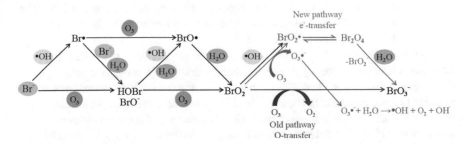

Fig. 9.2 General pathway in the production of bromate from bromide. *Source* [178]

The international Agency for Research on Cancer (IARC) classified bromate as a level IIB carcinogenic compound. Ozonation has proven to be the most effective in disinfecting drinking water. However, it produces disinfection by-products (DBP) such as aldehydes, carboxylic acid, ketoacid, ketone, inorganic halogen and nitrile [179, 180]. This is problematic especially if bromide is present in water as it can be oxidized to bromate [181] as shown in Fig. 9.2. The maximum permitted concentration of bromate in drinking water is 10 μg/L [182, 183]. However, due to ozone treatment of water, bromate concentration in potable water ranges between 0.4 and 60 μg/L making it a major factor that limits the adoption of large-scale ozonation during water treatment processes [184]. Remediation of bromate from water using microbes is more advantageous as it is cost effective and requires low energy consumption in comparison with chemical and physical methods.

The mechanism of bromate bioremediation is not well-studied. Thus, biological technology leverage on bromate as the electron acceptor to generate bromide (Br^-) without the accumulation of stable Bromite ion (BrO_2^-), and hypobromite (BrO^-). Postulations of co-metabolism of bromate via nitrate reductase, (per)chlorate reductase and sulfate reductase have been made [185, 186]. However, in a bid to ascertain this claim, numerous studies nullify these claims as reduction of bromate is inhibited in the presence of nitrate. *Pseudomonas* sp., a denitrifying bacteria was confirmed capable in reducing Bromate to bromide but in the presence of nitrate, bromate reduction did not occur [187, 188]. It was also established that bromate reduction ability of an autohydrogenotrophic microbial community was also inhibited by high concentrations (50 mg/L) of nitrate in a rotating biofilm-electrode reactor [189]. Reasons being that nitrate is the more preferred electron acceptor in the presence of bromate when the electron donor, methane, was limited [190]. In another study, bromate was removed when measurable concentrations of dissolved oxygen (DO) and nitrate were discharged, but upon their introduction, bromate removal decreased [191]. It was established that in the presence of bromate, DO, and nitrate, DO competed against bromate as electron acceptor [192]. Bromate reduction was not hindered in the absence and presence of sulphate possibly due to the lower potential of sulphate to bromate and nitrate [178, 188, 193]. A diaphorase isoform belonging to a bromate-reducing bacterium *Rhodococcus* sp. Br-6, possess a bromate-reducing activity, but in an indirect way that involves the interplay of biotic and abiotic reactions, the latter was dependent of redox mediators including ferric iron and a redox dye 2,6-dichloroindophenol (DCIP) [194].

Microbes capable of reducing bromate include denitrifiers, sulfate-reducing and (per)chlorate reducing bacteria which are ubiquitous and phylogenetically diverse in the environment. Analysis of the 16S rRNA amplicons would provide genetic biomarkers used in identifying bromate-reducing organisms some of which include *Gammaproteobacteria, Firmicutes, Betaproteobacteria, Actinobacteria* [193], *Clostridium (Firmicutes), Citrobacter (Proteobacteria)* [185]. *Bacteroidetes, Alphaproteobacteria* [186]. *Sphingomonas* sp. 4721 (*Proteobacteria*) *Deinococcus* sp. 4710 (*Deinococcus–Thermus*) [192]. *pirochaetacea* spp. (*Spirochaetes*) and *Denitratisoma* spp. *Proteobacteria* [195]. *Exiguobacterium (Firmicutes), Arthrobacter (Actinobacteria), Chlorobium (Bacteroidetes)* [189].

Bromate-reducing biochemical mechanisms which could be by co-metabolism, respiratory or other specific pathways can be clarified and ascertained via the integration of metagenomics and metatranscriptomic technologies. This would provide genetic insight to the metabolic pathways which would be useful in improving bromate removal efficiencies under different environmental conditions.

9.3   Future Perspective

It is noteworthy that contaminants do not occur singly in water (oxyanions inclusive) but in mixture with others substances in the matrix they are found, having a wide range of concentrations. In this light, the successful removal of contaminants should be tackled from a multicontaminant removal approach. A few researches have addressed this challenge. For instance, Manna et al. [196] reported the successful removal of arsenate and chromate from an aquatic matrix but only with bioinspired silica nanoparticles assembled microsphere. Their design, mimicking silica structure of diatoms, showed efficient oxyanion-binding property with high recyclability. However, the use of organisms in multioxyanions removal in aquasystems is rare. Furthermore, the use of biological treatments is the right direction for removal of a wide range of pollutants from the environment since they are eco-friendly and relatively cheaper. However, their use for large scale water treatment is still minimal. There is need for more research addressing the use of microorganisms in oxyanion removal from aqua systems at this level.

References

1. Goutam SP, Saxena G, Roy D, Yadav AK, Bharagava RN (2020) Green synthesis of nanoparticles and their applications in water and wastewater treatment. In: Bioremediation of industrial waste for environmental safety. Springer, pp 349–379
2. Santos SC, Ungureanu G, Volf I, Boaventura RA, Botelho CM (2018) Macroalgae biomass as sorbent for metal ions. In: Biomass as renewable raw material to obtain bioproducts of high-tech value. Elsevier, pp 69–112
3. Gautam RK, Mudhoo A, Lofrano G, Chattopadhyaya MC (2014) Biomass-derived biosorbents for metal ions sequestration: adsorbent modification and activation methods and adsorbent regeneration. J Environ Chem Eng 2:239–259
4. Taylor BF, Oremland RS (1979, Mar 3) Depletion of adenosine triphosphate in Desulfovibrio by oxyanions of group VI elements. Curr Microbiol (2):101–103
5. Bridges CC, Zalups RK (2005) Molecular and ionic mimicry and the transport of toxic metals. Toxicol Appl Pharmacol 204:274–308
6. Karanasios K, Vasiliadou I, Pavlou S, Vayenas D (2010) Hydrogenotrophic denitrification of potable water: a review. J Hazard Mater 180:20–37
7. Liao C, Carlson BA, Paulson RF, Prabhu KS (2018) The intricate role of selenium and selenoproteins in erythropoiesis. Free Radical Biol Med 127:165–171
8. Axley MJ, Stadtman TC (1989) Selenium metabolism and selenium-dependent enzymes in microorganisms. Annu Rev Nutr 9:127–137

9. Roundhill DM, Koch HF (2002) Methods and techniques for the selective extraction and recovery of oxoanions. Chem Soc Rev 31:60–67
10. Fu F, Wang Q (2011) Removal of heavy metal ions from wastewaters: a review. J Environ Manage 92:407–418
11. Geering HR, Cary EE, Jones L, Allaway W (1968) Solubility and redox criteria for the possible forms of selenium in soils. Soil Sci Soc Am J 32:35–40
12. Maier KJ, Knight AW (1993) Comparative acute toxicity and bioconcentration of selenium by the midge Chironomus decorus exposed to selenate, selenite, and seleno-DL-methionine. Arch Environ Contam Toxicol 25:365–370
13. Stefaniak J, Dutta A, Verbinnen B, Shakya M, Rene ER (2018) Selenium removal from mining and process wastewater: a systematic review of available technologies. J Water Supply Res Technol Aqua 67:903–918
14. Oremland RS, Hollibaugh JT, Maest AS, Presser TS, Miller LG, Culbertson CW (1989) Selenate reduction to elemental selenium by anaerobic bacteria in sediments and culture: biogeochemical significance of a novel, sulfate-independent respiration. Appl Environ Microbiol 55:2333–2343
15. Macy JM, Lawson S, DeMoll-Decker H (1993) Bioremediation of selenium oxyanions in San Joaquin drainage water using Thauera selenatis in a biological reactor system. Appl Microbiol Biotechnol 40:588–594
16. Cantafio AW, Hagen KD, Lewis GE, Bledsoe TL, Nunan KM, Macy JM (1996) Pilot-scale selenium bioremediation of San Joaquin drainage water with Thauera selenatis. Appl Environ Microbiol 62:3298–3303
17. Oremland RS, Blum JS, Bindi AB, Dowdle PR, Herbel M, Stolz JF (1999) Simultaneous reduction of nitrate and selenate by cell suspensions of selenium-respiring bacteria. Appl Environ Microbiol 65:4385–4392
18. Herbel MJ, Blum JS, Oremland RS, Borglin SE (2003) Reduction of elemental selenium to selenide: experiments with anoxic sediments and bacteria that respire Se-oxyanions. Geomicrobiol J 20:587–602
19. Kessi J, Ramuz M, Wehrli E, Spycher M, Bachofen R (1999) Reduction of selenite and detoxification of elemental selenium by the phototrophic bacterium Rhodospirillum rubrum. Appl Environ Microbiol 65:4734–4740
20. Pearce CI, Pattrick RA, Law N, Charnock JM, Coker VS, Fellowes JW, Oremland RS, Lloyd JR (2009) Investigating different mechanisms for biogenic selenite transformations: Geobacter sulfurreducens, Shewanella oneidensis and Veillonella atypica. Environ Technol 30:1313–1326
21. Maltman C, Yurkov V (2018) Bioremediation potential of bacteria able to reduce high levels of selenium and tellurium oxyanions. Arch Microbiol 200:1411–1417
22. Staicu LC, van Hullebusch ED, Lens PN (2017) Industrial selenium pollution: wastewaters and physical–chemical treatment technologies. In: Bioremediation of selenium contaminated wastewater. Springer, pp 103–130
23. Hunter WJ, Kuykendall LD (2005) Removing selenite from groundwater with an in situ biobarrier: laboratory studies. Curr Microbiol 50:145–150
24. Luek A, Brock C, Rowan DJ, Rasmussen JB (2014) A simplified anaerobic bioreactor for the treatment of selenium-laden discharges from non-acidic, end-pit lakes. Mine Water Environ 33:295–306
25. Bakircioglu Y, Bakircioglu D, Akman S (2010) Biosorption of lead by filamentous fungal biomass-loaded TiO_2 nanoparticles. J Hazard Mater 178:1015–1020
26. Fujita M, Ike M, Kashiwa M, Hashimoto R, Soda S (2002) Laboratory-scale continuous reactor for soluble selenium removal using selenate-reducing bacterium, Bacillus sp. SF-1. Biotechnol Bioeng 80:755–761
27. Tan LC, Papirio S, Luongo V, Nancharaiah YV, Cennamo P, Esposito G, Van Hullebusch ED, Lens PN (2018) Comparative performance of anaerobic attached biofilm and granular sludge reactors for the treatment of model mine drainage wastewater containing selenate, sulfate and nickel. Chem Eng J 345:545–555

28. Altringer P, Larsen D, Gardner K (1989) Bench scale process development of selenium removal from wastewater using facultative bacteria. In: International symposium on biohydrometallurgy, Jackson Hole, pp 643–657
29. Pieniz S, Okeke BC, Andreazza R, Brandelli A (2011) Evaluation of selenite bioremoval from liquid culture by Enterococcus species. Microbiol Res 166:176–185
30. Kuyucak N (1990) Feasibility of biosorbents application. In: Biosorption of heavy metals, vol 4, pp 372–377
31. Kapoor A, Viraraghavan T (1995) Fungal biosorption—an alternative treatment option for heavy metal bearing wastewaters: a review. Bioresour Technol 53:195–206
32. Espinosa-Ortiz EJ, Rene ER, van Hullebusch ED, Lens PN (2015) Removal of selenite from wastewater in a Phanerochaete chrysosporium pellet based fungal bioreactor. Int Biodeterior Biodegrad 102:361–369
33. Gharieb M, Wilkinson S, Gadd G (1995) Reduction of selenium oxyanions by unicellular, polymorphic and filamentous fungi: cellular location of reduced selenium and implications for tolerance. J Ind Microbiol 14:300–311
34. Marinescu G, Stoicescu AG, Teodorof L (2011) Industrial nutrient medium use for yeast selenium preparation. In: Annals of the University Dunarea de Jos of Galati Fascicle VI—food technology, vol 35
35. Nettem K, Almusallam AS (2013) Equilibrium, kinetic, and thermodynamic studies on the biosorption of selenium (IV) ions onto Ganoderma lucidum biomass. Sep Sci Technol 48:2293–2301
36. Leong YK, Chang J-S (2020) Bioremediation of heavy metals using microalgae: recent advances and mechanisms. Bioresour Technol 303:122886
37. Tuzen M, Sarı A (2010) Biosorption of selenium from aqueous solution by green algae (Cladophora hutchinsiae) biomass: equilibrium, thermodynamic and kinetic studies. Chem Eng J 158:200–206
38. Mane P, Bhosle A, Jangam C, Vishwakarma C (2011) Bioadsorption of selenium by pretreated algal biomass. Adv Appl Sci Res 2:202–207
39. Johansson CL, Paul NA, de Nys R, Roberts DA (2015) The complexity of biosorption treatments for oxyanions in a multi-element mine effluent. J Environ Manage 151:386–392
40. Silver S, Phung LT (2005) Genes and enzymes involved in bacterial oxidation and reduction of inorganic arsenic. Appl Environ Microbiol 71:599–608
41. Li F, Guo H, Zhou X, Zhao K, Shen J, Liu F, Wei C (2017) Impact of natural organic matter on arsenic removal by modified granular natural siderite: evidence of ternary complex formation by HPSEC-UV-ICP-MS. Chemosphere 168:777–785
42. Smith AH, Lopipero PA, Bates MN, Steinmaus CM (2002) Arsenic epidemiology and drinking water standards. American Association for the Advancement of Science
43. Prasad KS, Ramanathan A, Paul J, Subramanian V, Prasad R (2013) Biosorption of arsenite (As + 3) and arsenate (As + 5) from aqueous solution by Arthrobacter sp. biomass. Environ Technol 34:2701–2708
44. Mukhopadhyay R, Rosen BP, Phung LT, Silver S (2002) Microbial arsenic: from geocycles to genes and enzymes. FEMS Microbiol Rev 26:311–325
45. Ramírez-Solís A, Mukopadhyay R, Rosen BP, Stemmler TL (2004) Experimental and theoretical characterization of arsenite in water: insights into the coordination environment of As⁻O. Inorg Chem 43:2954–2959
46. Teclu D, Tivchev G, Laing M, Wallis M (2008) Bioremoval of arsenic species from contaminated waters by sulphate-reducing bacteria. Water Res 42:4885–4893
47. Norman N (1998) Chemistry of arsenic, antimony and bismuth. Blackie Academic and Professional, London, p 403
48. Hamamura N, Itai T, Liu Y, Reysenbach AL, Damdinsuren N, Inskeep WP (2014) Identification of anaerobic arsenite-oxidizing and arsenate-reducing bacteria associated with an alkaline saline lake in Khovsgol, Mongolia. Environ Microbiol Rep 6:476–482
49. Murphy T, Guo J (2003) An introduction of arsenic toxicity and its management. In: Aquatic arsenic toxicity and treatment

50. Bhakta JN, Ali MM (2020) Biosorption of arsenic: an emerging eco-technology of arsenic detoxification in drinking water. In: Arsenic water resources contamination. Springer, pp 207–230
51. Kostal J, Yang R, Wu CH, Mulchandani A, Chen W (2004) Enhanced arsenic accumulation in engineered bacterial cells expressing ArsR. Appl Environ Microbiol 70:4582–4587
52. Wang S, Zhao X (2009) On the potential of biological treatment for arsenic contaminated soils and groundwater. J Environ Manage 90:2367–2376
53. Leiva ED, Rámila CdP, Vargas IT, Escauriaza CR, Bonilla CA, Pizarro GE, Regan JM, Pasten PA (2014) Natural attenuation process via microbial oxidation of arsenic in a high Andean watershed. Sci Total Environ 466:490–502
54. Yamamura S, Amachi S (2014) Microbiology of inorganic arsenic: from metabolism to bioremediation. J Biosci Bioeng 118:1–9
55. Johnson DL (1972) Bacterial reduction of arsenate in sea water. Nature 240:44–45
56. Biswas R, Majhi AK, Sarkar A (2019) The role of arsenate reducing bacteria for their prospective application in arsenic contaminated groundwater aquifer system. Biocatal Agric Biotechnol 20:101218
57. Sahabi DM, Takeda M, Suzuki I, Koizumi J-I (2009) Removal of Mn^{2+} from water by "aged" biofilter media: the role of catalytic oxides layers. J Biosci Bioeng 107:151–157
58. Ike M, Miyazaki T, Yamamoto N, Sei K, Soda S (2008) Removal of arsenic from groundwater by arsenite-oxidizing bacteria. Water Sci Technol 58:1095–1100
59. Wan J, Klein J, Simon S, Joulian C, Dictor M-C, Deluchat V, Dagot C (2010) As^{III} oxidation by Thiomonas arsenivorans in up-flow fixed-bed reactors coupled to As sequestration onto zero-valent iron-coated sand. Water Res 44:5098–5108
60. Yang L, Li X, Chu Z, Ren Y, Zhang J (2014) Distribution and genetic diversity of the microorganisms in the biofilter for the simultaneous removal of arsenic, iron and manganese from simulated groundwater. Bioresour Technol 156:384–388
61. Corsini A, Zaccheo P, Muyzer G, Andreoni V, Cavalca L (2014) Arsenic transforming abilities of groundwater bacteria and the combined use of Aliihoeflea sp. strain 2WW and goethite in metalloid removal. J Hazard Mater 269:89–97
62. Akhter M, Tasleem M, Alam MM, Ali S (2017) In silico approach for bioremediation of arsenic by structure prediction and docking studies of arsenite oxidase from Pseudomonas stutzeri TS44. Int Biodeterior Biodegrad 122:82–91
63. Manirethan V, Raval K, Balakrishnan RM (2020) Adsorptive removal of trivalent and pentavalent arsenic from aqueous solutions using iron and copper impregnated melanin extracted from the marine bacterium Pseudomonas stutzeri. Environ Pollut 257:113576
64. Pandey PK, Choubey S, Verma Y, Pandey M, Chandrashekhar K (2009) Biosorptive removal of arsenic from drinking water. Bioresour Technol 100:634–637
65. Sarı A, Tuzen M (2009) Biosorption of As (III) and As (V) from aqueous solution by macrofungus (Inonotus hispidus) biomass: equilibrium and kinetic studies. J Hazard Mater 164:1372–1378
66. Aryal M, Ziagova M, Liakopoulou-Kyriakides M (2010) Study on arsenic biosorption using Fe (III)-treated biomass of Staphylococcus xylosus. Chem Eng J 162:178–185
67. Battaglia-Brunet F, Crouzet C, Burnol A, Coulon S, Morin D, Joulian C (2012) Precipitation of arsenic sulphide from acidic water in a fixed-film bioreactor. Water Res 46:3923–3933
68. Sun J, Hong Y, Guo J, Yang J, Huang D, Lin Z, Jiang F (2019) Arsenite removal without thioarsenite formation in a sulfidogenic system driven by sulfur reducing bacteria under acidic conditions. Water Res 151:362–370
69. Florentino AP, Weijma J, Stams AJ, Sánchez-Andrea I (2015) Sulfur reduction in acid rock drainage environments. Environ Sci Technol 49:11746–11755
70. Zhang L, Zhang Z, Sun R, Liang S, Chen G-H, Jiang F (2018) Self-accelerating sulfur reduction via polysulfide to realize a high-rate sulfidogenic reactor for wastewater treatment. Water Res 130:161–167
71. Aguilar NC, Faria MC, Pedron T, Batista BL, Mesquita JP, Bomfeti CA, Rodrigues JL (2020) Isolation and characterization of bacteria from a Brazilian gold mining area with a capacity of arsenic bioaccumulation. Chemosphere 240:124871

72. Escalante G, Campos V, Valenzuela C, Yañez J, Zaror C, Mondaca M (2009) Arsenic resistant bacteria isolated from arsenic contaminated river in the Atacama Desert (Chile). Bull Environ Contam Toxicol 83:657–661

73. Shakya S, Pradhan B, Smith L, Shrestha J, Tuladhar S (2012) Isolation and characterization of aerobic culturable arsenic-resistant bacteria from surfacewater and groundwater of Rautahat District, Nepal. J Environ Manage 95:S250–S255

74. Majumder A, Bhattacharyya K, Bhattacharyya S, Kole S (2013) Arsenic-tolerant, arsenite-oxidising bacterial strains in the contaminated soils of West Bengal, India. Sci Total Environ 463:1006–1014

75. Huang K, Chen C, Zhang J, Tang Z, Shen Q, Rosen BP, Zhao F-J (2016) Efficient arsenic methylation and volatilization mediated by a novel bacterium from an arsenic-contaminated paddy soil. Environ Sci Technol 50:6389–6396

76. Das S, Barooah M (2018) Characterization of siderophore producing arsenic-resistant Staphylococcus sp. strain TA6 isolated from contaminated groundwater of Jorhat, Assam and its possible role in arsenic geocycle. BMC Microbiol 18:104

77. Fazi S, Amalfitano S, Casentini B, Davolos D, Pietrangeli B, Crognale S, Lotti F, Rossetti S (2016) Arsenic removal from naturally contaminated waters: a review of methods combining chemical and biological treatments. Rend Lincei 27:51–58

78. Kim N, Park M, Yun Y-S, Park D (2019) Removal of anionic arsenate by a PEI-coated bacterial biosorbent prepared from fermentation biowaste. Chemosphere 226:67–74

79. Ahmad A, Heijnen L, de Waal L, Battaglia-Brunet F, Oorthuizen W, Pieterse B, Bhattacharya P, van der Wal A (2020) Mobility and redox transformation of arsenic during treatment of artificially recharged groundwater for drinking water production. Water Res 115826

80. Amoozegar MA, Ghasemi A, Razavi MR, Naddaf S (2007) Evaluation of hexavalent chromium reduction by chromate-resistant moderately halophile, Nesterenkonia sp. strain MF2. Process Biochem 42:1475–1479

81. Narayani M, Shetty KV (2013) Chromium-resistant bacteria and their environmental condition for hexavalent chromium removal: a review. Crit Rev Environ Sci Technol 43:955–1009

82. Xu C-H, Zhu L-J, Wang X-H, Lin S, Chen Y-M (2014) Fast and highly efficient removal of chromate from aqueous solution using nanoscale zero-valent iron/activated carbon (NZVI/AC). Water Air Soil Pollut 225:1845

83. Gruber J, Jennette KW (1978, May 30) Metabolism of the carcinogen chromate by rat liver microsomes. Biochem Biophys Res Commun 82(2):700–706

84. Tzou Y, Chen Y, Wang M (1998) Chromate sorption by acidic and alkaline soils. J Environ Sci Health Part A 33:1607–1630

85. Losi M, Amrhein C, Frankenberger W Jr (1994) Biodegradation of chromate-contaminated groundwater by reduction and precipitation in surface soils. J Environ Qual 23:1141–1150

86. Luli G, Talnagi J, Strohl WR, Pfister R (1983) Hexavalent chromium-resistant bacteria isolated from river sediments. Appl Environ Microbiol 46:846

87. Bopp L, Chakrabarty A, Ehrlich H (1983) Chromate resistance plasmid in Pseudomonas fluorescens. J Bacteriol 155:1105–1109

88. Horitsu H, Nishida H, Kato H, Tomoyeda M (1978) Isolation of potassium chromate-tolerant bacterium and chromate uptake by the bacterium. Agric Biol Chem 42:2037–2043

89. Sau G, Chatterjee S, Sinha S, Mukherjee SK (2008) Isolation and characterization of a Cr (VI) reducing Bacillus firmus strain from industrial effluents. Pol J Microbiol 57:327–332

90. Shakoori A, Tahseen S, Haq R (1999) Chromium-tolerant bacteria isolated from industrial effluents and their use in detoxication of hexavalent chromium. Folia Microbiol 44:50–54

91. Alvarez AH, Moreno-Sánchez R, Cervantes C (1999) Chromate efflux by means of the ChrA chromate resistance protein from Pseudomonas aeruginosa. J Bacteriol 181:7398–7400

92. Cervantes C, Campos-García J, Devars S, Gutiérrez-Corona F, Loza-Tavera H, Torres-Guzmán JC, Moreno-Sánchez R (2001) Interactions of chromium with microorganisms and plants. FEMS Microbiol Rev 25:335–347

93. Ohtake H, Cervantes C, Silver S (1987) Decreased chromate uptake in Pseudomonas fluorescens carrying a chromate resistance plasmid. J Bacteriol 169:3853–3856

94. Wang Y-T, Xiao C (1995) Factors affecting hexavalent chromium reduction in pure cultures of bacteria. Water Res 29:2467–2474
95. Cervantes C, Silver S (1992) Plasmid chromate resistance and chromate reduction. Plasmid 27:65–71
96. Thatoi H, Das S, Mishra J, Rath BP, Das N (2014) Bacterial chromate reductase, a potential enzyme for bioremediation of hexavalent chromium: a review. J Environ Manage 146:383–399
97. Wang C, Chen J, Hu W-J, Liu J-Y, Zheng H-L, Zhao F (2014) Comparative proteomics reveal the impact of OmcA/MtrC deletion on Shewanella oneidensis MR-1 in response to hexavalent chromium exposure. Appl Microbiol Biotechnol 98:9735–9747
98. Han J-C, Chen G-J, Qin L-P, Mu Y (2017) Metal respiratory pathway-independent Cr isotope fractionation during Cr (VI) reduction by Shewanella oneidensis MR-1. Environ Sci Technol Lett 4:500–504
99. Yu X, Jiang Y, Huang H, Shi J, Wu K, Zhang P, Lv J, Li H, He H, Liu P (2016) Simultaneous aerobic denitrification and Cr (VI) reduction by Pseudomonas brassicacearum LZ-4 in wastewater. Bioresour Technol 221:121–129
100. Baldiris R, Acosta-Tapia N, Montes A, Hernández J, Vivas-Reyes R (2018) Reduction of hexavalent chromium and detection of chromate reductase (ChrR) in Stenotrophomonas maltophilia. Molecules 23:406
101. Polti MA, Amoroso MJ, Abate CM (2007) Chromium (VI) resistance and removal by actinomycete strains isolated from sediments. Chemosphere 67:660–667
102. Megharaj M, Avudainayagam S, Naidu R (2003) Toxicity of hexavalent chromium and its reduction by bacteria isolated from soil contaminated with tannery waste. Curr Microbiol 47:0051–0054
103. Kalsoom A, Batool R, Jamil N (2020) An integrated approach for safe removal of chromium (VI) by Brevibacterium sp. Pak J Sci 72:18
104. Sugiyama M, Tsuzuki K, Hidaka T, Ogura R, Yamamoto M (1991) Reduction of chromium (VI) in Chinese hamster V-79 cells. Biol Trace Elem Res 30:1–8
105. Cohen RR, Ozawa T (2013) Microbial sulfate reduction and biogeochemistry of arsenic and chromium oxyanions in anaerobic bioreactors. Water Air Soil Pollut 224:1732
106. Schroeder DC, Lee GF (1975) Potential transformations of chromium in natural waters. Water Air Soil Pollut 4:355–365
107. Ibrahim AS, El-Tayeb MA, Elbadawi YB, Al-Salamah AA (2011) Bioreduction of Cr (VI) by potent novel chromate resistant alkaliphilic Bacillus sp. strain KSUCr5 isolated from hypersaline Soda lakes. Afr J Biotechnol 10:7207–7218
108. Sathishkumar K, Murugan K, Benelli G, Higuchi A, Rajasekar A (2017) Bioreduction of hexavalent chromium by Pseudomonas stutzeri L1 and Acinetobacter baumannii L2. Ann Microbiol 67:91–98
109. Jin R, Liu Y, Liu G, Tian T, Qiao S, Zhou J (2017) Characterization of product and potential mechanism of Cr (VI) reduction by anaerobic activated sludge in a sequencing batch reactor. Sci Rep 7:1–12
110. Ma L, Xu J, Chen N, Li M, Feng C (2019) Microbial reduction fate of chromium (Cr) in aqueous solution by mixed bacterial consortium. Ecotoxicol Environ Saf 170:763–770
111. Pattanapipitpaisal P, Brown N, Macaskie L (2001) Chromate reduction by Microbacterium liquefaciens immobilised in polyvinyl alcohol. Biotechnol Lett 23:61–65
112. Humphries A, Nott K, Hall L, Macaskie L (2005) Reduction of Cr (VI) by immobilized cells of Desulfovibrio vulgaris NCIMB 8303 and Microbacterium sp. NCIMB 13776. Biotechnol Bioeng 90:589–596
113. Tucker M, Barton L, Thomson B (1998) Reduction of Cr, Mo, Se and U by Desulfovibrio desulfuricans immobilized in polyacrylamide gels. J Ind Microbiol Biotechnol 20:13–19
114. Poopal AC, Laxman RS (2008) Hexavalent chromate reduction by immobilized Streptomyces griseus. Biotechnol Lett 30:1005–1010
115. Xu L, Luo M, Li W, Wei X, Xie K, Liu L, Jiang C, Liu H (2011) Reduction of hexavalent chromium by Pannonibacter phragmitetus LSSE-09 stimulated with external electron donors under alkaline conditions. J Hazard Mater 185:1169–1176

116. Farag S, Zaki S (2010) Identification of bacterial strains from tannery effluent and reduction of hexavalent chromium. J Environ Biol 31:877
117. Gao J, Zhang X, Yu J, Lei Y, Zhao S, Jiang Y, Xu Z, Cheng J (2020) Cr (VI) removal performance and the characteristics of microbial communities influenced by the core-shell Maifanite/ZnAl-layered double hydroxides (LDHs) substrates for chromium-containing surface water. Biochem Eng J 107625
118. Hussain S, Aziz HA, Isa MH, Ahmad A, Van Leeuwen J, Zou L, Beecham S, Umar M (2011) Orthophosphate removal from domestic wastewater using limestone and granular activated carbon. Desalination 271:265–272
119. Loganathan P, Vigneswaran S, Kandasamy J, Bolan NS (2014) Removal and recovery of phosphate from water using sorption. Crit Rev Environ Sci Technol 44:847–907
120. Yeoman S, Stephenson T, Lester JN, Perry R (1988, Jan 1) The removal of phosphorus during wastewater treatment: a review. Environm Pollut 49(3):183–233
121. Awual MR (2019) Efficient phosphate removal from water for controlling eutrophication using novel composite adsorbent. J Clean Prod 228:1311–1319
122. Long F, Gong J-L, Zeng G-M, Chen L, Wang X-Y, Deng J-H, Niu Q-Y, Zhang H-Y, Zhang X-R (2011) Removal of phosphate from aqueous solution by magnetic Fe–Zr binary oxide. Chem Eng J 171:448–455
123. USEPA (2009) Valuing the protection of ecological systems and services: a report of the EPA science advisory board, EPA-SAB-09-012
124. Weiner ER (2012) Applications of environmental aquatic chemistry: a practical guide. CRC Press
125. Glick RE, Schlagnhaufer CD, Arteca RN, Pell E (1995) Ozone-induced ethylene emission accelerates the loss of ribulose-1, 5-bisphosphate carboxylase/oxygenase and nuclear-encoded mRNAs in senescing potato leaves. Plant Physiol 109:891–898
126. He ZH, Chillingworth RK, Brune M, Corrie JE, Trentham DR, Webb MR, Ferenczi MA (1997) ATPase kinetics on activation of rabbit and frog permeabilized isometric muscle fibres: a real time phosphate assay. J Physiol 501:125–148
127. Illmer P, Schinner F (1992) Solubilization of inorganic phosphates by microorganisms isolated from forest soils. Soil Biol Biochem 24:389–395
128. Wani PA, Zaidi A, Khan AA, Khan MS (2005) Effect of phorate on phosphate solubilization and indole acetic acid releasing potentials of rhizospheric microorganisms. Ann Plant Prot Sci 13:139–144
129. Chen Y, Rekha P, Arun A, Shen F, Lai W-A, Young CC (2006) Phosphate solubilizing bacteria from subtropical soil and their tricalcium phosphate solubilizing abilities. Appl Soil Ecol 34:33–41
130. Kumar V, Behl RK, Narula N (2001) Establishment of phosphate-solubilizing strains of Azotobacter chroococcum in the rhizosphere and their effect on wheat cultivars under green house conditions. Microbiol Res 156:87–93
131. De Freitas J, Banerjee M, Germida JJ (1997) Phosphate-solubilizing rhizobacteria enhance the growth and yield but not phosphorus uptake of canola (Brassica napus L.). Biol Fertil Soils 24:358–364
132. Chung H, Park M, Madhaiyan M, Seshadri S, Song J, Cho H, Sa T (2005) Isolation and characterization of phosphate solubilizing bacteria from the rhizosphere of crop plants of Korea. Soil Biol Biochem 37:1970–1974
133. Vazquez P, Holguin G, Puente M, Lopez-Cortes A, Bashan Y (2000) Phosphate-solubilizing microorganisms associated with the rhizosphere of mangroves in a semiarid coastal lagoon. Biol Fertil Soils 30:460–468
134. Bass JIF, Reece-Hoyes JS, Walhout AJ (2016) Zymolyase-treatment and polymerase chain reaction amplification from genomic and plasmid templates from yeast. Cold Spring Harb Protoc
135. Nilsson T, Rova M, Bäcklund AS (2013) Microbial metabolism of oxochlorates: a bioenergetic perspective. Biochim Biophys Acta Bioenerg 1827:189–197

136. Kounaves SP, Stroble ST, Anderson RM, Moore Q, Catling DC, Douglas S, McKay CP, Ming DW, Smith PH, Tamppari LK (2010) Discovery of natural perchlorate in the Antarctic Dry Valleys and its global implications. Environ Sci Technol 44:2360–2364

137. Mastrocicco M, Di Giuseppe D, Vincenzi F, Colombani N, Castaldelli G (2017) Chlorate origin and fate in shallow groundwater below agricultural landscapes. Environ Pollut 231:1453–1462

138. Knight BA, Shields BM, He X, Pearce EN, Braverman LE, Sturley R, Vaidya B (2018) Effect of perchlorate and thiocyanate exposure on thyroid function of pregnant women from South-West England: a cohort study. Thyroid Res 11:9

139. Logan BE (2001) Peer reviewed: assessing the outlook for perchlorate remediation. ACS Publications

140. Matsubara T, Fujishima K, Saltikov CW, Nakamura S, Rothschild L (2017) Earth analogues for past and future life on Mars: isolation of perchlorate resistant halophiles from Big Soda Lake. Int J Astrobiol 16:218–228

141. Okeke BC, Giblin T, Frankenberger WT Jr (2002) Reduction of perchlorate and nitrate by salt tolerant bacteria. Environ Pollut 118:357–363

142. Niziński P, Błażewicz A, Kończyk J, Michalski R (2020) Perchlorate–properties, toxicity and human health effects: an updated review. Rev Environ Health 1

143. Murray C, Bolger P (2014) Environmental contaminants: perchlorate

144. Xu J, Logan BE (2003) Measurement of chlorite dismutase activities in perchlorate respiring bacteria. J Microbiol Methods 54:239–247

145. Smith PN, Severt SA, Jackson WA, Anderson TA (2006) Thyroid function and reproductive success in rodents exposed to perchlorate via food and water. Environ Toxicol Chem Int J 25:1050–1059

146. Coates JD, Achenbach LA (2004) Microbial perchlorate reduction: rocket-fuelled metabolism. Nat Rev Microbiol 2:569–580

147. Carlström CI, Loutey DE, Wang O, Engelbrektson A, Clark I, Lucas LN, Somasekhar PY, Coates JD (2015) Phenotypic and genotypic description of Sedimenticola selenatireducens strain CUZ, a marine (per) chlorate-respiring gammaproteobacterium, and its close relative the chlorate-respiring Sedimenticola strain NSS. Appl Environ Microbiol 81:2717–2726

148. Balk M, Mehboob F, van Gelder AH, Rijpstra WIC, Damsté JSS, Stams AJ (2010) (Per)chlorate reduction by an acetogenic bacterium, Sporomusa sp., isolated from an underground gas storage. Appl Microbiol Biotechnol Bioeng 88:595–603

149. Ricardo AR, Carvalho G, Velizarov S, Crespo JG, Reis MA (2012) Kinetics of nitrate and perchlorate removal and biofilm stratification in an ion exchange membrane bioreactor. Water Res 46:4556–4568

150. Liebensteiner MG, Pinkse MW, Schaap PJ, Stams AJ, Lomans BP (2013) Archaeal (per) chlorate reduction at high temperature: an interplay of biotic and abiotic reactions. Sci Total Environ 340:85–87

151. Waller AS, Cox EE, Edwards EA (2004) Perchlorate-reducing microorganisms isolated from contaminated sites. Environ Microbiol Rep 6:517–527

152. Acevedo-Barrios R, Bertel-Sevilla A, Alonso-Molina J, Olivero-Verbel J (2019, Feb 17) Perchlorate-reducing bacteria from hypersaline soils of the Colombian caribbean. Int J Microbiol 2019

153. Xiao Y, Roberts DJ (2013) Kinetics analysis of a salt-tolerant perchlorate-reducing bacterium: effects of sodium, magnesium, and nitrate. Environ Sci Technol 47:8666–8673

154. Youngblut MD, Tsai C-L, Clark IC, Carlson HK, Maglaqui AP, Gau-Pan PS, Redford SA, Wong A, Tainer JA, Coates JD (2016) Perchlorate reductase is distinguished by active site aromatic gate residues. J Biol Chem 291:9190–9202

155. Coleman ML, Ader M, Chaudhuri S, Coates JD (2003) Microbial isotopic fractionation of perchlorate chlorine. Appl Environ Microbiol 69:4997–5000

156. Brundrett M, Horita J, Anderson T, Pardue J, Reible D, Jackson WA (2015) The use of chlorate, nitrate, and perchlorate to promote crude oil mineralization in salt marsh sediments. Environ Sci Pollut Res 22:15377–15385

157. Coates JD, Bruce RA, Patrick J, Achenbach LA (1999) Hydrocarbon bioremediative potential of (per) chlorate-reducing bacteria. Bioremediat J 3:323–334
158. Coates JD, Michaelidou U, Bruce RA, O'Connor SM, Crespi JN, Achenbach LA (1999) Ubiquity and diversity of dissimilatory (per) chlorate-reducing bacteria. Appl Environ Microbiol Rep 65:5234–5241
159. Giblin T, Frankenberger W (2001) Perchlorate and nitrate reductase activity in the perchlorate-respiring bacterium perclace. Microbiol Res 156:311–315
160. Zu Y, Zhao Y, Xu K, Tong Y, Zhao F (2016) Preparation and comparison of catalytic performance for nano $MgFe_2O_4$, GO-loaded $MgFe_2O_4$ and GO-coated $MgFe_2O_4$ nanocomposites. Ceram Int 42:18844–18850
161. Hord NG, Tang Y, Bryan NS (2009) Food sources of nitrates and nitrites: the physiologic context for potential health benefits. Am J Clin Nutr 90:1–10
162. Espejo-Herrera N, Gracia-Lavedan E, Pollan M, Aragonés N, Boldo E, Perez-Gomez B, Altzibar JM, Amiano P, Zabala AJ, Ardanaz E (2016) Ingested nitrate and breast cancer in the Spanish Multicase-Control Study on Cancer (MCC-Spain). Environ Health Perspect 124:1042–1049
163. Kuai L, Verstraete W (1998) Ammonium removal by the oxygen-limited autotrophic nitrification-denitrification system. Appl Environ Microbiol Rep 64:4500–4506
164. Li Y, Go YK, Ooka H, He D, Jin F, Kim SH, Nakamura R (2020) Enzyme mimetic active intermediates for nitrate reduction in neutral aqueous media. Angew Chem Int Ed 59:9744–9750
165. Sliekers AO, Derwort N, Gomez JC, Strous M, Kuenen J, Jetten M (2002) Completely autotrophic nitrogen removal over nitrite in one single reactor. Water Res 36:2475–2482
166. Van Dongen U, Jetten MS, Van Loosdrecht MJ (2001) The SHARON®-Anammox® process for treatment of ammonium rich wastewater. Water Sci Technol 44:153–160
167. Lu L, Guo X, Zhao J (2017) A unified nonlocal strain gradient model for nanobeams and the importance of higher order terms. Int J Eng Sci 119:265–277
168. Neissi A, Rafiee G, Farahmand H, Rahimi S, Mijakovic I (2020) Cold-resistant heterotrophic ammonium and nitrite-removing bacteria improve aquaculture conditions of rainbow trout (Oncorhynchus mykiss). Microbial Ecol 1–12
169. Haroon MF, Hu S, Shi Y, Imelfort M, Keller J, Hugenholtz P, Yuan Z, Tyson GW (2013) Anaerobic oxidation of methane coupled to nitrate reduction in a novel archaeal lineage. Nature 500:567–570
170. Xie T, Xia Y, Zeng Y, Li X, Zhang Y (2017) Nitrate concentration-shift cultivation to enhance protein content of heterotrophic microalga Chlorella vulgaris: over-compensation strategy. Bioresour Technol 233:247–255
171. Nie W-B, Xie G-J, Ding J, Peng L, Lu Y, Tan X, Yue H, Liu B-F, Xing D-F, Meng J (2020) Operation strategies of n-DAMO and Anammox process based on microbial interactions for high rate nitrogen removal from landfill leachate. Environ Int 139:105596
172. Wang Y, Zhou W, Jia R, Yu Y, Zhang B (2020) Unveiling the activity origin of a copper-based electrocatalyst for selective nitrate reduction to ammonia. Angew Chem Int Ed 132:5388–5392
173. Gao J, Jiang B, Ni C, Qi Y, Bi X (2020) Enhanced reduction of nitrate by noble metal-free electrocatalysis on P doped three-dimensional Co_3O_4 cathode: mechanism exploration from both experimental and DFT studies. Chem Eng J 382:123034
174. Navada S, Sebastianpillai M, Kolarevic J, Fossmark RO, Tveten A-K, Gaumet F, Mikkelsen Ø, Vadstein O (2020) A salty start: brackish water start-up as a microbial management strategy for nitrifying bioreactors with variable salinity. Sci Total Environ 139934
175. Pinkernell U, Von Gunten U (2001) Bromate minimization during ozonation: mechanistic considerations. Environ Sci Technol 35:2525–2531
176. Winid B (2013) Bromine as a potential threat to the aquatic environment in areas of mining operations. Gospodarka Surowcami Mineralnymi-Mineral Resources Management
177. von Gunten U, Salhi E (2003) Bromate in drinking water a problem in Switzerland? Ozone Sci Eng 25:159–166

178. Pinkernell U, von Gunten U (2001) Bromate minimization during ozonation: mechanistic considerations. Environ Sci Technol 35(12):2525–2531
179. Xu P, Janex M-L, Savoye P, Cockx A, Lazarova V (2002) Wastewater disinfection by ozone: main parameters for process design. Water Res 36:1043–1055
180. Plewa MJ, Wagner ED, Richardson SD, Thruston AD, Woo Y-T, McKague AB (2004) Chemical and biological characterization of newly discovered iodoacid drinking water disinfection byproducts. Environ Sci Technol 38:4713–4722
181. Glaze WH, Weinberg HS, Cavanagh JE (1993) Evaluating the formation of brominated DBPs during ozonation. J Am Water Works Assoc 85:96–103
182. Fang J-Y, Shang C (2012) Bromate formation from bromide oxidation by the UV/persulfate process. Environ Sci Technol 46:8976–8983
183. Liu Y, Yang Y, Pang S, Zhang L, Ma J, Luo C, Guan C, Jiang J (2018) Mechanistic insight into suppression of bromate formation by dissolved organic matters in sulfate radical-based advanced oxidation processes. Chem Eng J 333:200–205
184. Butler R, Godley A, Lake R, Lytton L, Cartmell E (2005) Reduction of bromate in groundwater with an ex situ suspended growth bioreactor. Water Sci Technol 52:265–273
185. Assunção A, Martins M, Silva G, Lucas H, Coelho MR, Costa MC (2011) Bromate removal by anaerobic bacterial community: mechanism and phylogenetic characterization. J Hazard Mater 197:237–243
186. Davidson AN, Chee-Sanford J, Lai HYM, Ho C-H, Klenzendorf JB, Kirisits MJ (2011) Characterization of bromate-reducing bacterial isolates and their potential for drinking water treatment. Water Res 45:6051–6062
187. Hijnen W, Voogt R, Veenendaal H, Van der Jagt H, Van Der Kooij D (1995) Bromate reduction by denitrifying bacteria. Appl Environ Microbiol 61:239–244
188. Hijnen W, Jong R, Van der Kooij D (1999) Bromate removal in a denitrifying bioreactor used in water treatment. Water Res 33:1049–1053
189. Zhong Y, Yang Q, Fu G, Xu Y, Cheng Y, Chen C, Xiang R, Wen T, Li X, Zeng G (2018) Denitrifying microbial community with the ability to bromate reduction in a rotating biofilm-electrode reactor. J Hazard Mater 342:150–157
190. Lai C-Y, Lv P-L, Dong Q-Y, Yeo SL, Rittmann BE, Zhao H-P (2018) Bromate and nitrate bioreduction coupled with poly-β-hydroxybutyrate production in a methane-based membrane biofilm reactor. Environ Sci Technol 52:7024–7031
191. Kirisits MJ, Snoeyink VL, Inan H, Chee-Sanford JC, Raskin L, Brown JC (2001) Water quality factors affecting bromate reduction in biologically active carbon filters. Water Res 35:891–900
192. Liu J, Yu J, Li D, Zhang Y, Yang M (2012) Reduction of bromate in a biological activated carbon filter under high bulk dissolved oxygen conditions and characterization of bromate-reducing isolates. Biochem Eng J 65:44–50
193. Kirisits MJ, Snoeyink VL, Chee-Sanford JC, Daugherty BJ, Brown JC, Raskin L (2002) Effect of operating conditions on bromate removal efficiency in BAC filters. J Am Water Works Assoc 94:182–193
194. Tamai N, Ishii T, Sato Y, Fujiya H, Muramatsu Y, Okabe N, Amachi S (2016) Bromate reduction by Rhodococcus sp. Br-6 in the presence of multiple redox mediators. Environ Sci Technol 50:10527–10534
195. Demirel S (2017) Denitrification performance and microbial community dynamics in a denitrification reactor as revealed by high-throughput sequencing. Water Sci Technol Water Supply 17:940–946
196. Manna J, Shilpa N, Bandarapu AK, Rana RK (2019 Feb 22) Oxyanion-binding in a bioinspired nanoparticle-assembled hybrid microsphere structure: effective removal of arsenate/chromate from water. ACS Applied Nano Materials 2(3):1525–1532
197. Thomas Sims J, Pierzynski GM (2005) Chemistry of phosphorus in soils. In: Chemical processes in soils, vol 8, pp 151–192

Chapter 10
The Halogen-Oxyanion Derivatives as Contaminants of Concern in Water

Moses O. Alfred, Daniel T. Koko, Ahmad Hosseini-Bandegharaei, Artur J. Motheo, and Emmanuel I. Unuabonah

Abstract Due to the adverse health and environmental effects of halogen, based on oxyanions in water, the occurrence and distribution have been of great concern, worldwide. Therefore, in this chapter, the occurrence, chemical structure, the natural and anthropogenic sources of these oxyanions were appraised. The impacts on human health and the environment and their fate in water are carefully enunciated. Different techniques that have been developed for the determination of halogen-oxyanions in aqueous solutions are discussed. Finally, a perspective for future research on halogen-based oxyanions is provided, and the possible research gaps, which are begging for answers, are also highlighted.

Keywords Halogen oxyanions · Water · Pollution · Environmental management · Perchlorate · Bromate

10.1 Introduction

Oxyanions are polyatomic negatively charged ions, containing oxygen with the generic formula $A_xO_y^{z-}$ (where A represents a chemical element, O represents an oxygen atom and z represents the overall charge of the ion) [1]. Some of these anions are generated and leached into water bodies through anthropogenic activities and

M. O. Alfred (✉) · D. T. Koko · E. I. Unuabonah
African Centre of Excellence for Water and Environmental Research (ACEWATER), Redeemer's University, PMB 230, Ede, Osun State, Nigeria
e-mail: alfredm@run.edu.ng

Department of Chemical Sciences, Redeemer's University, PMB 230, Ede, Osun State, Nigeria

A. Hosseini-Bandegharaei
Faculty of Health, Sabzevar University of Medical Sciences, Sabzevar, Iran

A. J. Motheo
São Carlos Institute of Chemistry, University of São Paulo, Av. Trabalhador Sãocarlense 400, São Carlos 13566-590, Brazil

© Springer Nature Switzerland AG 2021
N. A. Oladoja and E. I. Unuabonah (eds.), *Progress and Prospects in the Management of Oxyanion Polluted Aqua Systems*, Environmental Contamination Remediation and Management, https://doi.org/10.1007/978-3-030-70757-6_10

natural means. Research into the removal of these anions from the environment is relatively very low when compared with those for cationic pollutants [2].

Among these oxyanions, the most toxic and mostly available in considerable concentration in water bodies are the halogen-based types [3]. Halogens, being strong oxidants, combined easily with oxygen to form polyatomic oxyanions. Of the six halogens elements, which include fluorine (F), chlorine (Cl), bromine (Br), iodine (I), astatine (At) and tennessine (Ts), in the periodic table, the oxyanions of chlorine, bromine and iodine are the most readily available. Polyatomic oxyanions with halogens can exist in four different forms, depending on the number of oxygen atoms present.

To name the most common form of the anion, use the stem of the halogen's name and add the ending "-*ate*". If the oxyanion has one less oxygen than the most common form, the ending changes from "-*ate*" to "-*ite*". If there is one less oxygen than the "-*ite*" anion, add the prefix "*hypo-*" to the beginning of the "-*ite*" anion's name. If there is one more oxygen than the most common anion (the "-*ate*" anion), add the prefix "*per-*" to the beginning of the "-*ate*" anion's name. Examples of these oxyanions include nitrate (NO_3^-), nitrite (NO_2^-), sulphate (SO_4^{2-}), sulphite (SO_3^{2-}), perchlorate (ClO_4^-), chlorate (ClO_3^-), chlorite (ClO_2^-), hypochlorite (ClO^-), perbromate (BrO_4^-), bromate (BrO_3^-), bromite (BrO_2^-), hypobromite (BrO^-), periodate (IO_4^-), iodate (IO_3^-), iodite (IO_2^-), hypoiodite (IO^-), tellurate (TeO_4^{2-}), tellurite (TeO_3^{2-}), selenate (SeO_4^{2-}), selenite (SeO_3^{2-}), arsenate (AsO_4^{3-}), arsenite (AsO_3^{3-}), phosphate (PO_4^{3-}), phosphite (PO_3^{3-}), carbonate (CO_3^{2-}) and borate (BO_3^{3-}).

Fluorine is the first member of the halogen family. It is the most reactive and the lightest, found as toxic pale yellow diatomic gas. Because of its reactivity, it forms compounds with all other elements apart from He, Ne and Ar. It is not found in elemental state but exists as minerals like fluorite, apatite and cryolite [4]. It is a strong oxidizing agent and has the ability to form compounds with oxygen, such as oxydifluoride (F_2O or OF_2) and dioxygen fluoride (F_2O_2) [5]. The formation of fluorine oxyanions in water is very difficult, because the contact of oxygen difluoride with water results in explosion, which makes them unstable [5]. Another reason F does not form oxyanions is its lack of energetically accessible orbitals for octet expansion [5].

The most common oxyanion of chlorine is chlorate. Its major source in water is from the use of chlorine dioxide disinfectants and oxidants. Chlorite is also found in water, due to the same application of hypochlorite as disinfectants and oxidants [6, 7]. Perchlorates usually get to the aquatic environment from the manufacture and use of rockets, fireworks and ammunition [8]. The negative impacts of perchlorate led to publishing a reference dose (RfD) in 2002 for perchlorate as 0.00003 mg/kg/d, with a drinking water equivalent level (DWEL) of approximately 1 µg/L [9, 10]. This was later updated in 2005 by the National Academy of Sciences and USEPA to 0.0007 mg/kg/d, with a DWEL of 24.5 µg/L [11], a value based on the no observable effect level (NOEL) of 0.007 mg/kg/d for inhibition of iodide uptake [11, 12].

Bromine occupy the fourth period in the periodic table. It can be oxidized to form oxyanions of environmental concern, such as hypobromite, bromite, bromate and perbromate [3]. Based on its negative impact on human health and the environment, bromate is considered to be the most important oxyanions of bromine. It contains bromine in its highest oxidation state +5, so it is an oxidizing agent, especially under acidic conditions. They are formed in water when ozonation is employed as disinfection method for water that contains bromine [13]. The existing guidelines on bromate contamination are based on its maximum acceptable concentration (MAC) of 0.01 mg/L (10 µg/L) on renal cell tumours in rats, taking into consideration limitations in analytical methodology and treatment technology [14]. The maximum contaminant level (MCL) of bromate was proposed to be 0.010 mg/L by the United States Environmental Protection Agency (USEPA) [14]. Salts of bromate include sodium bromate ($NaBrO_3$) and potassium bromate ($KBrO_3$), both of which are white crystalline substances that readily dissolve in water. Both salts are used in industrial dyeing processes, for hair treatments and as dough conditioners in bread formulation. Bromate has been extensively used as an oxidizer in food processing, such as flour milling, beer malting and cheese making [15].

Iodine is the least reactive element of the halogens (besides astatine) because of its large atomic size. It exists in low-temperature geochemical environment as elemental iodine (I_2), iodide (I^-), iodate (IO_3^-) and periodate (IO_4^-). Iodine is a trace element that exists in natural waters, such as seawater, freshwater and rain water [16]. It has several synthetic isotopes, including some that are radioactive with short half-lives. It is chemically similar to chlorine and bromine, but much less reactive as an oxidizer. Iodine-129 (^{129}I) has been reported to be one of the top risk radioactive contaminants in the environment, due to its long half-life (1.6×10^7 years), toxicity and mobility [17]. Iodine exists in large quantity, as iodate, in diet enhancement tablets and in iodized salts, for the treatment of hypothyroidism [18]. This results in large amount of iodates in the environment [18]. The existence of the different iodine oxyanions depend on the environmental pH value. However, iodate tend to be the dominant specie alongside iodide [19]. As the environmental pH value increases, naturally occurring iodides are easily oxidized to iodate (IO_3^-), periodate and iodide [20].

Astatine (At) is considered to be the rarest element in the earth's crust, because it is only formed from the radioactive disintegration of heavier elements. Other than exhibiting more metallic character than other halogens, it is similar to iodine in properties. Its isotopes are stable, apart from At-210, which disintegrates easily to the deadly Po-210 [21]. It is known to concentrate in delicate parts of the human body, such as thyroid gland, lungs, spleen and liver [22]. The effects of At in the environment cannot be over emphasized, bearing in mind that some isotopes are used in nuclear medicine for the treatment of cancer. Hence, there is a possibility that they leach into the environment [23]. However, the chemistry of oxyanions of At is not well established; therefore, it would not be further considered in this chapter.

Tennessine (Ts) was recently discovered in the year 2010, but was formally added to the periodic table in the year 2016 [24]. Considering the position of Ts on the periodic table, it is expected to behave like the other halogens. Scientific explanations observed that relativistic effects from the Ts's valence electrons will inhibit it from producing anions or achieving higher oxidation states. Therefore, in some sense, it behaves like a metalloid or post-transition element. However, being radioactive in nature, it posed health risk to animals/human.

This chapter highlights the occurrence, distribution and effects of halogen-based oxyanions in water. The impact of these oxyanions on human health and the environment, the different techniques that have been explored for their determination in aquatic environment and methods that have been employed for their removal in aqueous system are presented. Considering the fact that there are few/no report(s) confirming the presence of fluorine, iodine, astatine and tennessine oxyanions in water, the focus of the chapter is limited to the oxyanions of chlorine and bromine. The chapter closes with a peep into future research, regarding halogen-based oxyanions in water and possible research gaps begging for further understanding especially in the behaviour and management of these oxyanions in water. A brief review of some of these research works is presented in Table 10.1.

10.1.1 Research Trends

The results of the analysis of published research articles on the occurrence of halogen oxanions in water in the last 100 years, are presented in Fig. 10.1. Scopus® Analyse document search tool, with the search term "halogen oxyanions in water" for each of the oxyanions of halogen (e.g. perchlorate, chlorate, chlorite and hypochlorite) was used; to get the total contribution of each of the anions, the sum of the individual halogen oxyanions was used.

According to the search results, oxyanions of chlorine are the most popular among the halogen oxyanions (Fig. 10.1). A yearly comparison of the publications mentioning the halogen oxyanions in the past 100 years (Fig. 10.1a), shows that there are more mentions for the oxyanion of chlorine than for any other halogen oxyanion per year. However, there is paucity of report on the oxyanions of fluorine, astatine and tennessine, which accounted for why these oxyanions are not represented on the charts. Chlorine oxyanions contributes 83.74% (15,557 articles) of total number of publications mentioning any halogen oxyanions, while bromine and iodine contribute only 10.94 (2032 articles) and 5.32% (988 articles), respectively (Fig. 10.1b).

Oxyanions of chlorine started gaining research interest around 1960 (Fig. 10.2a). For oxyanions of bromine, there was not much attention until 1980 (Fig. 10.2c), while for iodine oxyanions, reports have only focused on iodate. Ever since, the number of publications reporting the oxyanions of chlorine and bromine has grown (Fig. 10.2a).

For the oxyanions of chlorine, hypochlorite has been the most reported, which may be due to its use as bleach and disinfectant. With 5357 published articles (i.e. 34.43% of the total publications related to oxyanions of Chlorine), hypochlorite is the most reported of all the oxyanions of chlorine. Chlorate is the least reported, contributing only 7.44% of the total number of articles on oxyanions of chlorine (1158 articles in the last 100 years) (Fig. 10.1b).

Table 10.1 Occurrence of halogen oxyanions in water

Anions	Matrix	Analytical method	Concentration (mg/L)	References
ClO_3^-	Drinking water	Spectrophotometry	4.5×10^{-6}	[25]
ClO_2^-	Drinking water	In-electrode coulometric titration	0.25	[26]
ClO_2^-	Raba river Rudawa Dlubina	UV/vis spectrophotometry	220 230 170	[27]
ClO_3^-	Drinking water	Liquid chromatography-electrospray ionization-mass spectrometry (LC-ESI-MS/MS)	0.19	[28]
ClO_3^-	NaClO solution	Iodometric titration	2–50	[29]
ClO_2^- ClO_3^-	Drinking water Drinking water	Flow injection A and ion chromatography	0.03 0.04	[30]
ClO_2^- ClO_3^-	Water Water	Potentiometric titration	13.0 3.38	[31]
ClO_4^-	River water Drinking water Groundwater Mineral water Swimming pool	Ion chromatography	16.1 14.8 5.41 2.58 205	[32]
ClO_4^-	Local city hardwater	Ion chromatography	5.0	[33]
ClO_4^-	Drinking water	Liquid–liquid extraction followed by flow injection electrospray mass spectrometry (ESI/MS)	1×10^{-5}	[34]
ClO_4^-	Groundwater for human consumption	Ion chromatography-mass spectroscopy-mass spectroscopy (IC-MS/MS)	1.5×10^{-4}	[35]
ClO_4^-	Groundwater	Ion chromatography-electrospray ionization-mass spectrometry (IC-ESI-MS)	24.4×10^{-6}	[36]
BrO_3^-	Drinking water	Dispersive liquid–liquid extraction and gas chromatography-electron capture detection	50×10^{-5}	[37]
BrO_3^-	Bottled drinking water	Ion chromatography with spectrophotometric detection after post column reaction	$5–169 \times 10^{-3}$	[38]
BrO_3^-	Waters	Diffusion reflection spectroscopy	5.0×10^{-4}	[39]

(continued)

Table 10.1 (continued)

Anions	Matrix	Analytical method	Concentration (mg/L)	References
BrO_3^-	Water from centralized supply Sift drinks Purified water Table water Table water Medicinal water	Photometric redox determination	0.4 1.7 3.4 5.6 3.7 9.1	[40]
BrO_3^-	Tap water	Spectrophotometry	25–750	[41]
BrO_3^-	Drinking water	Spectrophotometry using phenothiazines	0.67 and 2.25	[42]
IO_3^-	Sea water and evaporates	Determined spectrophotometrically as the starch-iodine complex without prior separation or concentration of the iodine	$50–84 \times 10^{-3}$	[43]
IO_3^-	Sea water	Determination of iodate using methylene blue as a chromogenic reagent	0.5–14	[44]
IO_3^-	Spring water Groundwater Sea water Stream water	High performance liquid chromatography (HPLC) with amperometric and spectrophotometric detection, and off-line UV irradiation	16.7 234 50–51.5 0.58–2.07	[45]

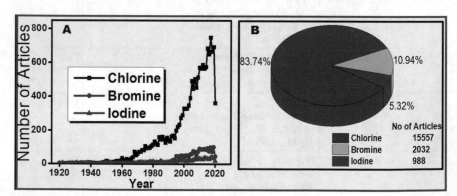

Fig. 10.1 Charts showing **a** the trend in research for the management of halogen oxyanions and **b** the contribution of each member of the halogen family to these studies

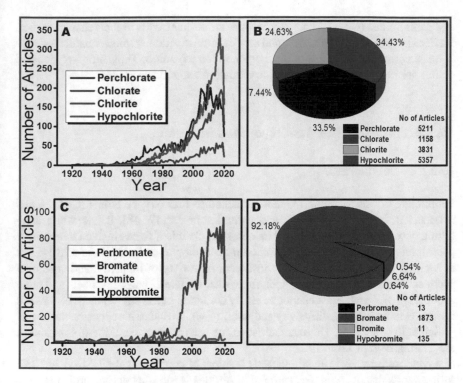

Fig. 10.2 Charts showing **a** the trend in research for the management of chlorine oxyanions, **b** the contribution of each oxyanions of chlorine to the management of studies, **c** the trend in research for the management of bromine oxyanions and **d** the contribution of each oxyanions of chlorine to the management of studies

For the oxyanions of bromine and iodine, bromate (Fig. 10.1d) and iodate have been the most reported in literatures. Bromate contributed 92.18% of the total reports on the bromine oxyanions (Fig. 10.1d), and iodate contributed about 98% for iodine.

10.2   Occurrence, Sources and Fate of Halogen Oxyanion in Aqua Systems

10.2.1   Oxyanions of Chlorine

Chlorine oxyanions are negatively charged polyatomic ions consisting of chlorine and oxygen. The chlorine atom is in an odd-number positive oxidation state, and oxygen atoms with lone pairs of electrons surround the atom. All chlorine oxyanions possess oxidizing ability, but variation exists in their stability. There are four types of chlorine oxyanions which are hypochlorites (ClO^-), chlorites (ClO_2^-), chlorates (ClO_3^-)

and perchlorates (ClO_4^-), and they exhibit unfamiliar trends in their characteristics. Their oxidizing ability decreases, and they become kinetically inferior oxidizers, with increasing number of atoms of the oxygen in the oxyanion. Thus, more oxygenated anions are more stable with respect to oxidation and reduction [46].

10.2.2 Fate of Chlorine-Oxyanions in Waters

10.2.2.1 Perchlorates

Perchlorate consists of chloride atom attached to four oxygen atoms (ClO_4^-), and it occurs in both anthropogenic and natural sources [47, 48]. It is a white crystalline powder or clear liquid, and its common chemical forms include ammonium perchlorate, sodium perchlorate, potassium perchlorate and perchloric acid [49]. It is a strong oxidizing agent and a constituent of variety of industrial and consumer products, which include missile fuel, fireworks, vehicle airbags, fertilizers and other products [50]. Perchlorate cannot be easily degraded in the environment because of its high water solubility, diffusivity and stability, which makes it a persistent inorganic pollutant in water [51]. The characteristics and uses of some perchlorate compounds are summarized in Table 10.2.

Perchlorate was first identified as a chemical of concern by the USEPA in 1985, following its discovery in wells around hazardous waste sites in California [51, 52]. In an aqueous systems, its fate is relative to the chemical and physical properties of the system [49]. Their salts are highly soluble in aqueous medium, but the ClO_4^- is very stable in the environment. It undergoes biodegradation by anaerobic bacteria, releasing significant level of carbon and electron donors, like oxygen and nitrate [53].

In the soil, ClO_4^- movement depends on the presence and circulation of water, since it has poor sorption ability which is why it does not bind to soil particles [54]. In the presence of adequate water, it may be completely leached from the soil [55]. Terrestrial and aquatic plants can take up perchlorate and accumulate it in various tissues or degrade it in their leaves and branches. At low concentrations, usually in groundwater, it is transferred along the groundwater gradient towards the elimination point [56]. These observations demonstrate that the knowledge about water flow paths and aquifer properties are essential prerequisites to understand groundwater ClO_4^- contamination [53].

10.2.3 Sources of Perchlorate

The sources of perchlorate are classified as natural and anthropogenic (man made manufactured) [57]. Naturally, occurring perchlorate has mainly been associated with the geographies of extremely arid climates. Artificial sources result predominantly in

Table 10.2 Characteristics and uses of some perchlorate compounds

Compound	Molecular weight (g/mol)	Melting Point (°C)	Boiling Pointing (°C)	Density (g/cm³)	Physical appearance	Aqueous solubility (g/100 mL)	Uses
Ammonium perchlorate (NH_4ClO_4)	117.49	>200	N.A	1.95	White crystalline orthorhombic and regular crystals	11.56 @ 0 °C 20.85 @ 20 °C 57.01 @ 100 °C	Used in rocket propellants (explosives and pyrotechnics), as an etching and engraving agents and in analytical chemistry
Sodium perchlorate ($NaClO_4$) (monohydrate)	122.44	468	462	2.50	White crystalline solid	209.6 @ at 25 °C (anhydrous) 209 @ 15 °C (monohydrate)	Used in making explosives, jet fuels and in chemical industries. Also used as medication and for preparation of oxygen
Potassium perchlorate ($KClO_4$)	138.55	610	600	2.52	Colourless/white crystalline powder, rhombohedral	0.76 @ 0 °C 1.5 @ 25 °C 21.08 @ 100 °C	Used as oxidizer, disinfectant, rocket propellant and antithyroid agent
Lithium perchlorate ($LiClO_4$)	106.39	236	430	2.43	Deliquescent, white crystal	42.3 @ 0 °C 59.8 @ 25 °C 119.5 @ 80 °C	A source of oxygen and as electrolyte in voltaic cells. Used in the synthesis of organic chemicals
Perchloric acid ($HClO_4.2H_2O$)	223.21	−17	203	1.77	White, hygroscopic powder	Very miscible	Used in preparing perchlorate salts and an important component for rocket fuel

areas where the manufacture, use and storage of ammunitions and rockets propellants took place over a period of time [58].

10.2.3.1 Natural Sources

In general, the atmospheric source of ClO_4^- is the UV-mediated photo-oxidation of inorganic chlorine in the presence of ozone [53]. Other natural sources of ClO_4^- include volcanic eruptions [57, 59] and from the conversion of organic chlorine to the inorganic forms [60]. Recent study suggested that the volatilization of chlorinated solvents could also give rise to the formation of ClO_4^- [59].

Another natural source of perchlorate is from atmospheric sources around arid and semi-arid deserts [53]. Perchlorate from the atmosphere is deposited on the earth's surface by a wet and dry process, which is subsequently transferred to surface or groundwater because of its high solubility and poor sorption [53]. This shows why most natural ClO_4^- occurrences are restricted to arid and semi-arid environments, as the evapotranspiration is strong and the amount of atmospheric deposition exceeds the rate of dissolution by the on-going precipitation [61]. Study has shown that the concentration of ClO_4^- increases with aridity index and the period of arid conditions [61]. Perchlorate is also found as a natural impurity in nitrate salts from Chile, which are imported and used to produce nitrate fertilizers, explosives and other products [62, 63]. The type of soil may also play an important role in ClO_4^- accumulation [53], and site-specific biological activities (e.g. plants or bacteria) can also influence ClO_4^- accumulation [64].

Paleogeochemical deposits are rich sources of naturally occurring perchlorate that are intensively mined and used for agricultural and industrial purposes.

10.2.3.2 Anthropogenic Sources

The main anthropogenic source of perchlorate is from areas where ammonium perchlorate is manufactured [65]. Synthetic ClO_4^- was first manufactured in commercial quantities in Masebo, Sweden in the 1890s [55]. Being a strong oxidizer, ammonium perchlorate is widely used in solid propellants for rockets explosives, fireworks and missiles [66, 67].

Perchlorate also exists in several products as impurities. For instance, it is found in sodium hypochlorite and sodium chlorate as a breakdown product and manufacturing by-product, respectively [68]. Sodium hypochlorite (NaOCl) has several applications, which include; surface purification, bleaching, odor removal and water disinfection. Hypochlorite solutions contain trace amounts of ClO_4^- that is formed during and after manufacture [11]. Most of the perchlorate contaminated sites are found in or close to military establishments, where wastewater containing perchlorate was discharged on the ground without proper treatment [69]. Electrochemical production of sodium chlorate can also generate ClO_4^- impurities at 50–230 mg kg^{-1} ClO_3^- [68].

10.2.4   Effects of Perchlorate on Human Health and the Environment

The effects of perchlorate can be studied based on their impact on the environment and the direct effects on human. Intensive industrial development has contributed to the increased pollution of the environment with hazardous oxyanions such as perchlorates, which are leached from different sources [2]. Perchlorates is very poisonous at very low concentrations and can be transferred into living organisms via inhalation, ingestion and skin adsorption, causing irreversible effects [70].

Perchlorates are also promising effective thyroid disruptors. Several investigations have proven that perchlorate contaminates the ground and surface water, basically at fireworks manufacturing and display sites [49, 50, 71]. The health and ecological effects of perchlorate released from fireworks have been assessed and reported to lead to direct or indirect perchlorate water contamination [72]. Environmental perchlorate contamination may also happen in areas used for missile recycling, propellant or ammunitions disposal, or other military operations [55].

A major toxic effect of perchlorate on human is thyroid disruption. Perchlorate competitively inhibits the sodium iodide symporter (NIS), an intrinsic membrane glycoprotein responsible for the uptake of iodide into the thyroid and other organs [50]. Iodide uptake inhibition is considered the mode of action for perchlorate [49]. Perchlorate is observed to be an endocrine disruptor because it can interfere with iodide uptake by the thyroid gland and thus result in decreased thyroid hormone production [47, 73]. Perchlorate also affects other tissues, where NIS is known to be present, such as lactating breast epithelium, gastrointestinal tract, placenta, skin and mammary gland [49]. The presence of perchlorate decreases the synthesis of circulating thyroid hormones in the adult and decreases their placental transfer to the foetus in pregnant women [74].

The Agency for Toxic Substances and Disease Registry (ATSDR) has established a minimal risk level (MRL) of 0.0007 mg/kg/day for chronic duration oral exposure (365 days or more) to perchlorate. An MRL is an estimate of the daily human exposure to a hazardous substance that is likely to be without appreciable risk of adverse non-cancer health effects over a specified duration of exposure [50].

In 2011, United States of America Environmental Protection Agency (USEPA) determined that perchlorate meets the Safe Drinking Water Act criteria for regulation as a contaminant. EPA then worked with the Food and Drug Administration (FDA) to develop a dose-response model for determining the effects of perchlorate on thyroid hormone production in humans. In 2017, EPA completed a peer review to evaluate EPA's draft dose-response model. It is proposed that a future peer review will evaluate EPA's draft approach for deriving a maximum contaminant level goal (MCLG) for perchlorate in drinking water [65].

10.2.4.1 Chlorates and Chlorites

Chlorate (ClO_3^-) is a monovalent inorganic anion obtained by the deprotonation of chloric acid ($HClO_3$), with the chlorine in the +5-oxidation state. It is a conjugate base of a chloric acid. Meanwhile, chlorite (ClO_2^-) is a conjugate base of chlorous acid ($HClO_2$). The most commonly used of these is sodium chlorite ($NaClO_2$). When dissolved in water, it forms positively charged sodium cations (Na^+) and negatively charged chlorite anions (ClO_2^-).

Chlorite and chlorate are disinfection by-products, resulting from the use of chlorine dioxide as a disinfectant and odour/taste control in water [7]. The WHO guideline values for chlorite and chlorate (0.7 mg/L each) are designated as provisional because the use of chlorine dioxide as disinfectant may result in their guideline values being exceeded [75]. European Council Directive 98/83/EC do not indicate a limit value for chlorite. However, the current Italian regulation for chlorite in drinking water is set at 0.7 mg/L by the Decree of the Ministry of Health of 5th September 2006 [75].

Chlorate was not included in the final Stage 1 Disinfectants and Disinfection By-products (D/DBP) rule, because the health effects data at that time were inadequate to establish a maximum contaminant level goal (MCLG) [76]. Thus, managing chlorate was an issue for water systems until the US Environmental Protection Agency (USEPA) established a maximum contaminant level (MCL) for this compound in the year 2015 [77]. This results from the facts that the use of hypochlorite has increased as a result of safety, security, regulatory or community concerns [11].

10.2.5 Sources of Chlorate and Chlorite

The major sources of chlorate and chlorite are water treatment plants that utilize hypochlorites for disinfection purposes [78]. Chlorate is produced, spontaneously, from the disproportionation reaction of hypochlorites (Eqs. 10.1 and 10.2) [79]. In this reaction, the amount of chlorate continues to appreciate as a function of the storage temperature, pH, time and concentration of the hypochlorites [79].

$$2ClO^- \rightarrow Cl_2 + O_2 \tag{10.1}$$

$$3ClO^- \rightarrow Cl_2 + ClO_3^- \tag{10.2}$$

Chlorite can be sourced from rock minerals, modified in the course of burial, plate collisions, hydrothermal activity or contact metamorphism [80]. Examples of such rock minerals include clinochlore, pennantite, chamosite, greenschist, phyllite and greenstone.

10.2.6   Effects of Chlorate and Chlorite on Human Health and the Environment

Chlorate causes oxidative destruction to red blood cells in mammals leading to haemolytic anaemia and methemoglobin formation [81]. In addition, $NaClO_3$ can lead to nephrotoxicity via redox imbalance, causing DNA and membrane damage, metabolic alterations and brush border membrane enzyme dysfunction [82]. Due to its toxicity, chlorate has been included in the third USEPA Contaminant Candidate List and it is under examination as a part of the current microbial and disinfection by-product regulations review [83].

The health impact of chlorate on human is the toxicity, through ingestion and inhalation, which typically lead to goitrogens, methemoglobin [7] and the enlargement of the thyroid gland [6]. Exposures to chlorites give rise to oxidative stress, resulting in changes in the red blood cells [60, 61], cause anaemia and affect the human nervous system [1].

Chlorite is not considered to have carcinogenic effect to humans, but may lead to haemolytic anaemia and allergic dermatitis [7, 84]. In some toxicological research, no definite toxicological effect associated with chlorite was observed, but it was recommended that prolonged exposure to 120 and 360 mg/L of chlorite may cause weight loss in mice and cerebellar lesions in the offspring of rats [75].

10.3   Oxyanions of Bromine

The presence of bromate (BrO_3^-) in drinking water can be attributed to two principal sources:

1.   Electrolysis of NaCl: The electrolytic reactions of sodium chloride that result in the production of hypochlorite solutions produce residues of bromate in the presence of some bromide.
2.   As disinfection by-product (DBP): The main source of bromate in water results from the use of ozone to disinfect drinking water, and the ozone reacts with the naturally occurring bromide in the source water [13, 85].

$$Br^- + O_3 \rightarrow BrO_3^- \tag{10.3}$$

Bromate formation in disinfected drinking water is influenced by factors such as bromide ion concentration, pH of the source water (maximally at 8.8), the concentration of ozone and the reaction time used to disinfect the water.

The formation of bromate results from different oxidation processes of bromine, which include chlorination in the presence of sunlight/UV [86], ozonation [87] and sulphate-based oxidation [88, 89]. However, it has been reported that the bromate formed during the chlorination treatment rapidly reacts with organic carbon present in the water treatment system to form other brominated disinfection by-products like bromoform [90, 91]. The mechanism of bromate formation can be studied under two stages. The first stage involves the transformation of Br^- to $HOBr/^-OBr$; and the subsequent stage involves the transformation of $HOBr/^-OBr$ formed to bromate [88].

The most commonly used bromate is that of potassium. Potassium bromate is a colourless, odourless white crystal or powder, highly soluble in water and less soluble in acetone, dimethyl sulphoxide, ethanol, methanol and toluene [92]. It has a melting point of 350 °C and decomposes at 370 °C [93, 94]. In distilled water, potassium bromate degrades (350–400 °C) to form potassium bromide (KBr) and oxygen (O_2). Potassium bromate was an active ingredient in baking industries [95] before it was banned, due to some acute health implications, such as abdominal pain, diarrhoea, irritation to the mucous membrane of the upper aero-digestive tract and vomiting [93].

10.3.1 Fate of Bromates in Waters

Bromate is non-volatile; thus, it is slightly adsorbed onto soil or sediment. Because it is a strong oxidant, its fate in the environment is strongly dependent on the interaction with organic matter, which results in the formation of bromide ion. Bromate remains very stable in solution at ambient temperature, which makes it difficult to be removed by simple treatment method like boiling, filtration, etc. [91]. Despite the fact that bromate is a thermodynamically strong oxidant, it degrades abiotically within air-dried soil by up to 64% in 14 days, under aerobic and anaerobic conditions [96], but the reaction rates are not significant in a natural context [91]. The relatively high solubility enhances its contamination of water bodies following any industrial spillage. Its low chemical reduction rates suggest that it is conservative while in surface and groundwaters, acting as analogue to bromide, which has been extensively used as a tracer in aquifers due to its unreactive nature.

10.3.2 Effects of Bromine Oxyanions in Water and the Environment

The European Union has classified bromate as one of the substances that causes cancer under the category Group 1B [93], while it is classified as a Group 2B or "possible human" carcinogen by the International Agency for Research on Cancer (IARC), because of the tumour induction in rats and mice and lack of evidence of carcinogenicity inhuman [97].

The first toxicological study on $KBrO_3$ was conducted in Japan in the year 1978, which reported that potassium bromate exerted toxicological effects in experimental rats after two years of oral administration [98]. From both acute and sub-acute test using oral administration, it showed some significant toxicity symptoms like hypothermia, dullness, weakness, nasal discharge, lacrimation, diarrhoea and decreased appetite [93]. It is also had significant effects on the kidney as well as nervous system and can cause hearing loss [13].

Bromate has been proven to be an animal carcinogen, resulting from the high dose testing, causing DNA damage [99]. Hence, it is a probable human carcinogen under apt dose conditions. Nevertheless, recent researches have proven bromate to decompose in stomach acid and in blood. The particular cancer type found in male rats, resulting from laboratory experiment, is not applicable to human.

The Environmental Protection Agency (EPA) and the World Health Organization (WHO) have established the maximum contaminant level (MCL) for bromate in public water systems at 10 ppb [100]. These recommendations were determined based on the conservative assumptions concerning human cancer risks (approximately 1/10,000 per lifetime) at low doses, as well as the analytical procedure capabilities at the time. Analytical methods can now achieve quantification limits below 1 ppb [101].

Other bromine oxyanions include bromites (BrO_2^-), hypobromites (BrO^-) and perbromates (BrO_4). Hypobromites are chemical compounds similar to hypochlorites, found to be present in the immune cells and used as germicide and antiparasitic substance [102]. Perbromate is a compound that contains BrO_4^- functional group, with bromine having an oxidation state of $+7$. Perbromate has been identified as a new potential disinfection by-product (DBP), but the concentration detected was negligibly small, under drinking water conditions [103]. Generally, there is a dearth of information on the presence and management of bromites, hypobromites and perbromates in water.

10.4   Oxyanions of Iodine

In comparison with other halogens, iodine oxyanions are scarce in the environment [104]. However, the presence in the earth crust has been explained using the iodine cycle, which involves different geological and biological stages. The ocean sediments have been reported to account for about 68% of the iodine in the natural environment and 27.7% is obtainable from sedimentary rocks. The most common forms of iodine found in the ocean are iodate, iodide and elemental iodine [104].

The nuclear power process gave rise to iodine oxyanions in the atmosphere and terrestrial environment before World War II [105]. For instance, the I-131 and I-129 produced artificially, through the fission of uranium and plutonium results in the generation of iodine oxyanions [106]. It can also be generated from fallout from natural disasters like Chernobyl, etc. [107].

10.4.1   Fate of Iodine Oxyanions in Waters

Iodate (IO_3^-) is one of the naturally occurring forms of iodine in the environment. The presence of this oxyanion of iodine in the environment insinuates their obvious transfer into water bodies. It is observed that the high mobility of iodate in the environment is facilitated by their ability to adsorb to ordinary common sediment minerals or synthetic materials in aqueous environment rendering them even difficult to removed [108]. The stable iodine isotope (^{127}I) is ubiquitous throughout the earth's surface in igneous rocks and soils, most commonly as impurities in saltpetre and natural brines [109]. These anions can be released into the environment from both natural sources and anthropogenic activities. The natural sources include volatilization of iodine from the oceans, weathering of rocks and volcanic activity. Anthropogenic activities, such as oxyanion waste stream effluent from municipal plants and burning of waste and fossil fuels, are the sources of iodine oxyanions [110]. In the atmosphere, iodine undergoes extensive photochemical changes and can exist as gaseous inorganic, gaseous organic or particulate forms [111].

Iodine oxyanions are mainly introduced into surface and groundwater via rainfall, for non-coastal regions, and the combination of rainwater and ocean spray in coastal regions [112]. For elemental iodine, which is diatomic molecule (I_2), it is not particularly soluble in water; however, it is able to form more soluble anion species (e.g. iodate (IO_3^-)) [113]. The structures of halogen compounds is clear to scientists, except for the structure of iodate [104]. Studies have found that there are large intermolecular forces at play in between the oxygen and iodine [18]. These interactions give the molecule a trigonally distorted octahedral area around the iodine [114]. This configuration can also be seen in lithium iodate, ammonium iodate and cerium iodate. Some of the iodates are soluble in water, while some, like calcium iodate, are insoluble and can be removed by filtration [115].

In aqueous solution, using the hard soft acid base (HSBA) system, iodide acts as a soft base, but iodate act as a hard base. Iodate commonly pairs with hard acids such as potassium (K^+), sodium (Na^+) and lithium, while iodide (I^-) commonly pairs with soft acids such as silver (Ag^+), copper (Cu^+) and gold (Au^+). The oxidation and reduction reactions of iodine in the environment are very complex. Iodate dominates at higher pH and higher activity of electron or electrode potential measurement (Eh), suggesting that iodide is an easily oxidized species [116].

10.5   Measurement of Halogen Oxyanions in Aqua Systems

Until recently more focus was placed on the determination of organic halogens, due to their persistent and bio-accumulative nature. Recently, attention has been shifted towards the inorganic forms, to understand their environmental roles [117]. The increasing interest in the monitoring of halides and halogen oxyanions has led to the development of sensitive and selective analytical methodologies for their

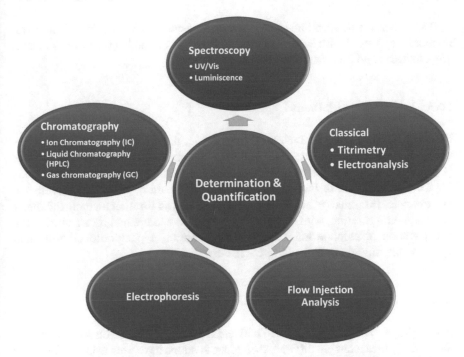

Scheme 10.1 Measurement technique for the determination and quantification of halide-based oxyanions

determination [117]. These methods can be classified into; classical, chromatography, spectroscopy, flow injection analysis and electrophoresis [117] (Scheme 10.1).

The selection of the method to be used in a particular analysis depends on: the availability of appropriate equipment; the number of samples to be analysed; expected content of the analyte; and the required detection and quantification limits [118].

Ion chromatography (IC) with suppressed conductivity detection remains the mostly used method for the determination of halide-based oxyanions. Nevertheless, indirect or hyphenated methods, such as liquid chromatography mass spectrophotometry (LC/MS), gas chromatography mass spectrometry (GC/MS) and so on are gradually gaining popularity and may replace the popular IC method in a matter of time [118, 119].

10.5.1 Classical Methods

Classical methods such as iodometry and potentiometry titrations have been used for the determination of halide-based oxyanions. These methods are based, almost exclusively, on their oxidizing characteristics [118]. However, most of the classical methods are laborious, lack sensitivity and suffer from interferences [118].

The general classification of the classical methods is: Titration (iodometric methods); spectrophotometric; and electroanalytical methods (i.e. potentiometric, amperometric and polarographic) [118].

10.5.1.1   Titration Methods

Different titration methods such as amperometric, direct potentiometric and iodometric titrations have been used for the determination of halogen oxyanions in water [120]. Some of these methods have been chosen as standard and reference methods for the determination of these oxyanions and validation of other newer methods. For example, Pisarenko et al. [121] used the iodometric titration method and direct potentiometric titration, with sulphite ion, as a reference method for measuring the concentration of chlorate ion in concentrated sodium hypochlorite solutions and selective quantitation of hypochlorite ion, respectively.

10.5.1.2   Spectrophotometric Methods

Although lack of chromophore(s) in halide-oxyanions makes the direct spectrophotometrically measurement difficult, their identifications have been achieved by interacting the oxyanion with other chemical species that are capable of absorbing photons, thereby improving the absorbance of the electromagnetic spectrum.

Based on this technique, several spectrophotometric applications for the measurement of halide-based oxyanions have been reported. For instance, a method, which involves the reduction of bromate in the presence of metabisulphite, to form bromine, followed by the reaction of the bromine with reduced fuchsin, to form a bromurated red coloured product that absorbs at 530 nm has been reported for the determination of trace amount of bromate in drinking water [122]. Using this method, Cl^-, ClO_3^- and SO_4^{2-} interferences that affect the commonly used ion chromatography method have no impact on the determination of bromate [122]. The linear response of the method is up to 20 mg/L of bromate, while the limit of detection is 1 mg/L (40 mm path length). The precision is 6% at 5 mg/L level of bromate ($n = 10$) [122].

Similarly, an indirect spectrophotometric determination of bromate was proposed by Oliveira et al. [123]. In their experiment, several phenothiazines, such as chlorpromazine, trifluoperazine and thioridazine were tested. However, observations showed that chlormomoromazine was more sensitive and had a lower limit of detection than the other phenothiazines. The methodology gives a linear range, between 25 and 750 µg/L, with limit of detection (LOD) of 6 µg/L, good precision (RSD < 1.6%, n ¼ 10), and determination frequency for bromate measurement [123].

Salami et al. [124] explored a spectrometry multi-commutation flow system for the measurement of hypochlorite in bleaching products. This method was also adapted for the determination of hypochlorite in water. Hypochlorite was determined, based on the spectrometric measurement of its reaction product with N, N-diethyl-p-phenylenediamine (DPD at 515 nm). The linear range for the method was between

2.68×10^{-5} and 1.88×10^{-4} mol L^{-1} (2–14 mg L^{-1}), with the limit of detection of 6.84×10^{-6} mol L^{-1} (0.51 mg L^{-1}). Analyte recovery ranged between 97.2 and 102.5% and the statistical analysis of the proposed method agreed with the reference iodometric method at 95% confidence level [124].

Another study reported a simple kinetic-spectrophotometric method for the simultaneous determination of binary mixtures of iodate and bromate in water samples [125]. The method was based on the kinetic profiles of different ratio of the analytes, thereby, allowing for the rapid and accurate simultaneous determination of bromate and iodate. The method was capable of simultaneous determination of $0.05–1.50$ g mL^{-1} each, with the ratio 30:1–1:30 for iodate–bromate [125].

Hosseini et al. [126] reported a method for a highly selective spectrometric determination of chlorate ions in the presence of chlorite ions. The method was based on the production of a coloured haloquinone, through the reaction of chlorate with benzidine in a hydrochloric acid medium (Eqs. 10.4 and 10.5).

$$ClO_3^- + 5Cl^- + 6H^+ \rightarrow 3Cl_2 + 3H_2O \tag{10.4}$$

$$\tag{10.5}$$

The method has a detection limit of 8×10^{-7} g/L chlorate, which is comparable with detection limit of ion chromatography, at 4×10^{-7} g/L chlorate [126].

Pena-Pereira et al. [117] gave a critical overview of contributions devoted to spectrophotometric determination of relevant halogen-containing species, with focus on the utilization of optical sensing of target analytes. The review focused on different nanoparticles that have been synthesized for the sensing of these oxyanions in different matrix, providing analytical validation of the reported methods and compliance with validation requirements [117].

10.5.1.3 Electroanalytical Methods

A selective potentiometric technique was developed for the simultaneous determination of OCl^- at high concentration (e.g. 1–3 M) alongside low concentrations of ClO_2^- and ClO_3^- (e.g. 50–150 mg/L), when these species are present together in solution. The method used sulphite ion as a mask, to selectively and quantitatively remove OCl^- at pH 10.5. This is so because, SO_4^{2-} does not react with ClO_2^- and ClO_3^- at pH 10.5. The remaining SO_4^{2-} is quantitatively removed with tri-iodide ion. Following the masking procedure, the sequential determination of ClO_2^- and ClO_3^- is carried out either by iodometric titration at the appropriate pH or by ion chromatography. The results from the direct potentiometric titration using sulphate are comparable with those of an indirect reference method [127].

A method has been proposed for the electrochemical determination of perchlorate ion by voltammetry, at the interface between two immiscible phases (water–o-nitrophenyloctyl ether). A demountable original-design amperometric ion-selective electrode, based on a laser-micro-perforated polymeric membrane, was fabricated for voltammetric measurements. The conditions of analytical signal recording in the determination of ClO_4^- were determined. The effect of interfering ions was assessed and amperometric selectivity coefficients were calculated. The accuracy of the procedure was verified by the added–found method. The developed electrode was applied to the determination of perchlorate in natural and drinking waters [128].

Salimi et al. [129] prepared a modified electrode used as a chronoamperometric detector for chlorate, iodate and bromate determination in flow systems or chromatographic instruments. The modification procedure was used for sensor and biosensor fabrications, using porphyrin derivatives as electron transfer mediators [129]. Ordeig et al. [130] described the direct detection of bromate, chlorate and iodate, using modified arrays of platinum ultra microelectrodes. The detection limits for these ions were IO_3^- 0.76 μM; BrO_3^-, 2.34 μM and ClO_3^-, 133.2 mM [130].

10.5.2   Chromatography

Chromatography is a technique that involves the separation of a mixture of components, thus allowing for their individual quantification. Separation of the components is usually achieved based on their relative mobility on a stationary phase while being carried along in a mobile phase. The main advantage of these methods of analysis is their ability to simultaneously measure different analytes at the same time. Different chromatographic techniques such as ion chromatography (IC), high performance liquid chromatography, gas chromatography (GC), and the hyphenated forms, with mass spectrometer (LCMS and GCMS), have been used for the measurement of halide-based oxyanions [28, 131–135].

10.5.2.1   Ion Chromatography

Ion chromatography is the routine method used for the measurement of halide-based oxyanions [118]. This is majorly because of the possibility for the simultaneous determination of several ions within minutes. These ions include: Ions of the same element with different oxidation states; those with different charges (cations and anions); and different species (organic and inorganic anions) [118]. Other advantages of ion chromatography include good limits of detection and quantification (at the level of μg/dm^3), ability to analyse a small amount of the sample, the possibility of using various detectors and a simple preparation method of analysis [136]. These have made it evolve as the standard method of choice for the determination of ionic species by many standard organizations such as the APHA methods 4110 for various anions and oxyanions [120, 135], ASTM method D6581-18 and USEPA Methods 321.8 for determination of BrO_3^- [137].

10.5.2.2   High Performance Liquid Chromatography

Constantinou et al. [28] reported a liquid chromatography-electrospray ionization-mass spectrometry-mass spectrometry (LC-ESI-MS/MS) method for the simultaneous determination of ClO_3^-, ClO_4^- and BrO_3^- in water and food samples. The study reported that 69% of the 284 water samples analysed contained chlorate above the limit of quantitation (LOQ) of 0.01 mg/L, with maximum amount of 1.1 mg/L. Bromate was detected in 5 drinking water samples at levels above the LOQ, at concentrations up to 0.026 mg/L. While for the determination of these anions in food, 247 food samples comprising of 19 different commodities, including fruits, vegetables, cereals and wine, were analysed. Chlorate was found to be present at maximum concentration of 0.83 mg/kg in a sample of cultivated mushrooms, while perchlorate was determined to be at concentrations below the LOQ of 0.05 mg/kg [28].

10.5.2.3   Gas Chromatography

Reddy-Noone et al. [138] developed a hyphenated gas chromatography method for the determination of bromate, iodate, bromide and iodide. For bromate and iodate determination, free bromide and iodide were first removed by anion exchange, with silver chloride, exploiting the differences in silver salts solubility product, being AgCl, 1.8×10^{-10}, AgBr, 5.0×10^{-13}, AgI, 8.3×10^{-17}, $AgBrO_3$, 5.5×10^{-5} and $AgIO_3$, 3.1×10^{-8}, followed by the ascorbic acid reduction of the oxyhalides to halides. The halides are then converted to 4-bromo-2, 6-dimethylaniline and 4-iodo-2,6-dimethylaniline, by their reaction with 2-iodosobenzoate in the presence of 2,6-dimethylaniline, at pH 6.4 and 2–3, respectively. The haloanilines were then extracted via Single drop microextraction (SDME) in 2 µL of toluene or liquid-phase microextraction (LPME) into 50 µL of toluene and injection of 2 µL of extract into GC-MS. The LPME was found to be more robust, sensitive and gives better extraction in shorter period than SDME. Linearity range was obtained between 0.05 µg and 25 mg/L of bromate/bromide and iodate/iodide, with a limit of detection of 20 ng/L of bromate, 15 ng/L of iodate, 20 ng/L of bromide and 10 ng/L of iodide for LPME [139].

10.5.3   Flow Injection Method

Flow injection analysis is one of the earliest techniques used for the determination of halide-based oxyanions. For instance, as early as 1984, flow injection voltammetry at a glassy carbon electrode was used for the measurement of hypochlorite in water [140]. This was achieved by the injection of the solution into a dilute sulphuric acid eluent containing 1% of potassium bromide. The amount of hypochlorite in solution was determined through the quantity of bromide that was replaced. Method linearity range was between 0.08 and 2×10^{-3} M. Hypobromite solutions containing an excess of bromide can be analysed using a similar method, but with the omission of potassium bromide from the eluent [140].

Similarly, Abdalla and Al-Swaidan [141] determined iodate, bromate and hypochlorite as iodine by flow injection amperometry at a platinum or glassy carbon electrode by injecting them into an eluent 0.20 M in hydrochloric acid and 0.024 M in potassium iodide or an eluent 2 M in sulphuric acid and 0.12 M in potassium iodide. The methods linearity is from 10^{-3} to 10^{-7} M [141].

Gordon et al. [142] developed a flow injection non-ion chromatographic method for the determination of bromate ion in bromide containing ozone treated waters using chlorpromazine. The workers reported a high precision (better than 2.5%) and accuracy (± 1 μ/L BrO_3) at a linear range of 1.0–30.0 μ/L BrO_3^- and a limit of detection (LOD) of 0.8 μ/L [142].

10.6 Future Perspective

The occurrence, distribution and effects of presence of halogen-based oxyanions in water are of high concern. Acquiring knowledge about their structure, their fate in water and specially their adverse effect on the human health and environment are of great importance. Keeping this in view, the precise determination of these oxyanions in aqueous media and their removal from water bodies are two realms which need competent focus in future research. Therefore, priority should be placed on devising sensitive methods for the precise determination of these substances in water media. In addition, finding highly effective and cost-effective ways for removal of such pollutants should be as a matter of urgency.

References

1. Yin YB, Guo S, Heck KN, Clark CA, Coonrod, CL, Wong MS (2018) Treating water by degrading oxyanions using metallic nanostructures. ACS Sustain Chem Eng 6:11160–11175
2. Weidner E, Ciesielczyk F (2019) Removal of hazardous oxyanions from the environment using metal-oxide-based materials. Materials 12:927
3. Theiss FL, Couperthwaite SJ, Ayoko GA, Frost RL (2014) A review of the removal of anions and oxyanions of the halogen elements from aqueous solution by layered double hydroxides. J Colloid Interface Sci 417:356–368
4. Fuge R (2019) Fluorine in the environment, a review of its sources and geochemistry. Appl Geochem 100:393–406
5. Henderson PB, Woytek AJ (2000) Fluorine compounds, inorganic, nitrogen. In: Kirk-Othmer encyclopedia of chemical technology

6. McCarthy WP, O'Callaghan TF, Danahar M, Gleeson D, O'Connor C, Fenelon MA, Tobin JT (2018) Chlorate and other oxychlorine contaminants within the dairy supply chain. Compr Rev Food Sci Food Saf 17:1561–1575

7. Alfredo K, Stanford B, Roberson JA, Eaton A (2015) Chlorate challenges for water systems. J Am Water Works Assoc 107:E187–E196

8. Sijimol M, Gopikrishna V, Dineep D, Mohan M (2017) Perchlorate in drinking water around rocket manufacturing and testing facilities and firework manufacturing sites in Kerala, India. Energy Ecol Environ 2:207–213

9. Tiemann M (2008) Perchlorate contamination of drinking water: regulatory issues and legislative actions. Library of Congress, Congressional Research Service, Washington

10. Yu L, Canas JE, Cobb GP, Jackson WA, Anderson TA (2004) Uptake of perchlorate in terrestrial plants. Ecotoxicol Environ Saf 58:44–49

11. Stanford BD, Pisarenko AN, Snyder SA, Gordon G (2011) Perchlorate, bromate, and chlorate in hypochlorite solutions: guidelines for utilities. J Am Water Works Assoc 103:71–83

12. Greer MA, Goodman G, Pleus RC, Greer SE (2002) Health effects assessment for environmental perchlorate contamination: the dose response for inhibition of thyroidal radioiodine uptake in humans. Environ Health Perspect 110:927–937

13. Barlokova D, Ilavsky J, Marko I, Tkacova J (2017) Removal of bromates from water. IOP Conf Ser Earth Environ Science, IOP Publishing 012021

14. Zhang Y, Li J, Li L, Zhou Y (2019) Influence of parameters on the photocatalytic bromate removal by F-graphene-TiO$_2$. Environ Technol 1–9

15. Dongmei L, Zhiwei W, Qi Z, Fuyi C, Yujuan S, Xiaodong L (2015) Drinking water toxicity study of the environmental contaminant—Bromate. Regul Toxicol Pharmacol 73:802–810

16. Gilfedder B, Lai S, Petri M, Biester H, Hoffmann T (2008) Iodine speciation in rain, snow and aerosols. Atmos Chem Phys 8

17. Parker KE, Golovich EC, Wellman DM (2014) Iodine adsorption on ion-exchange resins and activated carbons: batch testing. Pacific Northwest National Lab (PNNL), Richland

18. Gong T, Zhang X (2013) Determination of iodide, iodate and organo-iodine in waters with a new total organic iodine measurement approach. Water Res 47:6660–6669

19. Szecsody JE, Pearce CI, Cantrell KJ, Qafoku N, Wang G, Gillispie EC, Lawter AR, Gartman BN, Brown CF (2018) Evaluation of remediation technologies for iodine-129: FY18 bench scale results. Pacific Northwest National Lab (PNNL), Richland

20. Zhang S, Schwehr K, Ho Y-F, Xu C, Roberts K, Kaplan D, Brinkmeyer R, Yeager C, Santschi P (2010) A novel approach for the simultaneous determination of iodide, iodate and organo-Iodide for [127]I and [129]I in environmental samples using gas chromatography—mass spectrometry. Environ Sci Technol 44:9042–9048

21. Aten A, Doorgeest T, Hollstein U, Moeken H (1952) Section 5: radiochemical methods. Analytical chemistry of astatine. Analyst 77:774–777

22. Yordanov A, Pozzi O, Carlin S, Akabani G, Wieland B, Zalutsky M (2004) Wet harvesting of no-carrier-added [211]At from an irradiated [209]Bi target for radiopharmaceutical applications. J Radioanal Nucl Chem 262:593–599

23. Jadiyappa S (2018) Radioisotope: applications, effects, and occupational protection. In: Principles and applications in nuclear engineering: radiation effects, thermal hydraulics, radionuclide migration in the environment

24. de Farias RF (2017) Estimation of some physical properties for tennessine and tennessine hydride (TsH). Chem Phys Lett 667:1–3

25. Hosseini S, Pourmortazavi S, Gholivand K (2009) Spectrophotometric determination of chlorate ions in drinking water. Desalination 245:298–305

26. Tkáčová J, Božíková J (2014) Determination of chlorine dioxide and chlorite in water supply systems by verified methods. Slovak J Civ Eng 22:21–28

27. Herman MG, Wieczorek M, Matuszek M, Tokarczyk J, Stafiñski M, Koscielniak P (2006) Determination of chlorite in drinking water and related aspects of environment protection. J Elementol 11

28. Constantinou P, Louca-Christodoulou D, Agapiou A (2019) LC-ESI-MS/MS determination of oxyhalides (chlorate, perchlorate and bromate) in food and water samples, and chlorate on household water treatment devices along with perchlorate in plants. Chemosphere 235:757–766

29. Girenko DV, Gyrenko AOA, Nikolenko NV (2019) Potentiometric determination of chlorate impurities in hypochlorite solutions. Int J Anal Chem 2019

30. Dietrich AM, Ledder TD, Gallagher DL, Grabeel MN, Hoehn RC (1992) Determination of chlorite and chlorate in chlorinated and chloraminated drinking water by flow injection analysis and ion chromatography. Anal Chem 64:496–502

31. Tang T-F, Gordon G (1980) Quantitative determination of chloride, chlorite, and chlorate ions in a mixture by successive potentiometric titrations. Anal Chem 52:1430–1433

32. Seiler MA, Jensen D, Neist U, Deister UK, Schmitz F (2017) Determination of trace perchlorate in water: a simplified method for the identification of potential interferences. Environ Sci Europe 29:30

33. Lamb JD, Simpson D, Jensen BD, Gardner JS, Peterson QP (2006) Determination of perchlorate in drinking water by ion chromatography using macrocycle-based concentration and separation methods. J Chromatogr A 1118:100–105

34. Magnuson ML, Urbansky ET, Kelty CA (2000) Determination of perchlorate at trace levels in drinking water by ion-pair extraction with electrospray ionization mass spectrometry. Anal Chem 72:25–29

35. Jackson WA, Böhlke JK, Gu B, Hatzinger PB, Sturchio NC (2010) Isotopic composition and origin of indigenous natural perchlorate and co-occurring nitrate in the southwestern United States. Environ Sci Technol 44:4869–4876

36. Parker DR, Seyfferth AL, Reese BKJ (2008) Perchlorate in groundwater: a synoptic survey of "pristine" sites in the coterminous United States. Environ Sci Technol 42:1465–1471

37. Nabi M, Ghoreishi SM, Behpour M (2020) Determination of bromate ions in drinking water by derivatization with 2-methyl-2-butene, dispersive liquid-liquid extraction and gas chromatography-electron capture detection. J AOAC Int

38. Musa M, Ahmed IM, Atakruni I (2010) Determination of bromate at trace level in Sudanese bottled drinking water using ion chromatography. J Chem 7

39. Maznaya YI, Zuy O, Vasilchuk T, Goncharuk V (2014) Determination of bromate ions in waters by diffusion reflection spectroscopy. J Water Chem Technol 36:174–179

40. Nayanova E, Elipasheva E, Sergeev G (2015) Photometric redox determination of bromate ions in drinking water. J Anal Chem 70:143–147

41. Oliveira SM, Segundo MA, Rangel AO, Lima JL, Cerda V (2011) Spectrophotometric determination of bromate in water using multisyringe flow injection analysis. Anal Lett 44:284–297

42. Farrell S, Joa J, Pacey G (1995) Spectrophotometric determination of bromate ions using phenothiazines. Anal Chim Acta 313:121–129

43. Schnepfe MMJ (1972) Determination of total iodine and iodate in sea water and in various evaporites. Anal Chim Acta 58:83–89

44. Narayana B, Pasha C, Cherian T, Mathew M (2006) Spectrophotometric method for the determination of iodate using methylene blue as a chromogenic reagent. Bull Chem Soc Ethiop 20:143–147

45. Takeda A, Tsukada H, Takaku Y, Satta N, Baba M, Shibata T, Hasegawa H, Unno Y, Hisamatsu S (2016) Determination of iodide, iodate and total iodine in natural water samples by HPLC with amperometric and spectrophotometric detection, and off-line UV irradiation. Anal Sci 32:839–845

46. Weidner E, Ciesielczyk FJM (2019) Removal of hazardous oxyanions from the environment using metal-oxide-based materials. Materials 12:927

47. Dozier M, Melton R, Hare M, Porter D, Lesikar BJ (2005) Drinking water problems: perchlorate. Texas FARMER Collection

48. Urbansky ET, Gu B, Magnuson ML, Brown GM, Kelty CA (2000) Survey of bottled waters for perchlorate by electrospray ionization mass spectrometry (ESI-MS) and ion chromatography (IC). J Sci Food Agric 80:1798–1804

49. Srinivasan A, Viraraghavan T (2009) Perchlorate: health effects and technologies for its removal from water resources. Int J Environ Res Public Health 6:1418–1442
50. Steinmaus CM (2016) Perchlorate in water supplies: sources, exposures, and health effects. Curr Environ Health Rep 3:136–143
51. Xie T, Yang Q, Winkler MK, Wang D, Zhong Y, An H, Chen F, Yao F, Wang X, Wu J (2018) Perchlorate bioreduction linked to methane oxidation in a membrane biofilm reactor: performance and microbial community structure. J Hazard Mater 357:244–252
52. Ginsberg GL, Hattis DB, Zoeller RT, Rice DC (2007) Evaluation of the US EPA/OSWER preliminary remediation goal for perchlorate in groundwater: focus on exposure to nursing infants. Environ Health Perspect 115:361–369
53. Cao F, Jaunat J, Sturchio N, Cancès B, Morvan X, Devos A, Barbin V, Ollivier P (2019) Worldwide occurrence and origin of perchlorate ion in waters: a review. Sci Total Environ 661:737–749
54. Hřibová Š, Gargošová HZ, Vávrová M (2014) Sorption ability of the soil and its impact on environmental contamination. Interdiscip Toxicol 7:177–183
55. Trumpolt CW, Crain M, Cullison GD, Flanagan SJ, Siegel L, Lathrop S (2005) Perchlorate: sources, uses, and occurrences in the environment. Remediat J 16:65–89
56. Bennett ER, Clausen J, Linkov E, Linkov I (2009) Predicting physical properties of emerging compounds with limited physical and chemical data: QSAR model uncertainty and applicability to military munitions. Chemosphere 77:1412–1418
57. Srinivasan R, Sorial GA (2009) Treatment of perchlorate in drinking water: a critical review. Sep Purif Technol 69:7–21
58. Goldhaber S (2019) Perchlorate: the two-decade journey to a proposed rule. Int J Sci Res Environ Sci Toxicol 4:1–8
59. Furdui VI, Zheng J, Furdui A (2018) Anthropogenic perchlorate increases since 1980 in the Canadian High Arctic. Environ Sci Technol 52:972–981
60. Jiang S, Cox TS, Cole-Dai J, Peterson KM, Shi G (2016) Trends of perchlorate in Antarctic snow: implications for atmospheric production and preservation in snow. Geophys Res Lett 43:9913–9919
61. Jackson WA, Böhlke JK, Andraski BJ, Fahlquist L, Bexfield L, Eckardt FD, Gates JB, Davila AF, McKay CP, Rao B (2015) Global patterns and environmental controls of perchlorate and nitrate co-occurrence in arid and semi-arid environments. Geochim Cosmochim Acta 164:502–522
62. Bardiya N, Bae J-H (2011) Dissimilatory perchlorate reduction: a review. Microbiol Res 166:237–254
63. Sijimol M, Jyothy S, Pradeepkumar A, Chandran MS, Ghouse SS, Mohan M (2015) Review on fate, toxicity, and remediation of perchlorate. Environ Forensics 16:125–134
64. Urbansky ET (1998) Perchlorate chemistry: implications for analysis and remediation. Bioremediat J 2:81–95
65. Calderón R, Godoy F, Escudey M, Palma P (2017) A review of perchlorate (ClO_4^-) occurrence in fruits and vegetables. Environ Monit Assess 189:82
66. Zheng S, Liu J, Wang Y, Li F, Xiao L, Ke X, Hao G, Jiang W, Li D, Li Y (2018) Effect of aluminum morphology on thermal decomposition of ammonium perchlorate. J Therm Anal Calorim 134:1823–1828
67. Kucharzyk KH, Crawford RL, Cosens B, Hess TF (2009) Development of drinking water standards for perchlorate in the United States. J Environ Manage 91:303–310
68. Aziz CE, Hatzinger PB (2008) Perchlorate sources, source identification and analytical methods. In: In situ bioremediation of perchlorate in groundwater. Springer, pp 55–78
69. Kannan K, Praamsma ML, Oldi JF, Kunisue T, Sinha RK (2009) Occurrence of perchlorate in drinking water, groundwater, surface water and human saliva from India. Chemosphere 76:22–26
70. Shahbazi-Manshadi Z, Kaykhaii M, Dadfarnia S, Haji Shabani AM (2019) Co-microprecipitation/flotation of trace amounts of cadmium from environmental samples through its complexation with iodide and neutralization with cetyltrimethylammonium bromide in the presence of perchlorate ions. Int J Environ Anal Chem 99:1365–1374

71. Du Z, Xiao C, Furdui VI, Zhang W (2019) The perchlorate record during 1956–2004 from Tienshan ice core, East Asia. Sci Total Environ 656:1121–1132

72. Sijimol M, Mohan M (2014) Environmental impacts of perchlorate with special reference to fireworks—a review. Environ Monit Assess 186:7203–7210

73. Interstate Technology and Regulatory Council Perchlorate Team (2008) Remediation technologies for perchlorate contamination in water and soil, PERC-2. Interstate Technology and Regulatory Council

74. Leung AM, Pearce EN, Braverman LE (2010) Perchlorate, iodine and the thyroid. Best Pract Res Clin Endocrinol Metab 24:133–141

75. Tang Y, Long X, Wu M, Yang S, Gao N, Xu B, Dutta S (2020) Bibliometric review of research trends on disinfection by-products in drinking water during 1975–2018. Sep Purif Technol 116741

76. Baribeau H, Prévost M, Desjardins R, Lafrance P, Gates DJ (2002) Chlorite and chlorate ion variability in distribution systems. J Am Water Works Assoc 94:96–105

77. Al-Otoum F, Al-Ghouti MA, Ahmed TA, Abu-Dieyeh M, Ali M (2016) Disinfection by-products of chlorine dioxide (chlorite, chlorate, and trihalomethanes): Occurrence in drinking water in Qatar. Chemosphere 164:649–656

78. Padhi R, Subramanian S, Satpathy K (2019) Formation, distribution, and speciation of DBPs (THMs, HAAs, ClO_2^-, and ClO_3^-) during treatment of different source water with chlorine and chlorine dioxide. Chemosphere 218:540–550

79. Kriem LS (2017) Chlorate formation in on-site hypochlorite generation facilities: effects of temperature, pH, and storage times

80. Worden R, Griffiths J, Wooldridge L, Utley J, Lawan A, Muhammed D, Simon N, Armitage PJ (2020) Chlorite in sandstones. Earth Sci Rev 204:103105

81. Siddiqui MS (1996) Chlorine-ozone interactions: formation of chlorate. Water Res 30:2160–2170

82. Ali SN, Arif H, Khan AA, Mahmood R (2018) Acute renal toxicity of sodium chlorate: redox imbalance, enhanced DNA damage, metabolic alterations and inhibition of brush border membrane enzymes in rats. Environ Toxicol 33:1182–1194

83. Breytus A, Kruzic AP, Prabakar S (2017) Chlorine decay and chlorate formation in two water treatment facilities. J Am Water Works Assoc 109:E110–E120

84. Sorlini S, Collivignarelli MC, Canato M (2015) Effectiveness in chlorite removal by two activated carbons under different working conditions: a laboratory study. J Water Supply Res Technol AQUA 64:450–461

85. Rivera-Utrilla J, Sánchez-Polo M, Polo AM, López-Peñalver JJ, López-Ramón MV (2019) New technologies to remove halides from water: an overview. In: Advanced research in nanosciences for water technology. Springer, pp 147–180

86. Liu C, Von Gunten U, Croué J-P (2013) Chlorination of bromide-containing waters: enhanced bromate formation in the presence of synthetic metal oxides and deposits formed in drinking water distribution systems. Water Res 47:5307–5315

87. Aljundi IH (2011) Bromate formation during ozonation of drinking water: a response surface methodology study. Desalination 277:24–28

88. Yang J, Dong Z, Jiang C, Wang C, Liu H (2019) An overview of bromate formation in chemical oxidation processes: occurrence, mechanism, influencing factors, risk assessment, and control strategies. Chemosphere 124521

89. Fang J-Y, Shang C (2012) Bromate formation from bromide oxidation by the UV/persulfate process. Environ Sci Technol 46:8976–8983

90. Wang F, van Halem D, Ding L, Bai Y, Lekkerkerker-Teunissen K, van der Hoek JP (2018) Effective removal of bromate in nitrate-reducing anoxic zones during managed aquifer recharge for drinking water treatment: Laboratory-scale simulations. Water Res 130:88–97

91. Butler R, Godley A, Lytton L, Cartmell E (2005) Bromate environmental contamination: review of impact and possible treatment. Crit Rev Environ Sci Technol 35:193–217

92. National Toxicology Program (1991) Chemical repository data sheet: potassium bromate. Research Triangle Park

93. Shanmugavel V, Santhi KK, Kurup AH, Kalakandan SK, Anandharaj A, Rawson A (2019) Potassium bromate: effects on bread components, health, environment and method of analysis: a review. Food Chem 125964

94. Elabass H, Suliman M (2017) instrumental measurement of potassium bromate residue in bread and flour samples in Khartoum state. Sudan University of Science and Technology

95. Ojeka E, Obidiaku M, Enukorah C (2006) Spectrophotometeric determination of bromate in bread by the oxidation of dyes. J Appl Sci Environ Manag 10:43–46

96. Kanan GJ, Al-Najjar HE (2010) Isolation and growth characterization of chlorate and/or bromate resistant mutants generated by spontaneous and induced foreword mutations at several gene loci in Aspergillus niger. Braz J Microbiol 41:1099–1111

97. Emeje M, Ofoefule S, Nnaji A, Ofoefule A, Brown S (2009) Assessment of bread safety in Nigeria: quantitative determination of potassium bromate and lead. Afr J Food Science

98. Kurokawa Y, Takayama S, Konishi Y, Hiasa Y, Asahina S, Takahashi M, Maekawa A, Hayashi Y (1986) Long-term in vivo carcinogenicity tests of potassium bromate, sodium hypochlorite, and sodium chlorite conducted in Japan. Environ Health Perspect 69:221–235

99. Abd El-Rahim AH, Abd-El-Moneim OM, Abd El-Kader HA, Abd El Raouf A (2018) Inhibitory effect of bee venom against potassium bromate causing genetic toxicity and biochemical alterations in mice. J Arab Soc Med Res 13:89

100. Soltermann F, Abegglen C, Gotz C, Von Gunten U (2016) Bromide sources and loads in Swiss surface waters and their relevance for bromate formation during wastewater ozonation. Environ Sci Technol 50:9825–9834

101. Cotruvo J (2018) Drinking water quality and contaminants guidebook. CRC Press

102. World Health Organization (2018) Alternative drinking-water disinfectants: bromine, iodine and silver

103. Bergmann MH, Iourtchouk T, Rollin J (2011) The occurrence of bromate and perbromate on BDD anodes during electrolysis of aqueous systems containing bromide: first systematic experimental studies. J Appl Electrochem 41:1109–1123

104. Cox EM, Arai Y (2014) Environmental chemistry and toxicology of iodine. In: Advances in agronomy. Elsevier, pp 47–96

105. Cox ML, Sturrock G, Fraser PJ, Siems ST, Krummel P (2005) Identification of regional sources of methyl bromide and methyl iodide from AGAGE observations at Cape Grim, Tasmania. J Atmos Chem 50:59–77

106. Matsuzaki H, Muramatsu Y, Ohno T, MaoW (2017) Retrospective reconstruction of Iodine-131 distribution through the analysis of Iodine-129. In: EPJ web of conferences, EDP sciences, pp 08014

107. Kaplan DI, Serne RJ, Parker KE, Kutnyakov IV (2000) Iodide sorption to subsurface sediments and illitic minerals. Environ Sci Technol 34:399–405

108. Li D, Kaplan DI, Price KA, Seaman JC, Roberts K, Xu C, Lin P, Xing W, Schwehr K, Santschi PH (2019) Iodine immobilization by silver-impregnated granular activated carbon in cementitious systems. J Environ Radioact 208:106017

109. Hou X, Aldahan A, Nielsen SP, Possnert G, Nies H, Hedfors J (2007) Speciation of 129I and 127I in seawater and implications for sources and transport pathways in the North Sea. Environ Sci Technol 41:5993–5999

110. Westby T, Cadogan A, Duignan G (2018) In vivo uptake of iodine from a Fucus serratus Linnaeus seaweed bath: does volatile iodine contribute? Environ Geochem Health 40:683–691

111. Saiz-Lopez A, Lamarque J-F, Kinnison DE, Tilmes S, Ordóñez C, Orlando JJ, Conley AJ, Plane J, Mahajan AS, Sousa Santos G (2012) Estimating the climate significance of halogen-driven ozone loss in the tropical marine troposphere. Atmos Chem Phys 12:3939–3949

112. Nisi B, Raco B, Dotsika E (2014) Groundwater contamination studies by environmental isotopes: a review. In: Threats to the quality of groundwater resources. Springer, pp 115–150

113. Theiss FL, Ayoko GA, Frost RL (2017) Removal of iodate (IO_3^-) from aqueous solution using LDH technology. Mater Chem Phys 202:65–75

114. Nielsen E, Greve K, Larsen J, Meyer O, Krogholm K, Hansen M (2014) Iodine, inorganic and soluble salts. The Danish Environmental Protection Agency, Copenhagen

115. Lyon BA, Milsk RY, DeAngelo AB, Simmons JE, Moyer MP, Weinberg HS (2014) Integrated chemical and toxicological investigation of UV-chlorine/chloramine drinking water treatment. Environ Sci Technol 48:6743–6753
116. Henson GL, Niemeyer L, Ansong G, Forkner R, Makkar HP, Hagerman AE (2004) A modified method for determining protein binding capacity of plant polyphenolics using radiolabelled protein. Phytochem Anal Int J Plant Chem Biochem Tech 15:159–163
117. Pena-Pereira F, García-Figueroa A, Lavilla I, Bendicho C (2020) Nanomaterials for the detection of halides and halogen oxyanions by colorimetric and luminescent techniques: a critical overview. Trends Anal Chem 115837
118. Michalski R, Łyko A (2013) Bromate determination: state of the art. Crit Rev Anal Chem 43:100–122
119. Laubli M, Proost R, Seifert N, Unger S, Wille A (2003) Bromate in drinking water—which method to use in ion chromatography. LC GC EUROPE, pp. 6–7
120. American Public Health Association, American Water Works Association, Water Environment Federation (2017) Standard methods for the examination of water and wastewater. American Public Health Association
121. Pisarenko AN, Stanford BD, Quiñones O, Pacey GE, Gordon G, Snyder SA (2010) Rapid analysis of perchlorate, chlorate and bromate ions in concentrated sodium hypochlorite solutions. Anal Chim Acta 659:216–223
122. Romele L (1998) Spectrophotometric determination of low levels of bromate in drinking water after reaction with fuchsin. Analyst 123:291–294
123. Oliveira SM, Segundo MA, Rangel AO, Lima JL, Cerdà V (2011) Spectrophotometric determination of bromate in water using multisyringe flow injection analysis. Anal Lett 44:284–297
124. Salami FH, Bonifácio VG, de Oliveira GG, Fatibello-Filho O (2008) Spectrophotometric multicommutated flow system for the determination of hypochlorite in bleaching products. Anal Lett 41:3187–3197
125. Afkhami A, Madrakian T, Bahram M (2005) Simultaneous spectrophotometric determination of iodate and bromate in water samples by the method of mean centering of ratio kinetic profiles. J Hazard Mater 123:250–255
126. Hosseini S, Pourmortazavi S, Gholivand K (2009) Spectrophotometric determination of chlorate ions in drinking water. Desalin Water Treat 245:298–305
127. Adam LC, Gordon G (1995) Direct and sequential potentiometric determination of hypochlorite, chlorite, and chlorate ions when hypochlorite ion is present in large excess. Anal Chem 67:535–540
128. Martynov LY, Mel'nikov AP, Astaf'ev AA, Zaitsev NK (2017) Voltammetric determination of perchlorate ion at a liquid–liquid microscopic interface. J Anal Chem 72:992–998
129. Salimi A, MamKhezri H, Hallaj R, Zandi S (2007) Modification of glassy carbon electrode with multi-walled carbon nanotubes and iron (III)-porphyrin film: application to chlorate, bromate and iodate detection. Electrochim Acta 52:6097–6105
130. Ordeig O, Banks CE, Del Campo FJ, Muñoz FX, Compton RG (2006) Electroanalysis of Bromate, iodate and chlorate at tungsten oxide modified platinum microelectrode arrays. Electroanalysis 18:1672–1680
131. Alomirah HF, Al-Zenki SF, Alaswad MC, Alruwaih NA, Wu Q, Kannan K (2020) Elevated concentrations of bromate in drinking water and groundwater from Kuwait and associated exposure and health risks. Environ Res 181:108885
132. Hamid KIA, Scales PJ, Allard S, Croue J-P, Muthukumaran S, Duke M (2020) Ozone combined with ceramic membranes for water treatment: impact on HO radical formation and mitigation of bromate. J Environ Manage 253:109655
133. Bichsel Y, Von Gunten U (1999) Determination of iodide and iodate by ion chromatography with postcolumn reaction and UV/visible detection. Anal Chem 71:34–38
134. Liu Y, Mou S (2004) Determination of bromate and chlorinated haloacetic acids in bottled drinking water with chromatographic methods. Chemosphere 55:1253–1258

135. American Public Health Association, Water Environment Federation (2005) Standard methods for the examination of water and wastewater. Washington
136. Michalski R (2009) Applications of ion chromatography for the determination of inorganic cations. Crit Rev Anal Chem 39:230–250
137. Creed JT, Magnuson ML, Pfaff JD, Brockhoff C (1996) Determination of bromate in drinking waters by ion chromatography with inductively coupled plasma mass spectrometric detection. J Chromatogr A 753:261–267
138. Reddy-Noone K, Jain A, Verma KK (2007) Liquid-phase microextraction–gas chromatography–mass spectrometry for the determination of bromate, iodate, bromide and iodide in high-chloride matrix. J Chromatogr A 1148:145–151
139. Reddy-Noone K, Jain A, Verma KK (2007) Liquid-phase microextraction–gas chromatography–mass spectrometry for the determination of bromate, iodate, bromide and iodide in high-chloride matrix. J Chromatogr A 1148:145–151
140. Fogg AG, Chamsi AY, Barros AA, Cabral J (1984) Flow injection voltammetric determination of hypochlorite and hypobromite as bromine by injection into an acidic bromide eluent and the indirect determination of ammonia and hydrazine by reaction with an excess of hypobromite. Analyst 109:901–904
141. Abdalla MA, Al-Swaidan HM (1989) Iodimetric determination of iodate, bromate, hypochlorite, ascorbic acid and thiourea using flow injection amperometry. Analyst 114:583–586
142. Gordon G, Bubnis B, Sweetin D, Kuo C-Y (1994) A flow injection, non-ion chromatographic method for measuring low level bromate ion in ozone treated waters. Ozone Sci Eng

Chapter 11
Monitoring and Management of Anions in Polluted Aqua Systems: Case Studies on Nitrate, Chromate, Pertechnetate and Diclofenac

Rana Ahmed, Philippe Moisy, Amitabh Banerji, Peter Hesemann, and Andreas Taubert

Abstract Anionic pollutants are widespread and pose severe environmental risks. This chapter focuses on four different anions as representative examples for various classes of anionic pollutants originating from various anthropogenic sources and activities: nitrate from agriculture, chromate from leather and tanning industries, pertechnetate from nuclear industry and military activities and finally, as an example of an organic anion, diclofenac from the pharmaceutical industry. All four anions pose different risks, but their management calls for similar remediation strategies. This chapter summarizes these strategies and discusses their challenges, their advantages and disadvantages.

Keywords Oxyanions · Diclofenac · Anionic pollutants · Adsorption · Water management · Surface chemistry

11.1 Introduction

Over the past decades, anion toxicity has been widely recognized as a tremendous environmental and health problem. Anion management and sequestration in water bodies have therefore been established as a central issue in most parts of the world. While there is a rather large and heterogeneous group of anions causing these problems, some anionic species are particularly important in terms of their health risks. As a result, tremendous efforts have focused on the removal of these anions from

R. Ahmed · A. Banerji · A. Taubert (✉)
Institute of Chemistry, University of Potsdam, Karl-Liebknecht-Strasse 24-25, 14476 Potsdam, Germany
e-mail: ataubert@uni-potsdam.de

P. Moisy
CEA, DES/ISEC/DMRC, Univ Montpellier, Marcoule, BP 17171, Bagnols sur Cèze, France

P. Hesemann (✉)
ICGM, Univ Montpellier - CNRS, Place Eugène Bataillon, Montpellier, France
e-mail: peter.hesemann@umontpellier.fr

water bodies such as rivers, lakes and groundwater. The current chapter focuses on four anions, two of which are classical oxyanions (chromate and nitrate), one is radioactive (pertechnetate) and one is an organic anion (diclofenac). All these anionic species pose similar health and environmental risks and can often be treated with similar methods as those used for metal oxyanion-contaminated water and soil.

Nitrate NO_3^- is of prime concern on a global scale, primarily due to problems associated with eutrophication [1, 2]. Chromate CrO_4^{2-} and related ions are common, for example, in regions with leather and tanning industries [3, 4]. Pertechnetate TcO_4^- is found in nuclear testing sites and in hospital wastewaters [5–7]. Diclofenac (DCF) is the only non-oxyanion in this series. We chose it as a representative example for pharmaceutical pollutants [8, 9] and as an illustration that often similar strategies can be used for water purification. These ions are different in that nitrate is based on a nonmetal, chromate and pertechnetate are based on a metal and DCF is an organic compound. However, quite some of their properties such as solubility and bioavailability are similar, leading to related strategies for water management.

Characteristically, oxyanions with a similar structure and with similar properties, for instance, SO_4^{2-}, NO_3^- or PO_4^{3-}, often coexist. As the concentrations of these individual anions can be quite high, and as anion exchange is often unspecific, the sequestration of anions can be more challenging than sequestration of cations [10, 11]. For example, Zhou et al. [12] found that ClO_4^- often coexists with NO_3^- in groundwater at very different concentration levels. Similarly, SeO_4^{2-} coexists with SO_4^{2-} in coal-mining wastewaters, again in very different concentrations. SeO_4^{2-} and SO_4^{2-} also coexist with NO_3^- in flue gas desulfurization wastewater. Additionally, some wastewater contains very high concentrations of some oxyanions (e.g., fertilizer industry wastewater at 1000 mg/L of NO_3^- or nuclear industries wastewater at 50,000 mg/L of NO_3^-). These high concentrations of one specific anion often hamper an effective sequestration. Additionally, the high solubility and high stability of nitrogen oxyanions make nitrogen compounds even more challenging to remove from wastewater. As a result, these processes are often energy and cost intensive [13]. For example, coprecipitation shows low potential for NO_3^- removal [14], while conventional ion exchange produces NO_3^-, SO_4^- and Cl^- containing brine that requires further disposal and triggers secondary pollution [15].

Occasionally, very specific properties of a particular oxyanion may also hinder effective sequestration. For example, the high volatility of Tc(VII) (in the form of Tc_2O_7) in wastewater makes the operation of its vitrification or glassification process inefficient [16]. Also, the low selectivity and sensitivity toward competing anions such as Cl^- and CO_3^{2-}, which are present in much higher concentration than pertechnetate, cause many methods for TcO_4^- removal methods to be rather poor. Similarly, Zhang et al. [17] reported that DCF removal efficiency varies from 0 to 80% in most wastewater treatment plants (WWTPs). Such a large variation may be attributed to multiple factors such as acidity of the water in the WWTP, aerobic/anaerobic ratios, sunlight irradiation and effectiveness of enzymatic cleavage of glucuronide conjugates of DCF. Katarina et al. [18] found that conventional WWTPs are ineffective for DCF removal and a significant fraction of DCF escapes into effluents unchanged. Designing a selective, efficient and stable means for CrO_4^{2-} reduction or adsorption is challenging as well [19]. One central reason for this issue is that chromium-based anions typically coexist with very high concentrations of CO_3^{2-}, HCO_3^-, SO_4^{2-}, PO_4^{3-} and Cl^- in cooling water and effluents from pigment and tannery industries. This reduces the overall effectiveness of the adsorbents for CrO_4^{2-} sequestration [19, 20].

Although some progress in the development of effective water treatment materials and processes has been made, a tremendous need for effective, yet cheap and simple, processes for water treatment worldwide remains. This chapter presents recent and promising approaches, with a focus on green and sustainable materials and processes. Before discussing these approaches, we will briefly summarize and classify the most common and important oxyanions.

11.2   Common Oxyanions: Sources, Abundance and Occurrence

Industrialization and industrial production rely on the use of many hazardous elements and compounds such as heavy metals, metalloids, oxyanions and derivatives [21–24]. Oxyanions are a particular challenge to water treatment because they are highly water-soluble, very mobile and non-biodegradable and therefore contaminate groundwater, natural aquatic systems and soils [11, 25].

Oxyanions are negatively charged polyatomic ions with the generic formula $A_xO_y^{z-}$, where A represents different metal or nonmetal elements and O are oxygen atoms. Oxyanions can be both monomeric and polymeric [26, 27]. Oxyanions can be produced from many reactions and processes. Usually, industries that generate alkaline wastes, mostly through high-temperature processes (thermal waste treatments, metal smelting, fossil fuel combustion), are considered as the common sources of oxyanions [28].

However, not all oxyanions are equally important in terms of availability and contamination. According to Adegoke et al. [27], ClO_4^-, AsO_4^{3-}, SeO_4^{2-}, BO_3^{3-}, CrO_4^{2-} and MoO_4^{2-} are common along with NO_3^-, SO_4^{2-}, PO_4^{3-} and CO_3^{2-}. Han et al. [29] reported that SbO_4^{3-} and TeO_4^{2-} are toxic even at trace levels as well as AsO_4^{3-} and BO_3^{3-}. Besides, MnO_4^- [30], MoO_4^- [31] and BrO_3^- [26, 32] have often been studied considering their sequestration from industrial wastewater because of their ability to pollute the natural aquatic system.

Furthermore, oxyanions such as VO_4^{3-} and WO_4^{2-} are emerging pollutants, even though their biogeochemistry is still poorly understood. Doğan and Aydın [33] reported that vanadium is released in large amounts into the aquatic environment from glass, rubber, inorganic chemical and pigment industries. Also, the production of phosphoric acid from vanadium-rich minerals results in large amounts of VO_4^{3-} ion extraction and release.

Tungsten has been identified as an emerging environmental contaminant by the United States Environmental Protection Agency (US EPA) [34, 35]. Indeed,

Petruzzelli and Pedron [36] showed that WO_4^{2-} is thermodynamically very stable in the environment.

Radiotoxic oxyanions such as TcO_4^- are gaining attention around some parts of the world, especially in regions where production of weapons-grade plutonium and nuclear testing sites are located. Sun et al. [37] showed that TcO_4^- is highly mobile and can be transported into groundwater. Due to its high mobility and toxicity, noticeable attention is also directed toward ReO_4^- sequestration [37, 38]. Ghosh et al. [38] classified ReO_4^- as a rare oxyanion, which is sometimes used as a substitute for TcO_4^- because of the structural likeness of the two anions in terms of hydration energy and charge density. Table 11.1 summarizes the oxyanions that frequently occur in the aquatic environment and their sources.

Furthermore, besides conventional oxyanions as defined above, organic drugs and metabolites bearing anionic groups such as carboxylate or sulfonate have also caused increasing concern. Their concentration has grown due to increased release from medical and pharmaceutical wastes to open waters and groundwater in North America and Europe [57–59]. Low-cost over-the-counter drugs like the nonsteroidal anti-inflammatory drugs (NSAIDs) ibuprofen, diclofenac, naproxen and indomethacin have therefore been identified as emerging contaminants [60]. Because of the heavy discharge by households, hospitals, pharmaceutical industries and agriculture, diclofenac is on the "Watch list of the EU Water Framework Directive" [9, 57]. As a result, DCF has been studied and reviewed in far more than 400 scientific articles in the previous two decades [9].

11.3 Oxyanions in Aquatic System: Public Health and Environment

In natural environments, oxyanions exist in a variety of species. Speciation depends on numerous factors including pH, electrochemical potential (Eh), redox conditions, electrolyte compositions, temperature and microbial activity [26, 61]. For example, SO_4^{2-} is thermodynamically more stable than SO_3^{2-} at pH 6.5–8.5 and Eh 0.1–0.4 V [26]. Also, SO_4^{2-} can be protonated at sufficiently low pH yielding HSO_4^-; both sulfate and sulfuric acid are major drivers in the formation of atmospheric aerosols, which have been held responsible for climate change, adverse public health effects and natural sulfur cycle misbalance [62]. Above 600 mg/L, the presence of SO_4^{2-} noticeably changes the taste of drinking water and causes laxative effects. Therefore, for industrial effluent and mine drainage, a maximum allowed concentration of 250–1000 mg/L SO_4^{2-} has been set [53]. Depending on the pH and concentration, different species such as $HSO3^-$, $S_2O_5^{2-}$, SO_3^- and dissolved SO_2 exist in the aquatic environment [63].

Nitrate is another very common and highly soluble water contaminant. It naturally occurs in the nitrogen cycle and is the most stable form of N for oxygenated systems [1]. In contrast, NO_2^- preferably exists in reducing environments and can be oxidized

Table 11.1 Important oxyanions as classified by the US EPA

Element	Oxyanion species	Uses and sources	References
Priority pollutants			
As	AsO_4^{3-}, AsO_3^{3-}	Wood preservation, metal adhesives, textile, pigments, electric plants, pharmaceuticals	[25, 39]
Cr	CrO_4^{2-}, $Cr_2O_7^{2-}$	Tannery, pigments, electroplating, cooling tower blowdown, chromium plating	[3, 39, 40]
Se	SeO_4^{2-}, SeO_3^{2-}	Glass, electronics, coal combustion, mining, petroleum industries, metal refineries	[41–44]
Sb	SbO_4^{3-}	Mining, smelting, waste incineration, fossil fuel combustion	[31, 41, 45]
Contaminant candidate list (CCL4)			
Cl	ClO_3^-	Paper and pulp, flour and oil processing, leather de-tanning, cleaning agent production	[26, 46, 47]
Mo	MoO_4^{2-}	Stainless steel production, metallurgy, pigment, catalyst industries	[25, 31, 46, 48]
Mn	MnO_4^-	Iron and steel alloy manufacturing, bleaching and disinfection by-product, catalysis	[46, 47]
Te	TeO_4^{2-}, TeO_3^{2-}	Gold mining, deep-sea hydrothermal vent systems	[46, 49, 50]
V	VO_4^{3-}	Photography, glass, rubber, ceramic, mining, oil refining	[25, 46, 48]
Emerging contaminants			
Cl	ClO_4^-	Solid rocket fuel propellant, pyrotechnics, explosives, fireworks, munitions	[26, 32, 51]
W	WO_4^{2-}	Military sites, fertilizer production, golf clubs, heat sinks, metal wires, turbine blades, television sets	[25, 35]

(continued)

Table 11.1 (continued)

Element	Oxyanion species	Uses and sources	References
Most abundant oxyanion species			
P	PO_4^{3-}, PO_3^{3-}	Food and beverage industry, agriculture, fertilizer, detergent production	[27, 52]
S	SO_4^{2-}, SO_3^{2-}	Beam house of tanning, fertilizer, chemical, textile, soap, paper	[53–55]
N	NO_3^-, NO_2^-	Nitrogenous fertilizer production, wastewater disposal, agricultural runoff	[1, 2, 26, 47]
Further hazardous pollutants			
B	BO_3^{3-}	Glass manufacturing, flame retardants, soap and detergent production	[25, 47]
Br	BrO_3^-	Textiles, flour milling, beer production, fish paste production	[26, 32, 47]
Tc	TcO_4^-	Plutonium production, military tests and waste, fission material production, radiopharmaceutical tracer	[56]

to NO_3^-. Gee Chai et al. [64] reported that one common biological NO_3^- formation process includes *Nitrosomonas* and *Nitrobacter* species. First, ammonium is oxidized to NO_2^- by *Nitrosomonas* and *Nitrobacter* and then further oxidizes NO_2^- to NO_3^-.

A high NO_3^- concentration in surface water increases algal growth and causes eutrophication and extinction of aquatic life [2, 65]. Vegetables, meat and drinking water are vectors for both NO_3^- and NO_2^- uptake by humans and animals. The World Health Organization (WHO) has categorized NO_3^- as a major drinking water contaminant with a maximum permitted limit (MPL) of 50 mg/L because it causes stomach cancer in adults and blue baby syndrome in infants [66].

NO_2^- is considerably more toxic than NO_3^- and causes respiratory tissue hypoxia and cell failure. In light of this, the European Food Safety Authority (EFSA) has set the MPL of 10–500 mg/kg and 50–175 mg/kg for NO_3^- and NO_2^-, respectively, for different types of food [65]. Rajmohan el al. [13] and Mohseni et al. [15] showed that high concentrations of NO_3^- in surface and groundwater are an acute problem in the UK, Europe, Japan, China, Saudi Arabia, Australia, North America, Morocco, western Iran and India, among others. Bijay et al. [67] suggested that developing countries in Asia or humid tropics of Africa are potentially more prone to NO_3^- pollution due to the high amounts of nitrogen-containing fertilizer used in agriculture. The study also revealed that countries with lower fertilizer use in north, central and west Africa (where soils are mostly aridisols and likely to stay dry almost all year) are less affected by NO_3^- pollution.

Besides nonmetal-based oxyanions, a wide variety of metal and metalloid oxyanions is found in the aquatic system in large quantities with a diverse speciation. For example, chromium is a toxic metal known to form many oxo species. Its most important oxidation states are Cr(III) and Cr(VI) [68]. Cr(III) is an essential nutrient for cell function and the metabolic system, relatively stable and less of a threat for public health and environment [69]. On the other hand, Cr(VI) affects the water ecosystem and the plant and animal kingdom adversely due to its high solubility mobility and lethal character [3]. Considering the different forms of Cr(VI) (e.g., $HCrO_4^-$, CrO_4^{2-}, $Cr_2O_7^{2-}$, H_2CrO_4), $HCrO_4^-$ is the dominant species at a pH 2.5–5 [70]. Rakhunde et al. [68] stated that the chances are higher to find $HCrO_4^-$ and CrO_4^{2-} in typical surface water when the concentration of chromium is less than 5 µg/L. $HCrO_4^-$ and CrO_4^{2-} are the prevalent species between pH 6 and 7, whereas concentrated Cr(VI) likely contains $Cr_2O_7^{2-}$ in highly acidic media [71]. Liu et al. [72] reported the presence of $HCrO_4^-$ at lower pH and the transformation to CrO_4^{2-} and $Cr_2O_7^{2-}$ with increasing pH.

These anions can bioaccumulate and cause damage to aquatic organisms such as mucous secretion, irregular swimming, discoloration, erosion of scales, increasing mortality and disorder in osmoregulatory functions [73]. Generally, Cr(VI) is taken up by the human body through the respiratory tract and skin, leading to teratogenicity, mutagenicity, carcinogenicity, neurotoxicity, liver and kidney failure, respiratory problems, skin disease and many other issues [71, 73, 74].

Several studies suggest that plants also take up CrO_4^{2-} and $Cr_2O_7^{2-}$ and subsequently suffer from lower plant yield, stem growth, root growth, germination and photosynthesis [74–76]. Probably, chromium also percolates through soils, undergoes various changes such as oxidation, reduction, precipitation and dissolution in the presence of dissolved oxygen and MnO_2, and ultimately contaminates the groundwater [74]. As a result, Cr(III) and Cr(VI) compounds are major contaminants and regulated by the WHO and US EPA. For drinking water, a maximum 0.05 mg/L is permissible by the WHO, while the US EPA only allows 0.01 mg/L [70].

Like chromium, vanadium forms a wide variety of oxyanions. A substantial increase of vanadium compounds in water and soil and the potential toxicity of these compounds is currently raising concerns around the world [25, 77]. According to the Forum of European Geological Surveys (FOREGS), the vanadium concentration in European soil ranges from 1.28 to 537 mg/kg [78]. Vanadium has been adopted in both US EPA Contaminant Candidate Lists (CCL 3 & 4) [77] based on the fact that high vanadium exposure can irritate the mucous membrane and lead to gastrointestinal disturbance, anemia, cough and bronchopneumonia, among others [33, 79]. Generally, a pH above 3 is best for the formation of various monovanadate (e.g., $VO_2(OH)^{2-}$, $VO_3(OH)^{2-}$, VO_4^{3-}), polyvanadate (e.g., $V_2O_6(OH)^{3-}$, $V_2O_7^{4-}$, $V_3O_9^{3-}$, $V_4O_{12}^{4-}$) and decavanadate (e.g., $V_{10}O_{26}(OH)_2^{4-}$, $V_{10}O_{26}(OH)^{5-}$, $V_{10}O_{28}^{6-}$) species [80].

Recently, an increasing use of tungsten has raised interest and concern. Although tungsten appears inert and less toxic, it impedes the growth of nitrogen-fixing bacteria and plants and is toxic for fish and other aquatic species [81]. Moreover, the death of ryegrass and red worms [25] and other observations make W an emerging contaminant [35]. Tetrahedrally coordinated WO_4^{2-} is the dominant form under neutral to alkaline conditions, while under acidic condition tungsten forms polytungstates like $W_7O_{24}^{6-}$ and $H_2W_{12}O_{43}^{10-}$ [25, 81].

Similar to tungsten, rising levels of tellurium oxyanions in seawater, mining areas and industrial areas have attracted enormous attention for quite some time. Tellurium is perceived as the least available stable element in the environment and may exhibit various oxidation states [82, 83]. Harada and Takahashi [84] found $HTeO_3^-$ as the main species at pH 2–4 and Eh 0–0.4 V. In contrast, $HTeO_4^-$, TeO_3^{2-} and TeO_4^{2-} are the dominant species at neutral to alkaline condition (pH 7.5–12) with an Eh range of 0.1–0.5 V [84]. Their solubility increases with temperature and alkalinity, and Te(IV) is ten times more toxic than Te(VI) [84]. Tellurium often stems from the rubber and metal industry. Further transport causes bioaccumulation in kidney, liver, heart and other tissues showing acute toxicity particularly in young children [85]. Furthermore, microorganisms and bacteria suffer from TeO_3^{2-} even at concentrations as low as 1 $\mu g/mL$ [86].

A completely different source of pollution, production of weapons-grade plutonium, nuclear fuel reprocessing facilities and research sites have been generating nuclear waste for decades. Among others, these sites produce radioactive technetium, a common surface and groundwater contaminant in nuclear testing sites [87] such as the United States Department of Energy (US DOE) sites [88]. Moreover, ^{99m}Tc is also used as a radiotracer in clinical settings [89].

Multiple studies have specified that technetium mostly occurs as Tc(VII) in the form of TcO_4^- under atmospheric or moderately oxidizing conditions[56, 87]. Under moderately reducing and anaerobic conditions, $TcO_2 * 2H_2O$ and TcS_2 are the dominant species as Tc(IV). Due to its high solubility and mobility, TcO_4^- is taken up by plants and thus accumulates in the food chain using the same mechanism as SO_4^{2-} and PO_4^{3-} [56]. Persistent accumulation of TcO_4^- in lung tissue may initiate significant risks, where technetium tainted dust or vapor inhalation promotes the chance of lung cancer and associated maladies [90]. Finally, TcO_4^- uptake causes cell death for blue-green algae [91] and the same study also finds TcO_4^- responsible for constraining photosynthesis and suffocation for many organisms.

Besides these purely inorganic oxyanions, organic anions represent another class of emerging contaminants. As a representative example, the presence and treatment of diclofenac (DCF), a common painkiller, in water has thoroughly been investigated. DCF and related compounds are widely spread over the globe. DCF was identified as ubiquitous organic pollutant in various water bodies, and it was therefore included in the EU watch list of emerging contaminants [92].

The fate of DCF has become a significant concern since most of the DCF (almost 75%) escapes the WWTPs unaltered and therefore contaminates aquatic environments and soil in its both parent and metabolite forms [9, 93, 94]. DCF has also been detected in drinking water in many countries including USA, Japan, Germany, France, Spain, Sweden or Taiwan [8, 9]. Long-term DCF exposure to the human body may cause gastrointestinal bleedings, renal papillary necrosis, peptic ulceration, hepatotoxicity and renal failure, among others [95]. Furthermore, DCF is responsible for renal failure in birds leading to their death [96]. Liver, heart, pulmonary and renal lesions in vultures, mice, rats and many other animals have also been linked to DCF [96–98].

11.4 Current Approaches to Oxyanion Sequestration

11.4.1 Nitrate as an Example of a Nonmetal-Based Oxyanion

Multiple solutions to address the problem of nitrate removal and reduction exist. As stated above, nitrate is a severe environmental problem [1, 2, 13, 15, 65, 67] and there is thus a need to effectively treat water contaminated with nitrate. These developments and approaches have been reviewed [99–107], and we will thus focus on a series of examples to illustrate the breadth of these approaches. They can be broken down using a materials perspective, but, unlike in the case of CrO_4^{2-} (see below), the variability of these different approaches seems somewhat broader and less centered around one or two core concepts.

Biomass and carbonaceous materials

Biomass, biochar and other carbonaceous materials have been studied for nitrate removal. For example, Heaney et al. [108] have shown that biochar can be modified with suitable components to improve nitrate sorption efficiency. They used low molecular weight organic acids (citric, malic, oxalic acids) to protonate the biochar which was otherwise not effective for nitrate removal. According to the authors, surface protonation provides enough positive charges on the biochar to remove nitrate by electrostatic interactions.

Exploiting waste materials that would otherwise be burned, Pan et al. [109] modified biogas residues with amine and ammonium groups. The sorption capacities of the materials were up to ca. 62 mg/g for nitrate and showed excellent reproducibility in recycling experiment. In a similar approach, Zhao et al. used agro-waste (corn cobs, peanut shells, cotton stalks) to produce carbonaceous adsorbents, but the highest capacity reported by these authors was around 15 mg/g.

Machida et al. [110, 111] showed that polyacrylonitrile (PAN) and rayon fibers can be transformed to carbonaceous adsorbents *via* steam activation or solvent treatment and subsequent carbonization. The two studies reported closely related approaches aiming at the introduction of nitrogen into the carbonaceous fiber materials. Nitrogen doping is intended to produce amine and ammonium sites that would then facilitate nitrate uptake *via* ionic interactions.

Taoufik et al. [112] used activated carbon and statistical analysis to evaluate the performance in nitrate removal with the highest fraction of nitrate removal being ca. 96%. While this work provides an interesting approach to minimize the actual number of experiments, there is essentially no information on the type of activated carbon used by the authors. As such, a valorization of the work is rather difficult.

As an extension of the work described above, Yin et al. [113] fabricated Mg–Al modified biochars from soybean waste. The addition of Mg and Al, which produced MgO, AlOOH and MgAl$_2$O$_4$ within the biochar matrix, increased the surface areas, up to ca. 350 m^2/g, and a maximum nitrate adsorption capacity of ca. 40 mg/g was achieved. One question that was not addressed in this study, however, is the loss of Al(III), which may pose environmental risks in itself.

In a somewhat different approach, Oyarzun et al. [114] demonstrated that carbon electrodes can be modified with suitable surfactants, and the application of suitable potentials to the electrodes further supports the adsorption or desorption of nitrate, due to mostly electrostatic interactions. This is an interesting approach because it enables an effective and simple adsorbent regeneration, in principle even in situ (Fig. 11.1).

Composites

Viglasova et al. [115] investigated the synthesis and properties of biochar/montmorillonite composites for nitrate removal. The biochar was produced from bamboo, and the main data showed that the addition of montmorillonite

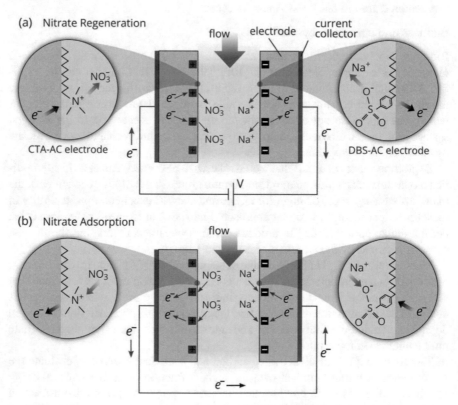

Fig. 11.1 Concept of nitrate removal using carbon electrode adsorbents. Figure reprinted from ref. [114] with permission

(yielding a carbon/mineral composite) more than doubled the adsorption capacity, when compared with plain biochar.

Cui et al. [116] used carbon materials obtained from crude oil waste modified with chitosan and iron oxide to remove nitrate. The resulting hybrid spheres showed an interesting adsorption behavior: over five recycling experiments, exactly the same rates of removal were recorded. In contrast, phosphate removal was less effective with continuous spent adsorbent recycling. The authors proposed that the interactions responsible for nitrate and (dihydrogen) phosphate removal are comprised of electrostatic interactions between various charged groups on the surface, but also through hydrogen bonding and exchange between nitrate or phosphate and surface-adsorbed anions such as chloride.

Again using chitosan, Zarei et al. [117] synthesized $Ag-TiO_2/\gamma-Al_2O_3$/chitosan nanocomposite thin films. The films exhibit some porosity, and their main principle of action is not adsorption but photocatalytic nitrate reduction. The catalyst removes ca. 50% of nitrate at catalyst doses of 1 g/L. Competing anions reduce the activity of the catalyst, and the highest activity was reported at pH over 7.

Using poly(ethylene imine) (PEI) rather than chitosan, Suzaimi et al. [118] modified porous rice husk silica with polyamine chains to enhance nitrate adsorption. The interesting feature of these materials is that the underlying silica matrix had an MCM-48 structure and therefore a high surface area (>700 m^2/g). Being porous, the materials contained large fractions of PEI after the modification step. This reduced the surface area but enhanced anion uptake through protonation, and hence electrostatic interaction. Indeed, the authors reported a nitrate sorption capacity of ca. 40 mg/g for neat rice husk silica and about 80 mg/g for the PEI-modified material. Using synthetic MCM-41 rather than biogenic MCM-48 type structures, Ebrahimi-Gatkash et al. [119] produced amino modified mesoporous materials with nitrate sorption capacities up to 38 mg/g. Moreover, Phan et al. [120] used an analogous approach by modifying rice husk ash with triamines. The resulting materials showed a very high adsorption capacity of over 160 mg/g and a reasonable recyclability over 10 cycles.

Karthikeyan and Meenakshi [121] reported the performance of magnetite/graphene oxide/carboxymethylcellulose nanocomposite beads prepared by a precipitation technique (Fig. 11.2). The beads are uniform in size and showed good adsorption capacities for phosphate and nitrate. The highest capacities were found at around neutral pH, at ca. 40 mg/g (even though the abstract and conclusion of this article states roughly twice this value).

Yang et al. [122] described a bifunctional adsorbent based on $Fe(OH)_3$ and an amine-modified polystyrene resin. The composite provided a large number of adsorption sites, and the authors reported that there is essentially no competition between nitrate and phosphate. Similar to another study [116], the authors found that the adsorption capacities of phosphate were reduced over ten recycle experiments, while the nitrate capacity essentially remained the same. High-angle annular dark field scanning transmission electron microscopy (HAADF-STEM), along with elemental mapping, further showed that both nitrogen (i.e., amine modification and nitrate)

Fig. 11.2 **a**, **b** Photographs of porous spheres for nitrate removal, **c–f** scanning electron microscopy images of beads and bead surface. Figure reprinted from ref. [121] and used with permission

and phosphorus (*i.e.*, phosphate) were homogeneously distributed throughout the materials (Fig. 11.3).

Using poly(amido amine) dendrimers rather than PEI to modify graphene oxide, Alighardashi et al. [123] produced powders with a flake-like morphology. The authors claimed that the nitrate removal is essentially independent of pH between pH 3.5 and 10.5 but as the data are given without error estimations, this is rather hard to quantify.

Jaworski et al. [124] used montmorillonite modified with various ammonium surfactants to introduce positive charges into the inorganic host and to open up the montmorillonite layer structure to improve the accessibility of internal sites. The importance of the interlayer opening is illustrated by the fact that in most cases no nitrate removal was observed with short-chain surfactants, while with longer-chain surfactants the exchange/removal capacity increased.

Finally, Omorogie et al. [125] described papaya seed/clay hybrids for the removal of nitrate. This work is based on earlier work [126] establishing that papaya/clay hybrids are effective for heavy metal removal but the more recent publication [125] demonstrates that these composites are in fact useful and effective for removal of a wide range of pollutants including anions.

Inorganics

Wang et al. [127] demonstrated that halide alloying in BiOX (X = Cl, Br) can be used to optimize the photocatalytic reduction of nitrate. The authors performed an in-depth analysis of the materials including measurements of the photocurrent generated upon

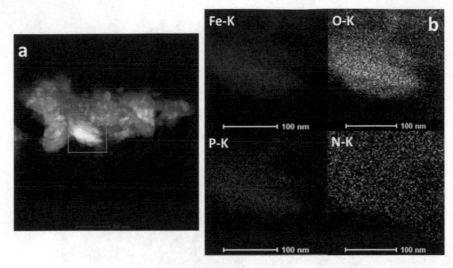

Fig. 11.3 a STEM-HAADF image and **b** elemental maps of Fe, O, P and N in the yellow frame in panel (**a**) of adsorbent prepared by Yang et al. Figure reprinted from ref. [122] and used with permission

irradiation and supported these measurements with analyses of defects. In particular, the authors showed that oxygen vacancies on the surface of the nanosheets were key reaction sites that favored the adsorption and reduction of nitrate, Fig. 11.4. Overall, this is a very interesting fundamental study, providing access to further concepts for catalyst design for nitrate reduction.

Li et al. [128] demonstrated highly effective photocatalytic reduction of nitrate with an inorganic/inorganic hybrid catalyst made from Pd/In clusters with different stoichiometries supported on FeO_x. Using a combination of experimental and theoretical methods, the authors demonstrated that the key step in the catalytic process is an electron transfer from the Pd/In cluster to the FeO_x support. This electron transfer is supposed to generate positively charged sites on the clusters, promoting the adsorption of nitrate and hydrogen, which then favors the catalytic nitrate reduction reaction at these sites.

Geng et al. [129] also studied the photocatalytic reduction of nitrate, but using a more conventional material, AgCl on TiO_2 nanotubes. The authors demonstrated that the chemical nature of the hole scavenger (formic acid *vs.* several other compounds) drastically affected the performance of the overall reaction. Moreover, the authors also showed that, with increasing irradiation time, the AgCl was transformed to Ag(0) nanoparticles that can clearly be distinguished in transmission electron microscopy analysis. Recycling experiments showed a reproducible performance over four cycles similar to other data [128], albeit with longer times needed for nitrate reduction.

Following the same general concept, Adamu et al. [130] showed that Cu-enriched TiO_2 was able to degrade both nitrate and oxalic acid at the same time. The authors highlighted the importance of the thermal treatment, as only some of the materials

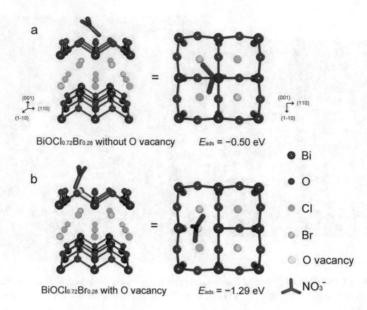

Fig. 11.4 Schematic representations of an NO_3^- ion adsorbed on the surface of $BiOCl_{0.72}Br_{0.28}$ **a** without and **b** with O vacancies (computational results). Figure reprinted from ref. [127] and used with permission

showed the necessary photochemical activity while others did not show any activity. In particular, the presence of Cu_2O appeared to be a key requirement for proper reaction of the catalyst with nitrate. The authors assigned this to the redox behavior of the Cu(I) ion, which was deemed necessary for the electron transfer reactions. Khatamian et al. [131] developed an ultrasound-assisted route for Fe_3O_4/bentonite hybrids. The materials effectively removed nitrate from aqueous solution, and the adsorbents were separated from the liquid phase by applying a magnetic field. In spite of this, it is difficult to compare the data to other work because the authors did not provide adsorption capacities.

Chu et al. [132] synthesized ammonium-modified zeolite Y by attaching aminosilanes to the zeolite solids. The resulting powders effectively removed nitrate, ammonium and phosphate from aqueous solution, and the materials can easily be recycled *via* washing with aqueous NaCl. In a similar study, Wu et al. [133] used $LaCl_3$ or hexadecyl trimethylammonium bromide-exchanged NaY zeolite. The La-modified zeolite showed high nitrate removal capacities.

Rezvani and Taghizadeh [134] used a simple raw material, clay (although they unfortunately did not specify the exact type of mineral), that was ground to below 150 μm in diameter, then hand formed into small spheres of ca. 5 mm in diameter and allowed to dry. A second type of material was prepared from montmorillonite clay with particle sizes below 20 nm. Both materials were heat-treated at 1000 °C to produce ceramic spheres. Adsorption experiments showed that nitrate was removed from solution by these particles, but as the authors did not provide errors of their experimental results, it is not possible to compare the performances of the two materials.

Resins and Gels

One of the major aspects in water treatment is cost. Duan et al. [135] have therefore studied the use of commercial trimethylamine-modified poly(styrene) (PS) anion exchange resins for nitrate removal. The authors identified a number of aspects that are critical for proper nitrate removal from nitrate-rich groundwater. In particular, the presence of sulfate and bicarbonate was reported as a major influence on the performance of the resin and the authors assigned the improved performance in some cases to the specific sulfate/chloride or bicarbonate/chloride ratios present in the system. Li et al. [136] followed the same approach, and the triethyl ammonium groups on the PS resins were found to be the most efficient moieties for nitrate removal. Moreover, these authors also stated that poly(acrylic) resins did not perform as well as the PS-based resins.

Li et al. [137] studied magnetic cationic hydrogels based on (3-acrylamidopropyl) trimethylammonium chloride producing a highly charged hydrogel network. The magnetic properties were introduced into the system by addition of γ-Fe_2O_3 during the synthesis. The authors reported a considerable pH dependence, and the highest nitrate sorption capacity of ca. 95.5 mg/g was observed between pH 5.2 and 8.8. Using a related approach, Sun et al. [138] synthesized gel adsorbents from an alkylammonium acrylate that were able to remove nitrate from real secondary treated wastewater.

Kalaruban et al. [139] described the combination of adsorption and electrodiffusion. A stainless steel mesh box was filled with Dowex ion exchange resin, and a piece of copper was used as the counter cathode. The combination of applying a potential to the steel/copper pair with the presence of the Dowex resin in the vicinity of the steel mesh box provided a better nitrate removal efficiency than either Dowex or potential difference alone. The authors also highlighted that there is a strong dependence on the performance from the electrode distance and the applied potential. The optimum settings are a distance of 1 cm, a potential of 3 V and a temperature of 30 °C.

Using a membrane approach, Maghsudi et al. [140] showed that pressure and the combination with a column setup can greatly enhance nitrate removal. The authors reported removal of over 90% of nitrate using appropriate combinations of pretreatment and membrane separation. Activated carbon improved the performance of the process.

Khalek et al. [141] investigated chitosan/gelatin/poly(2-(N,N-dimethylamino) ethyl methacrylate) mixed gels along with different modifications of this combination. The gels were prepared by gamma irradiation, and various modifications, such as the reaction with thiourea to provide thiol groups, were investigated. The authors highlighted that the presence of SiO_2 improved the adsorption capacity, when compared with the materials prepared without silica. While the authors did not provide detailed insights into why this is the case, they provided thermodynamic information, which allowed for the estimation of the contributions of the individual components to the overall sorption process.

Various processes

Finally, there are a few approaches that cannot be grouped in one of the above sections. First, Zhang et al. [142] reported a nitrate-dependent ferrous oxidation process. This process involves bacteria, and the authors initially attempted to couple nitrate reduction with As(III) oxidation to simultaneously remove nitrate and arsenic from solution. However, the authors found a rather strong pH dependence of this coupled process; thus, the reaction did not work properly in all pH regimes. Consequently, the authors employed a modified process, with a separation of the two steps, whereby first nitrate is reduced by the bacterial process followed by removal of the arsenic.

Saha et al. [143] fabricated a biofilm containing three bacterial strains that, in combination, simultaneously removed nitrate and phosphate to levels considered safe for drinking water. The authors highlighted the fact that the combination of several bacterial strains in the bioreactor is necessary, as individual strains were not able to individually perform the same reactions at the same level. According to the authors, "This is by far the fastest, energy efficient technology for wastewater treatment, which adds to sustainability of irrigation and aquaculture and reduced dependency on synthetic fertilizers and potable water for irrigation."

Gao et al. [144] introduced a new ultrafiltration process for simultaneous removal of nitrate and phosphate. The key component in this process is a tailor-made membrane based on amorphous zirconium hydroxide, quaternary ammonium salts and poly(vinylidene fluoride). Membranes were fabricated by phase inversion and had thicknesses in the order of 80 to 250 μm. According to the authors, the high nitrate and phosphate removal efficiencies were due to the high ion exchange capacities and the high number of surface hydroxyl groups introduced *via* the amorphous zirconium hydroxide.

Kalaruban et al. [145] used a combination of adsorbents (Dowex resins, amine-grafted corn cobs, amine-grafted coconut shells), in combination with a semi-continuous membrane process. Essentially, the method involved the adsorption of the contaminant onto the adsorbents, followed by membrane diffusion and separation of the purified water phase using a pump setup. The process is continuous in the sense that the water feed and effluent can be run in continuous mode but the adsorbents must be exchanged once their capacity is exhausted.

Using a completely different approach, Hosseini and Mahvi [146] used controlled freezing for nitrate removal. The authors highlighted the fact that freezing is among the simplest and most readily available processes even to non-experts. Moreover, their data suggested that a single freezing–melting run may in many cases be sufficient to lower the nitrate concentration to below the WHO limits considered safe for drinking water.

11.4.2 Chromate as an Example of a Metal-Based Oxyanion

As detailed in the introduction, chromium, especially Cr(VI), poses significant environmental threats and health hazards [71, 73–76]. As a result, there is a need for chromium removal from water bodies and drinking water sources. Given the fact that a large fraction of chromium contamination occurs in developing countries, these water treatment methods must be cheap, readily accessible and preferably be based on local resources. The following sections illustrate some of the developments that have been put forward over the last years. These developments have also been reviewed [147–152].

There have been numerous suggestions as to how Cr can be removed from water by adsorption. From a materials perspective, these can be grouped into: biomass and carbon-based adsorbents; composite materials, typically organic or carbon/inorganic composites; inorganic materials; and polymer resins and gels. Sometimes, these materials also contain moieties that provide photocatalytic or redox activity to further enhance the materials performance beyond mere adsorption.

Biomass and carbonaceous materials

Biomass is among the most interesting pools for raw materials because there is a large variety of biomaterials that is accessible worldwide. Moreover, biomass comes in a virtual unlimited supply and depending on the region, different biomasses are available. Thus, there is no need for long transport chains, and often, biomass is available as waste material that is not used otherwise. Therefore, there is no competition with food supplies.

Biomass and carbonized biomass have been studied for chromium removal [149, 151, 152]. To highlight a few recent approaches, we will focus on some studies that have appeared in the recent past and put forward interesting examples of this general approach.

Erabee et al. [153] studied the use of garden waste for the generation of activated carbons. A pool of agricultural waste was treated with $ZnCl_2$ as an activating agent, and fixed bed studies showed that Cr(VI) can be removed from aqueous solution using the resulting materials. Among other variables, the authors showed that the flow rates did not significantly influence the performance of the materials. However, increasing the column length in fixed bed experiments drastically improved the overall removal effectiveness from 42 to 95%, when the bed depth was increased from 3 to 7 cm.

Using a somewhat unusual biomass resource, *Sterculia villosa Roxb.*, Patra et al. [154] prepared sorbents for Cr(VI), using a combination of grinding (below 200 nm), acid activation and thermal treatment. The surface areas of the material ranged between 700 and 800 m^2/g, and the presence of Cr on the dry samples was shown using energy-dispersive X-ray spectroscopy. The authors also reported a strong pH dependence of Cr adsorption, with an essentially 100% removal efficiency at low pH (2–3) and ca. 10–20% at pH above 7.

Pakade et al. [155] studied macadamia nut shells as a source of activated carbon production. Macadamia is farmed in large volumes in Australia, Hawaii and South

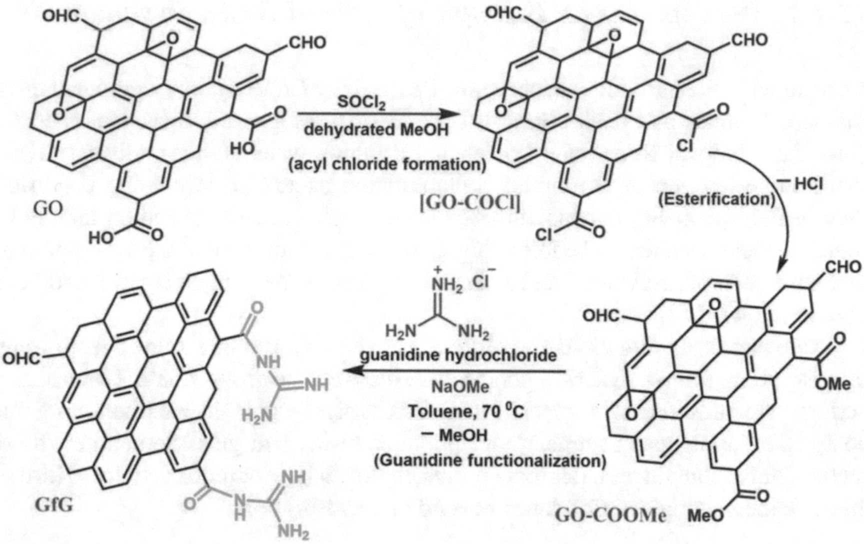

Fig. 11.5 Synthesis of guanidine-modified graphene oxide for chromate sorption. Reprinted from ref. [156] with permission

Africa, and macadamia nut farming produced large amounts of nut shells that are essentially unused agro-waste. The activated carbon was produced by a steam-activated heat treatment at 900°C producing carbons with surface areas >900 m^2/g and adsorption capacity of 23.3 mg/g at pH 2. According to the authors, the process also involved some reduction of Cr(VI) to Cr(III).

Using an approach that may not currently be accessible to mass applications due to high cost, Mondal et al. [156] presented an interesting fundamental study on how reduced graphene oxide can be functionalized with guanidine groups to enable Cr(VI) removal (Fig. 11.5). The authors performed an in-depth characterization of the materials and demonstrated a maximum adsorption capacity of 139 mg/g in acidic conditions and 51 mg/g in acid-free conditions. Importantly, these results can be achieved at low adsorbate concentrations and even increase further at higher concentrations of 400 ppm, to reach 173 mg/g. Clearly, while this approach is not suitable for large-scale applications, the general idea is certainly interesting for further technology development.

Using a closely related approach, Nkutha et al. [157] have made graphene-based sorbents by modification of graphene oxide with amino- and mercaptosilanes providing materials with –NH₂ and –SH functional groups suitable for metal binding. The main outcome of the article is the observation that, analogous to the work by Mondal et al. [156], the highest adsorption efficiencies were observed at pH 2. Again, the issue that could be raised here is that neither graphene nor specific silanes are necessarily very cheap; there is thus a need to further transfer these reactions and concepts to cheaper and more readily available carbon materials.

In a somewhat different approach, Bashir et al. [158] used carbon materials produced from sugarcane bagasse to directly bind heavy metals (Cd and Cr) in contaminated soil to prevent (or at least reduce) the uptake of heavy metals by crops and to improve microbial activity favoring plant growth. Biochar was produced *via* a rather simple process using pyrolysis at 500°C for 2 h. As the raw materials have a particle size in the millimeter range, the resulting products are rather coarse-grained. The authors demonstrated that the resulting biochar reduces the concentrations of free Cr and Cd by up to 85%. In particular, experiments in real soil samples demonstrated a high effectivity even at comparably low doses of added carbon to the soil of 15 g/kg of soil. The authors also measured the heavy metal uptake in plants grown on treated soils and found a significant decrease in Cd and Cr content in the respective plants. Moreover, biochar addition also positively affects the microbial activity. Overall, this is therefore a highly interesting approach toward removing heavy metals not only from open surface waters but also in food production.

Saranaya et al. [159] used *Annona reticulata Linn* fruit peels to remove Cr(VI) from aqueous solution. The somewhat unique aspect of this study is the fact that the materials were neither heat-treated nor carbonized; rather, the authors directly used the oven-dried materials. This is an interesting aspect because avoiding heat treatments reduces the energy and equipment cost involved in material preparation. Questions that need to be addressed, however, involve the long-term stability of such materials upon exposure to aqueous media. The authors highlighted that the materials were among the best performing fruit peel-based adsorbent for Cr(VI) reported so far, with a maximum capacity of 111 mg/g. Overall, biomass and biomass-derived materials are clearly a benchmark technology in low-cost, high-performance approaches for Cr removal from water and aqueous media. There are, however, alternative approaches that will be discussed next.

Composites

Besides carbon, composites are of interest because they can combine the advantageous properties of carbon or organic matter such as open pore structures and high surface areas with advantages of inorganics such as high sorption capacities or high mechanical stability. Indeed, numerous composites have been reported for Cr removal from water [133, 147].

Xu et al. [160] have prepared magnetic carbon/magnetite nanocomposites with high surface areas (Fig. 11.6). The synthesis of the materials was achieved by electrospinning of polystyrene in a fibrous mesh followed by calcination in the presence of ferric nitrate, producing robust magnetic carbon composites. The main features of these materials are the capacity to rapidly adsorb Cr(VI) as CrO_4^{2-} and the high adsorption rate constants.

Kumar et al. [161] developed a similar system, through the combination of graphitic carbon nitride (g-C_3N_4) with magnetite nanoparticles. The material simultaneously adsorbs Cr(VI) and reduces p-nitroaniline through a photocatalytic process. The authors reported a very high adsorption capacity of 555 mg/g for Cr(VI). This

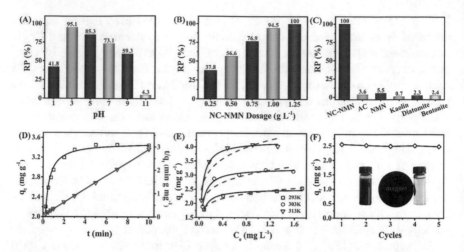

Fig. 11.6 **a** Removal percentage (RP%) of Cr(VI) versus pH, **b** effect of different adsorbent dosages, **c** comparison of Cr(VI) adsorption performance, **d** Cr(VI) adsorption rate qt versus t for and corresponding kinetic plot, **e** Cr(VI) adsorption isotherm at different temperatures, **f** Cr(VI) adsorption retention on magnetic carbon composite. Figure and caption reprinted from ref. [160] with permission

value was assigned to the presence of numerous –OH, –NH$_2$, –NHR and other coordinating groups on the material surface. Interestingly, the surface areas of these materials were below 100 m^2/g but the materials showed excellent Cr(VI) and Cr(III) removal efficiency, approaching 99% at higher pH. This is an interesting difference to carbon-based materials, such as those discussed above [153–159], which showed high adsorption capacities at low pH but performed rather poorly at high pH.

Xing et al. [162] synthesized iron oxide nanoparticles with an aminosilane surface coating. The combination of amino groups and nanoscale dimension of the material led to rapid Cr(VI) removal kinetics such that solutions containing up to 100 mg/L were Cr-free within one minute. Moreover, the materials can easily be recycled by a simple washing process. On the other hand, the Cr adsorption capacity was rather low at 8 mg/g.

Greenstein et al. [163] have developed iron oxide/polymer composites, including core–shell nanofibers for heavy metal removal. To produce the composite filter materials, the authors used electrospun poly(acrylonitrile) (PAN) fiber mats with embedded iron oxide nanoparticles to generate PAN/hematite composites with different iron oxide loadings. The resulting fiber mats, Fig. 11.7, remove As(V), Cr(VI), Cu(II) and Pb(II) and, according to the authors, "generally matched expectations from more traditional iron oxide sorbents." The article highlighted two interesting aspects that extend beyond more classical adsorbents: core–shell hybrid fibers prepared by the authors show a significantly higher performance, and the materials, being filters, were able to remove suspended solids along with dissolved metal ions. This is certainly an interesting combination of capabilities with tremendous application potential.

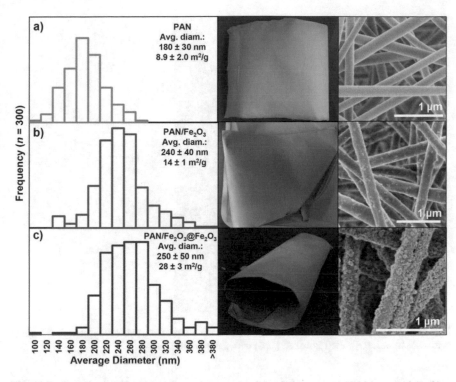

Fig. 11.7 Size distribution histograms, photograph of the fiber mat and SEM image of the fiber mats for **a** PAN, **b** embedded PAN/Fe$_2$O$_3$, and **c** core–shell PAN/Fe$_2$O$_3$@Fe$_2$O$_3$ nanofibers. Figure reprinted from ref. [163] with permission

Finally and using a different approach, Sathvika et al. [164] used *Saccharomyces cerevisiae* and *Rhizobium* in combination with carbon nanotubes to remove Cr(VI) from aqueous solution. The cell wall of the biological species was held responsible for the effective adsorption of the Cr(VI) ions to the materials. As the authors highlighted in their article, the work illustrated the "confluence of biotechnology and nanomaterials as an emerging area toward heavy metal remediation."

Inorganics

A further extension of adsorbent materials is the use of fully inorganic materials. Again, these studies have been reviewed [147, 150] but a few new developments should be highlighted. Hajji et al. [165] studied the adsorption of chromate onto siderite, ferrihydrite and goethite. The focus of this study was not on the optimization of adsorption capacities or the adsorption rates. Rather, the authors performed an in-depth analysis of the (competitive) adsorption of Cr(VI) and related species. The data provided clear evidence that the material (adsorbent) architecture is a key parameter to control for adjusting not only adsorption but also desorption behavior.

Similarly, Wang et al. [166] studied the role of calcium polysulfide (CPS) in CPS/iron oxide composite materials (Fig. 11.8). Again, the study focused on the role

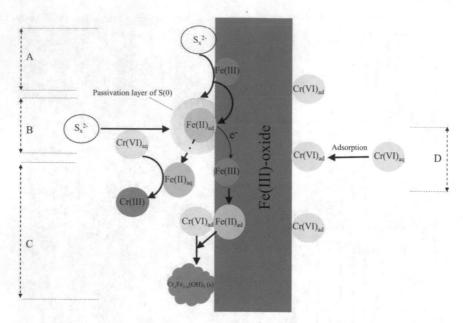

Fig. 11.8 Proposed mechanism of iron oxide-promoted Cr(VI) sequestration using CPS under acidic conditions. Reprinted from ref. [166] with permission

of the iron species and the polysulfide in these adsorbents and the authors proposed that an electron shuttling processes between all constituents were responsible for effective Cr(VI) adsorption along with the reduction of Cr(VI) to Cr(III). The sulfur species and the redox couple Fe(II)/Fe(III) were assigned key roles in this process. This is an interesting and detailed account of how the combination of adsorption and redox chemistry could provide new approaches to water treatment.

Wang et al. [167] reported on the use of sulfidated zerovalent iron for Cr(VI) and Sb(III) removal. Similar to the previous study [166], these authors reported a clear influence of the sulfur present in the adsorbents on the performances. Especially, the Fe/S ratio controlled the efficiency of Cr(VI) and Sb(III) removal and the presence of Sb(III) promoted Cr(VI) removal, while the inverse effect was not observed.

Whitaker et al. [168] explored the use of biogenic iron oxides (BIOS) for Cr(VI) uptake. The rationale behind this approach is that BIOS are nanocrystalline, have a high surface area and are often partly porous. Moreover, they contain many defects in their crystal structure to accommodate foreign ions like Cr(VI) and they often also contain organic components that may further enhance the adsorption capacities and/or the adsorption rates. Similar to ref. [166], Whitaker et al. [168] did not focus on improving adsorption behavior. Rather, they focused on understanding the mechanisms and single processes that happened during adsorption and reduction of Cr(VI) to Cr(III). They again proposed an electron transfer from Fe(II) to Cr(VI), similar to ref. [166], but through a less complicated pathway.

Suzuki et al. [169] followed similar ideas, but using fungal Mn oxides (BMOS) rather than BIOS. In contrast to all studies discussed so far, this work focused on Cr(III) and the authors showed that BMOS are effective oxidizers for Cr(III) to Cr(VI) under aerobic conditions. Moreover, BMOS were effective adsorbents for Cr(III), reaching a maximum of ca. 120% of Cr(III) with respect to the molar Mn content. While the oxidation and release of Cr(VI) is an issue, the very high Cr binding capacity of these materials is clearly an asset.

Thomas et al. [170] studied a series of green rusts for Cr(VI) remediation. The particular focus of their study was the investigation of cation substitution within the green rusts and the effect on Cr uptake. The most obvious effect was that Mg(II) effectively stabilized the green rust and thus provided a solution to a common problem with green rust, namely the low stability if exposed to environmental conditions. The authors suggested that Mg-stabilized green rusts could be a viable adsorbent for Cr(VI) remediation and could also be sinks for bivalent metal ions such as Ni(II) or Cu(II).

Gels and Resins

While there are numerous organic adsorbents [147] including algae [148], gels and resins have attracted attention as well. To just highlight a few recent approaches, Zhang et al. [171] used Fe-cross-linked alginate hydrogels to perform a synergistic redox reaction between Cr(VI) and As(III), resulting in Cr(III) and As(V), both of which can then be removed to over 80% from aqueous solutions using the as-prepared hydrogels. The key responsible functional component is Fe(II), which formed upon irradiation and enabled the reduction of Cr(VI) to Cr(III), to start the reaction cascade.

Kou et al. [172] used recombinant proteins to produce protein hydrogels containing specific metal or oxyanion binding domains (*e.g.*, to bind chromate). While the synthesis is rather complex, the power of the approach lies in the fact that protein-binding motifs can be combined to even address complex situations with water containing numerous heavy metal species at the same time.

Xie et al. [173] used hyper-cross-linked polymeric ionic liquids to remove Cr(VI) in highly competitive conditions, with Cl^-, NO_3^-, $H_2PO_4^-$ and SO_4^{2-} present in the solution. The cross-linked resin can be recycled and—exploiting an advantage that is intrinsic to all ionic liquids—can be adjusted to account for changes in the application by exchange of ions.

Finally, Hesemann and coworkers have used ionosilicas to remove chromate from aqueous solution [174]. The high number of positive charges in these materials provided solid adsorbents with high capacities of up to 2.6 mmol/g.

11.4.3 Pertechnetate as an Example of a Radioactive Oxyanion

Technetium presents many oxidation levels in aqueous solution, from $+VII$ to $-I$ [175]. The most stable oxidation state in aerated solution is $+VII$, in the form of the oxyanion TcO_4^-. Correspondingly, the redox chemistry of this element is very complex, and several disproportionation reactions result in the coexistence of several oxidation states in equilibrium in solution [175]. The chemical forms are very different with coordination numbers between 4 and 9, and Tc species come in the form of oxyanion, cation, oxocation and neutral species with low solubility, such as TcO_2, or volatile as Tc_2O_7 or the salt of $CsTcO_4$ [176]. While this chapter focuses on the oxyanion TcO_4^-, it is important to note that several reservoirs at different oxidation states are likely to exist in the environment. Technetium is a radioelement, meaning that all isotopes (from 90 to 110) are radioactive [177]; therefore, the toxicity is both chemical (that of a heavy metal) and radiochemical (radiotoxicity). Technetium is a natural, non-primordial element formed by spontaneous fission of the uranium isotope 238, and the quantities in the earth's crust are very small, below $\ll 10^{-15}$ mol.L^{-1} [178]. As a result, most of the Tc on earth is anthropological. It was first produced ($^{95m}Tc_{43}$ and $^{97m}Tc_{43}$) by man using a cyclotron at the University of Berkeley by Lawrence, following the irradiation of a natural molybdenum target by deuterons, according to the nuclear reaction:

$$^{A}Mo_{42}(d, n) \rightarrow {}^{A+1}Tc_{43}$$

Synthesis by artificial means led to its name, which means "artificial" in Greek. It has been isolated and characterized, first by Perrier and Segré in 1937 [175]. The use of techniques of trace chemistry developed in radiochemistry and the analogy with the other elements of group VII (with Mn and Re) of Mendeleïev's periodic classification made it possible to initially describe its chemical properties [177]. Only three isotopes have a radioactive decay period longer than 61 days. These are the isotopes 97, 98 and 99 with a half-life greater than 10^5 y ($97 - T_{1/2} = 2.6 \ 10^6$ y, $98 - T_{1/2} = 4.2 \ 10^6$ y and $99 - T_{1/2} = 2.13 \ 10^5$ y) [175]. As these isotopes are radioactive, they must be handled in facilities dedicated to the handling of radioactive substances [179]. The one present in the largest quantity is the isotope ^{99}Tc, which is a soft beta emitter ($E_{max} = 293.5$ keV and $E_{avg} = 84.6$ keV), with no associated gamma radiation emission. The specific activity is $6.36 \ 10^8$ Bq.g^{-1}. It is mainly produced by the fission reaction of fissile isotopes such as ^{235}U or ^{239}Pu. The quantity depends on the nuclear reaction, the fuel (^{235}U, ^{238}U, ^{239}Pu, ...) and the burnup. For example, there is about 1 kg of Tc per 1 ton of uranium in spent nuclear fuel [180]. The best-known isotope, especially for medical applications, is the metastable ^{99m}Tc isotope (radiopharmaceuticals isotope). It is mainly produced by irradiation of ^{99}Mo by neutrons according to the following nuclear reactions:

$$^{98}Mo_{42}(n, \gamma)^{99}Mo^{\beta-} \rightarrow {}^{99m}Tc^{\gamma-} \rightarrow {}^{99}Tc_{43}$$

The half-life of ^{99}Mo is $T_{1/2} = 2.75$ d. The transition from ^{99m}Tc to ^{99}Tc leads to the emission of gamma radiation of the order of 140 keV with a very short half-life of $T_{1/2} = 6.02$ h [175].

Technetium is present in the environment mainly due to human activities [181]. There are two particular areas: nuclear testing and preparation of plutonium for these weapons, and nuclear reactor accidents. Concerning the atmospheric testing of nuclear weapons, it should be noted that the fission of the ^{235}U and ^{239}Pu nuclei produces ^{99}Tc, in a volatile form of Tc(VII) (Tc$_2$O$_7$) at high temperature. Hydrolysis in the atmosphere leads to the formation of the anion TcO$_4^-$. Concerning the production of plutonium for military applications, the joint production of Tc and its subsequent separation from other fission products leads to the accumulation of a very large amount of Tc, a fraction of which is released into the environment from the fission product storage vessels. The release of ^{99}Tc during the accident at the Chernobyl (Ukraine) reactor in 1986 and Fukushima (Japan) reactor in 2011 also led to the release of the TcO$_4^-$ anions into the atmosphere from volatile Tc(VII) species (Tc$_2$O$_7$ and CsTcO$_4$, mostly). The amount of artificial Tc present in the environment has been estimated to be in the order of 2 tons ($1.3 \ 10^3$ TBq), and the chemical form in aqueous media under aerobic conditions is indeed Tc(VII) in the form of the TcO$_4^-$ anion [178].

It is in the spent fuel cycle of nuclear power that the quantity of technetium (^{99}Tc) accumulated is the highest [182]. During the reprocessing, the release into the environment is controlled. Depending on whether the spent fuel cycle is closed (as in France, Japan, etc.) or opened (as in the USA, etc.), the Tc is present either in a dedicated containment matrix or in the fuel. The chemical form is multiple, ranging from zero oxidation state (Tc metal) to a pertechnetate salt, *via* TcO$_2$. Regardless of the conditioning (containment matrix or spent fuel), this nuclear waste will be stored in suitable geological disposal sites. However, it should be noted that the transmutation of ^{99}Tc by nuclear reactions (neutron) leads to stable Rh and Pd nuclides. Concerning the use of the ^{99m}Tc isotope in medical applications, the quantity involved is very small but will most likely increase with the development of nuclear medicine in the world [183]. For medical applications, the chemical form is also extremely variable. The most common forms are the TcO$_4^-$ anion and complexes in solution of the oxidation state +V and +III, like TcO^{3+} and Tc^{3+}. The speciation of the Tc element in environmental aqueous media generally depends on aerobic or non-aerobic conditions. In an aerobic environment, Tc will mainly be in the oxidation state +VII, in the anionic form TcO$_4^-$, whereas in an anaerobic environment, Tc can exist in different oxidation states: the oxidized form Tc(VII), therefore TcO$_4^-$, and the reduced form Tc(IV), TcO$_2$ [184, 185]. This will also be linked to the presence of redox agents in the environment and the presence of complexing agents such as carbonate anions, humic acids or chloride ions depending on the geological site [175, 176]. Given the low concentration of Tc, radioactivity (the specific activity of ^{99}Tc is 6.36 Bq.g^{-1}) is generally used as a unit of mass (Bq.g^{-1}) and volume (Bq. L^{-1}) concentration.

The pertechnetate anion in aqueous solution, often after a redox reaction, has a strong capacity to sorb onto mineral surfaces (rocks, sediment and soils) and to accumulate in plants and microorganisms [178]. These natural compartments therefore constitute an important reservoir of the element Tc in the environment, and it is important to consider them for the remediation of aquatic environments. The concentration of ^{99}Tc in aquatic environments is highly variable (several orders of magnitude) and greatly depends on the location of the sampling: near or far from a source of Tc. It also depends on the date of the measurement: before or after nuclear test shutdowns, accidents at the Chernobyl and Fukushima reactors or the reduction of Tc release in spent nuclear fuel processing operations. The orders of magnitude range from 0.02 Bq.L^{-1} to 20 Bq.L^{-1} (close to Monaco in 1986, May) [186] in rainwater, and from 10^{-5} Bq.L^{-1} to 10^{-1} Bq.L^{-1} in seas and oceans [181]. These concentrations are well below 10^{-10} mol.L^{-1}, which is extremely low on the chemical scale. The concepts of radiochemistry (trace scale chemistry) are generally used to describe the speciation of ions in solution [187]. Even though these concentrations are very low, the radiotoxicity of ^{99}Tc is very high and the long-term risk associated with humans is very significant.

Very few studies concern the remediation of natural waters (groundwater, rainwater, lakes, rivers and seas and oceans) by technetium. The vast majority of the work concerns the treatment of industrial solutions resulting from the production of plutonium for military applications [188] and to a lesser extent the management of waste from medical applications [89, 189]. However, it should be noted that there is vast literature concerning either the separation of technetium in PUREX separation processes (separation of uranium and plutonium from fission products such as ^{99}Tc) or the remediation of storage tanks for fission products from the preparation of "military plutonium." In these processes, liquid–liquid extraction and separation on ion exchange resin are often used. In the case of separation by liquid–liquid extraction, ammonium extracts or molecular or supramolecular complexing agents are commonly used [190].

While liquid–liquid extraction techniques are very well suited for the separation of constituents in macro-concentration, this technique is poorly suited for species that are very diluted in the environment. Moreover, in the case of radioactive pollutants, the subsequent management of organic solvent or ion exchange resin contaminated with a radioactive substance is an additional constraint. This work will not be further developed here, but it was routinely presented at the annual specific Tc and Re chemistry conference (ISTR—International Symposium on Technetium and Rhenium—Science and Utilization). These more conventional approaches were not described in this chapter, but many publications are available on the subject [16, 188–194]. However, recent developments concerning the use of bi-functionalized anion exchange resins for the recovery of TcO$_4^-$ from groundwater should be noted [195].

Furthermore, in the case of the separation of a radioactive element, it is necessary to consider the effect of ionizing radiation on the separation system. For example, mineral supports are often preferred over organic supports, due to the radiolytic degradation of the more fragile organic bonds [196]. Finally, it should be considered that selectivity is extremely important for trace elements because other contaminants in the environment, which are not always radioactive, may be present at much higher concentrations (10^3 to 10^6 times higher). Moreover, it is also important to consider the management of the radioactive waste after separation. In the most recent work, the separative system developed is also intended to be transformed into a containment matrix [88, 197, 198]. In addition, it should be noted that the constraints linked to the handling of a radioelement such as Tc often lead to the use of stable chemical analogues such as ReO_4^- and ClO_4^- or the ^{99m}Tc isotope in the ultra-trace state, which is easy to detect thanks to its gamma emission.

The most recent scientific approaches developed for the separation of Tc are of two types: separation by ion exchange on metal–organic framework (MOF) supports and adsorption on active surfaces of nanoparticles. Other systems, such as organic adsorbents, layered double hydroxide or dendrimer systems will not be discussed here [16, 199].

Recent works showed the practical interest and promise of MOF materials for the treatment of solutions contaminated with radioactive anions such as $^{129}I^-$, $^{79}SeO_3^{2-}$, $^{79}SeO_4^{2-}$ and $^{99}TcO_4^-$ [190, 200]. These cationic MOFs are built from neutral soft donor ligand and transition metal. The few main criteria that guide the synthesis of MOFs, in order to amplify the interaction with the TcO_4^- anion, include: the exchange anion used for the construction of the MOF must be compatible with the medium and must constitute pores/channels suitable for anion exchange (ClO_4^-, Cl^-, NO_3^- and SO_4^{2-}); the hydrophobic nature of the surface of the material and of the pores/channels is an asset for selectivity; and the possibility of creating a network of hydrogen bonds.

The choice of transition metal is also important. For the capture of the TcO_4^- anion, two MOF structures have been specifically studied: SCU-100 and cationic Zr-based MOF [200, 201]. Recently, an eightfold interpenetrated three-dimensional cationic MOF material SCU-100 has shown high immobilization capabilities for TcO_4^- anion with fast kinetics. Sorption isotherms were measured at room temperature with TcO_4^- or ReO_4^-. The good selectivity towards other anions, which are present in higher concentrations than Tc, and the radiostability of the structure are further attractive factors in these new materials. The proposed mechanism is based on anion exchange and strong interaction between the Ag^+ sites constituting MOF SCU-100 with the tetradentate neutral nitrogen-donor ligand. The single-crystal X-ray diffraction on the ReO_4^--based structure highlights the coordination between the two metals, Re and Ag (Re–O–Ag), and the existence of several hydrogen bonds.

Hesemann and coworkers have developed a strategy using a platform made from mesoporous silica [202], to covalently bind reactive chemical functions. These properties have been exploited by using an imidazolium unit and different alkyl chain lengths between the mesoporous network and the imidazolium cation, and also the one fixed on the second nitrogen of the imidazolium. These ion exchange material showed a very good extraction capacity for the TcO_4^- anion and especially an excellent selectivity against other anions (Fig. 11.9). These results were explained by the hydrophobicity of both the mesoporous silica support and the presence of alkyl chains. The alkyl chains also allowed the structuring of the mesoporous silica/water interface by a network of hydrogen bonds.

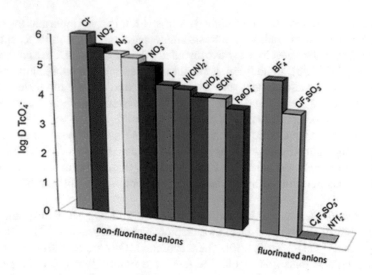

Fig. 11.9 Distribution coefficients of TcO_4^- in the presence of different competing ions ($c = 0.1 \, mol \, L^{-1}$). Reprinted from [202] with permission

Shen et al. [203] studied technetium removal from water. The separation is based on the sorption of the anion ReO_4^- (as TcO_4^- surrogate) on an ordered urea-functionalized mesoporous polymeric nanoparticle (urea-MPN). The nanosized spherical morphology and 3D interconnected ordered cubic mesopore structure were obtained by the surfactant directed self-assembly process with the low-concentration urea–phenol–formaldehyde resol oligomers under hydrothermal conditions. The sorption capacity of this material was good, with fast kinetics (less than 30 min). Analysis by NMR and XPS spectroscopies showed that the mechanism involved the effective hydrogen bond provided by the urea functional groups bonding the tetra-hedral ReO_4^- anion. In addition, the nanometric dimensions of the channels and the large specific surface favored the diffusion of the anion. Finally, the hydrophobic pore surface of the urea-MPN was also beneficial for decreasing the ReO_4^- charge. The stability of the ordered mesoporous structure is maintained after an irradiation with 200 kGy dose radiation.

In a recent study [204], the recovery of the ReO_4^- anion (used as a simulant of TcO_4^-) from nanoparticle-anchored biochar composite (ZBC) ZnO was observed. In this study, the selectivity toward competitive anions (such as I^-, NO_2^-, NO_3^-, SO_4^{2-}, PO_4^{3-}) and the stability toward gamma radiolysis were evaluated and found to be highly favorable. It has been shown that the high sorption capacity of the ReO_4^- anion was due to the synergistic effect of the hydrophobicity (superhydrophobicity) of the surface and the nature of the sorbent (and in particular according to the different crystal planes of the ZnO nanoparticles). The electron transfer between ReO_4^- and the (002) surface of ZnO appears to be greater than with the (100) plane, leading to a stronger interaction for the (002) plane.

11.4.4 Diclofenac as an Example of an Organic Anion

Due to enormous anthropogenic activities (pharmaceutical industry and agriculture), pollution is not limited on purely inorganic species, but organic ionic species also have a strong ecological impact on various natural compartments. The incorporation of ionic groups is a common strategy to modify the physical properties of organic compounds; consequently, numerous organic compounds exist as ionic species. This is due to the fact that water solubility is often required for an efficient use of organic compounds as pesticide, pharmaceutical drug or dye. Indeed, various types of organic anions such as anionic pharmaceuticals and anionic dyes are extensively used in different industries and some of them are produced on a large ton-scale [8, 205, 206].

These organic anions display interesting properties, such as biological activity and optical properties, which renders them interesting for various applications in medicine, agriculture, catalysis, sensors, solar cells, printing, textiles, bioimaging and energetic materials. Various pharmaceutically active compounds (PhACs) and, more generally, pharmaceutical and personal care products (PPCPs) [207] are ionic organic compounds. The presence of organic anions in water bodies is therefore closely related to anthropogenic activities, and these compounds are widely spread all over the planet. The high water solubility of these compounds, which is often a key property for the use as pharmaceuticals or pesticides, further increases their mobility in soil and water bodies. The water solubility therefore favors rapid distribution of these compounds and hampers their elimination *via* water treatment processes.

Synthetic anionic organic species have been present in the environment since decades, but recently, significant progress in analytical methods (such as GC–MS and LC–MS) allowed for a detailed monitoring. This is particularly true for water bodies, where they exist at concentrations down to the ng/L scale. Anionic organic compounds are therefore ubiquitous and particularly abundant in surface, ground and drinking water. They can also be found in other environmental compartments (soils), and due to the water treatment and upgrading processes, also in sewage sludge and even the effluents of WWTPs. The presence of these compounds in the environment can cause direct and indirect harm to humans, animals and plants. Besides the purely inorganic oxyanions, such as chromate and nitrate, organic anionic species therefore have a tremendous impact on water management aspects. Their monitoring, separation and removal from diverse water bodies and particularly from wastewater are therefore significant challenges and of essential concern to human health.

Pharmaceuticals are an important class of emerging contaminants [208] because of their inherent biological activity and the potential to enter the aquatic environment. One compound in particular, (2-[2,6-dichlorophenylamino]phenylethanoic acid sodium salt) (diclofenac, DCF), is among the most prominent examples of such a species. Even though DCF is a classical organic anion, we chose to discuss its dissemination and monitoring in this chapter because it is among the most studied and most often detected pharmaceuticals worldwide [209], and huge data indicating DCF pollution are available. DCF is an excellent model compound illustrating the challenges related to the management of anionic organic species and their metabolites in water treatment and upgrading. It represents a large panel of organic carboxylates, sulfonates, phosphonates, etc., which are widely used as water-soluble drugs, dyes and pesticides and present in nearly all natural compartments.

Due to its intensive use, in both human medicine and veterinary applications, DCF is detected in nearly all water bodies (wastewater, surface water, effluents). It was included in the EU watch list of pharmaceuticals requiring detailed environmental monitoring in 2014 (Directive 2014/80 of the EU) [92] but has recently been removed from the list, as sufficiently monitoring data are now available. Today, DCF is omnipresent in nearly all-natural compartments (aquatic environment and soil). This part gives a short overview on the use of DCF, its fate in different natural compartments and the consequences and repercussions of its presence in these ecologic systems (Fig. 11.10). The fate, effects and risks related to the presence of DCF in the environment have recently been reviewed by Sathishkumar et al. [9].

DCF is a nonsteroidal anti-inflammatory drug (NSAID), similar to other widely used pharmaceuticals, such as ibuprofen, naxopren, ketoprofene and aspirin. NSAIDs are "over-the-counter" drugs in most countries; thus, they are easily accessible. Nowadays, DCF is one of the most frequently administered agents for pain relief and inflammatory diseases in humans and animals. DCF was introduced in the 1960s by Ciba-Geigy [210], but it has been commercialized under different brand names (Athrifen, Diclodoc, Diclofenbeta, Voltaren and many others). It is usually commercialized as sodium or potassium salt and applied *via* oral or topical administration. Some physical properties of diclofenac (free acid) and diclofenac sodium salt are summarized in Table 11.2.

Although DCF is widely used, there is a lack of reliable data about its annual production and consumption. The estimated global annual consumption was ca. 940 tons in the early 2000s and approximately 1500 tons ten years later [213]. However, these values only concern human health applications and neglected veterinary use. This difference is important due to intensive veterinary use and particularly to the

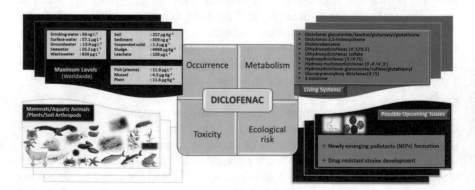

Fig. 11.10 Diclofenac: metabolism, occurrence in the natural environment and ecological risk on human and animal health. Reprinted with permission from ref. [9]

Table 11.2 Comparison of some physical and chemical properties of diclofenac and diclofenac sodium

	Diclofenac-*free acid*	Diclofenac *sodium*
Structure		
Formula	$C_{14}H_{11}Cl_2NO_2$	$C_{14}H_{10}Cl_2NNaO_2$
Molecular weight	296.15 g/mol	318.13 g/mol
CAS number	15307-86-5	15307-79-6
Water solubility	2.43 mg/L [211]	20.4 g/L [212]
Melting point	158–159 °C	283–285 °C

different fates of drugs used for human and animal diseases. Whereas drugs used for human health applications mostly enter the water treatment circuit *via* WWTPs, drugs originating from veterinary uses mostly move directly into the aquatic ecosystem, *i.e.*, in surface water, without further treatment.

As DCF is a low-cost over-the-counter drug, it is extensively used in both emerging and developed countries. This is highlighted on the one side by the estimated data for Brazil and the Indian subcontinent, with annual consumptions of more than 60 tons in each country and estimated consumption in Germany (82 t in 1999), France (16 t in 2003) and England (26 tons in 2000) [213]. Unfortunately, no precise recent data exist. In spite of this, the available data clearly indicate that DCF is a widely used and globally present drug. It is therefore a valuable model compound to monitor the spread and fate, and to evaluate the effects of the presence of anionic organic species and in particular of pharmaceuticals in the ecosystem.

The presence of DCF in the environment sometimes has serious consequences. First studies, indicating deep impact of DCF on the environment, go back to the early 2000s. DCF was held responsible for a sharp decline in the population of three vulture species in the Indian subcontinent, due to the consumption of carcasses of cattle treated with DCF. The intolerance of these birds toward DCF led to renal failure and finally to their death [98]. A similar scenario led to the breakdown of other vulture and eagle populations in Kenya, ten years later [214]. To prevent further decline in vulture population, the use of DCF for veterinary use has been banned in India, Nepal and Pakistan since 2006.

Detailed data on the human, animal and plant metabolism of DCF (Fig. 11.11) revealed the presence of its metabolites in various environmental compartments [8, 9, 17].

The human metabolism of DCF produces a variety of compounds, mainly hydroxylated DCF derivatives, and the half-life of DCF is 2 h [216]. After this relatively

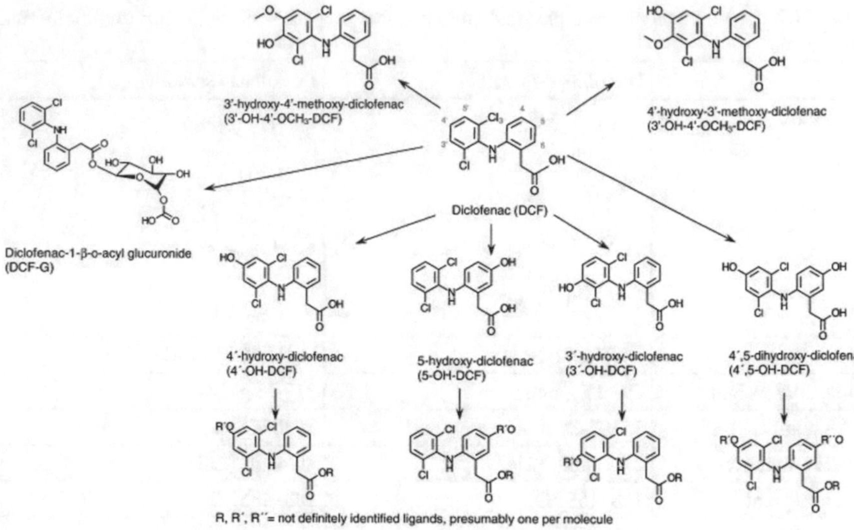

Fig. 11.11 DCF and its main metabolites. Reprinted from ref. [215] with permission

short period, it is excreted, either as such or as one of its metabolites, *via* urine (2/3) or feces (1/3) [17]. DCF is mainly metabolized *via* hydroxylation, with 4′-hydroxo (OH) diclofenac as major product [215], either as pure compounds or their conjugated forms with glucuronic acid, glucuronide, taurine or sulfate. Usually, DCF is nearly completely metabolized, and only a minor part is excreted as the un-metabolized compound. Approximately 10% of the administered dose is eliminated as taurine, glucuronide or sulfate conjugates, and more than 80% as DCF derivatives, mostly hydroxylated species and their conjugates (Fig. 11.12). Hence, the monitoring of DCF should not be limited to the pure drug, but must be extended to its metabolites.

After excretion, DCF enters the aquatic environment, either directly or after its transit *via* WWTPs. The various entries of DCF and its metabolites are illustrated in Fig. 11.13. DCF is initially introduced in surface water, but can later be found in groundwater and even in drinking and mineral water, due to inefficient elimination during wastewater treatment. Due to the continuous influx of DCF and its metabolites into the natural environment, it is considered a pseudo-persistent pollutant. Despite its relatively short half-life time in freshwater of 8 days [217], the pseudo-persistent nature of DCF makes in an important case study for organic anionic contaminants.

We focus here on the monitoring of DCF in water bodies originating from France and Germany as representative examples of developed countries. The presence of DCF and its metabolites has been observed since the beginning of the 2000s. Selected data of DCF concentrations in surface, ground and wastewater together with some data from seawater, are given in Table 11.3. The first report on the presence of DCF in surface water appeared as early as 2002, when Heberer et al. reported a DCF concentration of approximately 1 μg/L in surface water in Berlin/Germany [218]. Since then, enormous amounts of data attested to the omnipresence of DCF and

Fig. 11.12 Identified metabolites of DCF and their percentages of oral dosage. Reprinted from ref. [17] with permission

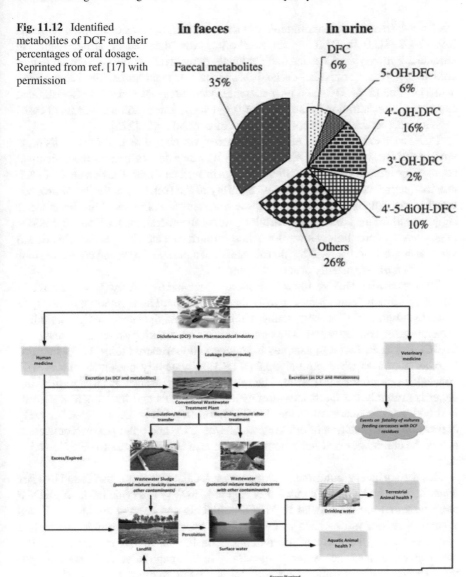

In faeces

In urine

Faecal metabolites
35%

DFC
6%

5-OH-DFC
6%

4'-OH-DFC
16%

3'-OH-DFC
2%

4'-5-diOH-DFC
10%

Others
26%

Fig. 11.13 Entry routes of DCF into the environment. Reprinted from ref. [8] with permission

its metabolites in almost all water bodies worldwide [9]. Remarkably, DCF was even found in surface water in very isolated regions, such as the northern Antarctic Peninsula [219].

DCF residues could be found in nearly all types of surface water such as canals, lakes and rivers. The presence of DCF is clearly related to dense population or industrial or agricultural activities. It originates from pharmaceutical production sites, wastewater treatment plant effluents of urban and tourist areas or hospital effluents [59, 220–223]. Both DCF and its metabolites were detected in surface water, in particular hydroxylated species and conjugate derivatives, resulting from human or animal excretion. Some other derivatives such as nitro and nitroso derivatives were also identified [224]. DCF and its metabolites were present in nearly all investigated samples in concentration in the range of 0.1–1 µg/L, but in some exceptional cases even in considerably higher concentrations of over 50 µg/L [225].

DCF has also been identified in groundwater, which is directly linked to its presence in surface water [226]. DCF in groundwater therefore originates from identical sources (agricultural, pharmaceutical, etc.) as in surface water. The presence of DCF residues in groundwater highlights its mobility in the soil. The pollution of groundwater is particularly problematic, as these resources are often used for direct water supplies and for the production of drinking water. In general, the DCF concentrations in groundwater are slightly lower than those in surface water. This is probably due to a partial degradation of DCF by microorganisms in the soil. However, concentrations of 10–500 ng/L are usually observed (Table 11.3).

The presence of DCF residues in drinking water and especially tap water directly relates to human health issues, and is therefore of particular importance. Despite water treatment, DCF has been found in different drinking water samples, including mineral water and tap water. The contamination of drinking water is widespread, as DCF has been found in samples in France and Germany (Table 11.3), but also in other countries (Sweden and Spain). DCF is generally present in traces with concentrations on the ng/L scale. The reason for the presence of DCF in drinking water is probably the inefficient elimination in WWTPs and inadequate treatment in drinking water treatment plants (DWTPs), together with other reasons. It clearly appears that the pollution of drinking water with DCF and other pharmaceuticals is one of the challenges in water treatment in the near future and has to be followed in detail.

Besides surface, ground and drinking water, DCF residues are also found in other water bodies. As DCF is not efficiently degraded in WWTPs (vide infra), high DCF concentrations are also found in WWTP effluents and sewage sludge. DCF and several of its derivatives such as hydroxylated species or glucuronide and sulfate conjugates could be detected, originating from human or animal excretion. The concentrations in surface water are generally in the range of several hundred ng/L up to several µg/L, in some exceptional cases even up to more than 50 µg/L.

The presence of DCF in the marine environment has recently been investigated [97]. Despite the relatively short half-life time of DCF in freshwater of approximately 8 days [217], DCF could also be found in various marine environments due to its continuous input from surface water (Fig. 11.14). DCF is therefore particularly present in coastal waters and sediments although in lower concentration. However, the presence of traces of DCF has a serious impact on marine organisms, due to the bioaccumulation of DCF. Mussels (*Mytilus galloprovincialis* and *Mytilus trossulus*) are particularly sensitive; this is due to their high filtration capacity and their ability to bioaccumulate pollutants [227]. The accumulation of DCF by mussels then generates

Table 11.3 Selected data of DCF occurrence in various water bodies in France and Germany

Entry	Country	Sample nature	DCF conc.	Sampling point	References
1	France	Surface water	300 ng/L	Doubs River	[228]
2	France	Surface water	166 ng/L	Loue River	[228]
3	France	Surface water	33,2 ng/L	Lergue River	[229]
4	Germany	Surface water	435 ng/L	Lake Tegel	[230]
5	Germany	Surface water	245 ng/L	Saale River	[231]
6	Germany	Surface water	15.0 µg/L	Erft River (near Cologne)	[232]
7	Germany	Surface water	2.1 µg/L	Teltow Canal (Berlin)	[233]
8	France	Groundwater	2.5 ng/L	Wells of Hérault Basin	[229]
9	France	Groundwater	9.7 ng/L	Rhone-Alpes region	[234]
10	Germany	Groundwater	590 ng/L	Monitoring wells	[235]
11	Germany	Groundwater	45 ng/L	Monitoring wells	[230]
12	France	Drinking water	56 ng/L	Rhone-Alpes water supply	[234]
13	France	Drinking water	2.5 ng/L	Hérault watershed	[236]
14	France	Wastewater (eff)	486 ng/L	WWTP	[229]
15	France	Wastewater (eff)	2.5 µg/L	WWTP	[228]
16	Germany	Wastewater (inf)	6.3 µg/L	WWTP in Berlin	[233]
17	Germany	Wastewater (eff)	3.0 µg/L	WWTP in Berlin	[233]
18	Germany	Wastewater (eff)	5.1 µg/L	WWTP	[237]
19	Germany	Primary sludge	7020 µg/kg	WWTP	[238]
20	Germany	Secondary sludge	310 µg/kg	WWTP	[238]
21	France	Seawater	7.7–63.4 ng/L	Seine Estuary	[239]
22	France	Seawater	1.5 µg/L	Mediterranean Sea	[236]
23	Germany	Seawater	0–9.2 ng/L	Baltic Sea	[240]
24	Germany	Seawater	0 - 6.2 ng/L	North Sea	[241]

oxidative stress and genotoxic effects [97]. DCF therefore seems to be responsible for various effects on marine organisms, including the alteration of biological functions.

The management of pharmaceuticals, such as DCF in wastewater, is particularly challenging as these pollutants are often only partially eliminated in WWTPs, and studies on the fate in WWTPs are often inconclusive or contradictory [215]. In fact, DCF elimination rates have been found in the range of 0–80%, with most values in the 20–40% interval. The reasons for these low elimination rates are poor biodegradability of DCF on one side and the low DCF adsorption ability of sewage sludge on the other side. Although high concentrations of DCF can be found adsorbed on sludge (Table 11.3, entries 19/20), a considerable fraction is still present in the

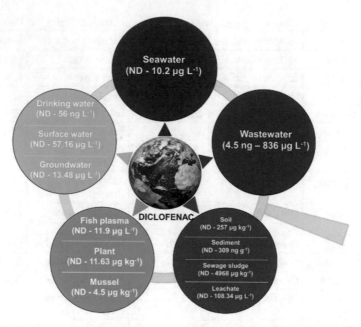

Fig. 11.14 Global occurrence of diclofenac range in various environmental matrices. ND: not detected. Reprinted from ref. [9] with permission

effluents of WWTPs. Both factors result in the presence of considerable amounts of DCF in WWTP effluents. In this way, DCF continuously enters the aquatic system *via* the surface water input (Table 11.3, entries 16/17), thus resulting in a permanent contamination of surface and groundwater.

These results show that innovative water treatment techniques have to be developed in order to reduce the presence of pharmaceuticals in the effluents of WWTPs. These techniques can be, for example, membrane bioreactor technologies. Additionally, the process parameters for wastewater treatment can be adapted in view of more efficient DCF mineralization. As an example, longer solid retention times (SRT) and longer hydraulic retention times (HRT) result in higher DCF elimination rates, but are hardly compatible with standard treatment processes applied in WWTPs. Biodegradation can also be enhanced using microorganisms capable to degrade DCF. These results indicate that the management of pharmaceuticals in the aquatic environment is a complex issue and further research is needed to find valuable strategies allowing for an efficient and complete transformation and/or degradation of these compounds in WWTPs and DWTPs [215]. DCF is neither efficiently mineralized nor efficiently adsorbed on sewage sludge. However, regarding this latter aspect, pH is a crucial factor. Elimination of DCF on sludge is strongly enhanced in slightly acidic media, compared with the neutral media. The adsorption of DCF in its protonated neutral form seems to be favored, when compared with the separation of the anionic carboxylate species [238, 242].

The omnipresence of DCF represents a high environmental risk, particularly due to its interactive effects with other types of environmental pollutants such as pharmaceuticals, pesticides, personal health products and metallic contaminants. The occurrence of all these pollutants, together with DCF in particular, in wastewater may give rise to synergistic effects affecting the whole aquatic life [8, 243]. The monitoring of DCF, together with other emerging contaminants, will therefore become a challenge in water treatment in the near future.

Effective adsorbents for DCF removal

DCF is an omnipresent compound; thus, efficient strategies for its removal are required. Various novel strategies from wastewater have recently been investigated, which include physical processes (adsorption and coagulation flocculation), biological (i.e., microbial or enzymatic) degradation and biotransformation, and chemical strategies, such as electrochemical degradation and photolytic/photocatalytic degradation [8, 215]. As an example, DCF can efficiently be degraded photolytically [244] *via* irradiation with UV light. The photodegradation can be assisted by the presence of photocatalytic materials, such as titanium dioxide [245, 246] or TiO_2-WO_3 mixed oxides [247].

Photodegradation is often combined with advanced oxidation processes (AOP), to transform organic pollutants into biodegradable intermediates. In this context, ozonation, using strong oxidizing agents such as H_2O_2, appeared as a suitable strategy for the mineralization [248]. However, both the degradations of DCF *via* photolytic or oxidation pathways result in the formation of a large panel of organic degradation products [249] whose ecotoxicity needs to be assessed as well (Fig. 11.15). On the other side, DCF can be separated from wastewater *via* solid–liquid extraction processes, using a large variety of adsorbents. Sequestration is an important method for the removal of DCF from water bodies, as other methods either are inefficient (biological degradation, adsorption on sewage sludge) or result in the formation of degradation products (photodegradation and oxidation processes).

A multitude of adsorbents that have been studied for DCF adsorption include metal–organic frameworks [250], activated carbons [251], graphene oxide [252], biomass [253] and biopolymers [254], clays and double-layered hydroxides [255, 256] and silicas [257–260]. Due to the chemical characteristics of DCF, its adsorption can be triggered by hydrophobic interactions, π–π stacking interaction or ionic interactions. Anion exchange materials seem to be particularly well suited for the adsorption [261].

Clays only show limited adsorption capacity for DCF but the affinity can be significantly increased *via* the incorporation of organic compounds. Various clays, such as bentonite and halloysite, have been tested as adsorbents for DCF. The intercalation of organic compounds, such as CTAB, into the interlayer space of the clay structures allows the synthesis of organoclay nanocomposites, which led to highly efficient DCF adsorbing materials [262, 263]. A similar approach was reported by Krajisnik, who described cetylpyridinium-modified natural zeolites for DCF adsorption [264]. In all cases, hydrophobic interactions, combined with increasing surface

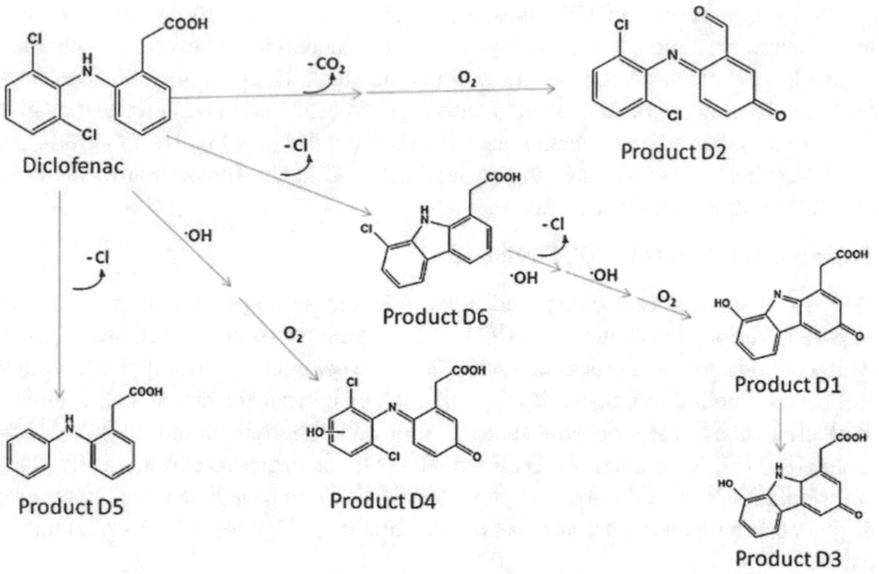

Fig. 11.15 Photodegradation products of DCF. Reprinted from ref. [244] with permission

charge, are the reasons for the improved adsorption properties of these organic–inorganic nanocomposites. The incorporation of surfactants within the interlayer spaces of clays or zeolites is crucial for the adsorption of organic pollutants and allows for a dramatic increase of the adsorption efficiency.

Due to their morphological and textural properties, metal–organic frameworks (MOFs) appear as very promising adsorbents for DCF adsorption. Indeed, a ZIF-67/CTAB nanocomposite has been used for the removal of DCF from aqueous media. The presence of CTAB increases the adsorption capacity of the composites as it enhances the surface charge of the material. Compared to the pure ZIF-67 material, the nanocomposite with CTAB shows a ten times higher adsorption capacity for DCF [265]. Other MOFs have also been used for DCF removal, and high adsorption capacities, up to 263 mg/g of MOF, were observed. The chemical nature of the adsorbent, i.e., the functionalization with ionic groups, together with the adsorption conditions (pH of the liquid media, contact time, etc.) is crucial to achieve high adsorption efficiencies. For example, the highest adsorption capacities have been reported for amine and sulfonate-functionalized UiO-66. These results may be explained by some particular adsorbent–adsorbate interactions, such as base–base repulsion or acid–base attraction involving the amine and carboxylate groups of DCF and the amine and sulfonate groups of the MOF adsorbent [266]. More recently, a porous Cu-based MOF showed extraordinary high adsorption capacity of up to 900 mg/g [267].

Another important class of adsorbents is carbon-based materials, such as activated carbons (ACs), graphene and graphene oxide. ACs are widely used as adsorbents, at both the laboratory and industrial scale, and have been investigated for the removal of DCF from aqueous solution. Due to the different chemical natures, the adsorption mechanism is mostly based on weak interactions, such as van der Waals bonding or $\pi-\pi$ stacking interactions. Activated carbons sometimes show high adsorption capacity of DCF up to 400 mg/g [205, 226]. A combination of ACs and MOFs was used to create porous carbonaceous materials *via* pyrolysis, so-called porous carbons derived from MOFs (PCDMs). These solids displayed higher adsorption capacities for DCF than the conventional porous carbons [268–270]. Graphene oxide is another promising adsorbent for DCF, due to the $\pi-\pi$ stacking interactions with the aromatic rings of DCF [252].

Porous silicas, ionosilicas and silica-supported ionic liquid phases (SILPs) are among the most polyvalent and tunable functional materials. Silica offers a large variety of possibilities for processing and morphology control, and ionic liquids (ILs) allow for the fine-tuning of the physicochemical properties of the materials, such as hydrophilicity/hydrophobicity and ionic interactions. These materials are therefore particularly suitable for the synthesis of ion exchange materials. In fact, porous thiol and amine-functionalized silica adsorb DCF more efficiently than the parent pristine silica, and this effect was attributed to mainly hydrophobic interactions [271].

On the other hand, the incorporation of ionic sites within the material gives rise to silica-based ion exchange materials. These ionosilicas and SILPs are versatile adsorbents for anionic drugs, displaying high adsorption capacities up to 600 mg/g [257, 260]. The polyvalence of these materials is highlighted by the fact that the adsorption process can be controlled by the counter anion [257]. This feature shows that ionosilicas and SILPs are universal adsorbent and offer multiple ways of chemical modification.

Renewable materials, such as bagasse, or biopolymers, such as chitosan, have also been successfully used for DCF removal from aqueous solution. For example, Isabel grape (*Vitis labrusca* × *Vitis vinifera*) bagasse was tested as a low-cost adsorbent. Despite the fact that the adsorption performances of bagasse are considerably lower (maximum capacity <50 mg/g), it appears as a good solution for DCF sequestration. This is due to its large-scale availability and insignificant commercial value, thus compensating for its lower efficiency [253]. Recently, bagasse-based PET/Fe(III) containing nanocomposites were successfully used for DCF adsorption [272]. Regarding functional biopolymers, Liang et al. reported a chitosan/iron oxide nanocomposite, displaying high DCF adsorption capacity of nearly 470 mg DCF/g [273]. The material is recoverable using magnetic field, due to the presence of magnetic Fe_3O_4, and can be reused in at least four adsorption–regeneration cycles, while maintaining a DCF removal efficiency of 70%. All these examples show that multiple types of materials with good to excellent DCF adsorption exist. The observed adsorption capacities are sometimes very high and can reach more than 600 mg DCF/g, but these results are often compromised by the presence of competing anionic species.

In view of an efficient reuse of these adsorbents, closed adsorption–desorption cycles and regeneration of these often-costly adsorbents have to be investigated. Furthermore, it has to be mentioned that most of these studies are carried out under laboratory conditions. The efficiencies of these materials under real-life conditions are still a rather unexplored field and have to be addressed before transposition on wastewater treatment under real large-scale applications. It also needs to be pointed out that rather minor changes in wastewater treatment have enormous effects on the elimination rates of DCF in WWTPs. For example, slight pH changes allow for a 20-fold increase in the DCF elimination rate. Other parameters, such as resident times, need to also be addressed for a more efficient elimination of DCF and other organic anionic species from wastewater [215].

All these findings clearly indicate that continuous monitoring of DCF and other pharmaceutically active compounds (PhACs) and, more generally, pharmaceutical and personal care products (PPCPs) in the aquatic environment are necessary [274]. Furthermore, novel techniques and strategies for efficient removal of these compounds from all water bodies are required. Regarding DCF, three main strategies are of importance for a more efficient removal from wastewater during conventional wastewater treatment:

(i)   process optimization in wastewater treatment, namely longer hydraulic residence times, leading to longer contact times of DCF with biomass
(ii)   enrichment of DCF-metabolizing microorganisms in the bioreactor
(iii)   pre-treatment of wastewater using (photo)oxidation methods.

However, as all these methods require drastic process modifications, because they are often time consuming and not always efficient and result in the production of secondary contamination from DCF degradation. The adsorption techniques are of particular importance and require further investigation to reach acceptable DCF concentrations in WWTP effluents.

11.5   Final Observations

At this point, it is quite fair to say that only innovation in oxyanion sequestration is not enough to combat environmental pollution. While many industries are responsible for significant pollution, they also provide employment and resources that are needed, especially in developing countries [275]. Therefore, it is the environmental authorities of many countries, especially the least developed and developing countries, which are under enormous pressure to control and reduce pollution [276]. It is clear that besides technological advances, environmental education plays a vital role in improving the overall conditions of the populations [277]. Therefore, raising awareness through continuous consultation, focus group discussion and environmental education has to be put in place to ensure that the people involved and local stakeholders recognize the alarming threat [278, 279]. It is also necessary that employers, laborers and management of these industries consider adopting certain strategies that reduce the overall health risk that are the consequences of untreated wastewater discharge. Educational workshops and training must be put in place to educate the relevant industries to understand and implement necessary cleanup methods. These efforts must be backed by systematic counseling with the local stakeholders to further support the development of a deeper understand of the threats of uncontrolled wastewater pollution. The same efforts must also highlight the benefits of feasible treatment planning and reorganization of water treatment [280].

Clearly, changing the approaches of industrial setting in the developing world will not take place overnight. However, it is absolutely necessary that continuous training at all levels (schools, universities, industries, pubic stakeholders) is established and that better pollutant management plans are put in action. Local pressure groups, environmental activists, governments and other environmental authorities are obvious partners in these developments [281].

Environmental education has been recognized as silver bullet to build more environmentally aware populations [282]. Typically, the problem of environmental pollution mitigates with well-informed citizens and schools and universities are highly suitable platforms to educate but also train scientists, engineers and decision makers [278]. Among others, demonstration experiments and small-scale live demonstration of pollutant removal using hands-on in schools, universities and any other public event that is accessible for education should be a useful tool for establishing an understanding of the issues and chances of water treatment for any society [283].

References

1. Bhatnagar A, Sillanpää M (2011) A review of emerging adsorbents for nitrate removal from water. Chem Eng J 168:493–504
2. Loganathan P, Vigneswaran S, Kandasamy J (2013) Enhanced removal of nitrate from water using surface modification of adsorbents–a review. J Environ Manage 131:363–374
3. Owlad M, Aroua MK, Daud WAW, Baroutian S (2009) Removal of hexavalent chromium-contaminated water and wastewater: a review. Water Air Soil Pollut 200:59–77
4. Salmani MH, Sahlabadi F, Eslami H, Ghaneian MT, Balaneji IR, Zad TJ (2019) Removal of Cr(VI) oxoanion from contaminated water using granular jujube stems as a porous adsorbent. Groundwater Sustain Develop 8:319–323
5. Gerland S, Lind B, Dowdall M, Kolstad AK (2002) Recent levels of technetium-99 in seawater at the west coast of Svalbard. Sci World J 2:1507–1513
6. Leonard KS, McCubbin D, Brown J, Bonfield R, Brooks T (1997) Distribution of technetium-99 in UK coastal waters. Mar Pollut Bull 34:628–636
7. Darab JG, Amonette AB, Burke DSD, Orr RD, Ponder SM, Schrick B, Mallouk TE, Lukens WW, Caulder DL, Shuh DK (2007) Removal of pertechnetate from simulated nuclear waste streams using supported zerovalent iron. Chem Mater 19:5703–5713
8. Lonappan L, Brar SK, Das RK, Verma M, Surampalli RY (2016) Diclofenac and its transformation products: environmental occurrence and toxicity—a review. Environ Int 96:127–138
9. Sathishkumar P, Meena RAA, Palanisami T, Ashokkumar V, Palvannan T, Gu FL (2020) Occurrence, interactive effects and ecological risk of diclofenac in environmental compartments and biota—a review. Sci Total Environ 698:134057
10. Verbinnen B, Block C, Vandecasteele C (2016) Adsorption of oxyanions from industrial wastewater using perlite-supported magnetite. Water Environ Res 88:408–414
11. Montes-Hernandez G, Concha-Lozano N, Renard F, Quirico E (2009) Removal of oxyanions from synthetic wastewater via carbonation process of calcium hydroxide: applied and fundamental aspects. J Hazard Mater 166:788–795

12. Zhou C, Ontiveros-Valencia A, Nerenberg R, Tang Y, Friese D, Krajmalnik-Brown R, Rittmann BE (2018) Hydrogenotrophic microbial reduction of oxyanions with the membrane biofilm reactor. Front Microbiol 9:3268

13. Rajmohan L, Margavelu G, Chetty R (2016) Review on challenges and opportunities in the removal of nitrate from wastewater using electrochemical method. J Environ Biol 37:1519–1528

14. Archna A, Sharma S, Sobti R (2012) Nitrate removal from ground water: a review. J Chem 9:1667–1675

15. Mohseni-Bandpi A, Elliott DJ, Zazouli MA (2013) Biological nitrate removal processes from drinking water supply—a review. J Environ Health Sci Eng 11:35

16. Banerjee D, Elsaidi SK, Aguila B, Li B, Kim D, Schweiger MJ, Kruger AA, Doonan CJ, Ma S, Thallapally PK (2016) Removal of pertechnetate-related oxyanions from solution using functionalized hierarchical porous frameworks. Chem A Eur J, 22:17581–17584

17. Zhang Y, Geißen S-U, Gal C (2008) Carbamazepine and diclofenac: removal in wastewater treatment plants and occurrence in water bodies. Chemosphere 73:1151–1161

18. Katarína G, Emília K, Ján H, Tomáš M (2017) Degradation of anti-inflammatory drug diclofenac in sewage water. Acta Chimica Slovaca 10:1–5

19. Aregay GG, Jawad A, Du Y, Shahzad A, Chen Z (2019) Efficient and selective removal of chromium (VI) by sulfide assembled hydrotalcite compounds through concurrent reduction and adsorption processes. J Mol Liq 294:111532

20. Sengupta AK, Clifford D (1986) Some unique characteristics of chromate ion exchange. React Polym, Ion Exch, Sorbents 4:113–130

21. Speight JG (2020) Sources of water pollution. In: Natural water remediation: chemistry and technology. Butterworth-Heinemann, Oxford, pp 165–198

22. Sutton R, Xie YN, Moran KD, Teerlink J (2019) Occurrence and sources of pesticides to urban wastewater and the environment. In: Goh KS, Gan J, Young DF, Luo Y (eds) Pesticides in surface water: monitoring, modeling, risk assessment, and management. Amer Chemical Soc, Washington, pp 63–88

23. Goel PK (2006) Water pollution: causes, effects and control. New Age International

24. Inyinbor AA, Adebesin BO, Oluyori AP, Adelani-Akande TA, Dada AO, Oreofe TA (2018) Water pollution: effects, prevention, and climatic impact. In: Glavan M (ed) Water pollution: effects, prevention, and climatic impact, water challenges of an urbanizing world. IntechOpen

25. Weidner E, Ciesielczyk F (2019) Removal of hazardous oxyanions from the environment using metal-oxide-based materials, materials (Basel), vol 12

26. Yin YB, Guo S, Heck KN, Clark CA, Coonrod CL, Wong MS (2018) Treating water by degrading oxyanions using metallic nanostructures. ACS Sustain Chem Eng 6:11160–11175

27. Adegoke HI, Adekola FA, Fatoki OS, Ximba BJ (2013) Sorptive interaction of oxyanions with iron oxides: a review. Polish J Environ Studies 22:7–24

28. Cornelis G, Johnson CA, Gerven TV, Vandecasteele C (2008) Leaching mechanisms of oxyanionic metalloid and metal species in alkaline solid wastes: a review. Appl Geochem 23:955–976

29. Han J, Kim M, Ro H-M (2020) Factors modifying the structural configuration of oxyanions and organic acids adsorbed on iron (hydr)oxides in soils. A Rev Environ Chem Lett

30. Al-Aoh HA (2019) Adsorption of MnO_4^- from aqueous solution by Nitraria retusa leaves powder; kinetic, equilibrium and thermodynamic studies. Mater Res Express 6:115102

31. Verbinnen B, Block C, Lievens P, Van Brecht A, Vandecasteele C (2013) Simultaneous removal of molybdenum, antimony and selenium oxyanions from wastewater by adsorption on supported magnetite. Waste Biomass Valorization 4:635–645

32. Theiss FL, Couperthwaite SJ, Ayoko GA, Frost RL (2014) A review of the removal of anions and oxyanions of the halogen elements from aqueous solution by layered double hydroxides. J Colloid Interface Sci 417:356–368

33. Doğan V, Aydın S (2014) Vanadium(V) removal by adsorption onto activated carbon derived from starch industry waste sludge. Sep Sci Technol 49:1407–1415

34. Cui M, Johannesson KH (2017) Comparison of tungstate and tetrathiotungstate adsorption onto pyrite. Chem Geol 464:57–68
35. USEPA (2017) EPA—Technical fact sheet—Tungsten, in, United States government
36. Petruzzelli G, Pedron F (2017) Tungstate adsorption onto Italian soils with different characteristics. Environ Monit Assess 189:379
37. Sun Q, Zhu L, Aguila B, Thallapally PK, Xu C, Chen J, Wang S, Rogers D, Ma S (2019) Optimizing radionuclide sequestration in anion nanotraps with record pertechnetate sorption. Nat Commun 10:1646
38. Ghosh R, Ghosh TK, Ghosh P (2020) Selective and efficient removal of perrhenate by an imidazolium based hexapodal receptor in water medium. Dalton Trans 49:3093–3097
39. Marinho BA, Cristóvão RO, Boaventura RAR, Vilar VJP (2019) As(III) and Cr(VI) oxyanion removal from water by advanced oxidation/reduction processes—a review. Environ Sci Pollut Res Int 26:2203–2227
40. Ma L, Islam SM, Liu H, Zhao J, Sun G, Li H, Ma S, Kanatzidis MG (2017) Selective and efficient removal of toxic oxoanions of As(III), As(V), and Cr(VI) by layered double hydroxide intercalated with MoS_4^{2-}. Chem Mater 29:3274–3284
41. USEPA (2014) EPA—Priority pollutant list, in, United States government
42. Tan LC, Nancharaiah YV, van Hullebusch ED, Lens PNL (2016) Selenium: environmental significance, pollution, and biological treatment technologies. Biotechnol Adv 34:886–907
43. Losi ME, Frankenberger WT (1997) Reduction of selenium oxyanions by enterobacter cloacae SLD1a-1: isolation and growth of the bacterium and its expulsion of selenium particles. Appl Environ Microbiol 63:3079–3084
44. Holmes AB, Gu FX (2016) Emerging nanomaterials for the application of selenium removal for wastewater treatment. Environ Sci: Nano 3:982–996
45. Ren JH, Ma LQ, Sun HJ, Cai F, Luo J (2014) Antimony uptake, translocation and speciation in rice plants exposed to antimonite and antimonate. Sci Total Environ 475:83–89
46. Kaszycki ME (2016) Drinking water contaminant candidate list 4—final, in, United States environmental protection agency, Federal Register, pp 9–13
47. WHO (2017) Guidelines for drinking-water quality, 4th edn, incorporating the 1st addendum (chapters), In: Chemical fact sheet. World Health Organization
48. Kailasam V (2009) The removal and recovery of oxo-anions from aqueous systems using nano-porous silica polyamine composites. In: Department of chemistry and biochemistry. University of Montana, pp 1–213
49. Maltman C, Donald LJ, Yurkov V (2017) Tellurite and tellurate reduction by the aerobic anoxygenic phototroph erythromonas ursincola, Strain KR99 is carried out by a novel membrane associated enzyme, Microorganisms, vol 5
50. Maltman C, Yurkov V (2019) Extreme environments and high-level bacterial tellurite resistance, Microorganisms, vol 7
51. USEPA (2017) EPA—Technical fact sheet—Perchlorate. In: United States government
52. Xie Q, Li Y, Lv Z, Zhou H, Yang X, Chen J, Guo H (2017) Effective adsorption and removal of phosphate from aqueous solutions and eutrophic water by fe-based MOFs of MIL-101. Sci Rep 7:3316
53. Runtti H, Tuomikoski S, Kangas T, Kuokkanen T, Rämö J, Lassi U (2016) Sulphate removal from water by carbon residue from biomass gasification: effect of chemical modification methods on sulphate removal efficiency. BioResources 11(2):2016
54. Muruganathan M, Raju GB, Prabhakar S (2004) Removal of sulfide, sulfate and sulfite ions by electro coagulation. J Hazard Mater 109:37–44
55. WHO (2004) Sulfate in drinking-water, Background document for development of WHO guidelines for drinking-water quality. In: World Health Organization
56. Huo L, Xie W, Qian T, Guan X, Zhao D (2017) Reductive immobilization of pertechnetate in soil and groundwater using synthetic pyrite nanoparticles. Chemosphere 174:456–465
57. Schröder P, Helmreich B, Škrbić B, Carballa M, Papa M, Pastore C, Emre Z, Oehmen A, Langenhoff A, Molinos M, Dvarioniene J, Huber C, Tsagarakis KP, Martinez-Lopez E,

Pagano SM, Vogelsang C, Mascolo G (2016) Status of hormones and painkillers in wastewater effluents across several European states—considerations for the EU watch list concerning estradiols and diclofenac. Environ Sci Pollut Res 23:12835–12866

58. Koumaki E, Mamais D, Noutsopoulos C (2017) Environmental fate of non-steroidal antiinflammatory drugs in river water/sediment systems. J Hazard Mater 323:233–241

59. Marsik P, Rezek J, Židková M, Kramulová B, Tauchen J, Vaněk T (2017) Non-steroidal anti-inflammatory drugs in the watercourses of Elbe basin in Czech Republic. Chemosphere 171:97–105

60. Tyumina EA, Bazhutin GA, Gomez ADPC, Ivshina IB (2020) Nonsteroidal anti-inflammatory drugs as emerging contaminants. Microbiology 89:148–163

61. Markelova E, Couture R-M, Parsons CT, Markelov I, Madé B, Van Cappellen P, Charlet L (2018) Speciation dynamics of oxyanion contaminants (As, Sb, Cr) in argillaceous suspensions during oxic-anoxic cycles. Appl Geochem 91:75–88

62. Sheng F, Jingjing L, Yu C, Fu-Ming T, Xuemei D, Jing-yao L (2018) Theoretical study of the oxidation reactions of sulfurous acid/sulfite with ozone to produce sulfuric acid/sulfate with atmospheric implications. RSC Advances 8:7988–7996

63. Müller IA, Brunner B, Breuer C, Coleman M, Bach W (2013) The oxygen isotope equilibrium fractionation between sulfite species and water. Geochim Cosmochim Acta 120:562–581

64. Gee Chai S, Pfeffer John T, Suidan Makram T (1990) Nitrosomonas and nitrobacter interactions in biological nitrification. J Environ Eng 116:4–17

65. Gouvêa LFC, Moreira AJ, Freschi CD, Freschi GPG (2018) Speciation of nitrite, nitrate and p-nitrophenol by photochemical vapor generation of NO using high-resolution continuum source molecular absorption spectrometry. J Food Compos Anal 70:28–34

66. Battas A, Gaidoumi AE, Ksakas A, Kherbeche A (2019) Adsorption study for the removal of nitrate from water using local clay. ScientificWorldJournal 2019:9529618

67. Bijay S, Yadvinder S, Sekhon GS (1995) Fertilizer-N use efficiency and nitrate pollution of groundwater in developing countries. J Contam Hydrol 20:167–184

68. Rakhunde R, Deshpande L, Juneja HD (2012) Chemical speciation of chromium in water: a review. Crit Rev Environ Sci Technol 42:776–810

69. Peng H, Leng Y, Cheng Q, Shang Q, Shu J, Guo J (2019) Efficient removal of hexavalent chromium from wastewater with electro-reduction, processes, vol 7

70. Diniz KM, Tarley CRT (2015) Speciation analysis of chromium in water samples through sequential combination of dispersive magnetic solid phase extraction using mesoporous amino-functionalized Fe_3O_4/SiO_2 nanoparticles and cloud point extraction. Microchem J 123:185–195

71. Zhitkovich A (2011) Chromium in drinking water: sources, metabolism, and cancer risks. Chem Res Toxicol 24:1617–1629

72. Liu W, Yang L, Xu S, Chen Y, Liu B, Li Z, Jiang C (2018) Efficient removal of hexavalent chromium from water by an adsorption–reduction mechanism with sandwiched nanocomposites. RSC Advan 8:15087–15093

73. Mitra S, Sarkar A, Sen S (2017) Removal of chromium from industrial effluents using nanotechnology: a review. Nanotechnol Environ Eng 2:11

74. Oliveira H (2012) Chromium as an environmental pollutant: insights on induced plant toxicity. J Bot 2012:375843

75. Appenroth KJ, Stöckel J, Srivastava A, Strasser RJ (2001) Multiple effects of chromate on the photosynthetic apparatus of Spirodela polyrhiza as probed by OJIP chlorophyll a fluorescence measurements. Environ Pollut 115:49–64

76. Castro RO, Trujillo MM, López Bucio J, Cervantes C, Dubrovsky J (2007) Effects of dichromate on growth and root system architecture of arabidopsis thaliana seedlings. Plant Sci 172:684–691

77. USEPA (2017) Contaminant candidate list (CCL) and regulatory determination. In: Chemical contaminant—CCL4, United States government

78. Larsson MA, Hadialhejazi G, Gustafsson JP (2017) Vanadium sorption by mineral soils: development of a predictive model. Chemosphere 168:925–932

79. Leiviskä T, Khalid MK, Sarpola A, Tanskanen J (2017) Removal of vanadium from industrial wastewater using iron sorbents in batch and continuous flow pilot systems. J Environ Manage 190:231–242
80. Hollemann AF, Wiberg N (2001) Lehrbuch der anorganischen Chemie. Walter de Gruyter
81. Iwai T, Hashimoto Y (2017) Adsorption of tungstate (WO_4) on birnessite, ferrihydrite, gibbsite, goethite and montmorillonite as affected by pH and competitive phosphate (PO_4) and molybdate (MoO_4) oxyanions. Appl Clay Sci 143:372–377
82. Grundler PV, Brugger J, Etschmann BE, Helm L, Liu W, Spry PG, Tian Y, Testemale D, Pring A (2013) Speciation of aqueous tellurium(IV) in hydrothermal solutions and vapors, and the role of oxidized tellurium species in Te transport and gold deposition. Geochim Cosmochim Acta 120:298–325
83. Keith M, Smith DJ, Jenkin GRT, Holwell DA, Dye MD (2018) A review of Te and Se systematics in hydrothermal pyrite from precious metal deposits: Insights into ore-forming processes. Ore Geol Rev 96:269–282
84. Harada T, Takahashi Y (2008) Origin of the difference in the distribution behavior of tellurium and selenium in a soil–water system. Geochim Cosmochim Acta 72:1281–1294
85. Yarema MC, Curry SC (2005) Acute tellurium toxicity from ingestion of metal-oxidizing solutions. Pediatrics 116:E319–E321
86. Maltman C, Yurkov V (2015) The effect of tellurite on highly resistant freshwater aerobic anoxygenic phototrophs and their strategies for reduction. Microorganisms 3:826–838
87. Farrell J, Bostick WD, Jarabek RJ, Fiedor JN (1999) Electrosorption and reduction of pertechnetate by anodically polarized magnetite. Environ Sci Technol 33:1244–1249
88. Li D, Seaman JC, Kaplan DI, Heald SM, Sun C (2019) Pertechnetate (TcO_4^-) sequestration from groundwater by cost-effective organoclays and granular activated carbon under oxic environmental conditions. Chem Eng J 360:1–9
89. Villar M, Borras A, Avivar J, Vega F, Cerda V, Ferrer L (2017) Fully automated system for Tc-99 monitoring in hospital and urban residues: a simple approach to waste management. Anal Chem 89:5858–5864
90. Icenhower JP, Qafoku N, Martin WJ, Zachara JM (2008) The geochemistry of technetium: a summary of the behavior of an artificial element in the natural environment. In: United States
91. Gearing P, Van Baalen C, Parker PL (1975) Biochemical effects of technetium-99-pertechnetate on microorganisms. Plant Physiol 55:240–246
92. Jurado A, Walther M, Diaz-Cruz MS (2019) Occurrence, fate and environmental risk assessment of the organic microcontaminants included in the watch lists set by EU decisions 2015/495 and 2018/840 in the groundwater of Spain. Sci Total Environ 663:285–296
93. Scheurell M, Franke S, Shah RM, Hühnerfuss H (2009) Occurrence of diclofenac and its metabolites in surface water and effluent samples from Karachi. Pakistan, Chemosphere 77:870–876
94. Wiesenberg-Boettcher I, Pfeilschifter J, Schweizer A, Sallmann A, Wenk P (1991) Pharmacological properties of five diclofenac metabolites identified in human plasma. Agents Actions 34:135–137
95. Gomaa S (2018) Adverse effects induced by diclofenac, ibuprofen, and paracetamol toxicity on immunological and biochemical parameters in Swiss albino mice. J Basic Appl Zool 79:5
96. Green RE, Newton I, Shultz S, Cunningham AA, Gilbert M, Pain DJ, Prakash V (2004) Diclofenac poisoning as a cause of vulture population declines across the Indian subcontinent. J Appl Ecol 41:793–800
97. Bonnefille B, Gomez E, Courant F, Escande A, Fenet H (2018) Diclofenac in the marine environment: a review of its occurrence and effects. Mar Pollut Bull 131:496–506
98. Oaks JL, Gilbert M, Virani MZ, Watson RT, Meteyer CU, Rideout BA, Shivaprasad HL, Ahmed S, Chaudhry MJI, Arshad M, Mahmood S, Ali A, Khan AA (2004) Diclofenac residues as the cause of vulture population decline in Pakistan. Nature 427:630–633
99. Xu X, Gao BY, Jin B, Yue QY (2016) Removal of anionic pollutants from liquids by biomass materials: a review. J Mol Liq 215:565–595

100. Loganathan P, Vigneswaran S, Kandasamy J (2013) Enhanced removal of nitrate from water using surface modification of adsorbents—a review. J Environ Manage 131:363–374
101. Ruiz-Bevia F, Fernandez-Torres MJ (2019) Effective catalytic removal of nitrates from drinking water: an unresolved problem? J Clean Prod 217:398–408
102. Ramos-Ruiz A, Field JA, Wilkening JV, Sierra-Alvarez R (2016) Recovery of elemental tellurium nanoparticles by the reduction of tellurium oxyanions in a methanogenic microbial consortium. Environ Sci Technol 50:1492–1500
103. Tyagi S, Rawtani D, Khatri N, Tharmavaram M (2018) Strategies for nitrate removal from aqueous environment using nanotechnology: a review. J Water Process Eng 21:84–95
104. Huno SKM, Rene ER, van Hullebusch ED, Annachhatre AP (2018) Nitrate removal from groundwater: a review of natural and engineered processes. J Water Supply Res Technol-Aqua 67:885–902
105. Jermakka J, Wendling L, Sohlberg E, Heinonen H, Vikman M (2015) Potential technologies for the removal and recovery of nitrogen compounds from mine and quarry waters in subarctic conditions. Crit Rev Environ Sci Technol 45:703–748
106. Kumar T, Mandlimath TR, Sangeetha P, Revathi SK, Kumar SKA (2018) Nanoscale materials as sorbents for nitrate and phosphate removal from water. Environ Chem Lett 16:389–400
107. Dai YJ, Wang WS, Lu L, Yan LL, Yu DY (2020) Utilization of biochar for the removal of nitrogen and phosphorus. J Cleaner Prod 257
108. Heaney N, Ukpong E, Lin CX (2020) Low-molecular-weight organic acids enable biochar to immobilize nitrate. Chemosphere 240
109. Pan JW, Gao BY, Song W, Xu X, Yue QY (2020) Modified biogas residues as an eco-friendly and easily-recoverable biosorbent for nitrate and phosphate removals from surface water. J Hazard Mater 382
110. Machida M, Yoo P, Amano Y (2019) Adsorption of nitrate from aqueous phase onto nitrogen-doped activated carbon fibers (ACFs). Sn Appl Sci 1
111. Machida M, Sakamoto T, Sato K, Goto T, Amano Y (2018) Adsorptive removal of nitrate from aqueous phase using steam activated and thermal treated polyacrylonitrile (PAN) fiber. J Fiber Sci Technol 74:158–164
112. Taoufik N, Elmchaouri A, Korili SA, Gil A (2020) Optimizing the removal of nitrate by adsorption onto activated carbon using response surface methodology based on the central composite design. J Appl Water Eng Res 8:66–77
113. Yin QQ, Wang RK, Zhao ZH (2018) Application of Mg-Al-modified biochar for simultaneous removal of ammonium, nitrate, and phosphate from eutrophic water. J Clean Prod 176:230–240
114. Oyarzun DI, Hemmatifar A, Palko JW, Stadermann M, Santiago JG (2018) Adsorption and capacitive regeneration of nitrate using inverted capacitive deionization with surfactant functionalized carbon electrodes. Sep Purif Technol 194:410–415
115. Viglasova E, Galambos M, Dankova Z, Krivosudsky L, Lengauer CL, Hood-Nowotny R, Soja G, Rompel A, Matik M, Briancin J (2018) Production, characterization and adsorption studies of bamboo-based biochar/montmorillonite composite for nitrate removal. Waste Manage 79:385–394
116. Cui XQ, Li H, Yao ZY, Shen Y, He ZL, Yang XE, Ng HY, Wang CH (2019) Removal of nitrate and phosphate by chitosan composited beads derived from crude oil refinery waste: sorption and cost-benefit analysis. J Clean Prod 207:846–856
117. Zarei S, Farhadian N, Akbarzadeh R, Pirsaheb M, Asadi A, Safaei Z (2020) Fabrication of novel 2D Ag-TiO$_2$/gamma-Al$_2$O$_3$/Chitosan nano-composite photocatalyst toward enhanced photocatalytic reduction of nitrate. Int J Biol Macromol 145:926–935
118. Suzaimi ND, Goh PS, Malek N, Lim JW, Ismail AF (2019) Performance of branched polyethyleneimine grafted porous rice husk silica in treating nitrate-rich wastewater via adsorption. J Environ Chem Eng 7
119. Ebrahimi-Gatkash M, Younesi H, Shahbazi A, Heidari A (2017) Amino-functionalized meso-porous MCM-41 silica as an efficient adsorbent for water treatment: batch and fixed-bed column adsorption of the nitrate anion. Appl Water Sci 7:1887–1901

120. Phan PT, Nguyen TT, Nguyen NH, Padungthon S (2019) Triamine-bearing activated rice husk ash as an advanced functional material for nitrate removal from aqueous solution. Water Sci Technol 79:850–856

121. Karthikeyan P, Meenakshi S (2020) In situ fabrication of magnetic particles decorated biopolymeric composite beads for the selective remediation of phosphate and nitrate from aqueous medium. J Environ Chem Eng 8

122. Yang WL, Shi XX, Wang JC, Chen WJ, Zhang LL, Zhang WM, Zhang XL, Lu JL (2019) Fabrication of a novel bifunctional nanocomposite with improved selectivity for simultaneous nitrate and phosphate removal from water. ACS Appl Mater Interfaces 11:35277–35285

123. Alighardashi A, Esfahani ZK, Najafi F, Afkhami A, Hassani N (2018) Development and application of graphene oxide/poly-amidoamines dendrimers (GO/PAMAMs) nano-composite for nitrate removal from aqueous solutions. Environ Process—An Int J 5:41–64

124. Jaworski MA, Flores FM, Fernandez MA, Casella M, Sanchez RMT (2019) Use of organo-montmorillonite for the nitrate retention in water: influence of alkyl length of loaded surfactants. Sn Appl Sci 1

125. Omorogie MO, Agunbiade FO, Alfred MO, Olaniyi OT, Adewumi TA, Bayode AA, Ofomaja AE, Naidoo EB, Okoli CP, Adebayo TA, Unuabonah EI (2018) The sequestral capture of fluoride, nitrate and phosphate by metal-doped and surfactant-modified hybrid clay materials. Chem Pap 72:409–417

126. Unuabonah EI, Gunter C, Weber J, Lubahn S, Taubert A (2013) Hybrid clay: a new highly efficient adsorbent for water treatment. ACS Sustain Chem Eng 1:966–973

127. Wang LY, Guo SQ, Chen YT, Pan ML, Ang EH, Yuana ZH (2020) A Mechanism investigation of how the alloying effect improves the photocatalytic nitrate reduction activity of bismuth oxyhalide nanosheets. Chemphotochem 4:110–119

128. Li JC, Li M, Yang X, Wang S, Zhang Y, Liu F, Liu X (2019) Sub-nanocatalysis for efficient aqueous nitrate reduction: effect of strong metal-support interaction. ACS Appl Mater Interfaces 11:33859–33867

129. Geng Z, Chen ZT, Li ZY, Qi X, Yang X, Fan W, Guo YN, Zhang LL, Huo MX (2018) Enhanced photocatalytic conversion and selectivity of nitrate reduction to nitrogen over AgCl/TiO$_2$ nanotubes. Dalton Trans 47:11104–11112

130. Adamu H, McCue AJ, Taylor RSF, Manyar HG, Anderson JA (2019) Influence of pretreatment on surface interaction between Cu and anatase-TiO$_2$ in the simultaneous photoremediation of nitrate and oxalic acid. J Environ Chem Eng 7

131. Khatamian M, Divband B, Shahi R (2019) Ultrasound assisted co-precipitation synthesis of Fe$_3$O$_4$/bentonite nanocomposite: performance for nitrate, BOD and COD water treatment. J Water Process Eng 31

132. Chu YB, Xu Y, Liu H, Yuan ZW, Zhao HZ (2018) Novel Na-zeolite covalently bonded with quaternary ammonium for simultaneous removal of phosphate, ammonium and nitrate. Desalin Water Treat 106:153–164

133. Wu DL, Sun Y, Wang LL, Zhang ZM, Gui JX, Ding AQ, Modification of NaY zeolite by lanthanum and hexadecyl trimethyl ammonium bromide and its removal performance for nitrate. Water Environ Res

134. Rezvani P, Taghizadeh MM (2018) On using clay and nanoclay ceramic granules in reducing lead, arsenic, nitrate, and turbidity from water. Appl Water Sci 8

135. Duan SP, Tong TZ, Zheng SK, Zhang XY, Li SD (2020) Achieving low-cost, highly selective nitrate removal with standard anion exchange resin by tuning recycled brine composition. Water Res 173

136. Li QM, Lu XY, Shuang CD, Qi CD, Wang GX, Li AM, Song HO (2019) Preferential adsorption of nitrate with different trialkylamine modified resins and their preliminary investigation for advanced treatment of municipal wastewater. Chemosphere 223:39–47

137. Li JY, Dong SX, Wang YL, Dou XM, Hao HT (2020) Nitrate removal from aqueous solutions by magnetic cationic hydrogel: effect of electrostatic adsorption and mechanism. J Environ Sci 91:177–188

138. Sun Y, Zheng WS, Ding XC, Singh RP (2019) Adsorption of nitrate by a novel polyacrylic anion exchange resin from water with dissolved organic matters: batch and column study. Appl Sci-Basel 9

139. Kalaruban M, Loganathan P, Kandasamy J, Naidu R, Vigneswaran S (2017) Enhanced removal of nitrate in an integrated electrochemical-adsorption system. Sep Purif Technol 189:260–266

140. Maghsudi S, Mirbagheri SA, Fard MB (2018) Removal of nitrate, phosphate and COD from synthetic municipal wastewater treatment plant using membrane filtration as a post-treatment of adsorption column. Desalin Water Treat 115:53–63

141. Khalek MAA, Mahmoud GA, Shoukry EM, Amin M, Abdulghany AH (2019) Adsorptive removal of nitrate ions from aqueous solution using modified biodegradable-based hydrogel. Desalin Water Treat 155:390–401

142. Zhang ML, Li YF, Long XX, Chong YX, Yu GW, He ZH (2018) An alternative approach for nitrate and arsenic removal from wastewater via a nitrate-dependent ferrous oxidation process. J Environ Manage 220:246–252

143. Saha A, Bhushan S, Mukherjee P, Chanda C, Bhaumik M, Ghosh M, Sharmin J, Datta P, Banerjee S, Barat P, Thakur AR, Gantayet LM, Mukherjee I, Chaudhuri SR (2018) Simultaneous sequestration of nitrate and phosphate from wastewater using a tailor-made bacterial consortium in biofilm bioreactor. J Chem Technol Biotechnol 93:1279–1289

144. Gao Q, Wang CZ, Liu S, Hanigan D, Liu ST, Zhao HZ (2019) Ultrafiltration membrane microreactor (MMR) for simultaneous removal of nitrate and phosphate from water. Chem Eng J 355:238–246

145. Kalaruban M, Loganathan P, Kandasamy J, Vigneswaran S (2018) Submerged membrane adsorption hybrid system using four adsorbents to remove nitrate from water. Environ Sci Pollut Res 25:20328–20335

146. Hosseini SS, Mahvi AH (2018) Freezing process—a new approach for nitrate removal from drinking water. Desalin Water Treat 130:109–116

147. Wu YH, Pang HW, Liu Y, Wang XX, Yu SJ, Fu D, Chen JR, Wang XK (2019) Environmental remediation of heavy metal ions by novel-nanomaterials: a review. Environ Pollut 246:608–620

148. Mazur LP, Cechinel MAP, de Souza S, Boaventura RAR, Vilar VJP (2018) Brown marine macroalgae as natural cation exchangers for toxic metal removal from industrial wastewaters: a review. J Environ Manage 223:215–253

149. Rangabhashiyam S, Balasubramanian P (2019) The potential of lignocellulosic biomass precursors for biochar production: performance, mechanism and wastewater application-a review. Ind Crops Prod 128:405–423

150. Ezzatahmadi N, Ayoko GA, Millar GJ, Speight R, Yan C, Li JH, Li SZ, Zhu JX, Xi YF (2017) Clay-supported nanoscale zero-valent iron composite materials for the remediation of contaminated aqueous solutions: a review. Chem Eng J 312:336–350

151. Saravanan A, Kumar PS, Yashwanthraj M (2017) Sequestration of toxic Cr(VI) ions from industrial wastewater using waste biomass: a review. Desalin Water Treat 68:245–266

152. Ugwu EI, Agunwamba JC (2020) A review on the applicability of activated carbon derived from plant biomass in adsorption of chromium, copper, and zinc from industrial wastewater. Environ Monit Assess 192

153. Erabee IK, Ahsan A, Imteaz M, Alom MM (2019) Adsorption of hexavalent chromium using activated carbon prepared from garden wastes. Desalin Water Treat 164:293–299

154. Patra C, Medisetti RMN, Pakshirajan K, Narayanasamy S (2019) Assessment of raw, acid-modified and chelated biomass for sequestration of hexavalent chromium from aqueous solution using Sterculia villosa Roxb. shells. Environ Sci Pollution Res 26:23625–23637

155. Pakade VE, Nchoe OB, Hlungwane L, Tavengwa NT (2017) Sequestration of hexavalent chromium from aqueous solutions by activated carbon derived from macadamia nutshells. Water Sci Technol 75:196–206

156. Mondal MK, Roy D, Chowdhury P (2018) Designed functionalization of reduced graphene oxide for sorption of Cr(VI) over a wide pH range: a theoretical and experimental perspective. New J Chem 42:16960–16971

157. Nkutha CS, Diagboya PN, Mtunzi FM, Dikio ED Application of eco-friendly multifunctional porous graphene oxide for adsorptive sequestration of chromium in aqueous solution. Water Environ Res

158. Bashir S, Hussain Q, Akmal M, Riaz M, Hu H, Ijaz SS, Iqbal M, Abro S, Mehmood S, Ahmad M (2018) Sugarcane bagasse-derived biochar reduces the cadmium and chromium bioavailability to mash bean and enhances the microbial activity in contaminated soil. J Soils Sediments 18:874–886

159. Saranya N, Nakeeran E, Nandagopal MSG, Selvaraju N (2017) Optimization of adsorption process parameters by response surface methodology for hexavalent chromium removal from aqueous solutions using annona reticulata linn peel microparticles. Water Sci Technol 75:2094–2107

160. Xu XJ, Zhang HY, Ma C, Gu HB, Lou H, Lyu SY, Liang CB, Kong J, Gu JW (2018) A superfast hexavalent chromium scavenger: magnetic nanocarbon bridged nanomagnetite network with excellent recyclability. J Hazard Mater 353:166–172

161. Kumar ASK, You JG, Tseng WB, Dwivedi GD, Rajesh N, Jiang SJ, Tseng WL (2019) Magnetically separable nanospherical g-C_3N_4@Fe_3O_4 as a recyclable material for chromium adsorption and visible-light-driven catalytic reduction of aromatic nitro compounds. ACS Sustain Chem Eng 7:6662–6671

162. Xing MC, Xie Q, Li XD, Guan T, Wu DY (2020) Monolayers of an organosilane on magnetite nanoparticles for the fast removal of Cr(VI) from water. Environ Technol 41:658–668

163. Greenstein KE, Myung NV, Parkin GF, Cwiertny DM (2019) Performance comparison of hematite (alpha-Fe_2O_3)-polymer composite and core-shell nanofibers as point-of-use filtration platforms for metal sequestration. Water Res 148:492–503

164. Sathvika T, Soni A, Sharma K, Praneeth M, Mudaliyar M, Rajesh V, Rajesh N (2018) Potential application of saccharomyces cerevisiae and rhizobium immobilized in multi walled carbon nanotubes to adsorb hexavalent chromium. Sci Rep 8

165. Hajji S, Montes-Hernandez G, Sarret G, Tordo A, Morin G, Ona-Nguema G, Bureau S, Turki T, Mzoughi N (2019) Arsenite and chromate sequestration onto ferrihydrite, siderite and goethite nanostructured minerals: Isotherms from flow-through reactor experiments and XAS measurements. J Hazard Mater 362:358–367

166. Wang T, Wang W, Liu Y, Li W, Li Y, Liu B (2020) Roles of natural iron oxides in the promoted sequestration of chromate using calcium polysulfide: pH effect and mechanisms. Sep Purif Technol 237

167. Wang YH, Shao QQ, Huang SS, Zhang BL, Xu CH (2018) High performance and simultaneous sequestration of Cr(VI) and Sb(III) by sulfidated zerovalent iron. J Clean Prod 191:436–444

168. Whitaker AH, Pena J, Amor M, Duckworth OW (2018) Cr(VI) uptake and reduction by biogenic iron (oxyhydr)oxides. Environ Sci Process Impacts 20:1056–1068

169. Suzuki R, Tani Y, Naitou H, Miyata N, Tanaka K (2020) Sequestration and oxidation of Cr(III) by fungal Mn oxides with Mn(II) oxidizing activity. Catalysts 10

170. Thomas AN, Eiche E, Gottlicher J, Steininger R, Benning LG, Freeman HM, Tobler DJ, Mangayayam M, Dideriksen K, Neumann T (2020) Effects of metal cation substitution on hexavalent chromium reduction by green rust. Geochem Trans 21

171. Zhang WF, Liu F, Sun YG, Zhang J, Hao ZP (2019) Simultaneous redox conversion and sequestration of chromate(VI) and arsenite(III) by iron(III)-alginate based photocatalysis. Appl Catal B-Environ 259:11

172. Kou SZ, Yang ZG, Luo JR, Sun F (2017) Entirely recombinant protein-based hydrogels for selective heavy metal sequestration. Polymer Chem 8:6158–6164

173. Xie YQ, Lin J, Liang J, Li MH, Fu YW, Wang HT, Tu S, Li J (2019) Hypercrosslinked mesoporous poly(ionic liquid)s with high density of ion pairs: efficient adsorbents for Cr(VI) removal via ion-exchange. Chem Eng J 378

174. Thach UD, Prelot B, Pellet-Rostaing S, Zajac J, Hesemann P (2018) Surface properties and chemical constitution as crucial parameters for the sorption properties of ionosilicas: the case of chromate adsorption. ACS Appl Nano Mater 1:2076–2087

175. Kugler HKK, Cornelius K (1982) Gmelin handbook of inorganic and organometallic chemistry, 8th edn. element T-C Tc. technetium (system-NR. 69) supplement 1–2 gmelin tc. technetium su. vol general properties. Isotopes. Production. Biology, Springer
176. Rard JA, Sandino MCA, Östhols E (1999) Chemical thermodynamics of technetium. North-Holland, OECD Nuclear Energy Agency
177. Boyd JE (1959) Technetium and promethium. J Chem Educ 36:1–3
178. Schwochau K (2000) Technetium: chemistry and radiopharmaceutical applications. Wiley
179. Sources and effects of ionizing radiation, UNSCEAR 2008, United Nations scientific committee on the effects of atomic radiation, in, 2008
180. Poineau F, Mausolf E, Jarvinen GD, Sattelberger AP, Czerwinski KR (2013) Technetium chemistry in the fuel cycle: combining basic and applied studies. Inorg Chem 52:3573–3578
181. Popova NN, Tananaev IG, Rovnyi SI, Myasoedov BF (2003) Technetium: behaviour during reprocessing of spent nuclear fuel and in environmental objects. Russ Chem Rev 72:101–121
182. Choppin G, Liljenzin J-O, Rydberg J, Ekberg C (2013) Radiochemistry and nuclear chemistry, 4th edn. Academic Press
183. (2010) Technetium and other metals in coordination chemistry. Nucl Med Biol 37:677–726
184. Yalcintas E, Scheinost AC, Gaona X, Altmaier M (2016) Systematic XAS study on the reduction and uptake of Tc by magnetite and mackinawite. Dalton Trans 45:17874–17885
185. Lukens WW, Saslow SA (2018) Facile incorporation of technetium into magnetite, magnesioferrite, and hematite by formation of ferrous nitrate in situ: precursors to iron oxide nuclear waste forms. Dalton Trans 47:10229–10239
186. Holm E, Rioseco J, Ballestra S, Walton A (1988) Radiochemical measurements of Tc-99—sources and environmental levels. J Radioanal Nucl Chem-Art 123:167–179
187. Adloff J-P, Guillaumont R (2018) Fundamentals of radiochemistry. CRC Press
188. Wilmarth WR, Lumetta GJ, Johnson ME, Poirier MR, Thompson MC, Suggs PC, Machara NP (2011) Review: waste-pretreatment technologies for remediation of legacy defense nuclear wastes. Solvent Extr Ion Exch 29:1–48
189. Mukhopadhyay B, Lahiri S (1999) Separation of the carrier free radioisotopes of second transition series elements. Solvent Extr Ion Exch 17:1–21
190. Banerjee D, Kim D, Schweiger MJ, Kruger AA, Thallapally PK (2016) Removal of TcO_4^- ions from solution: materials and future outlook. Chem Soc Rev 45:2724–2739
191. Long KM, Goff GS, Ware SD, Jarvinen GD, Runde WH (2012) Anion exchange resins for the selective separation of technetium from uranium in carbonate solutions. Ind Eng Chem Res 51:10445–10450
192. Liu Z-W, Han B-H (2020) Evaluation of an imidazolium-based porous organic polymer as radioactive waste scavenger. Environ Sci Technol 54:216–224
193. Da H-J, Yang C-X, Yan X-P (2019) Cationic covalent organic nanosheets for rapid and selective capture of perrhenate: an analogue of radioactive pertechnetate from aqueous solution. Environ Sci Technol 53:5212–5220
194. Luo W, Huang Q, Antwi P, Guo B, Sasaki K (2020) Synergistic effect of ClO_4^- and Sr^{2+} adsorption on alginate-encapsulated organo-montmorillonite beads: implication for radionuclide immobilization. J Colloid Interface Sci 560:338–348
195. Gu BH, Brown GM, Bonnesen PV, Liang LY, Moyer BA, Ober R, Alexandratos SD (2000) Development of novel bifunctional anion exchange resins with improved selectivity for pertechnetate sorption from contaminated groundwater. Environ Sci Technol 34:1075–1080
196. Thach UD, Hesemann P, Yang G, Geneste A, Le Caer S, Prelot B (2016) Ionosilicas as efficient sorbents for anionic contaminants: radiolytic stability and ion capacity. J Colloid Interface Sci 482:233–239
197. McKeown DA, Buechele AC, Lukens WW, Muller IS, Shuh DK, Pegg IL (2007) Research program to investigate the fundamental chemistry of technetium. In: U.S. Department of Energy, Office of Scientific and Technical Information
198. Reinig KM, Seibert R, Velazquez D, Baumeister J, Khosroshahi FN, Wycoff W, Terry J, Adams JE, Deakyne CA, Jurisson SS (2017) Pertechnetate-induced addition of sulfide in small olefinic acids: formation of $[TcO(dimercaptosuccinate)_2]^{5-}$ and $[TcO(mercaptosuccinate)_2]^{3-}$ Analogues. Inorg Chem 56:13214–13227

199. Stephan H, Spies H, Johannsen B, Klein L, Vögtle F (1999) Lipophilic urea-functionalized dendrimers as efficient carriers for oxyanions. Chem Commun 1875–1876
200. Xiao C, Silver MA, Wang S (2017) Metal-organic frameworks for radionuclide sequestration from aqueous solution: a brief overview and outlook. Dalton Trans 46:16381–16386
201. Sheng D, Zhu L, Xu C, Xiao C, Wang Y, Wang Y, Chen L, Diwu J, Chen J, Chai Z, Albrecht-Schmitt TE, Wang S (2017) Efficient and selective uptake of TcO$_4^-$ by a cationic metal-organic framework material with open Ag$^+$ sites. Environ Sci Technol 51:3471–3479
202. Petrova M, Guigue M, Venault L, Moisy P, Hesemann P (2015) Anion selectivity in ion exchange reactions with surface functionalized ionosilicas. Phys Chem Chem Phys 17:10182–10188
203. Shen J, Chai W, Wang KX, Zhang F (2017) Efficient removal of anionic radioactive pollutant from water using ordered urea-functionalized mesoporous polymeric nanoparticle. ACS Appl Mater Interfaces 9:22440–22448
204. Hu H, Sun L, Gao Y, Wang T, Huang Y, Lv C, Zhang Y-F, Huang Q, Chen X, Wu H (2020) Synthesis of ZnO nanoparticle-anchored biochar composites for the selective removal of perrhenate, a surrogate for pertechnetate, from radioactive effluents. J Hazard Mater 387
205. Amouzgar P, Wong M, Horri B, Salamatinia B (2016) Advanced material for pharmaceutical removal from wastewater. In: Mishra AK (ed) Smart materials for waste water application. Scrivener Publishing LLC, pp 179–212
206. Lellis B, Fávaro-Polonio CZ, Pamphile JA, Polonio JC (2019) Effects of textile dyes on health and the environment and bioremediation potential of living organisms. Biotechnol Res Innov 3:275–290
207. Liu J-L, Wong M-H (2013) Pharmaceuticals and personal care products (PPCPs): a review on environmental contamination in China. Environ Int 59:208–224
208. Khetan SK, Collins TJ (2007) Human pharmaceuticals in the aquatic environment: a challenge to green chemistry. Chem Rev 107:2319–2364
209. He B-S, Wang J, Liu J, Hu X-M (2017) Eco-pharmacovigilance of non-steroidal anti-inflammatory drugs: necessity and opportunities. Chemosphere 181:178–189
210. Sallmann AR (1986) The history of diclofenac. Am J Med 80:29–33
211. Foan L, Vignoud S, Ricoul F (2017) Device and methode for extracting aromatic-ring compounds contained in a liquid sample. EP 16191669A 20160930, US 2017097325
212. Jesus AR, Soromenho MRC, Raposo LR, Esperanca JMSS, Baptista PV, Fernandes AR, Reis PM (2019) Enhancement of water solubility of poorly water-soluble drugs by new biocompatible N-acetyl amino acid N-alkyl cholinium-based ionic liquids. Eur J Pharm Biopharm 137:227–232
213. Acuna V, Ginebreda A, Mor JR, Petrovic M, Sabater S, Sumpter J, Barcelo D (2015) Balancing the health benefits and environmental risks of pharmaceuticals: diclofenac as an example. Environ Int 85:327–333
214. Sharma AK, Saini M, Singh SD, Prakash V, Das A, Dasan RB, Pandey S, Bohara D, Galligan TH, Green RE, Knopp D, Cuthbert RJ (2014) Diclofenac is toxic to the steppe eagle aquila nipalensis: widening the diversity of raptors threatened by NSAID misuse in South Asia. Bird Conserv Int 24:282–286
215. Vieno N, Sillanpaa M (2014) Fate of diclofenac in municipal wastewater treatment plant—a review. Environ Int 69:28–39
216. Wishart DS, Knox C, Guo AC, Shrivastava S, Hassanali M, Stothard P, Chang Z, Woolsey J (2006) DrugBank: a comprehensive resource for in silico drug discovery and exploration. Nucleic Acids Res 34:D668–D672
217. Tixier C, Singer HP, Oellers S, Muller SR (2003) Occurrence and fate of carbamazepine, clofibric acid, diclofenac, ibuprofen, ketoprofen, and naproxen in surface waters. Environ Sci Technol 37:1061–1068
218. Heberer T, Reddersen K, Mechlinski A (2002) From municipal sewage to drinking water: fate and removal of pharmaceutical residues in the aquatic environment in urban areas. Water Sci Technol 46:81–88

219. Gonzalez-Alonso S, Merino LM, Esteban S, de Alda ML, Barcelo D, Duran JJ, Lopez-Martinez J, Acena J, Perez S, Mastroianni N, Silva A, Catala M, Valcarcel Y (2017) Occurrence of pharmaceutical, recreational and psychotropic drug residues in surface water on the northern antarctic peninsula region. Environ Pollut 229:241–254

220. Lindim C, de Zwart D, Cousins IT, Kutsarova S, Kuehne R, Schueuermann G (2019) Exposure and ecotoxicological risk assessment of mixtures of top prescribed pharmaceuticals in Swedish freshwaters. Chemosphere 220:344–352

221. Williams M, Kookana RS, Mehta A, Yadav SK, Tailor BL, Maheshwari B (2019) Emerging contaminants in a river receiving untreated wastewater from an Indian urban centre. Sci Total Environ 647:1256–1265

222. Lin H, Chen L, Li H, Luo Z, Lu J, Yang Z (2018) Pharmaceutically active compounds in the Xiangjiang river, China: distribution pattern, source apportionment, and risk assessment. Sci Total Environ 636:975–984

223. Riva F, Zuccato E, Davoli E, Fattore E, Castiglioni S (2019) Risk assessment of a mixture of emerging contaminants in surface water in a highly urbanized area in Italy. J Hazard Mater 361:103–110

224. Osorio V, Sanchis J, Abad JL, Ginebreda A, Farre M, Perez S, Barcelo D (2016) Investigating the formation and toxicity of nitrogen transformation products of diclofenac and sulfamethoxazole in wastewater treatment plants. J Hazard Mater 309:157–164

225. Olaitan JO, Anyakora C, Bamiro T, Tella AT (2014) Determination of pharmaceutical compounds in surface and underground water by solid phase extraction-liquid chromatography. J Environ Chem Ecotoxicol 6:20–26

226. Lopez-Serna R, Jurado A, Vazquez-Sune E, Carrera J, Petrovic M, Barcelo D (2013) Occurrence of 95 pharmaceuticals and transformation products in urban groundwaters underlying the metropolis of barcelona. Spain, Environ Pollut 174:305–315

227. Swiacka K, Szaniawska A, Caban M (2019) Evaluation of bioconcentration and metabolism of diclofenac in mussels mytilus trossulus—laboratory study. Mar Pollut Bull 141:249–255

228. Chiffre A, Degiorgi F, Bulete A, Spinner L, Badot P-M (2016) Occurrence of pharmaceuticals in WWTP effluents and their impact in a karstic rural catchment of Eastern France. Environ Sci Pollut Res 23:25427–25441

229. Rabiet M, Togola A, Brissaud F, Seidel J-L, Budzinski H, Elbaz-Poulichet F (2006) Consequences of treated water recycling as regards pharmaceuticals and drugs in surface and ground waters of a medium-sized Mediterranean catchment. Environ Sci Technol 40:5282–5288

230. Heberer T, Adam M (2004) Transport and attenuation of pharmaceutical residues during artificial groundwater replenishment. Environ Chem 1:22–25

231. Moeder M, Braun P, Lange F, Schrader S, Lorenz W (2007) Determination of endocrine disrupting compounds and acidic drugs in water by coupling of derivatization, gas chromatography and negative chemical ionization mass spectrometry. Clean-Soil Air Water 35:444–451

232. Jux U, Baginski RM, Arnold HG, Kronke M, Seng PN (2002) Detection of pharmaceutical contaminations of river, pond, and tap water from cologne (Germany) and surroundings. Int J Hyg Environ Health 205:393–398

233. Schmidt S, Hoffmann H, Garbe L-A, Schneider RJ (2018) Liquid chromatography-tandem mass spectrometry detection of diclofenac and related compounds in water samples. J Chromatogr A 1538:112–116

234. Vulliet E, Cren-Olive C (2011) Screening of pharmaceuticals and hormones at the regional scale, in surface and groundwaters intended to human consumption. Environ Pollut 159:2929–2934

235. Sacher F, Lang FT, Brauch HJ, Blankenhorn I (2001) Pharmaceuticals in groundwaters—analytical methods and results of a monitoring program in Baden-Wurttemberg, Germany. J Chromatogr A 938:199–210

236. Togola A, Budzinski H (2008) Multi-residue analysis of pharmaceutical compounds in aqueous samples. J Chromatogr A 1177:150–158

237. Stuelten D, Zuehlke S, Lamshoeft M, Spiteller M (2008) Occurrence of diclofenac and selected metabolites in sewage effluents. Sci Total Environ 405:310–316
238. Ternes TA, Herrmann N, Bonerz M, Knacker T, Siegrist H, Joss A (2004) A rapid method to measure the solid-water distribution coefficient (K_d) for pharmaceuticals and musk fragrances in sewage sludge. Water Res 38:4075–4084
239. Togola A, Budzinski H (2007) Analytical development for analysis of pharmaceuticals in water samples by SPE and GC-MS. Anal Bioanal Chem 388:627–635
240. Noedler K, Voutsa D, Licha T (2014) Polar organic micropollutants in the coastal environment of different marine systems. Mar Pollut Bull 85:50–59
241. Weigel S, Kuhlmann J, Hühnerfuss H (2002) Drugs and personal care products as ubiquitous pollutants: occurrence and distribution of clofibric acid, caffeine and DEET in the North Sea. Sci Total Environ 295:131–141
242. Urase T, Kikuta T (2005) Separate estimation of adsorption and degradation of pharmaceutical substances and estrogens in the activated sludge process. Water Res 39:1289–1300
243. Stancova V, Plhalova L, Bartoskova M, Zivna D, Prokes M, Marsalek P, Blahova J, Skoric M, Svobodova Z (2014) Effects of mixture of pharmaceuticals on early life stages of tench (Tinca tinca). Biomed Research International
244. Salgado R, Pereira VJ, Carvalho G, Soeiro R, Gaffney V, Almeida C, Vale Cardoso V, Ferreira E, Benoliel MJ, Ternes TA, Oehmen A, Reis MAM, Noronha JP (2013) Photodegradation kinetics and transformation products of ketoprofen, diclofenac and atenolol in pure water and treated wastewater. J Hazard Mater 244:516–527
245. Sousa MA, Goncalves C, Vilar VJP, Boaventura RAR, Alpendurada MF (2012) Suspended TiO_2-assisted photocatalytic degradation of emerging contaminants in a municipal WWTP effluent using a solar pilot plant with CPCs. Chem Eng J 198:301–309
246. Calza P, Sakkas VA, Medana C, Baiocchi C, Dimou A, Pelizzetti E, Albanis T (2006) Photocatalytic degradation study of diclofenac over aqueous TiO_2 suspensions. Appl Catal B-Environ 67:197–205
247. Mugunthan E, Saidutta MB, Jagadeeshbabu PE, Visible light assisted photocatalytic degradation of diclofenac using TiO_2-WO_3 mixed oxide catalysts. Environ Nanotechnol, Monit Manage 10:322–330
248. Ribeiro AR, Nunes OC, Pereira MFR, Silva AMT (2015) An overview on the advanced oxidation processes applied for the treatment of water pollutants defined in the recently launched directive 2013/39/EU. Environ Int 75:33–51
249. Iovino P, Chianese S, Canzano S, Prisciandaro M, Musmarra D (2017) Photodegradation of diclofenac in wastewaters. Desalin Water Treat 61:293–297
250. Dhaka S, Kumar R, Deep A, Kurade MB, Ji S-W, Jeon B-H (2019) Metal-organic frameworks (MOFs) for the removal of emerging contaminants from aquatic environments. Coord Chem Rev 380:330–352
251. Ahmed MJ (2017) Adsorption of non-steroidal anti-inflammatory drugs from aqueous solution using activated carbons: review. J Environ Manage 190:274–282
252. Nam SW, Jung C, Li H, Yu M, Flora JRV, Boateng LK, Her N, Zoh KD, Yoon Y (2015) Adsorption characteristics of diclofenac and sulfamethoxazole to graphene oxide in aqueous solution. Chemosphere 136:20–26
253. Antunes M, Esteves VI, Guegan R, Crespo JS, Fernandes AN, Giovanela M (2012) Removal of diclofenac sodium from aqueous solution by Isabel grape bagasse. Chem Eng J 192:114–121
254. Desbrieres J, Guibal E (2018) Chitosan for wastewater treatment. Polym Int 67:7–14
255. Xiong T, Yuan X, Wang H, Wu Z, Jiang L, Leng L, Xi K, Cao X, Zeng G (2019) Highly efficient removal of diclofenac sodium from medical wastewater by Mg/Al layered double hydroxide-poly(m-phenylenediamine) composite. Chem Eng J 366:83–91
256. Maia GS, de Andrade JR, da Silva MGC, Vieira MGA (2019) Adsorption of diclofenac sodium onto commercial organoclay: kinetic, equilibrium and thermodynamic study. Powder Technol 345:140–150
257. Almeida HFD, Neves MC, Trindade T, Marrucho IM, Freire MG (2020) Supported ionic liquids as efficient materials to remove non-steroidal anti-inflammatory drugs from aqueous media. Chem Eng J 381:122616

258. Barczak M, Borowski P (2019) Silica xerogels modified with amine groups: influence of synthesis parameters on porous structure and sorption properties. Microporous Mesoporous Mater 281:32–43
259. Fontanals N, Ronka S, Borrull F, Trochimczuk AW, Marce RM (2009) Supported imidazolium ionic liquid phases: a new material for solid-phase extraction. Talanta 80:250–256
260. Bouchal R, Miletto I, Thach UD, Prelot B, Berlier G, Hesemann P (2016) Ionosilicas as efficient adsorbents for the separation of diclofenac and sulindac from aqueous media. New J Chem 40:7620–7626
261. Landry KA, Boyer TH (2013) Diclofenac removal in urine using strong-base anion exchange polymer resins. Water Res 47:6432–6444
262. Ghemit R, Makhloufi A, Djebri N, Flilissa A, Zerroual L, Boutahala M (2019) Adsorptive removal of diclofenac and ibuprofen from aqueous solution by organobentonites: study in single and binary systems. Groundwater Sustain Dev 8:520–529
263. Salaa F, Bendenia S, Lecomte-Nana GL, Khelifa A (2020) Enhanced removal of diclofenac by an organohalloysite intercalated via a novel route: performance and mechanism. Chem Eng J 396:125226
264. Krajisnik D, Dakovic A, Milojevic M, Malenovic A, Kragovic M, Bogdanovic DB, Dondur V, Milic J (2011) Properties of diclofenac sodium sorption onto natural zeolite modified with cetylpyridinium chloride. Colloids Surf B-Biointerfaces 83:165–172
265. Lin K-YA, Yang H, Lee W-D (2015) Enhanced removal of diclofenac from water using a zeolitic imidazole framework functionalized with cetyltrimethylammonium bromide (CTAB). Rsc Advan 5:81330–81340
266. Hasan Z, Khan NA, Jhung SH (2016) Adsorptive removal of diclofenac sodium from water with Zr-based metal-organic frameworks. Chem Eng J 284:1406–1413
267. Luo Z, Fan S, Liu J, Liu W, Shen X, Wu C, Huang Y, Huang G, Huang H, Zheng M (2018) A 3D stable metal-organic framework for highly efficient adsorption and removal of drug contaminants from water. Polymers 10
268. Bhadra BN, Jhung SH (2017) A remarkable adsorbent for removal of contaminants of emerging concern from water: porous carbon derived from metal azolate framework-6. J Hazard Mater 340:179–188
269. Bhadra BN, Ahmed I, Kim S, Jhung SH (2017) Adsorptive removal of ibuprofen and diclofenac from water using metal-crossmark organic framework-derived porous carbon. Chem Eng J 314:50–58
270. An HJ, Bhadra BN, Khan NA, Jhung SH (2018) Adsorptive removal of wide range of pharmaceutical and personal care products from water by using metal azolate framework-6-derived porous carbon. Chem Eng J 343:447–454
271. Suriyanon N, Punyapalakul P, Ngamcharussrivichai C (2013) Mechanistic study of diclofenac and carbamazepine adsorption on functionalized silica-based porous materials. Chem Eng J 214:208–218
272. Salomao GR, Americo-Pinheiro JHP, Isique WD, Torres NH, Cruz IA, Ferreira LFR Diclofenac removal in water supply by adsorption on composite low-cost material. Environ Technol 17
273. Liang XX, Omer AM, Hu ZH, Wang YG, Yu D, Ouyang XK (2019) Efficient adsorption of diclofenac sodium from aqueous solutions using magnetic amine-functionalized chitosan. Chemosphere 217:270–278
274. Ebele AJ, Abou-Elwafa Abdallah M, Harrad S (2017) Pharmaceuticals and personal care products (PPCPs) in the freshwater aquatic environment. Emerging Contam 3:1–16
275. Siddique HMA, Kiani AK (2020) Industrial pollution and human health: evidence from middle-income countries. Environ Sci Pollut Res 27:12439–12448
276. Van Rooij B, McAllister L (2012) Environmental challenges in middle-income countries: a comparison of enforcement in Brazil, China, Indonesia, and Mexico. Avoiding the Middle-Income Trap, Law and Development of Middle-Income Countries, pp 288–306
277. Raut P (2014) Role of higher education institutions in environmental conservation and sustainable development: a case study of Shivaji University, Maharashtra, India. J Environ Earth Sci 4:30–34

278. Miyan M, Salam A, Nuruzzaman M, Naznin S (2016) Investment analysis of environment pollution in educational institutions. Int J Sci Technol Res 5:231–234
279. Stavropoulos S, Wall R, Xu Y (2018) Environmental regulations and industrial competitiveness: evidence from China. Appl Econ 50:1378–1394
280. EETAP (2018) Environmental education—a tool for the prevention of water pollution? In: EETAP (Environmental education and training partnership)
281. Binder S, Neumayer E (2005) Environmental pressure group strength and air pollution: an empirical analysis. Ecol Econ 55:527–538
282. Edsand H-E, Broich T (2020) The Impact of environmental education on environmental and renewable energy technology awareness: empirical evidence from Colombia. Int J Sci Math Educ 18:611–634
283. Ateia M, Yoshimura C (2016) In-situ biological water treatment technologies for environmental remediation: a review. J Bioremed Biodegradation 7:1–8

Chapter 12
Removal of Nitrogen Oxyanion (Nitrate) in Constructed Wetlands

Fidelis O. Ajibade, Nathaniel A. Nwogwu, Kayode H. Lasisi,
Temitope F. Ajibade, Bashir Adelodun, Awoke Guadie, Adamu Y. Ugya,
James R. Adewumi, Hong C. Wang, and Aijie Wang

Abstract The increasing levels of nitrogen oxyanion pollution especially nitrate in water environments have become a critical issues of concern because of the potential risk on ecology and human health. Owing to its distinctive merits of sustainability, lesser operational and maintenance expenditure, the utilization of constructed wetland systems for the treatment of wastewater has turned out to be predominant worldwide. Its nitrogen oxyanion removal performance has received significant attention in the last two decades. This chapter presents a comprehensive outline of the application of constructed wetlands (CW) for nitrogen oxyanion removal from water

F. O. Ajibade (✉) · K. H. Lasisi · T. F. Ajibade · J. R. Adewumi
Department of Civil and Environmental Engineering, Federal University of Technology, PMB 704, Akure, Nigeria
e-mail: foajibade@futa.edu.ng

F. O. Ajibade · N. A. Nwogwu · A. Guadie · H. C. Wang · A. Wang (✉)
Research Centre for Eco-Environmental Sciences, Chinese Academy of Sciences, Beijing 100085, PR China
e-mail: ajwang@rcees.ac.cn

F. O. Ajibade · N. A. Nwogwu · K. H. Lasisi · T. F. Ajibade
University of Chinese Academy of Sciences, Beijing 100049, PR China

N. A. Nwogwu
Department of Agricultural and Bioresources Engineering, Federal University of Technology, Owerri, Nigeria

K. H. Lasisi · T. F. Ajibade
Institute of Urban Environment, Chinese Academy of Sciences, Xiamen 361021, PR China

B. Adelodun
Department of Agricultural and Biosystems Engineering, University of Ilorin, PMB 1515, Ilorin, Nigeria

Department of Agricultural Civil Engineering, Kyungpook National University, Daegu, South Korea

A. Y. Ugya
Department of Environmental Management, Kaduna State University, Kaduna State, Nigeria

College of New Energy and Environment, Jilin University, Changchun, PR China

© Springer Nature Switzerland AG 2021
N. A. Oladoja and E. I. Unuabonah (eds.), *Progress and Prospects in the Management of Oxyanion Polluted Aqua Systems*, Environmental Contamination Remediation and Management, https://doi.org/10.1007/978-3-030-70757-6_12

and wastewater. The removal mechanisms and transformations of nitrogen are also discussed. In addition, the major factors that influence the removal performances in CWs are elucidated, especially the types of carbon sources commonly used, and how it affects the denitrification process. This chapter would be useful to engineers and researchers in the field of water and wastewater engineering.

Keywords Carbon sources · Constructed wetlands · Nitrate · Nitrogen pollution · Wastewater

12.1 Background

12.1.1 Nitrate in the Environment

Nitrate (NO_3^-) is one of the major generic forms of nitrogen oxyanions that exist naturally in moderate concentrations in different environmental media. The oxidation of nitrites (NO_2^-) majorly generates nitrates during nitrification process of the nitrogen cycle. The nitrogen cycle is the biogeochemical cycle by which organic protein from animals and plants origin is converted into ammonia (NH_3) and then NO_2^-, and NO_3^- in the environment. The transformations of the different nitrogen forms are carried out via physicochemical and biological processes (Fig. 12.1). Owing to its high solubility in water, the presence of NO_3^- has adverse effects on the environment, as it greatly accounts for the pollution of soil, surface water and the groundwater [1]. Several wastewater types such as urban drainage, landfill leachate, industrial and agricultural wastewater that contain nitrogenous compounds initiate undesirable phenomena (e.g. eutrophication and methemoglobinemia (i.e. blue baby syndrome)) when they are released into water bodies [2–4]. The concentrations of NO_3^- in these wastewaters vary from low to high, and thus, demand an appropriate technique for the removal. According to Rajmohan et al. [5], the usual NO_3^- level in polluted water ranges from 200 to 500 mg/L, based on the nature of the source (Table 12.1), but wastewater from nuclear industries contain up to 50,000 mg/L of NO_3^-.

Excess NO_3^-, discharged from the large-scale utilization of agricultural fertilizers, concentrated livestock feeding operations and disposal of partially treated sewage, that enters the groundwater, is among the priority pollutants of the groundwater system. Over 10,000 public water supply wells are estimated to have high levels of nitrate in the USA and thousands of wells were also ascertained with nitrate concentrations at or above the established health standards, across Western Europe and Asia [14]. Consequently, the maximum permissible concentration limit of NO_3^- in drinking water was set at 10 mg/L as nitrate-nitrogen (NO_3–N) by the US Environmental Protection Agency, while 50 mg/L NO_3^- was set by World Health Organization to address the concerns of methemoglobinemia in infants [14, 15]. High levels of NO_3^- is recognized to cause environmental and public health issues. The presence of this nitrogen oxyanion in water environments is a global challenge that needs urgent attention. To this end, various technological solutions, including electrodialysis, chemical

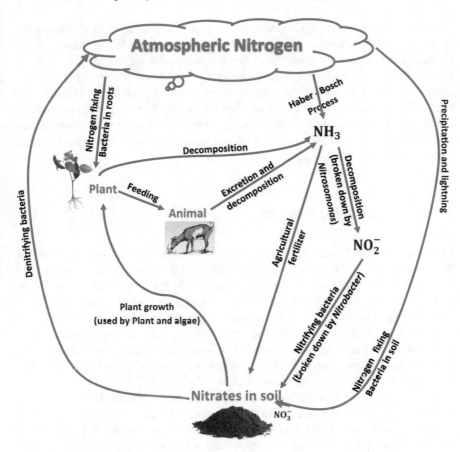

Fig. 12.1 Nitrogen cycle

Table 12.1 Nitrate levels at various sources as reported in the literature

Wastewater source	Nitrate level (mg/L)	References
Domestic wastewaters/septic tanks	70–85	[6, 7]
Fertilizer, Diaries, metal finishing industries	200	[8]
Tannery, Pisa, Italy	222	[9]
Glasshouses waste	325	[10]
Brackish water	1000	[11]
Explosives factory, China	3600	[12]
Nuclear industry	50,000	[13]

Adapted from (Rajmohan et al. 2017)

reduction, membrane separation, adsorption, sequencing batch reactor, moving bed bioreactors, electrochemical denitrification, reverse osmosis; ion exchange, photo-catalytic degradation and membrane bioreactors have been developed to solve this menace [3, 16–27]. However, these technologies are always limited by their costly installation and high operational cost, secondary pollution, sludge production that need disposal, incomplete removal efficiency [28, 29]. In this chapter, constructed wetland (CW) systems, which are generally cost effective, simple, environmentally non-disruptive, ecologically sound, with relatively low maintenance cost, will be expounded in relation to NO_3^- removal.

12.1.2 Constructed Wetlands

Constructed wetlands, also referred to as treatment wetlands, are engineered systems that are designed and fabricated to treat several kinds of wastewater with relatively low external energy requirements and operationally simple technology and main-tenance (Fig. 12.2) [30–35]. Milani et al. [36] defined CW as a "sustainable and efficient solutions used around the world to treat wastewater as an alternative or a supplement to intensively engineered treatment plants". They are complex, inte-grated systems that involve the interaction of soil, water, plants, animals, microbes and the environment. The CWs have become an essential alternative wastewater treatment system since the method combines relatively high performance of pollu-tant removal with low maintenance and simple operation [37]. The CWs are planned methods designed and constructed to apply the natural procedures involving wetland vegetation, soils and the associated microbial assemblages to assist in wastewater treatment. It can effectively remove suspended solids, organic pollutants and nutri-ents from wastewater [38–40]. The CWs provide an inexpensive and reliable method for treating a variety of wastewaters such as sewage, landfill leachate, mine leachate, urban storm-water and agricultural run-off. This system of treatment is very efficient for nutrient removal and comparatively simple to construct, operate, maintain and suitable for advanced and polishing treatment if water reuse is an option [41]. The main NO_3^- removal mechanisms in wetlands are seepage loss, plant uptake and deni-trification [42] which are further expatiated in Sect. 12.2. Table 12.2 summarizes research studies on NO_3^- removal using CWs.

12.2 Nitrogen Transformation in Constructed Wetlands

As an ecological treatment technology, CWs have been largely utilized in recent decades in wastewater treatment plants. Before the arrival of CWs technology, conventional activated sludge-type wastewater treatment plants have been used for nitrogen removal but only minimal quantity is removed, via the consumption of the organic matter fraction of the wastewater.

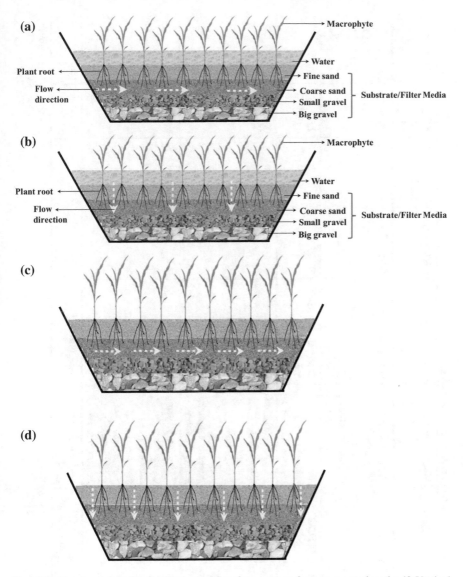

Fig. 12.2 Constructed wetlands **a** Horizontal flow free water surface constructed wetland **b** Vertical flow free water surface constructed wetland **c** Horizontal sub-surface flow constructed wetland **d** Vertical sub-surface flow constructed wetland

In wastewater treatment operations, nitrate is removed through a process known as denitrification. This is a process, where organic and ammonia nitrogen is converted, through a process known as nitrification, to NO_3^- in the absence of oxygen (an anaerobic environment). The NO_3^- produced through nitrification is further reduced to nitrogen gas in this same anoxic environment, thus completing the denitrification process [65, 66]. This process is carried out by several range of autotrophic and

Table 12.2 Nitrate removal performance from various constructed wetland types

Source of WW	NO_3–N Influent (mg/L)	CW types	Macrophyte type	Substrate type	HRT (days)	Carbon source	NO_3–N removal (%)	References
Sullage	14.40 ± 0.23	HSSFCW	Water hyacinth (*Eichhornia crassipes*), *Commelina cyanea*, *Phragmites australis*	Sandstone, gravel	6–12	–	50–94	[30]
Aquaculture	8.50 ± 0.06	VDFCW	*Sacciolepsis Africana; Commelina cyanae*	Charcoal, fine sand and gravel	5	–	2.50 ±0.09	[43]
Synthetic NO_3^-	21–47	VUFCW	*Phragmite australis, Commelina communis, Penniserum purpureum, Ipomoea aquatica, Pistia stratiotes*	Fine sand	4.2	Fructose	60–97	[44]
Synthetic	22 ± 2.89	VUFCW	*Phragmite australis, Commelina communis, Penniserum purpureum, Ipomoea aquatica, Pistia stratiotes*	Fine sand	4.2	Fructose	70–99	[45]
Industrial	13	Hybrid CW (FWS-VF)	*Schoenoplectus validus*	Sand, gravel	6	–	78	[46]

(continued)

Table 12.2 (continued)

Source of WW	NO$_3$-N Influent (mg/L)	CW types	Macrophyte type	Substrate type	HRT (days)	Carbon source	NO$_3$-N removal (%)	References
Secondary effluent from WWTP	15.12 ± 1.99	ICWS	*Vetiver zizanioides, Coix lacrymajobi L*	Sand, Steel slag, Peat	10	–	76	[47]
NO$_3^-$ contaminated groundwater	12 ± 0.5	Hybrid (unsaturated VF – HSSFCW)	*Phragmites australis*	Cork, granitic gravel	2	–	80–99	[48]
Synthetic	20	VFCW	*Iris pseudacorus*	Iron scraps, biochar, sand, gravel	3	Sucrose	87	[49]
Synthetic	12.01 ± 1.32	HSSFCW	*Canna indica L*	Sand, gravel	5	–	83	[50]
Synthetic agricultural run-off	7.4	FWSCW	–	Hydrilla verticillate —GS, PS, gravel; Vallisneria—GS, PS, gravel	–	Glucose	90–100	[51, 52]
Synthetic	50	VFCW	*Acorus calamus L*	Gravel	3	SCSs; SCSs combined with NZVI; Without SCSs and NZVI	47–91	[53]
Synthetic	65/100/150	VFCW	-	Peat soil	–	–	49–83	[54]
Domestic	8.6	VFCW	Water hyacinth (*Eichhornia crassipes*),	Hydroponic	–	–	100	[55]

(continued)

Table 12.2 (continued)

Source of WW	NO$_3$–N Influent (mg/L)	CW types	Macrophyte type	Substrate type	HRT (days)	Carbon source	NO$_3$–N removal (%)	References
Synthetic	10.83 ± 1.61	IVCW	*Arundo donax, Canna indica*	Gravel	2.4	–	59–79	[56]
Hospital	0.9 ± 0.2(dry) 1.1 ± 0.5 (rainy)	HSSFCW	Cattails (*Typha domingensis*), *Cyperus papyrus*, dark green bulrush (*scirpus atrovirens*) and sugar cane (*Saccharum officinarum*)	Brick, gravel	4	–	80 (dry season) 81 (rainy season)	[57]
Domestic	14.4 ± 1.6	HSSFCW	*Cyperus papyrus, Canna indica, Hedychium coronarium*	Porous stone, sand, tepezil	3	–	44	[58]
Synthetic	10–30	VFCW	*Canna indica L*	Quartz sand, Sponge iron particles (S-Fe^0)	0.25–0.5	Glucose	16–87	[59, 60]
Synthetic	100	HSSFCW	*Phragmites australis, Typha angustifolia*	River gravel, breakstone	–	Rice husk, glucose	96	[61]
Synthetic	10.04 ± 1.29	VUFCW	*Typha orientalis*	Ceramsite	5	PHBV	93–99	[62]
Synthetic	15 (TN)	EFTW	*Canna indica L*	Styrofoam floating mat	7	Glucose	83–98	[63]

(continued)

Table 12.2 (continued)

Source of WW	NO_3–N Influent (mg/L)	CW types	Macrophyte type	Substrate type	HRT (days)	Carbon source	NO_3–N removal (%)	References
Municipal	5.7 ± 2.4	VPSCW	*Canna hybrids, Zantedeschia aethiopica*	Tezontle	–	–	82–84	[64]

HRT: Hydraulic retention time; WW: Wastewater; WWTP: Wastewater treatment plant; VDFCW: Vertical downward flow constructed wetland; FWSCW: Free water surface constructed wetland; ICWS: Integrated constructed wetland system; VPSCW: Vertically partially saturated constructed wetland; EFTW: Enhanced floating treatment wetland; VF: Vertical flow; VFCW: Vertical flow constructed wetlands; GS: Granitic sulphur; PS: Pretreated soil; HSSFCW: horizontal sub-surface flow constructed wetland; VUFCW: vertical upward flow constructed wetland; SCSs: Solid carbon sources; IVCW: Integrated vertical constructed wetlands; PHBV: Poly(3-hydroxybutane-hydroxyvalerate) heterotrophic facultative anaerobic bacteria, which are capable of utilizing NO_3^- (and NO_2^-), under anoxic conditions, as an electron acceptor [67]. Some of these bacteria include *Pseudomonas, Micrococcus, Bacillus, Paracoccus denitrificans and Achromobacter.* For better nitrogen removal, an external organic carbon source is needed to act as an electron donor in the respiratory chain [38], and CWs are a better option to achieving this.

Although the eutrophication and the toxic effects of NO_3^- on aquatic organisms of both vertebrate and invertebrate species are sources of concerns [68], it also boosts plants' growth, which sequentially promotes the environmental biogeochemistry in the wetlands. The circulation of nitrogen in wetlands involves composite processes, while very straightforward chemical conversion of this element still poses a great task in environmental engineering. Such processes, which include bacterial actions, plant/microbial uptake, adsorption (interaction between ionized NH_3 and the media in sub-surface horizontal flow, (SSHF) CWs), and volatilization (i.e. transformation of aquatic NH_4^+ to gaseous NH_3, within the operating pH regime of the surface flow CW), mostly achieved nitrogen removals in wetlands [68–70].

Nitrogen transformation involves some processes and mechanisms, which lead to the transference of wetland nitrogen from one point to the other without any consequential molecular alteration [69]. As earlier noted, the physical processes of management of nitrogen oxyanion in CW include, settling of particles and re-suspension, dissolution and diffusion, plant translocation, litterfall, volatilization and sorption [68, 69]. Generally, nitrogen oxyanion removal in CW occurs through two processes that include biological and physicochemical treatment processes. The five major biological treatment process include denitrification, nitrification, ammonification (mineralization), assimilation and decomposition [69, 71, 72]. The physicochemical processes include, sedimentation, NH_3 stripping, breakpoint chlorination and ion exchange [70, 73]. It was suggested that low oxygen and organic matter contents in the root zone offers restriction to nitrification and denitrification processes [69]. However, an integration of partial nitrification and anaerobic NH_4^+ oxidation has equally been recommended to be resourceful in removing nitrogen from constructed wetlands. This is largely due to the autotrophic nature of anaerobic ammonia oxidation (Anammox) process, in which NH_4^+ is completely converted into nitrogen gas in the presence of NO_2^- and without the addition of organic matter [69].

12.2.1 Ammonification

The ammonification refers to the process by which the organic nitrogen fraction is transformed to NH_3, through a biological process [71]. The first stage of nitrification in sub-surface flow CW (SSFCW) systems is initiated by ammonification, if the inbound wastewater is highly loaded with organic nitrogen [74]. This biochemical process, where the amino acids fractions are exposed to oxidative deamination yielding NH_3 is acomplex and exergonic process, (Eq. 12.2.1) [74, 75].

$$\text{Amino acids} \rightarrow \text{Imino acids} \rightarrow \text{Keto acids} \rightarrow NH_3 \qquad (12.2.1)$$

Since the process of occurrence decreases with depth, it shows that ammonification is quickest within the upper zone of the wetlands, where the aerobic condition is prominent. It is time-consuming within the lower zone, where the environment moves from facultative anaerobic condition to obligate anaerobic condition [71, 76]. In CWs, the inorganic ammoniacal-nitrogen is mostly removed by nitrification–denitrification processes, but ammonification kinetically progresses faster than nitrification [71]. Kadlec and Knight [70] suggested that the ammonification process progresses quicker in higher temperature, doubling the rate with a temperature rise of 10 °C. The pH range observed to be ideal for ammonification is 6.5–8.5 [74, 77, 78]. The ammonification process is therefore generally affected by pH, temperature, carbon-to-nitrogen (C/N) ratio, soil structure and available nutrient [76]. Furthermore, processes such as adsorption, plant uptake and volatilization are suggested to be resourceful in ammonia–nitrogen removal [38], though the effectiveness of nitrification–denitrification processes is, in general, suggested to be the most resourceful in NH_4^+ removal [71].

12.2.2 Nitrification

Nitrification is the major transformation mechanism by which the level of ammonia nitrogen is reduced. This reduction is achieved through the conversion of the ammonia nitrogen into oxidized form of nitrogen (i.e. NO_2^- and NO_3^-). Graaf et al. [79] defined nitrification as the biological formation of nitrate or nitrite from compounds containing reduced nitrogen with oxygen (O_2) as their terminal electron receptor. Lee et al. [71] defined it as the chemolithoautotrophic oxidation of NH_3 to NO_3^- in the presence of adequate O_2, occurring in two successive oxidative steps, namely ammonia oxidation (NH_3 to NO_2^-) and nitrite oxidation (NO_2^- to NO_3^-), carried out by nitrifying bacteria. These bacteria use NH_3 or NO_2^- as an energy source, O_2 as the terminal electron recipient and carbon dioxide as the carbon source [71]. The first stage is the oxidation of NH_3 to NO_2^-, by ammonium oxidizing bacteria such as *Nitrosomonas* or *Nitrospira* or *Nitrosococcus* (Eq. 12.2.2) [71, 74].

$$NH_4^+ + 1.5O_2 \xrightarrow{\text{Nitroso - genus}} NO_2^- + H_2O + 2H^+ \qquad (12.2.2)$$

The above first stage is succeeded by the second stage which is the oxidation of NO_2^- by nitrite-oxidizing bacteria such as *Nitrobacter* or *Nitrospira*. The second stage is described by Eq. 12.2.3 [71, 74].

$$NO_2^- + 0.5O_2 \xrightarrow{\text{Nitro - genus}} NO_3^- \qquad (12.2.3)$$

The oxygen consumption of nitrification process is estimated to be 3.16 mg O_2 per mg $NH_4 - N$ oxidized, and 1.11 mg O_2 per mg $NO_2 - N$ oxidized, while *Nitrosomonas* and *Nitrobacter* produce 0.15 mg cells per mg $NH_4 - N$ oxidized and 0.02 mg cells per mg $NO_2 - N$ respectively [71]. Furthermore, alkalinity is necessary as 7.07 mg $CaCO_3$ per mg $NH_4 - N$ oxidized [80]. The acid formation (i.e. low pH value) during nitrification process causes alkalinity reduction and a deep reduction in pH [68, 80–82], and a swift decline in the nitrification rate below the neutral pH value [81]. Hence, it is important to replenish the alkaline level with lime during the process, when there is a drop in alkalinity [80]. Though nitrification is basically attributed to chemoautotrophic bacteria, it is suggested that heterotrophic nitrification takes place, which can be significant [68]. Aside from autotrophic nitrification, heterotrophic nitrifying bacteria are also capable of producing $NO_3 - N$. Some of these species (in bacteria, algae and fungi) are *Actinomycetes, Arthrobacter globiformis, Aerobacter aerogenes, Bacillus, Mycobacterium phlei, Streptomyces griseus, Theosphaera and Pseudomonas* [38, 74, 83]. Gerardi [83] affirmed that although these heterotrophic nitrifiers are resourceful, the nitrification rates achieved by *Nitrosomonas* and *Nitrobacter* groups are significantly greater (relatively greater by 1000 to 10,000 times) [74]. However, owing to constraints against nitrification and denitrification processes, offered by low oxygen and organic matter concentration in SSF, it has been affirmed that a combination of partial nitrification and Anammox is a resourceful means of removing nitrogen from CWs [69]. Moreover, since the Anammox process is autotrophic, the transformation of NH_4^+ to nitrogen could be possible without adding organic matter [69].

12.3   Denitrification

Kadlec and Wallace [68] defined denitrification as the process by which NO_3^- is transformed to dinitrogen (N_2) via intermediates such as NO_2^-, nitric oxide, and nitrous oxide, and finally nitrogen (Eq. 12.2.4). The denitrification process is also called NO_3^- dissimilation, and it is accomplished by facultative heterotrophic organisms that can use NO_3^- as the terminal electron receptor, and organic carbon as an electron donor under anoxic condition [71]. During the transformation, inorganic nitrogens such as NO_2^- and NO_3^- are usually reduced to harmless nitrogen gas by denitrifying bacteria [71, 84, 85]. Some denitrifiers require organic substrates to get their carbon source for growth and evolution, whereas others use inorganic substances as their energy sources and CO_2 as their carbon source [86]. Therefore, denitrifying bacteria are categorized into two main species, namely autotrophs and heterotrophs [71]. However, earlier studies have focussed on the heterotrophic denitrification process, due to its frequency in conventional wastewater treatment plants [71, 87], while the autotrophic denitrification process started gaining attention in recent studies [88–94].

Moreover, denitrification is led by some heterotrophic microorganisms like *Pseudomonas, Micrococcus, Achromobacter and Bacillus,* under anaerobic or low-oxygen conditions. Denitrificating microbes can be grouped as: organotrophs (e.g.

Pseudomonas, Alcaligenes, Bacillus, Agrobacterium, Flavobacterium, Propionibacterium and *Vibrio*), chemolithotrophs (e.g. *Thiobacillus, Thiomicrospira, Nitrosomonas*), photolithotrophs (e.g. *Rhodopsuedomonas*), diazotrophs (e.g. *Rhizobium, Azospirillum*), archaea (e.g. *Halobacterium*) and other microorganisms such as *Paracoccus* or *Neisseria* [68]. The fraction of total nitrogen removal through denitrification is normally 60–95%.

$$2NO_3^- \rightarrow 2NO_2^- \rightarrow 2NO \rightarrow 2N_2O \rightarrow N_2 \qquad (12.2.4)$$

12.3.1 Assimilation Process of Nitrogen

The uptake of nitrogen by plants or microbes is regarded as the assimilation process. Masclaux-Daubresse et al. [95] and Xu et al. [96] asserted that the usage of nitrogen by plants encompasses numerous stages, including uptake, assimilation, translocation and, when the plant is ageing, recycling and remobilization. The assimilation process occurs via the formation of organic nitrogen compounds such as amino acids from inorganic nitrogen compounds available in the environment. Organisms like plants, fungi and specific bacteria that cannot fix nitrogen gas (N_2) rely on the ability to assimilate NO_3^- or NH_3 for their needs. Animals also depend fully on the organic nitrogen form for their food. Several studies have affirmed the significance of the removal of NH_3 from water by wetland plants [97–104]. However, many of these studies are commonly seen to portray the measurement of gross nitrogen uptake, without deduction for consequential losses due to plant death and decomposition, with associated leaching as well as re-solubilization of nitrogen [68].

For nitrogen removal from the wetland water, the attention is usually on the net influence of the macrophytes (macroflora) on the water phase concentrations [68]. When discussing plant uptake as a process of nitrogen removal from wetland water, terms such as phytomass (the totality of vegetative materials, living and dead), biomass (all living vegetative materials) and necromass (all dead vegetative materials) are often used [68]. Macrophytes are vital in enhancing nitrogen removal from wetlands due to their functions such as providing surfaces and O_2 for the growth of microbes within the rhizosphere, thus improving nitrification [99, 105–107], and providing carbon from root secretions (due to photosynthetically fixed carbon, within a range of 5–25% C), enhancing organics removal and denitrification process [97, 108–111]. Various relative researches between unplanted and planted wetlands indicated good nitrogen and organics removal, with the latter yielding more significant results, hence indicating the necessity of macroflora for enhancing nitrogen removal operations in CWs [74].

Inorganic nitrogen forms are usually transformed into organic compounds through the uptake of NH_3 and NO_3^- by macrophytes. This serves as the building blocks for cells and tissues [78]. The ability of rooted plants to utilize sediment nutrients partly describes their massive yield in comparison with planktonic algae in many systems [112]. Different plant species have varying ability in their ideal nitrogen forms absorbed, and the nutrient concentration of plants tissues also influences the uptake and storage rate of nutrient [71]. However, NH_4^+ preference is conventional in macroflora within NH_4^+ -rich environments where restricted nitrification occurs [113]. In general, the uptake of nitrogen by plants varies along with system configurations, loading ranges, type of wastewater and environmental conditions [74]. In nitrogen removal, plants contribution is affirmed to be about 0.5–40.0% of the total nitrogen removal [74, 103, 104]. Plant biomass accumulates 60% of total nitrogen thus, enhancing nitrogen removal significantly [103]. For efficient nutrient assimilation and storage, plants with features such as high tissue nutrient content, rapid growth and ability to achieve high-standing crops are preferably desired. On the contrary, plants with immense biomass accumulation during autumn and winter have a likelihood of releasing a considerable amount of their stored nitrogen back into the water during the winter season [38]. Brodrick et al. [114] equally suggested that decaying plant materials could also raise the concentration of nutrients in the effluent through leaching [74].

Some selected plants have been employed in constructed wetlands; however, *Phragmites australis* remains the most typical plant used in SSFCW due to its capability to pass O_2 from its leaves through the stems and rhizomes and out of from its fine hair roots into the rhizosphere [115]. Reports from literature about the ability of the plant to convey oxygen (thereby fostering microbial conversion and nitrification) express various illustrations [74, 116–118]. Armstrong et al. [116] noted O_2 release (per unit wetland area) by phragmites species to be in the range of 5–12 g O_2 per square metre per day, while the O_2 release by phragmites in a study by Brix and Schierup [117] gives a record of only 0.02 g O_2 per square meter in soil substrate. The oxygen released by phragmites species recorded by Bavor et al. [118] is about 0.8 g O_2 per square meter in gravel substrate [74]. Figure 12.3 represents the major typical routes for nitrogen removal in SSFCWs.

12.4 Factors Affecting Nitrogen Removal Efficiency in CWs

Nitrogen oxyanion removal efficiency, especially NO_3^-, in CWs has been discussed to involve various biological and physicochemical processes. Therefore, various environmental factors are bound to affect the efficiencies of these processes, thereby limiting the oxyanion removal efficiency. Some of such factors include pH, temperature, hydraulic residence time (HRT), NO_2^- concentration, oxygen concentration, vegetation type (wetland plant species) and density, activity of microorganism, distribution of wastewater, climate, and attributes of influent[71, 74, 119–122]. It should be noted that most of these factors are interdependent; hence, a variation of one factor often leads to a consequent change in other factors [121]. Furthermore, Kuschk et al. [123] stressed that the two major factors affecting the nitrogen removal from CWs are temperature and HRT [71]. The following subsections give a concise analysis of the key factors influencing nitrogen removal efficiency in CWs.

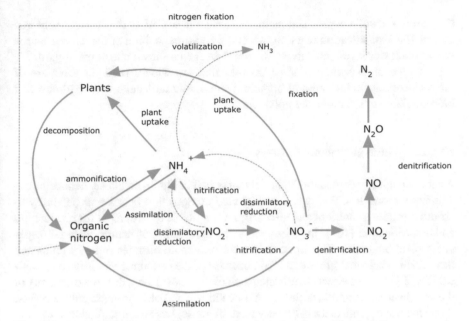

Fig. 12.3 Key classical nitrogen removal routes in sub-surface flow wetlands [74]

12.4.1 Carbon Source

Amidst all the aforementioned factors, the carbon source is one of the known dominant external factors which has disreputed the nitrogen oxyanion removal efficiency of CWs [56, 124]. Other factors to be considered in the choice of the carbon source include cost, handling and storage safety/stability, denitrification rate, degree of utilization, kinetics, sludge production, the content of unfavourable/toxic compounds. A commonly used carbon source, which is readily available, and with a high denitrification rate is methanol. Other closely related examples are ethanol and acetic acid. Although the nitrogen removal efficiency, using the methanol carbon source is desirable, the existence of NO_2^- accumulation in wastewater with high NO_3^- concentration often results in bacteria growth suppression [125]. Furthermore, the danger of overapplication of easily biodegradable materials of these liquid carbon sources through aerobic degradation can adversely impact nitrogen oxyanion removal [126]. Considering the aforementioned shortcomings, plant-based carbon sources have been considered [124, 127], and used to treat different wastewater, such as domestic sewage, agricultural run-off and industrial effluent. At present, most of the denitrifying bacteria in CWs are heterotrophic which require organic carbon sources for the substantive effect on denitrification process for nourishment and NO_3^- reduction [128]. In a study by Zhao and Chen [129], it was discovered that ammonia nitrogen, nitrous nitrogen and total nitrogen (when alkali-treated corn stover was used as additional carbon source material) were removed in the upper and middle layers, while nitrate is removed mainly at the bottom layer. Xiao et al. [130] showed the addition of solid carbon source to the vertical flow CW. In the system, there was almost 100% nitrification reaction, when the addition position of the carbon source was at the lower layer, thus giving the total nitrogen removal rate at the highest.

The type of exogenous carbon sources that are widely used for CWs are of three classes, including natural organic matters, low molecular carbohydrates and biodegradable macromolecule polymers.

12.4.1.1   Natural Organic Matters

Plant is a naturally degradable material that is rich in lignin, cellulose, hemicellulose and many more. In CWs, it is the most vital composition because of its ability to absorb nitrogen as nutrient and also to provide suitable environment for nitrification and denitrification [131]. In recent years, the application of natural material (especially plant biomass) as a carbon source for maximum nitrogen removal efficiency has gained substantial ground in CWs, because of the economic viability and practicability [132]. Their effects in nitrogen oxyanion removal vary, due to the diversity in the composition of lignin, cellulose, hemicellulose and other components in plants. Some natural organic matters (mostly plant biomass) have been studied to assess the effectiveness and efficiency of plant carbon source in CW denitrification rate [133] (Table 12.3). Conversely, these natural materials (especially for plant biomass) have some demerits, which include unstable carbon supply and discharge of coloured matter [134], which sometimes affect their applications.

12.4.1.2   Low Molecular Carbohydrate

Low molecular organic carbon sources have some desirable properties, which have also gained them recognition as an external carbon source. They are rich in carbon, which can easily be used up during decomposition. If classified in terms of physical form, they are liquid organic substances, which are termed liquid carbon sources. Examples are glucose [142–144], fructose [44], ethanol, methanol [145–147] and acetic acid [148].

12.4.1.3   Biodegradable Macromolecule Polymers

In recent time, a wide range of external carbon sources, which by the physical classification are solid organic substances, were checked in some laboratory studies, to function as physical support for biofilm formation in solid-phase denitrification system [149]. Some of the polymers were even blended. Examples of these polymer/polymer blends used so far are polybutylene succinate [150], polycaprolactone [151, 152], polyhydroxyalkanoates [153], polyvinyl alcohol [154], starch [155], starch/polyvinyl alcohol [156], poly(3-hydroxybutyrate-co-3-hydroxyvalerate)/poly(lactic acid) [157] and PHBV/starch [149].

Table 12.3 Plant biomass used as carbon source in CWs and their denitrification rates and removal efficiencies. Adapted from [133]

Carbon Source	Wastewater Treatment	Denitrification Rate (g m^{-3} d^{-1})	NO$_3^-$ − N Influent (mg L^{-1})	Effluent (mg L^{-1})	Removal efficiency (%)	References
Wheat straw	Drinking water	32–53	≈ 20	0	100	[135]
Sawdust	Groundwater	0.24–3.36	NA	NA	>95.0	[136]
Wheat straw	Simulated sewage	NA	200	>170	>15	[121]
G.verrucosa	Simulated sewage	13.2	100	0	100	[137]
Giant reed	Simulated sewage	3.36	100	0	100	[137]
Liquorice	Synthetic brackish water	20.64	100	0	100	[138]
Giant reed	Synthetic brackish water	85.92	100	13	87	[138]
Giant reed	Synthetic brackish water	101.52	100	0–13	87–100	[138]
Cotton wool	Aquaculture wastewater	NA	>200	<10	>95	[139]
Pine bark	Landfill leachates	33.6	600	0	100	[140]
Pine woodchip	Simulated sewage	2.4	17.2	11.6	32.6	[141]
Maize cobs	Simulated sewage	6.24	17.2	4.9	71.5	[141]
Wheat straw	Simulated sewage	4.56	17.2	8.9	48.3	[141]
Green waste	Simulated sewage	5.04	17.2	5.8	66.3	[141]
Sawdust	Simulated sewage	4.32	17.2	8.6	50.0	[141]
Eucalyptus	Simulated sewage	3.6	17.2	10.3	40.1	[141]
Maize cobs	Municipal portable water	19.8	141	0	100	[119]

(continued)

Table 12.3 (continued)

Carbon Source	Wastewater Treatment	Denitrification Rate (g m^{-3} d^{-1})	NO$_3^-$ − N Influent (mg L^{-1})	Effluent (mg L^{-1})	Removal efficiency (%)	References
Wheat straw	Municipal portable water	10.5	141	NA	NA	[119]
Softwood	Municipal portable water	5.8	141	NA	NA	[119]
Hardwood	Municipal portable water	3.0	141	NA	NA	[119]

NA: No data available

Hitherto, there have been some investigations of denitrification performance and microbial community structure in both liquid and solid carbon sources supported denitrification systems but their differences are scarcely studied. Srinandan et al. [158] reported that both liquid and solid organic carbon sources influence the nitrate removal activity, biofilm architecture and community structure although molecular weight and chemical structure of the biodegradable polymers were generally higher and more complicated when compared with the liquid carbon sources. Furthermore, denitrification performance and microbial diversity using starch/PCL and ethanol as an electron donor for nitrate removal were also investigated through comparison. The outcome revealed that the ethanol system displayed a higher denitrification rate while the blended starch/PCL system had richer microbial diversity [159].

Generally, when blended polymers and other external carbon sources were utilized, they yielded a good denitrification effect [160, 161] but the water of the solution of the blended polymer carbon source took some ample of time, causing some lag period. In addition, the morphology of the blended materials, surface properties and particle size posed great influence on the denitrification rate, with its biodegradability and denitrification performance decreasing with increasing molecular weight [162]. Thus, polymer/polymer blended carbon sources are not commonly used because of some factors such as high market price and slow release of carbon sources which requires a lot of time.

12.4.2 Selected Operating Parameters

12.4.2.1   pH

It has been established that nitrification process consumes alkalinity. Vymazal [38] affirmed that a pH >8.0 is capable of decreasing nitrification and denitrification processes to an insignificant level, with denitrification process occurring slowly at pH 5. Moreover, previous studies have suggested that high pH leads to a decline in dissolved oxygen (DO) in substrate [163], thus influencing nitrification and denitrification processes [32]. Also, some studies suggested that pH <6.0 and >8.0 hinder denitrification [32, 38, 81, 163], while the peak rate is observed at a pH range 7.0–7.5 [74, 164].

12.4.2.2 Temperature

Temperature is a significant environmental factor that controls the solid-phase denitrification process by hindering the activity of the associated enzymes in both hydrolysis of the solid substrate and reduction of NO_3^- [165]. In other words, temperature affects both microbial activities and diffusion rate of O_2 in constructed wetlands [166]. A temperature range between 16.5 and 32 °C is favourable for nitrification in CWs [74, 167], while the most efficient removal occurs at temperature that ranged between 20 and 25 °C [164, 166]. The nitrification- and denitrification-associated microbial activities decreased significantly at temperatures below 15 °C and above 30 °C [123].

Many studies have investigated the activities of denitrifiers in CW sediments during various climatic conditions and found that their activities are generally more robust in spring and summer than in autumn and winter [74, 168–171]. Oostrom and Russell [172] affirmed that, in general, the degree of removal of NO_3 is greater around summer than during winter [71]. Denitrification is usually believed to terminate at temperatures below 5 °C [71]. In soils, the optimal temperature limits for nitrification and ammonification are 30–40 °C and 40–60 °C, respectively [38].

12.4.2.3 Hydraulic Residence Time and Hydraulic Loading

Hydraulic residence time (HRT) is an important factor in nitrogen removal. The nitrogen removal efficiency is highly influenced by the flow condition and the residence time [4]. An increase in wastewater residence time leads to an intense decrease of ammonium and total Kjeldahl nitrogen concentrations in treated effluent [71]. This is because of the lengthier time of contact of nitrogen pollutant with microorganisms that gives advantage to the microbe to play a significant catabolic activity [74]. Lee et al. [71] also stated that lengthier HRT is necessary in nitrogen removal from wetlands than for BOD and COD removal. An eight-day HRT at a temperature above 15 °C is needed in SSFCWs [173]. However, if anaerobic conditions dominate in the wetlands, there is likelihood that an increase in HRT will not facilitate NO_3^- removal [174]. About 3–4 h HRT is required when NO_3^- concentration is not more than 40 mg L^{-1} and a minimum HRT of 6 h is necessary when NO_3^- concentration is more than 70 mg L^{-1} [175]. Hydraulic loading is also important in this regard, especially in SSFWs. Saeed and Sun [74] affirmed that the greater the hydraulic loading the faster the passage of wastewater through the media.

12.4.2.4 Dissolved Oxygen (DO)

As earlier discussed, most denitrifiers are facultative anaerobic organisms that use nitrate as a terminal electron recipient in the absence of oxygen or under anoxic condition. DO is a great and energetic electron recipient, and for that reason, it exhibits direct competition or inhibition of enzymes, which consequently results to suppression of the denitrification process [165]. Denitrification could happen at DO concentration to the level of 4.0–5.0 mg L^{-1} [165], though the denitrification rate declined with increase in DO levels [153]. Furthermore, it is vital to note that the presence of DO promotes upsurge in carbon source consumption as a portion of predisposed organic carbon is used up by aerobic respiration instead of denitrification [153, 176]. Since enzymatic actions in reducing nitrate can be inhibited by DO, there may occur nitrate accumulation [165]. The lower the oxygen concentration, the higher the denitrification becomes [177].

Denitrification rate of *Diaphorobacter nitroreducens* strain NA10B decreased as the DO concentration increased, when using poly(3-hydroxybutyrate-co-3-hydroxyvalerate) (PHBV) powders as carbon substrate, even when more than 3 mg $NO_3 - N\, g^{-1}\, h^{-1}$ is maintained under complete aerobic conditions [178]. Gutierrez-Wing et al. [153] reported that the denitrification rate decreased from 5.5 to 0.5 g $NO_3 - N\, L^{-1}\, d^{-1}$ when the DO concentration increased from 0.5 to 4.0 mg L^{-1} in a circulating aquaculture water system filled with polyhydroxybutyrate (PHB). At the DO levels of 4–5 mg L^{-1}, a least denitrification rate of 0.18 g $NO_3 - N\, L^{-1}\, d^{-1}$ was noticed for 6 days and thereafter declined to zero. Xu et al. [179] also reported that the nitrate removal increased to more than 85% with increasing DO levels in the influent from 1.5 to 4.0 mg L^{-1}, and decreased to 50% at DO levels > 4.0 mg L^{-1} in a solid-phase denitrification system, using corncobs as carbon source. Wang and Chu [165] thus suggested that controlling the DO levels in the denitrification reactor appeared to be needless, but then it could promote the efficiency of the process.

12.4.3 Vegetation Type

Macrophytes (also known as macroflora, phytoremediators, hydrophytes, wetland plants and aquatic plants) are those plant species naturally found thriving in wetlands of all sorts, either in or on the water. They play a significant role in CWs and have been extensively used for decontamination of water bodies. For instance, their roots provide surface areas for microbial activities and aerobic zones in the wetlands. The rhizosphere is the most active reaction zone in a CW as it promotes the relationship benefits that exit amongst plants, microbes, soil and contaminants, thereby enhancing physical and biochemical processes [71]. Studies revealed that parts (the above-ground and below-ground) of the macrophytes enhanced microbial diversity and offer enormous surface areas for biofilm development which is accountable for the majority of the microbial activities occurring in the CWs [180, 181]. The categories of macrophytes commonly used in CW are emergent plants (Arundo donax L.,

Juncus spp., Phragmites spp., Typha spp., Iris spp., and *Eleocharis spp),* submerged plants (*Myriophyllum verticillatum, Hydrilla verticillata, Ceratophyllum demersum, and Vallisneria natans*), floating leaved plants (water spinach (*Ipomoea aquatica*), water lettuce (*Pistia stratiotes*) *Nymphaea tetragona, Nymphoides peltata, Trapa bispinosa and Marsilea quadrifolia*), free-floating plants (Water hyacinth (*Eichhornia crassipes*), *Lemna minor, Hydrocharis dubia* and *Salvinia natans*) and other large wetland grass-like plants like Bulrushes (e.g. *Scirpus luviatilis, Scirpus validus, Scirpus cyperinus*). It has been substantiated that planting of more than one species of macrophytes enhances the removal performance of CWs because the presence of diverse kind of plant species offers a more favourable microbial activities and longer retention time [182, 183]. For optimum treatment efficiency and favourable CW design, a detailed understanding of plant species, uniqueness of microorganism groups, and the associations between biogenic matters and particular components in contaminants are required.

12.5 Conclusion

Nitrate pollution remains a vital problem in the pursuit of environmental sustainability in water environments. This chapter has shed light on the viable means of treating nitrate contaminated water using an ecologically based technology called constructed wetland. CWs have been proved to be a beneficial and promising technique in wastewater treatment because of their low-cost, environmental quality preservation and easy maintenance. This chapter also summarizes several factors responsible for nitrogen removal in CW treatment systems from water and wastewater, including the various transformations of nitrogen with a focus on nitrogen oxyanion (nitrate).

References

1. Patel RK (2016) Nitrates—its generation and impact on environment from mines: a review. In: National conference on sustainable mining practice. India. pp 2–3
2. Rossi F, Motta O, Matrella S, Proto A, Vigliotta G (2015) Nitrate removal from wastewater through biological denitrification with OGA 24 in a batch reactor. Water 7:51–62. https://doi.org/10.3390/w7010051
3. Ghafari S, Hasan M, Aroua MK (2008) Bio-electrochemical removal of nitrate from water and wastewater—a review. Biores Technol 99:3965–3974
4. Taylor GD, Fletcher TD, Wong THF, Breen PF (2005) Nitrogen composition in urban runoff—implications for stormwater management. Water Res 39:1982–1989
5. Rajmohan KS, Gopinath M, Chetty R (2018) Bioremediation of nitrate-contaminated wastewater and soil. In: Varjani S, Agarwal A, Gnansounou E, Gurunathan B (eds) Bioremediation: applications for environmental protection and management. energy, environment, and sustainability. Springer, Singapore, pp 387–409. https://doi.org/10.1007/978-981-10-7485-1_19
6. Oladoja NA, Ademoroti CMA (2006) The use of fortified soil-clay as on-site system for domestic wastewater purification. Water Res 40:613–620

7. Wu C, Chen Z, Liu X, Peng Y (2007) Nitrification–denitrification via nitrite in SBR using real-time control strategy when treating domestic wastewater. Biochem Eng J 36:87–92
8. Peyton BM, Mormile MR, Petersen JN (2001) Nitrate reduction with halomonas campisalis: kinetics of denitrification at pH 9 and 12.5% NaCl. Water Res 35:4237–4242
9. Munz G, Gori R, Cammilli L, Lubello C (2008) Characterization of tannery wastewater and biomass in a membrane bioreactor using respirometric analysis. Biores Technol 99:8612–8618
10. Park JBK, Craggs RJ, Sukias JPS (2009) Removal of nitrate and phosphorus from hydroponic wastewater using a hybrid denitrification filter (HDF). Biores Technol 100:3175–3179
11. Dorante T, Lammel J, Kuhlmann H, Witzke T, Olfs HW (2008) Capacity, selectivity, and reversibility for nitrate exchange of a layered double-hydroxide (LDH) mineral in simulated soil solutions and in soil. J Plant Nutr Soil Sci 171:777–784
12. Shen J, He R, Han W, Sun X, Li J, Wang L (2009) Biological denitrification of high-nitrate wastewater in a modified anoxic/oxic-membrane bioreactor (A/O-MBR). J Hazard Mater 172:595–600
13. Francis CW, Hatcher CW (1980) Biological denitrification of high nitrate wastes generated in the nuclear industry. Bio Fluid Bed Treat Water Waste 1:235–250
14. Reinsel M (n.d) Nitrate removal technologies: new solutions to an old problem. Guest column on December 10, 2014. https://www.wateronline.com/doc/nitrate-removal-technologies-new-solutions-to-an-old-problem-0001. Accessed 21 April 2020
15. Tsai HH, Ravindran V, Williams MD, Pirbazari M (2004) Forecasting the performance of membrane bioreactor process for groundwater denitrification. J Environ Eng Sci 3:507–521
16. Nujić M, Milinković D, Habuda-Stanić M (2017) Nitrate removal from water by ion exchange. Croatian J Food Sci Technol 9:182–186. https://doi.org/10.17508/CJFST.2017.9.2.15
17. Du R, Peng Y, Cao S, Wu C, Weng D, Wang S, He J (2014) Advanced nitrogen removal with simultaneous Anammox and denitrification in sequencing batch reactor. Biores Technol 162:316–322
18. Fu F, Dionysiou DD, Liu H (2014) The use of zero-valent iron for groundwater remediation and wastewater treatment: a review. J Hazard Mater 267:194–205
19. Anderson JA (2011) Photocatalytic nitrate reduction over Au/TiO$_2$. Catal Today 175:316–321
20. Bhatnagar A, Sillanpaa MA (2011) Review of emerging adsorbents for nitrate removal from water. Chem Eng J 168:493–504
21. Li M, Feng C, Zhang Z, Yang S, Sugiura N (2010) Treatment of nitrate contaminated water using an electrochemical method. Biores Technol 101:6553–6557
22. Samatya S, Kabay N, Yüksel Ü, Arda M, Yüksel M (2006) Removal of nitrate from aqueous solution by nitrate selective ion exchange resins. React Funct Polym 66:1206–1214
23. Wang JL, Kang J (2005) The characteristics of anaerobic ammonium oxidation (ANAMMOX) by granular sludge from an EGSB reactor. Process Biochem 40:1973–1978
24. Aslan S, Turkman A (2003) Biological denitrification of drinking water using various natural organic solid substrates. Water Sci Technol 48:489–495
25. Schoeman JJ, Steyn A (2003) Nitrate removal with reverse osmosis in a rural area in South Africa. Desalination 155:15–26
26. Elmidaoui A, Elhannouni F, Menkouchi Sahli AM, Chay L, Elabbassi H, Hafsi M, Largeteau D (2001) Pollution of nitrate in Moroccan ground water: removal by electrodialysis. Desalination 136:325–332
27. Kapoor A, Viraraghavan T (1997) Nitrate removal from drinking water—review. J Environ Eng 123:371–380
28. Della Rocca C, Belgiorno V, Meriç S (2007) Overview of in-situ applicable nitrate removal processes. Desalination 204:46–62
29. He Q, Feng C, Chen N, Zhang D, Hou T, Dai J, Hao C, Mao B (2019) Characterizations of dissolved organic matter and bacterial community structures in rice washing drainage (RWD)-based synthetic groundwater denitrification. Chemosphere 215:142–152
30. Ajibade FO, Adewumi JR (2017) Performance evaluation of aquatic macrophytes as a constructed wetland for municipal wastewater treatment. FUTA J Eng Eng Technol (FUTAJEET) 11:1–11

31. Guo L, Lv T, He K, Wu S, Dong X, Dong R (2017) Removal of organic matter, nitrogen and faecal indicators from diluted anaerobically digested slurry using tidal flow constructed wetlands. Environ Sci Pollut Res 24:5486–5496. https://doi.org/10.1007/s11356-016-8297-2

32. He K, Lv T, Wu S, Guo L, Ajmal Z, Luo H, Dong R (2016) Treatment of alkaline stripped effluent in aerated constructed wetlands: feasibility evaluation and performance enhancement. Water 8:386

33. Kizito S, Lv T, Wu S, Ajmal Z, Luo H, Dong R (2017) Treatment of anaerobic digested effluent in biochar-packed vertical flow constructed wetland columns: role of media and tidal operation. Sci Total Environ 592:197–205

34. Wu S, Lv T, Lu Q, Ajmala Z, Dong R (2016) Treatment of anaerobic digestate supernatant in microbial fuel cell coupled constructed wetlands: evaluation of nitrogen removal, electricity generation, and bacterial community response. Sci Total Environ 580:339–346

35. Ajibade FO, Wang H, Guadie AA, Ajibade TF, Fang Y, Sharif HMA, Liu W, Wang A (2021) Total nitrogen removal in biochar amended non-aerated vertical flow constructed wetlands for secondary wastewater effluent with low C/N ratio: Microbial community structure and dissolved organic carbon release conditions. Biores Technol 322:124430. https://doi.org/10.1016/j.biortech.2020.124430

36. Milani M, Marzo A, Toscano A, Consoli S, Cirelli GL, Ventura D, Barbagallo S (2019) Evapotranspiration from horizontal subsurface flow constructed wetlands planted with different perennial plant species. Water 11:2159. https://doi.org/10.3390/w11102159

37. Mena J, Rodriguez L, Nunez J, Fernández FJ, Villasenor J (2008) Design of horizontal and vertical subsurface flow constructed wetlands treating industrial wastewater. WIT Trans Ecol Environ 111:555–563

38. Vymazal J (2007) Removal of nutrients in various types of constructed wetlands. Sci Total Environ 380:48–65

39. Kadlec RH (2008) The effects of wetland vegetation and morphology on nitrogen processing. Ecol Eng 33(1):26–141

40. Peng L, Hua Y, Cai J, Zhao J, Zhou W, Zhu D (2014) Effects of plants and temperature on nitrogen removal and microbiology in a pilot-scale integrated vertical-flow wetland treating primary domestic wastewater. Ecol Eng 64:285–290

41. Białowiec A, Albuquerque A, Randerson PF (2014) The influence of evapotranspiration on vertical flow subsurface constructed wetland performance. Ecol Eng 67:89–94

42. International Water Association (2000) Constructed wetlands for pollution control. Processes, performance, design and operation. IWA Publishing, London, p 156

43. Omotade IF, Alatise MO, Olanrewaju OO (2019) Recycling of aquaculture wastewater using charcoal based constructed wetlands. Int J Phytorem 21:399–404. https://doi.org/10.1080/15226514.2018.1537247

44. Lin YF, Jing SR, Wang TW, Lee DY (2002) Effects of macrophytes and external carbon sources on nitrate removal from groundwater in constructed wetlands. Environ Pollut 119:413–420. https://doi.org/10.1016/S0269-7491(01)00299-8

45. Lin YF, Jing SR, Lee DY, Chang YF, Shih KC (2007) Nitrate removal and denitrification affected by soil characteristics in nitrate treatment wetlands. J Environ Sci Health, Part A: Toxic/Hazard Subst Environ Eng 42:471–479. https://doi.org/10.1080/10934520601187690

46. Domingos S, Germain M, Dallas S, Ho G (n.d) Nitrogen removal from industrial wastewater by hybrid constructed wetland systems. In: 2nd IWA-ASPIRE conference and exhibition, 28 October–31 November, 2007. Perth, Western Australia. https://researchrepository.murdoch.edu.au/4088/

47. Xiong J, Guo G, Mahmood Q, Yue M (2011) Nitrogen removal from secondary effluent by using integrated constructed wetland system. Ecol Eng 37:659–662

48. Aguilar L, Gallegos A, Arias CA, Ferrera I, Sánchez O, Rubio R, Saad MB, Missagia B, Caro P, Sahuquillo S, Pérez C, Morató J (2019) Microbial nitrate removal efficiency in groundwater polluted from agricultural activities with hybrid cork treatment wetlands. Sci Total Environ 653:723–734

49. Jia L, Liu H, Kong Q, Li M, Wu S, Wu H (2020) Interactions of high-rate nitrate reduction and heavy metal mitigation in iron-carbon-based constructed wetlands for purifying contaminated groundwater. Water Res 169:115285

50. Wang W, Song X, Li F, Jia X, Hou M (2020) Intensified nitrogen removal in constructed wetlands by novel spray aeration system and different influent COD/N ratios. Bioresources Technol 306:123008. https://doi.org/10.1016/j.biortech.2020.123008

51. Hang Q, Wang H, He Z, Dong W, Chu Z, Ling Y, Yan G, Chang Y, Li C (2020) Hydrilla verticillata–sulfur-based heterotrophic and autotrophic denitrification process for nitrate-rich agricultural runoff treatment. Int J Environ Res Public Health 17:1574. https://doi.org/10.3390/ijerph17051574

52. Hang Q, Wang H, Chu Z, Hou Z, Zhou Y, Li C (2017) Nitrate-rich agricultural runoff treatment by vallisneria-sulfur based mixotrophic denitrification process. Sci Total Environ 587–588:108–117. https://doi.org/10.1016/j.scitotenv.2017.02.069

53. Zhao Y, Song X, Cao X, Wang Y, Zhao Z, Si Z, Yuan S (2019) Modified solid carbon sources with nitrate adsorption capability combined with nZVI improve the denitrification performance of constructed wetlands. Biores Technol 294:122189. https://doi.org/10.1016/j.biortech.2019.122189

54. Kleimeier C, Liu H, Rezanezhad F, Lennartz B (2018) Nitrate attenuation in degraded peat soil-based constructed wetlands. Water 10:355. https://doi.org/10.3390/w10040355

55. Ajibade FO, Adeniran KA, Egbuna CK (2013) Phytoremediation efficiencies of water hyacinth in removing heavy metals in domestic sewage (a case study of university of Ilorin, Nigeria). Int J Eng Sci 2:16–27

56. Chang JJ, Wu SQ, Dai YR, Liang W, Wu ZB (2013) Nitrogen removal from nitrate-laden wastewater by integrated vertical-flow constructed wetland systems. Ecol Eng 58:192–201

57. Dires S, Birhanu T, Ambelu A (2019) Use of broken brick to enhance the removal of nutrients in subsurface flow constructed wetlands receiving hospital wastewater. Water Sci Technol 79:156–164. https://doi.org/10.2166/wst.2019.037

58. Zamora S, Marín-Muñíz JL, Nakase-Rodríguez C, Fernández-Lambert G, Sandoval L (2019) Wastewater treatment by constructed wetland eco-technology: influence of mineral and plastic materials as filter media and tropical ornamental plants. Water 11:2344. https://doi.org/10.3390/w11112344

59. Si Z, Song X, Wang Y, Cao X, Wang Y, Zhao Y, Ge X, Sand W (2020) Untangling the nitrate removal pathways for a constructed wetland sponge iron coupled system and the impacts of sponge iron on a wetland ecosystem. J Hazard Mater 393:122407. https://doi.org/10.1016/j.jhazmat.2020.122407

60. Si Z, Song X, Cao X, Wang Y, Wang Y, Zhao Y, Ge X, Tesfahunegn AA (2020) Nitrate removal to its fate in wetland mesocosm filled with sponge iron: impact of influent COD/N ratio. Frontiers Environ Sci Eng 2020(14):4. https://doi.org/10.1007/s11783-019-1183-7

61. Yu G, Peng H, Fu Y, Yan X, Du C, Chen H (2019) Enhanced nitrogen removal of low C/N wastewater in constructed wetlands with co-immobilizing solid carbon source and denitrifying bacteria. Biores Technol 280:337–344. https://doi.org/10.1016/j.biortech.2019.02.043

62. Sun H, Yang Z, Wei C, Wu W (2018) Nitrogen removal performance and functional genes distribution patterns in solid-phase denitrification sub-surface constructed wetland with micro aeration. Biores Technol 263:223–231

63. Zhang L, Sun Z, Xie J, Wu J, Cheng S (2018) Nutrient removal, biomass accumulation and nitrogen-transformation functional gene response to different nitrogen forms in enhanced floating treatment wetlands. Ecol Eng 112:21–25

64. Nakase C, Zurita F, Nani G, Reyes G, Fernández-Lambert G, Cabrera-Hernández A, Sandoval L (2019) Nitrogen removal from domestic wastewater and the development of tropical ornamental plants in partially saturated mesocosm-scale constructed wetlands. Int J Environ Res Public Health 16:4800. https://doi.org/10.3390/ijerph16234800

65. Tchobanoglous G, Burton FL, Stensel HD (2003) Wastewater engineering. McGraw-Hill, New York, p 56

66. Li B, Irvin S, Baker K (2007) The variation of nitrifying bacterial population sizes in a sequencing batch reactor (SBR) treating low, mid, high concentrated synthetic wastewater. J Environ Eng Sci 6:651–663
67. Harold L, Haunschild LK, Hopes G, Tchobanoglous G, Darbya JL (2010) Anoxic treatment wetlands for denitrification. Ecol Eng 36:1544–1551
68. Kadlec RH, Wallace S (2009) Treatment wetlands, 2nd edn. CRC Press, Boca Raton
69. Gajewska, M., Skrzypiec, K (2018) Kinetics of nitrogen removal processes in constructed wetlands. E3S Web Conf 26:1–4. https://doi.org/10.1051/e3sconf/20182600001
70. Kadlec RH, Knight RL (1996) Treatment wetlands. Boca Raton, FL 33431. CRC Press LLC, USA
71. Lee C, Fletcher TD, Sun G (2009) Nitrogen removal in constructed wetland systems. Eng Life Sci 9:11–22. https://doi.org/10.1002/elsc.200800049
72. Sonavane PG, Munavalli GR (2009) Modeling nitrogen removal in a constructed wetland treatment system. Water Sci Technol 60:301–309. https://doi.org/10.2166/wst.2009.319
73. US EPA (1993) Subsurface flow constructed wetlands for wastewater treatment: a technology assessment. Office of Water, Washington, D.C. EPA 832-R-93-008
74. Saeed T, Sun G (2012) A review on nitrogen and organics removal mechanisms in subsurface flow constructed wetlands: dependency on environmental parameters, operating conditions and supporting media. J Environ Manage 112:429–448
75. Savant NK, DeDatta SK (1982) Nitrogen transformations in wetland rice soils. Adv Agron 35:241–302
76. Reddy KR, Patrick WII Jr (1984) Nitrogen transformations and loss in flooded soils and sediments. CRC Crit Rev Environ Control 13:273
77. Patrick WH Jr, Wyatt R (1964) Soil nitrogen loss as a result of alternate submergence and dying. Proc—Soil Sci Soc Am 28:647–653
78. Vymazal J (1995) Algae and element cycling in wetlands. CRC Press Inc
79. Graaf AA, Bruijn P, Robertson LA, Jetten MS, Kuenen JG (1996) Autotrophic growth of anaerobic ammonium oxidizing micro-organisms in a fluidized bed reactor. Microbiology 142:2187–2196
80. Ahn YH (2006) Sustainable nitrogen elimination biotechnologies: a review. Process Biochem 41:1709–1721
81. Guadie A, Xia S, Zhang Z, Zeleke J, Guo W, Ngo HH, Hermanowicz SW (2014) Effect of intermittent aeration cycle on nutrient removal and microbial community in a fluidized bed reactor-membrane bioreactor combo system. Biores Technol 156:195–205
82. Guadie A, Xia S, Zhang Z, Guo W, Ngo HH, Hermanowicz SW (2013) Simultaneous removal of phosphorus and nitrogen from sewageusing a novel combo system of fluidized bed reactor–membranebioreactor (FBR–MBR). Biores Technol 149:276–285
83. Gerardi MH (2002) Nitrification and denitrification in the activated sludge process. Wiley Inc., New York
84. Prosnansky M, Sakakibarab Y, Kuroda M (2002) High-rate denitrification and SS rejection by biofilm-electrode reactor (BER) combined with microfiltration. Water Res 36:4801–4810
85. Szekeres S, Kiss I, Kalman M, Soares MI (2002) Microbial population in a hydrogen-dependent denitrification reactor. Water Res 36:4088–4094
86. Rijn JV, Tal Y, Schreier HJ (2006) Denitrification in recirculating systems: theory and applications. Aquacult Eng 34:364–376
87. Breisha GZ, Winter J (2010) Bio-removal of nitrogen from wastewaters—a review. J Am Sci 6:508–528
88. Chen D, Dai T, Wang H, Yang K (2015) Nitrate removal by a combined bioelectrochemical and sulfur autotrophic denitrification (CBSAD) system at low temperatures. Desalin Water Treat 57:1–7. https://doi.org/10.1080/19443994.2015.1101024
89. Chen D, Yang K, Wang H (2016) Effects of important factors on hydrogen-based autotrophic denitrification in a bioreactor. Desalin Water Treat 57:3482–3488. https://doi.org/10.1080/19443994.2014.986533

90. Chen D, Yang K, Wang H, Lv B (2014) Nitrate removal from groundwater by hydrogen-fed autotrophic denitrification in a bio-ceramsite reactor. Water Sci Technol 69:2417–2422. https://doi.org/10.2166/wst.2014.167

91. Chung J, Amin K, Kim S, Yoon S, Kwon K, Bae W (2014) Autotrophic denitrification of nitrate and nitrite using thiosulfate as an electron donor. Water Res 58:169–178. https://doi.org/10.1016/j.watres.2014.03.071

92. Wang X, Xing L, Qiu T, Han M (2013) Simultaneous removal of nitrate and pentachlorophenol from simulated groundwater using a biodenitrification reactor packed with corncob. Environ Sci Pollut Res 20:2236–2243. https://doi.org/10.1007/s11356-012-1092-9

93. Kim J, Park K, Cho K, Nam S, Park T, Bajpai R (2005) Aerobic nitrification–denitrification by heterotrophic Bacillus strains. Biores Technol 96:1897–1906

94. Kim S, Jung H, Kim KS, Kim IS (2004) Treatment of high nitrate containing wastewaters by sequential heterotrophic and autotrophic denitrification. J Environ Eng 130:1475–1480

95. Masclaux-Daubresse C, Daniel-Vedele F, Dechorgnat J, Chardon F, Gaufichon L, Suzuki A (2010) Nitrogen uptake, assimilation and remobilization in plants: challenges for sustainable and productive agriculture. Ann Bot 105:1141–1157. https://doi.org/10.1093/aob/mcq028

96. Xu G, Fan X, Miller AJ (2012) Plant nitrogen assimilation and use efficiency. Annu Rev Plant Biol 63:153–182. https://doi.org/10.1146/annurev-arplant-042811-105532

97. Bialowiec A, Janczukowicz W, Randerson PF (2011) Nitrogen removal from wastewater in vertical flow constructed wetlands containing LWA/gravel layers and reed vegetation. Ecol Eng 37:897–902

98. Dan TH, Quang LN, Chiem NH, Brix H (2011) Treatment of high-strength wastewater in tropical constructed wetlands planted with Sesbania sesban: horizontal subsurface flow versus vertical downflow. Ecol Eng 37:711–720

99. Cui L, Ouyang Y, Lou Q, Yang F, Chen Y, Zhu W, Luo S (2010) Removal of nutrients from wastewater with *Canna indica L.* under different vertical-flow constructed wetland conditions. Ecol Eng 36:1083–1088

100. Kantawanichkul S, Kladprasert S, Brix H (2009) Treatment of high-strength wastewater in tropical vertical flow constructed wetlands planted with typha angustifolia and cyperus involucratus. Ecol Eng 35:238–247

101. Landry GM, Maranger R, Brisson J, Chazarenc F (2009) Nitrogen transformations and retention in planted and artificially aerated wetlands. Water Res 43:535–545

102. Huett DO, Morris SG, Smith G, Hunt N (2005) Nitrogen and phosphorus removal from plant nursery runoff in vegetated and unvegetated subsurface flow wetlands. Water Res 39:3259–3272

103. Shamir E, Thompson TL, Karpiscak MM, Freitas RJ, Zauderer J (2001) Nitrogen accumulation in a constructed wetland for dairy wastewater treatment. J Am Water Resour Assoc 37:315–325

104. Drizo A, Frost CA, Smith KA, Grace J (1997) Phosphate and ammonium removal by constructed wetlands with horizontal subsurface flow, using shale as a substrate. Water Sci Technol 35:95–102

105. Bayley ML, Davison L, Headley TR (2003) Nitrogen removal from domestic effluent using subsurface flow constructed wetlands: influence of depth, hydraulic residence time and pre-nitrification. Water Sci Technol 48:175–182

106. Kaseva ME (2004) Performance of a sub-surface flow constructed wetland in polishing pre-treated wastewater e a tropical case study. Water Res 38:681–687

107. Langergraber G (2005) The role of plant uptake on the removal of organic matter and nutrients in subsurface flow constructed wetlands: a simulation study. Water Sci Technol 51:213–223

108. Brix H (1997) Do macrophytes play a role in constructed treatment wetlands? Water Sci Technol 35:11–17

109. Masi F (2008) Enhanced denitrification by a hybrid HF-FWS constructed wetland in a large-scale wastewater treatment plant. In: Vymazal J (ed) Wastewater treatment, plant dynamics and management in constructed and natural wetlands. pp 267–275

110. Osorio AC, Villafañe P, Caballero V, Manzan Y (2011) Efficiency of mesocosm scale constructed wetland systems for treatment of sanitary wastewater under tropical conditions. Water Air Soil Pollut 220:161–171
111. Wang R, Baldy V, Périssol C, Korboulewsky N (2012) Influence of plants on microbial activity in a vertical-downflow wetland system treating waste activated sludge with high organic matter concentrations. J Environ Manage 95:S158–S164
112. Wetzel RG (11–16 Nov, 2000) Fundamental processes within natural and constructed wetland ecosystems: short-term versus long-term objectives. 7th International conference on wetland systems for water pollution control. Lake Buena Vista pp 3–12
113. Garnett TP, Shabala SN, Smethurst PJ, Newman IA (2001) Simultaneous measurement of ammonium, nitrate and proton fluxes along the length of eucalyptus roots. Plant Soil 236:55–62
114. Brodrick SJ, Cullen P, Maher W (1988) Denitrification in a natural wetland receiving secondary treated effluent. Water Res 22:431–439
115. Biddlestone AJ, Gray KR, Job GD (1991) Treatment of dairy farm wastewaters in engineered reed bed systems. Process Biochem 26:265–268
116. Armstrong W, Armstrong J, Beckett RM (1990) Measurement and modeling of oxygen release from roots of phragmites Australis. Constructed wetlands for water pollution control. Pergamon Press, Oxford, UK
117. Brix H, Schierup H (1990) Soil oxygenation in constructed reed beds: the role of macrophyte and soil atmosphere interface oxygen transport. Constructed wetlands for water pollution control. Pergamon Press, Oxford, UK
118. Bavor HJ, Roser DJ, McKersie SA, Breen P (1988) Treatment of secondary effluent. Report to Sydney Water Board, Australia
119. Cameron SG, Schipper LA (2010) Nitrate removal and hydraulic performance of organic carbon for use in denitrification beds. Ecol Eng 36:1588–1595
120. Sirivedhin T, Gray KA (2006) Factors affecting denitrification rates in experimental wetlands: field and laboratory studies. Ecol Eng 26:167–181
121. Aslan Ş, Türkman A (2004) Simultaneous biological removal of endosulfan (α +β) and nitrates from drinking waters using wheat straw as substrate. Environ Int. 30:449–455
122. Ingersoll TL, Baker LA (1998) Nitrate removal in wetland microcosms. Water Res 32:677–684
123. Kuschk P, Wiessner A, Kappelmeyer U, Weissbrodt E, Kaestner M, Stottmeister U (2003) Annual cycle of nitrogen removal by a pilot-scale subsurface horizontal flow in a constructed wetland under moderate climate. Water Res 37:4236–4242
124. Zhong F, Huang S, Wu J, Cheng S, Deng Z (2019) The use of microalgal biomass as a carbon source for nitrate removal in horizontal subsurface flow constructed wetlands. Ecol Eng 127:263–267. https://doi.org/10.1016/j.ecoleng.2018.11.029
125. Glass C, Silverstein J (1998) Denitrification kinetics of high nitrate concentration water: pH effect on inhibition and nitrite accumulation. Water Res 32:831–839
126. Shen Z, Zhou Y, Liu J, Xiao Y, Cao R, Wu F (2015) Enhanced removal of nitrate using starch/PCL blends as solid carbon source in a constructed wetland. Biores Technol 175:239–244
127. Áséy A, Édegaard H, Bach K, Pujol R, Hamon M (1998) Denitrification in a packed bed biofilm reactor (Biofor)—experiments with different carbon sources. Water Res 32(5):1463–1470
128. Liu X, Fu X, Pu A, Zhang K, Luo H, Anderson BC, Li M, Huang B, Hu L, Fan L, Chen W, Chen J, Fu S (2019) Impact of external carbon source addition on methane emissions from a vertical subsurface-flow constructed wetland. Greenhouse Gases: Sci Technol 9:331–348. https://doi.org/10.1002/ghg.1847
129. Zhao Q, Chen Y (2015) Nitrogen removal effect in different locations of tidal flow constructed wetland and the effect of external carbon source on it. Water Supply Technol 9(4):32–37
130. Xiao L, He F, Liang X (2012) Effect of adding solid carbon source on the treatment effect of vertical flow constructed wetland sewage. Lake Sci 24:843–848
131. Nikolausza M, Kappelmeyera U, Szekelyb A, Rusznyakb A, Marialigetib K, Kastnera M (2008) Diurnal redox fluctuation and microbial activity in rhizosphere of wetland plants. Eur J Soil Biol 44:324–333

132. Fu G, Huangshen L, Guo Z, Zhou Q, Wu Z (2017) Effect of plant-based carbon sources on denitrifying microorganisms in a vertical flow constructed wetland. Biores Technol 224:214–221

133. Hang Q, Wang H, Chu Z, Ye B, Li C, Hou Z (2016) Application of plant carbon source for denitrification by constructed wetland and bioreactor: review of recent development. Environ Sci Pollut Res 23:8260–8274. https://doi.org/10.1007/s11356-016-6324-y

134. Chang JJ, Lu YF, Chen JQ, Wang XY, Luo T, Liu H (2016) Simultaneous removals of nitrate and sulfate and the adverse effects of gravel-based biofilters with flower straws added as exogenous carbon source. Ecol Eng 95(2016):189–197

135. Soares MIM, Abeliovich A (1998) Wheat straw as substrate for water denitrification. Water Res 32:3790–3794

136. Schipper LA, Vojvodić-Vuković M (2001) Five years of nitrate removal, denitrification and carbon dynamics in a denitrification wall. Water Res 35:3473–3477

137. Ovez B (2006) Batch biological denitrification using arundo donax, glycyrrhiza glabra, and gracilaria verrucosa as carbon source. Process Biochem 41:1289–1295

138. Ovez B, Ozgen S, Yuksel M (2006) Biological denitrification in drinking water using glycyrrhiza glabra and arunda donax as the carbon source. Process Biochem 41:1539–1544

139. Singer A, Parnes S, Gross A, Sagi A, Brenner A (2008) A novel approach to denitrification processes in a zero-discharge recirculating system for small-scale urban aquaculture. Aquacult Eng 39:72–77

140. Trois C, Pisano G, Oxarango L (2010) Alternative solutions for the biodenitrification of landfill leachates using pine bark and compost. J Hazard Mater 178:1100–1105

141. Warneke S, Schipper LA, Matiasek MG, Scow KM, Cameron S, Bruesewitz DA, McDonald IR (2011) Nitrate removal, communities of denitrifiers and adverse effects in different carbon substrates for use in denitrification beds. Water Res 45:5463–5475

142. Lu S, Hu H, Sun Y, Yang J (2009) Effect of carbon source on the denitrification in constructed wetlands. J Environ Sci 21:1036–1043. https://doi.org/10.1016/S1001-0742(08)62379-7

143. Zhang M, Zhao L, Mei C, Yi L, Hua G (2014) Effects of plant material as carbon sources on TN removal efficiency and N$_2$O flux in vertical-flow-constructed wetlands. Water Air Soil Pollut 225:11. https://doi.org/10.1007/s11270-014-2181-9

144. Wu S, Kuschk P, Brix H, Vymazal J, Dong R (2014) Development of constructed wetlands in performance intensifications for wastewater treatment: a nitrogen and organic matter targeted review. Water Res 57(5):40–55

145. Gomez MA, Gonzalez-Lopez J, Hontoria-Garcia E (2006) Influence of carbon source on nitrate removal of contaminated groundwater in a denitrifying submerged filter. J Hazard Mater 80:69–80

146. Hareendran RA (2010) Study of denitrification kinetics at low temperatures using methanol as the external carbon sources. The thesis, George Washington University. p 47

147. Chen X, He S, Zhang Y, Huang X, Huang Y, Chen D, Huang XC, Tang J (2015) Enhancement of nitrate removal at the sediment-water interface by carbon addition plus vertical mixing. Chemosphere 136:305–310. https://doi.org/10.1016/j.chemosphere.2014.12.010

148. Rustige H, Nolde E (2007) Nitrogen elimination from landfill leachates using an extra carbon source in subsurface flow constructed wetlands. Water Sci Technol 56(3):125–133

149. Chu L, Wang J (2016) Denitrification of groundwater using PHBV blends in packed bed reactors and the microbial diversity. Chemosphere 155:463–470

150. Wu W, Yang L, Wang J (2013) Denitrification using PBS as carbon source and biofilm support in a packed-bed bioreactor. Environ Sci Pollut Res 20:333–339

151. Chu L, Wang J (2011) Nitrogen removal using biodegradable polymers as carbon source and biofilm carriers in a moving bed biofilm reactor. Chem Eng J 170:220–225

152. Zhang Q, Ji F, Xu X (2016) Effects of physicochemical properties of poly-ε-caprolactone on nitrate removal efficiency during solid-phase denitrification. Chem Eng J 283:604–613

153. Gutierrez-Wing MT, Malone RF, Rusch KA (2012) Evaluation of polyhydroxybutyrate as a carbon source for recirculating aquaculture water denitrification. Aquacult Eng 51:36–43

154. Marusincova H, Husarova L, Ruzicka J, Ingr M, Navrátil V, Buňková L, Koutny M (2013) Polyvinyl alcohol biodegradation under denitrifying conditions. Int Biodeterior Biodegradation 84:21–28. https://doi.org/10.1016/j.ibiod.2013.05.023

155. Shen Z, Zhou Y, Liu J, Xiao Y, Cao R, Wu F (2014) Enhanced removal of nitrate using starch/PCL blends as solid carbon source in a constructed wetland. Biores Technol 175C:239–244

156. Li P, Zuo J, Xing W, Tang L, Ye X, Li Z, Yuan L, Wang K, Zhang H (2013) Starch/polyvinyl alcohol blended materials used as solid carbon source for tertiary denitrification of secondary effluent. J Environ Sci 25:1972–1979. https://doi.org/10.1016/S1001-0742(12)60259-9

157. Wu W, Yang F, Yang L (2012) Biological denitrification with a novel biodegradable polymer as carbon source and biofilm carrier. Biores Technol 118:136–140

158. Srinandan CS, D'souza, G., Srivastava, N., Nayak, B.B., Nerurkar, A.S, (2012) Carbon sources influence the nitrate removal activity, community structure and biofilm architecture. Bioresour Technol 117:292–299

159. Shen Z, Zhou Y, Wang J (2013) Comparison of denitrification performance and microbial diversity using starch/polylactic acid blends and ethanol as electron donor for nitrate removal. Bioresour Technol 131:33–39

160. Boley A, Müller WR, Haider G (2000) Biodegradable polymers as solid substrate and biofilm carrier for denitrification in recirculated aquaculture systems. Aquacult Eng 22:75–85

161. Huai J, Wu J, Zhong F, Cheng S (2018) Level. Advances in research on carbon source replenishment strategies for constructed wetlands. Environ Prot Frontiers 8(6):475–481. https://doi.org/10.12677/aep.2018.86059

162. Qian Z (2016) Research on nitrogen and phosphorus removal of low carbon source wastewater based on solid phase nitrification and adsorption phosphorus [D]: [Ph.D. Thesis]. Chongqing University, Chongqing

163. Rørslett B, Berge D, Johansen SW (1986) Lake enrichment by submersed macrophytes: a Norwegian whole-lake experience with Elodea canadensis. Aquat Bot 26:325–340

164. US EPA (1975) Process design manual for nitrogen control. Office of Technology Transfer, Washington, DC

165. Wang J, Chu L (2016) Biological nitrate removal from water and wastewater by solid-phase denitrification process. Biotechnol Adv 34:1103–1112

166. Phipps RG, Crumpton WG (1994) Factors affecting nitrogen loss in experimental wetlands with different hydrologic loads. Ecol Eng 3:399–408

167. Katayon S, Fiona Z, Noor MM, Halim GA, Ahmad J (2008) Treatment of mild domestic wastewater using subsurface constructed wetlands in Malaysia. Int J Environ Stud 65:87–102

168. Langergraber G, Prandtstetten Ch, Pressl A, Rohrhofer R, Haberl R (2007) Optimization of subsurface vertical flow constructed wetlands for wastewater treatment. Water Sci Technol 55:71–78

169. Nivala J, Hoos MB, Cross C, Wallace S, Parkin G (2007) Treatment of landfill leachate using an aerated, horizontal subsurface flow constructed wetland. Sci Total Environ 380:19–27

170. Tuncsiper B (2007) Removal of nutrient and bacteria in pilot-scale constructed wetlands. J Environ Sci Health, Part A 42:1117–1124

171. Herkowitz J (1986) Listowel artificial marsh project report. Ontario Ministry of the Environment, Water Resources Branch, Toronto

172. Oostrom AJ, Russell JM (1994) Denitrification in constructed wastewater wetlands receiving high concentrations of nitrate. Water Sci Technol 29:7–14

173. Akratos CS, Tsihrintzis VA (2007) Effect of temperature, HRT vegetation and porous media on removal efficiency of pilot scale horizontal subsurface flow constructed wetlands. Ecol Eng 29:173–191

174. Huang J, Reneau RB, Hagedorn C (2000) Nitrogen removal in constructed wetlands employed to treat domestic wastewater. Water Res 34:2582–2588

175. Zhou W, Sun Y, Wu B, Zhang Y, Huang M, Miyanaga T, Zhang Z (2011) Autotrophic denitrification for nitrate and nitrite removal using sulfur limestone. J Environ Sci 23:1761–1769. https://doi.org/10.1016/S1001-0742(10)60635-3

176. Boley A, Muller WR (2005) Denitrification with polycaprolactone as solid substrate in a laboratory-scale recirculated aquaculture system. Water Sci Technol 52:495–502
177. Viotti P, Collivignarelli MC, Martorelli E, Raboni M (2016) Oxygen control and improved denitrification efficiency by dosing ferrous ions in the anoxic reactor. Desalin Water Treat 57:18240–18247
178. Hiraishi A, Khan ST (2003) Application of polyhydroxyalkanoates for denitrification in water and wastewater treatment. Appl Microbiol Biotechnol 61:103–109
179. Xu ZX, Shao L, Yin HL, Chu HQ, Yao YJ (2009) Biological denitrification using corncobs as a carbon source and biofilm carrier. Water Environ Res 81:242–247
180. Button M, Nivala J, Weber KP, Aubron T, Müller RA (2015) Microbial community metabolic function in subsurface flow constructed wetlands of different designs. Ecol Eng 80:162–171
181. Chen Y, Wen Y, Zhou Q, Vymazal J (2014) Effects of plant biomass on nitrogen transformation in subsurface-batch constructed wetlands: a stable isotope and mass balance assessment. Water Res 63:158–167
182. Abou-Elela SI, Golinielli G, Abou-Taleb EM, Hellal MS (2013) Municipal wastewater treatment in horizontal and vertical flows constructed wetlands. Ecol Eng 61:460–468
183. Karathanasis AD, Potter CL, Coyne MS (2003) Vegetation effect on fecal bacteria BOD, and suspended solid removal in constructed wetlands treating domestic wastewater. Ecol Eng 20:157–169

Chapter 13
Global Laws and Economic Policies in Abatement of Oxyanion in Aqua Systems: Challenges and Future Perspectives

Ngozi C. Ole, Rasaki S. Dauda, and Emmanuel I. Unuabonah

Abstract Water security and sanitation are precursors for socio-economic development, survival of flora and fauna, food security and healthy ecosystems. However, when these are compromised, they tend to have an adverse effect on the health of the populace and the socio-economic development of the entire society. This chapter investigates global laws and economic policies aimed at the abatement of toxic oxyanions (e.g. nitrate, fluoride, perchlorate etc) in aqua systems. Using a non-doctrinal cum systematic analysis, the extent of legal and economic instruments in controlling, reducing and preventing toxic oxyanion pollutants in water was examined. Relevant international treaties and instruments were analysed including the Universal Declaration on Human Rights (UDHR) 1948, the International Covenant on Economic, Social and Cultural Rights (ICESCR) 1966, the United Nations Convention on the Laws of the Sea (UNCLOS) 1982 and the United Nations Convention on the Law of the Non-Navigational Uses of International Watercourses (UNWC) 1997. Moreover, the use of different command and control (CAC) and economic instruments (EI) were also studied. The findings revealed that the provisions of the legal instruments are not strong and clear enough to compel states to adopt adequate measures for the prevention of toxic oxyanion pollutants in marine areas. In addition, even though the CAC and EI approaches have been adopted for pollution abatement across countries, the latter appear to have gained wider acceptance, due to some of the advantages it offers over and above the former approach. Nevertheless, the chapter recommends the combination of regulatory and economic approaches as the way forward in achieving the abatement of toxic oxyanion in aqua systems. One of

N. C. Ole · R. S. Dauda (✉) · E. I. Unuabonah
African Centre of Excellence for Water and Environmental Research (ACEWATER), PMB 230, Ede, Nigeria

N. C. Ole
Faculty of Law, Redeemer's University, Ede, Nigeria

R. S. Dauda
Department of Economics, Redeemer's University, PMB 230, Ede, Nigeria

E. I. Unuabonah
Department of Chemical Sciences, Redeemer's University, Ede, Nigeria

© Springer Nature Switzerland AG 2021
N. A. Oladoja and E. I. Unuabonah (eds.), *Progress and Prospects in the Management of Oxyanion Polluted Aqua Systems*, Environmental Contamination Remediation and Management, https://doi.org/10.1007/978-3-030-70757-6_13

the recommended regulatory approaches is the amendment of existing treaties and instruments to incorporate stronger obligations on states, which will feasibly achieve effective measures for the reduction and control of toxic oxyanion pollutants. The justification for the eco-legal approach to control toxic oxyanion pollutants is to yield the best optimal outcome because none of the instruments can operate in isolation, especially in a dynamic and complex society. Both, complement and reinforce each other.

13.1 Introduction

Water security and sanitation are indomitably, a precursor for socio-economic development, survival of flora and fauna, food security and healthy ecosystems [1–3]. Given the nexus between water and human existence, issues of water security form topical and recurrent themes in various international instruments [4]. The United Nations Agenda 21 recognises that water is imperative for all aspects of human existence, and thus, urges countries of the world to make available quality and adequate water supplies to the entire population of the world [5]. Similarly, the United Nations Agenda 2030 stipulates that ensuring access to clean and affordable water for all by 2030 remains a common aspiration for countries of the world [6]. Thus, the preservation of water resources from pollutants is a common concern shared by all [7].

The pollution of water resources by the anthropological introduction of toxic oxyanions (also called oxoanions) undermines the quality of water available for flora and fauna. Toxic oxyanions are inorganic pollutants such as nitrate, selenate, perchlorate and chromate, which are introduced in groundwater and surface water through municipal, industrial, agricultural and mining wastewaters [8, 9]. The presence of oxyanions in inordinate quantities in water renders it unsafe for human consumption [10, 11]. Studies across literature have concluded that drinking oxyanion polluted water can lead to medical disorders such as goitre, nausea, kidney disorder, cancer and in extreme situation death [10–12]. It can also be poisonous to aquatic life like fishes, plants and mammals [11]. A recent incident of the presence of oxyanions (heavy metal compounds and sludge) in the Atlantic Ocean resulted in the death of thousands of fishes washed by tidal waves on the shores of some coastlines in the Niger Delta Area of Nigeria [13]. While the anthropological introduction of a toxic oxyanion may be from a single source, its spread can transcend from the point of pollution to other interconnected surface and groundwater [10, 14]. Thus, the adverse effects of pollution from oxyanions are not confined to the initial point of pollution.

In some instances, the pollution caused by oxyanion may have transboundary effects in the context of shared water resources between two or more countries [11]. The latter is highly the case because over 60% of the surface water across the globe crisscrosses the political boundaries of more than one state [15]. The transboundary surface waters provide jobs, food and support countless of economies. Pollution caused by oxyanion can undermine economic development, peace and stability, thereby straining relationship between and among states [15]. Put in the words of McCarthy, 'ecological effects know no boundaries; they can spill across geopolitical frontiers and cause cumulative externalities worldwide' [16].

The transboundary effects of oxyanion pollutants also apply to groundwater. It is estimated that groundwater constitutes about 97% of the freshwater on earth [17]. Thus, it is not out of place that most groundwater interacts with surface lakes, rivers, streams and oceans [17]. Therefore, the pollution of groundwater in one state can have a transboundary effect on the surface water and groundwater of another state [17]. Hence, it is imperative for the regulation of oxyanion pollutants with international laws.

Various instruments are employed to regulate the prevention and reduction of water pollution. While some approach it from a right-based and economic perspectives, others regulate it from the standpoint of its transboundary effects. The Universal Declaration on Human Rights (UDHR) 1948 and the International Covenant on Economic, Social and Cultural Rights (ICESR) 1966 are the major international laws that address the prevention of pollution from a right-based approach [18]. On the other hand, the United Nations Convention on the Laws of the Sea (UNCLOS) 1982 [16] and the Convention on the Law of the Non-Navigational Uses of International Watercourses (CLNUIL)1997, are the international treaties that approach the prevention of water pollution from a transboundary point of view [16].

The above-mentioned instruments have been the subject matter of several academic works with the aim of ascertaining the extent of their effectiveness in preventing water pollution and achieving water sanitation. McCarthy analyses the effectiveness of UNCLOS 1982 in preventing transboundary noise pollution [16] while Meier examines the regulation of pollution by international law from a human rights perspective [18]. McCaffrey on the other hand, discusses in detail the regulation of transboundary pollution and the extent to which it affects the non-navigational use of watercourses [17]. Similarly, Nollkaempre analysed the Convention on the Law of the Non-Navigational Uses of International Watercourses (CLNUIL) 1997 and the Convention on the Protection and Use of Transboundary Watercourses and International Lakes (CPTWIL) 1992 and concluded that its provisions are shredded with ambiguity; thus affecting its effectiveness [19]. Hall [20] however, examines the intersection between international laws and national laws that regulate the prevention of water pollution as applicable in the USA and Canada. These are some of the regulatory laws that have been applied to the control of pollutants in aqua systems that discussed extensively in the literature.

Regardless of the above, none of the existing legal literature contain an analysis of the extent to which the existing international instruments on water pollution regulates the prevention of pollution of water resources by oxyanion pollutants. Moreover, the use of economic instruments for abatement of oxyanions in water is still been explored in many countries, and so literature is not yet conclusive on this. Thus, the gap in the literature remains the justification for this work. In this regard, this chapter, using non-doctrinal methodology, coupled with review of existing literature, contains an analysis of the effectiveness of international law as well as economic instruments in regulating the prevention and control of the pollution of water resources by oxyanion pollutants. In furtherance to the latter mentioned theme, the chapter is divided into five in seriatim, i.e. an introduction, the analysis of the right-based approach to the regulation of oxyanion pollutants by international laws, the examination of regulation of oxyanion pollutants to the extent that it has a transboundary effect, analysis of economic instruments as oxyanion pollutants control measures, future perspectives and a reasoned conclusion.

13.2 The Global Regulation of Oxyanion Pollutants: A Right-Based Approach

The right to quality water and environment, albeit one devoid of pollution, is arguably a concomitant part of the realisation of human rights [21]. The rationale is that human beings cannot remain healthy and have a life if water and water deduced food is polluted [22]. The point has been previously made that the presence of oxyanion pollutants in drinking water can have dextrous effects on human health. It can also have adverse effects on food security, given that the death of aquatic life means less food available for coastal communities, and the world at large. Thus, there is a substantial nexus between the prevention of oxyanion caused pollution, and human existence made possible by the realisation of human rights [22]. The UDHR 1948 and the ICESCR 1966 are the foundational framework for human rights across the globe. These are analysed to ascertain their extent in regulating and preventing oxyanion pollution.

13.2.1 The Universal Declaration on Human Rights (UDHR) 1948

The UDHR 1948 is the first international instrument on human rights [23]. It was adopted by the United Nations, after the Second World War, to create a basis for the universal understanding of what human rights are, and the standard benchmark for the realisation of such rights by all people and nations [24]. Notably, the UDHR started as a soft instrument [25] but has metamorphosised into a hard law initiative [23]. For one, some of its provisions are part of customary international law [25]. The implication of being a customary international law is that the provisions are legally binding even on states that are non-party to it [26]. Also, it has been strengthened and made binding by the International Convention on the Elimination of All Forms of Racial Discrimination 1963 which provides that 'State Parties undertake to adopt immediate and effective measures... to propagating the purposes and principles of the... the Universal Declaration of Human Rights (UDHR)... [27]'

The UDHR instrument has provisions that are arguably relevant to the prevention of water pollution, including oxyanion pollutants. Article 3 of the UDHR provides that everyone has a right to life, liberty and security [24]. Article 25 grants that everyone has the right to a standard of living adequate for the health and well-being of himself and his family [24]. The UDHR further imposes the duty on states to strive to secure the universal and effective recognition of the mentioned rights through the adoption of progressive measures [24]. Notably, the word water pollution was not expressly mentioned in the UDHR instrument. However, some authors argue that the prevention of water pollution is embedded in the realisation of the right to life and a standard of living adequate to attaining good health [28]. The aforementioned is because the consumption of polluted water can and does interfere with the full enjoyment of the right to life, and the standard of living adequate to guarantee health. Moreover, oxyanion polluted water undermines the right to life and health, given that it causes illness such as cancer, goitre, kidney infections and death [10, 12]. What is more, oxyanion pollutants can also affect the quality of living by disrupting food security through the destruction of aquatic lives [11]. Accordingly, it can be argued that the provisions of UDHR on the right to life and standard of living create an obligation on states to adopt measures for the prevention of the pollution of waters by oxyanion pollutants.

While the above position is well established in academic literature, it is not expressly stated in the UDHR legal regime. The implication is that the nexus between the mentioned provisions of the UDHR, and the prevention of oxyanions pollution, as stated in academic literature, is merely persuasive [29]. Thus, it is entirely the prerogative of states to decide whether their definition of right to life and standard of living adequate to health, will be stretched to accommodate the prevention of water pollution including oxyanion pollutants. Some developed states have entirely accepted the academic position that the realisation of the right to life and health encompasses the prevention of water pollution [30]. For instance, the European Union (EU) states have adopted the water framework directive which contains provisions on the prevention of water pollution by chemical substances including oxyanions [31, 32]. Regrettably, some other regions like West African states are yet to adopt such measures. Undoubtedly, the role of the UDHR in the prevention of oxyanion pollution would have been more robust, if there is a clear expression of the synergy between the right to life and health on the one hand, and the prevention of water pollution on the other hand.

Regardless of the identified weakness of the UDHR, it creates a basis, albeit an imperfect one for legislative measures to be adopted for the prevention of water pollution by oxyanions to the extent that there is a political will to that effect. The fundamental human rights including the right to life and health declared in the UDHR has been included in over 100 Constitutions [23]. The incorporated provisions of the UDHR on the right to life and health create a basis for an argument to be made for a right-based approach to the regulation of oxyanion. As such, environmental right enthusiasts can leverage on the constitutional provisions on the right to life and health to agitate for the adoption of legislative measures prohibiting the pollution of water resources through the anthropological introduction of oxyanions.

13.2.2 The International Covenant on Economic, Social and Cultural Rights (ICESCR) 1966

The International Covenant on Economic, Social and Cultural Rights (ICESCR) 1966 is 'a multilateral human rights treaty adopted by the United Nations General Assembly' [33]. It is an instrument derived from the UDHR 1948. However, while the UDHR contains a skeletal declaration on human rights with little provisions on the duties of states in implementing it; the ICESCR details out the meaning of the declared rights and antecedent obligations on states [33]. It is worth mentioning that the conglomerate of the UDHR 1948, ICESCR 1966 and the International Covenant on Civil and Political Rights (ICCPR) 1966 is called the International Bill of Rights (IBR) [34]. Over 170 countries of the world have ratified the ICESCR 1966. It has some relevant provisions for the prevention of pollution of water by oxyanions.

Article 12 (1) of the ICESCR 1966 provides that 'the states parties to the present Covenant recognise the right of everyone to the enjoyment of the highest attainable standard of physical and mental health' [35]. It went further in subsection (2) to outline the steps to be taken by the State Parties to achieve a full realisation of this right including 'the improvement of all aspects of environmental and industrial hygiene' [35]. The ICESCR 1966 creates an obligation on member states to take steps in line with their maximum available capacity through individual and co-operative efforts, intending to 'achieve progressively the full realisation of the rights … by the adoption of legislative measures' [35]. The point has been made that oxyanion pollutants undermine the overall well-being of the environment [36]. Thus, the provisions of the ICESCR 1966 on the improvement of all aspects of environmental hygiene arguably creates an impetus for the prevention of oxyanion pollutants in water resources [28]. Such preventive measures as provided for in the ICESCR 1966 can be through the adoption of laws to prevent water pollution by oxyanions.

Regrettably, the absence of the parameters for defining 'environmental hygiene' in the ICESCR 1966 creates the perfect leeway for inaction by member states. The provision on 'all aspects of environmental hygiene' offers the basis for an argument to be made for the adoption of legislative measures to prevent water pollution by oxyanions under the ICESCR 1966. Unfortunately, while it is reasonably inferable, the ICESCR 1966 did not define 'all aspects of environmental hygiene' to expressly accommodate the prevention of water pollution. The implication is that it is entirely a prerogative of a member state to rightly stretch the meaning of all aspects of environmental hygiene to include the regulation of oxyanion pollutants [37]. In other words, any state which does not want to regulate water pollution by oxyanions can simply define the improvement of all aspects of environmental hygiene to accommodate their inaction. Thus, the provision on environmental hygiene is as strong as the political will of individual member states in this context [37]. Where such political will is lacking, the provision may be of no import in underpinning legislative measures for the prevention of pollution by oxyanions.

Debatably, the clarification offered by the United Nations Committee on Economic, Social and Cultural Rights (UNCESCR) strengthens the definition of 'all aspects of environmental hygiene' to cover the prevention of water pollution by oxyanions. The UNCESCR is a committee of independent experts established by the United Nations to monitor the general progression of members on the implementation of the ICESCR [38]. The UNCESCR in its General Comment No. 14 provides some indices on what is meant by 'the improvement of all aspects of environmental and industrial hygiene' as used in the ICESCR 1966 [39]. The indices include,

the requirement to ensure an adequate supply of safe and potable water and basic sanitation; the prevention and reduction of the population's exposure to harmful substances such as radiation and harmful chemicals and other detrimental environmental conditions that directly or indirectly impact upon human health [39].

As reiterated, oxyanion pollutants undermine the 'adequate supply of safe and potable water' [36]. It is also detrimental to human health when consumed to the extent that makes it a harmful substance [36]. As such, the indices provided by the UNCESCR as enunciated above cover the prevention and reduction of water pollution by oxyanions. Thus, achieving the highest standard of physical and mental health entails the use of legislative measures by individual states to prevent water pollution by oxyanions.

However, the UNCESCR clarification on what is meant by 'improvement of environmental hygiene' is merely persuasive. The Vienna Law of Treaties 1969 is the international law that provides the basis for the interpretation of other international law and treaties. It provides that the only instrument that will be of import in the interpretation of a treaty are those made by the member parties pursuant or in furtherance to such treaty [40]. The UNCESR is a committee of independent experts and as such their General Comment cannot qualify as the instrument provided for in the latter provision of the treaties [41]. Thus, the definition of 'improvement of environmental hygiene' in the UNCESCR General Comment is not binding but merely persuasive. As a result, a lot will depend on the political will of a state to transpose the indices in the General Comment into legislative measures for the prevention of oxyanions.

Regardless of the shortcoming of the ICESCR 1966, it is a stronger basis for the adoption of legislative measures for the prevention of oxyanion pollutants in water in comparison with the UDHR. It was reiterated that the UDHR provides for the right to life and the standard of living adequate to guarantee health. However, it was completely silent on the nexus between environmental protection and the realisation of the mentioned rights. On that note, it was concluded that it creates an imperfect foundation for the adoption of legislative measures for preventing water pollution by oxyanions from a right-based perspective. The ICESCR 1966 provides explicitly that environmental hygiene is included in the right to enjoy the highest attainable standard of physical and mental health. Thus, the ICESCR 1966 creates an even stronger basis in comparison with the UDHR for the prevention of oxyanion pollutants to the extent that it undermines the attainment of the right to the highest standard of physical health.

13.3   The Global Regulation of Toxic Oxyanion:
A Transboundary Approach

Transboundary pollution is defined as 'pollution whose physical origin is situated wholly or in part within the area under the jurisdiction of one [state] and which has adverse effects, other than effects of a global nature, in the area under the jurisdiction of [another state]' [20]. The pollution of marine waters by oxyanions can have transboundary effects where the polluted water or its effect transcends the political boundaries of the state where the pollution originated [42]. A simple example of the transboundary effects of polluted water was the case of *Ohio v. Wyandotte Chemical Corp* [43], where the U.S. State of Ohio succeeded in a suit against a Canadian company for damages caused by the mercury that was dumped by the company into the Canadian part of Lake Erie which has harmful effects on the entire flora and fauna in Ohio's part of Lake Erie. Monetary damages were awarded for the harm done to its fish, wildlife, vegetation and the affected citizens of Ohio. The United Nations Convention on the Laws of the Sea (UNCLOS) 1982 [16] and the Convention on the Law of the Non-Navigational Uses of International Watercourses (CLNUIL) 1997 are the instruments that cover the regulation of water pollution from a transboundary perspective. Thus, they will be analysed to determine their relevancy in the prevention of water pollution by oxyanions.

13.3.1   United Nations Convention on the Laws of the Sea (UNCLOS) 1982

The UNCLOS 1982 is the primary treaty that comprehensively covers every spectrum of human affairs and relationship with the ocean [44]. It creates the needed legal order for every matter on the use of ocean water resources to the extent that it is called the 'Constitution of the Ocean' [44]. UNCLOS 1982 is globally subscribed to with well over 170 states of the world being parties to it [44]. Even for the states that have not signed it, most of its provisions are still binding on them because they are considered as 'customary international law' [45]. UNCLOS regulates the ocean space as one entity and as expected, covers the duties of states with respect to the marine environment [44].

UNCLOS 1982 provides that states shall protect and preserve the marine environment [46]. As such, it mandates the states to 'take measures individually or collectively to prevent, control or reduce the pollution of the marine environment from any source using the best practical means in their disposal and in accordance with their disposal' [46]. The measures adopted by the states shall cover all sources of pollution of marine waters, including the release of toxic, harmful or noxious substances from land-based sources and dumping [47]. The benchmark for determining whether a substance is toxic or harmful under the mentioned provision is 'it can result or is likely to result in such deleterious effects as harm to living resources and marine life, hazards to human health...., impairment of quality for use of seawater...' [46] As reiterated, oxyanions are harmful substances that have dangerous effects on human health and living resources of the sea [36, 48]. It can distort the quality of the use of seawater [36] negatively. So, the duty of the state to protect and preserve the environment extends to toxic oxyanions.

Importantly, UNCLOS empowers states to adopt laws and regulations for the reduction, prevention and control of marine pollution from all sources, including oxyanions. It provides for the establishment of maritime zones, including territorial waters, exclusive economic zones and continental shelf [46, 49]. In the light of the later, it empowers coastal states to make laws and regulations for the prevention, reduction and control of marine pollution from all sources including land-based sources and dumping [46, 47]. Toxic oxyanions are introduced into the marine environment through the desposition of wastewater from land-based sources such as agricultural, mining and petroleum activities [9, 48]. They can also be introduced by the deliberate dumping of waste containing toxic oxyanions into the marine environment [11]. Whatever the source of oxyanion pollutants, the coastal states may make laws for its prevention.

However, there is no mandatory obligation on states to make laws and regulations for the prevention of marine pollution by oxyanions. The primary culprits in the pollution of marine waters through the introduction of toxic oxyanions are non-state actors within the jurisdictions of states [36]. Awkwardly, such individual actors are not parties to UNCLOS and, as such, its provisions do not directly apply to them [50]. The only way for its provisions to be relevant in the prevention of marine pollution by oxyanions is for states to adopt laws and regulations that reflect the spirit of UNCLOS [50]. UNCLOS provides that 'coastal states may adopt laws and regulations... in respect...the prevention of pollution thereof' in their maritime zones [46]. The use of the word 'may' in the mentioned provisions means that the powers are discretionary [51, 52]. Thus, it is entirely the discretion of states to exercise their powers to make laws and regulations that will bar individual entities within their political boundaries from introducing oxyanion pollutants into marine waters [51, 52]. As such, there is room for inaction given that the adoption of such laws are now contingent on the political will of coastal states. It is not surprising that some coastal states have not enacted any laws for the prevention or control of marine pollution by oxyanions.

In summary, the UNCLOS 1986 represents a good start for the prevention of marine pollution through the anthropological introduction of oxyanions. While states may not be mandated to adopt laws for the prevention of oxyanions, the regime establishes liability for damages caused to another state by the transboundary effects of pollution. As such, if the introduction of oxyanion pollutants in one state leads to the pollution of marine water in another state, the state responsible can be liable to pay damages to the affected state. One can argue that such liability should serve as an impetus for coastal states to adopt laws that will prevent such marine pollution by oxyanions. Regardless, the fact that some states are yet to have such laws are clear indications that more is needed. Thus, more is required in the form of a sturdy international framework that will nudge states to adopt laws for the prevention of the pollution of marine waters by toxic oxyanions.

13.3.2   The United Nations Convention on the Law of the Non-navigational Uses of International Watercourses (UNWC) 1997

The United Nations Convention on the Law of the Non-Navigational Uses of International Watercourses (UNWC) 1997 was adopted to establish 'an overarching framework that contains basic standards and rules for the cooperation between watercourse states on the use, management and protection of international watercourses' [53, 54]. The term 'international watercourses' has been defined as 'a system of surface waters and groundwaters constituting by virtue of their physical relationship a unitary whole' [55]. Thus, international watercourses consist of rivers, lakes, glaciers, aquifers, canals and reservoirs to the extent that they are shared by more than one state [17]. The existing international watercourses are said to be equivalent to almost half of the world's total land surface [56]. The UNWC 1997 also applies to the non-navigational uses of international watercourses [57] in contrast to the UNCLOS 1986 regime, which applies only to seas and the entire ocean. The UNWC 1997 has been signed and ratified by 36 parties [58]. It is apt to mention that the provisions of the UNWC 1997 constitute customary international law [59]. Thus, it applies to states of the world irrespective of whether they have signed the UNWC 1997 or not.

The UNWC 1997 contains a provision that is relevant to the prevention of the pollution of international watercourses by oxyanions. It creates a mandatory duty on State Parties to protect the environment of the international watercourses [55]. It creates a further obligation on State Parties to prevent the pollution of the international watercourses that 'may cause significant harm to other watercourse states or their environment, including harm to human health or safety' through the adoption of measures [55, 60]. The UNWC 1997 does not define the meaning of measures, but from the point of semantics, it covers legal initiatives like laws and regulations. Arguably, oxyanions have the propensity to cause significant harm to the aquatic environment, flora and fauna of other watercourse states [12]. Therefore, an environmental activist can argue that the provisions of the UNWC 1997 mandates states to adopt measures which may include actions for the prevention of oxyanion pollutants.

On the other hand, there is no specific definition of 'significant harm' as provided for in the convention [55, 61]. The UNWC did not also stipulate who has the final say on what is significant harm, especially in cases of transboundary pollution between two or more watercourse states [55, 61]. As such, there is room for a watercourse state to justify inaction in the adoption of such laws for the prevention of oxyanions by arguing that it does not consider it to be of such a nature as to cause significant harm to the watercourses.

Furthermore, the UNWC 1997 provides that a watercourse state shall, at the instance of another party consult to arrive at mutually accepted measures for the prevention of pollution [55]. The mutually accepted measures are defined to include establishing the lists of substances which may be prohibited, limited or controlled [55]. Thus, where the pollution of a watercourse by oxyanions has transboundary effects, the innocent state can initiate negotiations with the concerned state to adopt mutual agreement for its control and/or prevention [62, 63]. It is apt to mention that the UNWC provisions mandate the concerned state, where the pollution originated from to attend such negotiations prohibition or limiting of the quantity of oxyanions [55, 57]. Hence, there is a possibility that such negotiations can lead to the adoption of bilateral or multilateral agreements for the prevention and control of oxyanion pollutants. While there are no records of such agreements negotiated for the prevention of oxyanions pollution, other agreements have been negotiated under this provision [63]. For instance, the adoption of the Nile River Basin Development Agreement was influenced by the UNWC provisions [63, 64].

Unfortunately, while a watercourse state may be mandated to attend such consultations, they are not obligated to enter into a bilateral or multilateral agreement. It is apt to mention that the introduction of oxyanion pollutants into watercourses is a product of industrial, agricultural and mining activities geared towards the economic development of a state [8, 9]. Thus, a watercourse state which is keen on economic development at the expense of environmental protection can leverage on the lack of a mandatory obligation to enter into such agreements [65, 66]. A case in point is China which has refused to enter into any form of agreement for the protection of the watercourses with other states because of the mentality of economic development first at the expense of environmental protection [65]. What is more, most watercourse states may not be willing to sacrifice their economic benefits, spend time and resources solely for adopting agreements on the prevention of oxyanions. They may be more willing to adopt such bilateral or multilateral agreements for the prevention of a wider scope of pollutants, including oxyanions.

In conclusion, the UNWC 1997 contains provisions that can be stretched to accommodate the prevention of oxyanions. Regardless, it is not strong enough to lead to widespread adoption of needed legislative measures for the prevention of pollution of marine waters by oxyanions. Commenting on the weaknesses of the UNWC 1997 provisions on pollutions, McCaffery opines that the soft obligations contained therein do not 'convey the clear message necessary for states to take it seriously if indeed they are able to ascertain from its text exactly what it requires them to do or refrain from doing' [67].

13.4   The Global Regulation of Oxyanion Pollutants: Economic Instruments Approach

The cost of pollution to the society is huge and covers health, environment, economic, social and psychological costs. In fact, pollution imposes greater costs on the society that appear larger than its prevention [68, 69]. Thus, it is practically cost effective to control pollution than treating it. Therefore, definite actions are required to regulate and control pollutants in aqua systems.

Prior to the introduction of economic instruments to pollution abatement, regulatory approach/instrument otherwise known as command and control (CAC) approach, as discussed earlier in this chapter was very popular in several countries; thus, making it the main policy instrument used in most countries to control pollution and to curb activities that are harmful to the environment. This approach centres on policing power, which involves monitoring, inspecting and enforcement of laws against pollution activities in order to limit discharge of certain pollutants and restrict some polluting activities to specific times or areas [68, 69]. Its instruments are in form of environmental laws, emission standards, regulations, and enforcement mechanisms aimed at control of all forms of pollution by dictating abatement decisions [70, 71]. A proper implementation and enforcement of this approach "affords a reasonable degree of predictability about how much pollution will be reduced" [69]. However, the approach has been fraught with some drawbacks, which make countries to search for alternatives.

For instance, regulatory instruments are said to be economically inefficient and costly to implement; thus, increasing the overall cost of production of firms [69]. Moreover, the cost incurred by the government or its regulatory agencies to ensure compliance with pollution control laws and measures is huge and could continue to be bloated as the number of firms rise. Furthermore, enforcement could become difficult as the number of private firms to be regulated continue to increase.

Another argument is that the CAC approach employs force to make all firms 'adopt the same measures and practices for pollution control and thus to accept identical shares of the pollution control burden regardless of their relative costs and impacts' not minding the fact that marginal pollution abatement costs differ among firms [72]. This is capable of raising the total social costs of pollution abatement. In developing countries, CAC approach to the abatement of pollution is made more difficult due to financial, institutional and political constraints [71]. Therefore, the preference currently in most advanced and developing countries is to adopt economic instruments for pollution control.

Economic or market-based approach has to do with the use of economic instruments, which are policy tools that influence the monetary costs and benefits of polluters' actions in order to minimise pollution [73]. This is relatively recent and have gained wider acceptance due to some of the advantages it offers above the CAC approach. The approach employed policy instruments such as prices, charges, permits, taxes, deposit-refund systems, subsidies and incentives are employed for the control and management of pollutants in the environment [5, 68, 69, 74].

This approach (market-based) is hinged on the 'polluter pays' principle, which itself emanates from the notion of environmental economics, and argues for internalisation of the cost of pollution [69]. It is the Principle 16 of the 1992 Rio Declaration, which submits that polluter should bear the costs of pollution control and management without distorting international trade and investment [5, 75, 76]. The principle is not completely new because environmental economists borrowed it from the thoughts of welfare economists. Arthur Cecil Pigou, a welfare economist was the first to argue for the imposition of taxes on air pollution in the 1920s, and since then, several literatures have emerged on the use and the benefits of economic instruments for the control of all forms of pollution [72, 77–79]. However, the approach became more pronounced with most countries preferring it to CAC since the 1992 Rio Declaration.

Economic instruments in recent times have been canvassed as more effective and efficient pollution control measure. This is informed by their cost-effectiveness potential, revenue generation capacity that can be used to finance activities aimed at controlling pollution, their efficiency and the ability to internalise externalities and adjust the behaviour of polluters in line with environmental protection, as well as serving as continuous and control incentives for environmental improvement and sustainability [69, 73, 80, 81].

Moreover, the approach has been argued as more appropriate to address the failure of CAC policies to deal effectively with pollution [72] and also offers polluters the opportunity of more flexible ways of finding least-cost solutions to pollution reduction, which can stimulate technological innovation and help to improve resource use and environmental quality [70]. In addition, unlike CAC, economic instruments can easily be administered because they require less regulators and monitoring; and like taxes, established, effective and efficient tax collection institutions and agencies exist in most countries that make implementation easier than regulatory institutions; less information may be required to implement economic instruments than CAC policies [71].

Economic instruments like pollution charges or taxes serve as the cost for using the environment by polluting firms. In economics, every activity attracts a cost or price. So, the price any user of the environment pays for such service is the charge. The charge, however, manifests in various forms such as the tax levied on the quantity and quality of the pollutants discharged by the polluter, which is simply known as effluent charge. Others are product charge; which centres on taxes placed on products used (inputs into the production of polluting products), produced (final commodities associated with pollution) or discharged that can harm the environment (polluting substances that are contained in inputs used for production); the user charge, which is placed on the use of any collective treatment facility; and administrative charge that is 'paid to authorities for such purposes as chemical registration or financing licensing and pollution control activities' [69, 71]. The charges serve as a source of revenue to the state and can be spent on pollution control programmes. Moreover, charges of this nature can also deter the firms from polluting the environment because they increase the cost of production, which may be difficult to pass to consumers of their products, particularly if demand for the firm's final product is elastic. Thus, charges can serve dual purposes of revenue generation and regulation of pollution activities by polluters.

These instruments are discussed further as they apply to the abatement of oxyanions in aqua systems.

13.4.1   Pricing as Economic Instrument to Control Water Pollution

Pricing is based on the neoclassical economist's view, which presupposes that efficiency can be attained when the marginal cost is set equal to the marginal benefit of any activity. Pricing of pollution activities can ensure optimal resource allocation within the economy.

Citing Pigou's argument for the use of economic instruments for pollution abatement, Andersen maintains that a properly priced pollution leads to optimal allocation of resources; however, 'when pollution remains an unpriced externality to market transactions, it will result in a less optimal allocation of resources' [72]. The author noted specifically that 'by adding a tax to pollution, it is at least in theory, possible to equate net marginal private costs with net marginal social costs, and thus to assure that market transactions lead to outcomes that are Pareto optimum'. In addition, the use of this instrument can reduce the cost inquired in the abatement of pollution.

Water pricing policy can be adopted to control excessive water usage, water degradation and pollution. The instrument promotes water conservation and water saving technology, encourage recycling and also generate revenue which can, in turn, be used to support water pollution abatement programmes if effectively and appropriately implemented.

The unique nature of water makes determination of its price different from what obtains in the mainstream economic theory. For instance, the neoclassical approach to price determination depends on the market forces. However, the case of water differs because it has the characteristics of being a private good and at the same time public good. In fact, water possesses some unique features such as 'the potential for open access, ability for re-use and the existence of public environmental goods that are dependent on water' coupled with the social and political forces in operation within the market field, the environmental implication of water usage as well as the human right to water and sanitation aspect based on resolution 64/292 of the United Nations [82–86].

In view of these, water pricing is carried out using the combination of market-based mechanism (market forces) and regulatory (administrative) principles. The implication of this is that some specific pricing strategies are fixed by regulatory authorities while others are based on market forces. Thus, water pricing revolves around the use of economic instruments such as tariffs, fees, levies and charges, which can be adopted for the abatement of oxyanion in aqua systems. However, some of the instruments are more effective than others regarding the abatement of water pollutants.

Generally, it appears the price of water is low across most countries, and so, a combination of instruments is required to control water pollution. Hence, price setting must be done strategically to ensure that appropriate costs such as direct, indirect (opportunity costs) and environmental (externalities) costs are covered [81]. According to Muldianto et al. [87] 'Water price should be set optimally, high enough to complement subsidies, but within the ability of users to pay for the water'.

Economic instruments that raise price may succeed in abating water pollutants because it may induce behaviour change in consumers towards water usage thereby limiting water consumption, which will reduce production. Moreover, if an instrument raises cost of production that cannot be passed on to consumers, probably because demand is highly elastic, production can also reduce and this will to some extend curb oxyanion in water. Moreover, any price that will effectively control and manage water must 'reflect not only the costs of supply (i.e. service delivery), but also costs related to the scarcity of the resource itself (e.g. externalities and opportunity costs)' [82].

13.4.2   Product Charges

Product charges as economic instruments can be used to control water pollution by placing charges on production of any commodity that pollutes surface water and/or groundwater either before production or in the course of its production, or even after consumption. As the name implies, a product charge is an indirect economic instrument placed on any commodity 'associated with pollution rather than on pollutants' [82]. Examples of such products are fertiliser, chlorofluorocarbons, lubricant oil, tires, batteries, as well as disposable razors and cameras [88]. A charge of this nature is capable of limiting the production and consumption of commodities capable of introducing oxyanions into the surface water and groundwater; particularly, when their demand is highly elastic. When demand for a product is highly elastic, it will be difficult to pass charges on them or on their inputs to consumers in form of high price. Hence, only the producer should bear greater part of the charge. For this reason, when the commodity in question or its input constitutes pollutants in the aqua systems, charges can kerb its production, thereby, preventing water pollution. Whereas, if demand is inelastic, it may not serve the purpose; nevertheless, much revenue can be generated from sales of the product, which can be expended on pollution control programmes. However, in a situation of a highly toxic product, a partial or outright ban will be preferred to product charges [69].

13.4.3   User Charges/Fees

User charges/fees can be employed to control or prevent the introduction of inorganic pollutants into aqua systems. A user charge grants access to facilities or services for use. According to Bernstein 'user charges may be variable (i.e. linked to water consumption or property values), fixed or some combination of the two and they are assessed on both municipal and industrial discharges into public sewerage' [69]. The charge stands as payment for using the service and the revenue generated from this, stands as the costs for providing the service [88]. However, the charges can be raised to the level that it can be used to sponsor pollution control activities. According to

O'Connor [88] Netherlands employed this instrument as 'a cost-recovery device for water treatment' and as 'an incentive device to lower discharges' while France and Germany have also used 'user fees' to control water pollution [72].

13.4.4 Effluent Charges

Effluent charges can also be used to abate oxyanions in aqua systems. These are emission charges imposed as non-compliance fees on firms that fail to comply with environmental standards or emit pollutants into water. Apart from serving as a means to check emissions, they can be employed 'to finance necessary measures for wastewater collection and purification, and to provide financial incentives for reducing discharges of effluent' [69]. This instrument, however, is costly to administer because it requires large institutional capacity, regulation and monitoring for effectiveness. Moreover, it has been reported that the instruments have been effective for revenue generation (as in the case of Mexico) rather than for pollution control because they are always too low and eroded in real terms by inflation, which make polluters to prefer payment than changing their polluting behaviour [69, 88]. Charges of this nature have also been employed in countries such as Denmark, Norway, France, Germany, Sweden and the Netherlands to control all forms of pollution, especially, water pollution successfully.

13.4.5 Enforcement Incentive

Enforcement incentive as an economic instrument can be effective in the control of pollutants in water. As the name implies, it is a form of charge or levy introduced to make polluters comply with 'environmental standards and regulations'. The implication is that if a polluter fails to abide by such standard and regulation, the firm will be made to pay the fine. According to Bernstein, this type of fine covers non-compliance fees paid when a polluter's discharge exceeds certain accepted level, performance bonds, which is a form of payments by firms before undertaken any activity that may potentially pollute the environment (this charge may be returned to the firm if the activity is environmental compliance), and liability assignment, which is a form of incentive given 'to actual or potential polluters to protect the environment by making them liable for any damage they cause' [69].

Furthermore, the economic instrument through incentives can encourage polluting firms to reduce pollution activities, expand their tentacles, boost number of new entrants into the industry and reduce compliance and administrative costs on the part of the government and polluting firms [69].

13.4.6 Subsidies

Subsidy is another important economic instrument that can effectively control water pollution. A subsidy is an incentive either as a direct payment or tax concession given to an individual or private firm to promote policies that benefit the public. The government can provide some incentives to polluters such as tax cut or tax break, low interest loans or even grants to prevent water pollution. These could enable polluting firms to invest in water pollution control measures. If however, the subsidy is removed and taxes raised to discourage water pollution, the cost of production becomes unnecessarily high, thereby raising the prices of the products of such firms beyond what consumers can afford.

13.4.7 Marketable or Tradable Permits

Marketable or tradable permits are forms of direct instruments, which can as well be used to kerb water pollution. They are 'quantity-based instruments, which limit overall levels of pollutants in a defined pollution control area'; making their environmental outcome 'more certain than for price-based instruments like charges and taxes' [88]. The scheme provides firms legal right or permit of some levels of emission and firms which do not attain the permitted level can trade-off the remaining with another firm or use the remaining to make up any excess emissions elsewhere. This instrument, which has been used in several advanced and developing countries like China, India, South Korea and the USA to control all forms of pollutions covering water, air, etc., has been adjudged to be cost effective and also serves as incentive for pollution abatement [70, 71, 89, 90]. However, the effectiveness of the instrument requires the establishment of strong rules, procedures as well as regulatory and enforcement capacity to define the trading area or zone; distribute 'the initial set of permits for defining, managing and facilitating permissible trading after the initial allocation; and for carrying out monitoring and enforcement activities' [69, 91]. In addition, for tradable permit instrument to be effective, there is the need to 'incorporate clearly defined time-specific emission ceilings; and make arrangement 'for monitoring, record-keeping, reporting'... and 'a reliable database' [88].

13.4.8 Deposit-Refund Scheme

Deposit-refund scheme is another economic instrument that can be employed to abate oxyanion in aqua systems. The focus of this instrument is on the users of some products that could be termed polluting products, such as plastics, drink cans, pesticide containers, glass and metal beverage containers not consumed or that do not dissolve during usage. The approach is to surcharge consumers of such products, which will be refunded when the polluting items are returned to designated places for proper disposal recycling [69, 88]. The scheme has been used in the USA successfully 'to control the disposal of lead-acid batteries and products containing potential pollutants such as aluminium and glass cans, pesticide, containers, and tyres' as well as in South Korea [68, 69]. However, for the scheme to be run successfully, it is important to identify the products in focus, and consumers that are willing to participate as well as make proper arrangements to handle collection and recycling of the product, and management of the finances that are involved [69]. In most cases, deposit-refund system is largely managed by the private sector, which bears the costs while 'incentives are in place for third parties to establish return services when users do not participate' [69, 91].

From the foregoing, it is apparent that economic instruments appear to be highly advantageous and have been adopted in several countries to control pollution. The approach apart from its ability to generate revenue, which can be used for different projects within the economy, it is efficient, equitable and also helps to reduce the cost of pollution abatement. Similarly, it provides some forms of liberty for firms to operate within market regulations without compulsion. The approach, particularly those instruments that serve as incentives like subsidies will encourage firms to willingly reduce pollution by employing cost effective methods and technologies.

However, economic instruments are not without some challenges. For instance, it has been argued that the instruments may not abate pollutions completely but can only reduce it by limiting the procurement of commodities that are linked with pollution. According to Allen Blackman and Winston Harrington [71], the use of taxes as pollution control instruments 'may be less politically acceptable than some other regulatory instruments', exclude nontargeted activities that may pollute the environment and have negative distributional effects, particularly, where it is possible to shift greater burden of such taxes on the poor households [71].

Moreover, as pointed out by Bernstein [69], it may not be possible to predict the effects of market-based instruments on environmental quality since most polluters may prefer to choose their own solutions while some of them may find it easy to continue engaging in polluting activities and prefer to pay charges if such are not set at the appropriate level. Furthermore, the use of 'sophisticated institutions to implement and enforce' the instruments is considered a drawback. It is also observed that the use of the instruments have faced stiff resistance from government agencies, firms and individual polluters in some countries because they prevent absolute control over polluters by government agencies and negotiating power of firms.

13.5   Perspectives For the Future

The discussions above presuppose that no single approach is capable of yielding the optimal outcome in the abatement of oxyanion in the aqua system because none of the instruments can operate in isolation, especially in a dynamic and complex society. They complement and reinforce each other. Therefore, it is important to consider the combination of regulatory and economic approaches as the future perspectives for the abatement of oxyanion in aqua systems. Larsen and Ipsen [81] have argued that 'successful implementation of most economic instruments' requires 'appropriate standards, effective administrative, monitoring and enforcement capacities, institutional co-ordination and economic stability', most of, which are regulatory in nature.

Economic instruments require a regulatory environment to thrive. In a study carried out by Andersen [72] on the importance of institutions and policy design for the use of economic instrument to control water pollution, it was reported that even though economic instruments have been found to be very powerful for the implementation of public policies, institutions are also essential because market-based policies cannot operate in a vacuum. Thus, CAC cannot be ignored while the market is left to itself. According to the author, the conditions under which market mechanisms operate are defined by both formal and informal government institutions. Moreover, 'since market-like policy instruments are usually applied within existing rules, institutions, and policy processes, the policy and administrative contexts in which they operate become important'.

Furthermore, economic instruments such as effluent charges, product charges, enforcement incentive and tradable permits can help to drive compliance with regulations. Falco reiterates that the success of economic instruments depends largely on 'a well-functioning monitoring and command and control system (including properly functioning institutions)'. Therefore, a mixture of regulatory and economic instruments is highly critical for the abatement of oxyanion in aqua systems.

Finally, efficient regulatory instruments which incorporate economic mechanisms are more appropriate. However, an option may be to amend existing regulatory instruments in order to eliminate the lapses identified in this work. Conversely, existing instruments should be amended to incorporate stronger obligations which will translate into effective measures for the prevention and control of water pollution. Such amendment should be in the light of fizzling out the identified lapses detailed in the body of this work. There may be some difficulty in getting sovereign states to voluntarily subscribe to tighter regimes that has environmental benefits without a conspicuous economic one. However, the latter is defeated by the fact that pollution of water from oxyanions has the propensity to cause long term economic harm, which makes it economically wise to address them at the onset. It is for this reason that it is said in the context of environmental regulation that 'an ounce of prevention is worth a pound of cure'.

13.6   Concluding Remarks

This chapter contains an analysis of the effectiveness of international law and economic (market-based) instruments in regulating the prevention of oxyanion pollutants in water resources. It identifies that the regulation of oxyanion pollution is approached from a right-based approach, a transboundary one and the use of economic instruments. The primary instruments that are relevant to the global regulation of oxyanion pollutants from a right-based approach are the UDHR 1948 and the ICESCR 1966. It was noted that the provisions of UDHR on the right to life and health create the basis for an argument to be made for the adoption of laws for the prevention of oxyanion pollutants. Notwithstanding, it was observed that the lack of express provisions detailing substantial obligations on states to regulate pollution, including oxyanions, allows for inaction by states in this regard. The latter observation was also the case with the analysis of the ICESR 1966. However, it was concluded that in comparison with the UDHR, the ICESCR 1966 creates a more robust but imperfect foundation for the inference of an obligation to adopt laws on oxyanion pollutants from a right-based approach.

Furthermore, the UNCLOS 1992 and the UNWC 1997 were identified and analysed as the instruments that regulate water pollution by oxyanions to the extent that it has or will have transboundary effects. Unlike the UDHR 1948 and the ICESCR 1966, they both contain provisions that directly mandate states to protect and preserve the environment. Regardless, the UNCLOS 1992 does not empower member states to adopt legislative measures for the prevention of water pollution, including oxyanions. Thus, a coastal state may elect not to adopt such laws. With respect to the UNWC 1997, it was argued that there is no mandatory obligation on states to enter into a mutual agreement with other watercourse states to adopt measures for the prevention of oxyanion pollutants.

On the other hand, the chapter examines economic instruments as efficient and effective for the abatement of oxyanions in aqua system. Instruments such as water pricing, effluent charges, product charges, user fees, enforcement incentive, subsidies, tradable permits and deposit-refund scheme were analysed. The instruments were observed to have been preferred and employed in several countries globally due to their cost-effectiveness potential, revenue generation capacity, their efficiency and the ability to internalise externalities and adjust the behaviour of polluters in line with environmental protection, as well as serving as continuous and control incentives for environmental improvement and sustainability. It was however observed that economic instruments require regulatory instruments to be more effective and efficient.

Thus, the chapter concludes that international law (regulatory instrument) is not sturdy enough to the extent that will translate to national laws for the prevention of water pollution by oxyanions. While economic instruments alone cannot operate to prevent water pollution through oxyanions.

The failure of international law in this regard is because the analysed global instruments were not made specifically for the prevention of water pollution. Thus, its application to marine pollution, including oxyanion pollutants is incidental to other subject matters like the global protection of human rights and the equitable distribution of marine resources.

It is therefore recommended that a convention be adopted for the prevention and control of inorganic pollutants which will include oxyanion pollutants. There is currently a global treaty on persistent organic pollutants, i.e. *the Stockholm Convention on Persistent Organic Pollutants 2001* [92]. The convention mentioned above restricts and prohibits the anthropological introduction of mentioned organic pollutants in water resources [92]. Regrettably, it does not extend to inorganic pollutants such as oxyanions [92]. As such, a similar convention should be adopted globally for inorganic pollutants, including oxyanions. The proposed convention should mandatorily impose an obligation on member states to prevent the dumping of oxyanion pollutants in transboundary and internal waters. The common trait of water pollution is that the introduction might be small, but the consequences may be unprecedented in scale. For instance, a simple blowout of an oil well in an oil platform (Macondo) in the Gulf of Mexico resulted in damages to the marine environment worth over billions of dollars [51]. The most crucial lesson extrapolated from the Macondo accident and other related accidents is that marine pollution may start small but end in massive damages often more than the cost of prevention [51]. Thus, it is for this reason that environmental law enthusiasts commonly opine when it comes to water pollution that 'an ounce of prevention is better than a pound of cure' [93–95]. The adoption of a convention for the prevention of the pollution of internal, marine and transboundary waters is necessary to forestall future unprecedented damages.

Finally, a mixture of regulatory and economic instruments will be more effective in the control and abatement of oxyanion pollutants in water if both are properly employed and well implemented in accordance with the peculiarities of each country.

References

1. United Nations, 'Water' <https://www.un.org/en/sections/issues-depth/water/>. Accessed 4 Apr 2020
2. Devlaeminck D et al (2017) *The human face of water security*, Springer, p 82
3. Pahl-Wostl C et al (2016) *Handbook on water security*, Edward Elgar Publishing, p 64
4. Kuokkanen T (2017) 'Water security and international Law'. 20(2) PELJ 2, 8
5. Report of the United Nations Conference on Environment and Development, A/CONF.151/26/Rev.l, vol l. Ch. 18.2 1992
6. The United Nations Transforming Our World: 2030 Agenda for Sustainable Development A/RES/70/1, Goal 6
7. Kumar CP (2018) Water securities: challenges and needs. Int Educ Sci Res J 4(1):26
8. Zhou C et al (2019) Hydrogenotrophic microbial reduction of oxyanions with the membrane biofilm reactor. Frontiers Microbiol 9:1
9. Fowler J, Smets B (2017) Microbial biotechnologies for potable water production. Microbrial Biotechnol 10:1094
10. Jiang SY, Ma A, Ramachandran S (2018) Negative air ions and their effects on human health and air quality improvement. Int J Mol Sci 19(10):2966
11. Breida M, Younssi SA, Ouammou M (2018) 'Pollution of water sources from agricultural and industrial effluents: special attention to $NO3^-$, Cr(VI), and Cu(II)'. < DOI:http://dx.doi.org/10.5772/intechopen.86921>. Accessed 6 Apr 2020
12. Gafur NA et al (2018) 'A case study of heavy metal pollution in water of bone river by artisanal small-scale goldmine activities in eastern part of Gorontalo, Indonesia'. Water 10:1506
13. Obe E (2020) 'Dead fish on Atlanta contaminated with heavy metal, Bayelsa Government Warns' <https://tell.ng/18-2/>. Accessed 25 Apr 2020
14. Soliman M (2019) Comparing a review of heavy metal intake and their toxicity on plants and animal health. Int J Plants Animal Environ Sci 9:182–189

15. The United Nations (2018) 'The water convention: responding to global water challenge' <https://www.unece.org/environmentalpolicy/conventions/water/envwaterpublicationspub/brochures-about-the-water-convention/2018/the-water-convention-responding-to-global-water-challenges.html >. Accessed 25 Apr 2020

16. McCarthy EM (2001) International regulation of transboundary pollutants: the emerging challenge of ocean noise. Ocean Coastal Law J 6(2):257–258

17. McCaffrey SC (2001) The law of international watercourses, Oxford University Press, p 31

18. Meier BM (2013) Implementing an evolving human rights through water and sanitation policy. Water Policy 15(1):116

19. Nollkaemper A (1993) The legal regime for transboundary water pollution: between discretion and constraint, Martinus Nijhoff

20. Hall ND (2007) Transboundary pollution: harmonising international and domestic law. Univ Michigan J Law Reform 40:681–682

21. United Nations Environmental Protection (2019) 'A human rights based approach to preventing plastic pollution'. Sea Circular Issue Brief 2, 1

22. Odeku KO, Paulos BM (2017) 'Prohibition of pollution of marine environment: challenges and prospects'. Environ Econ 8(3):126,132

23. Brown G The universal declaration of human rights on the 21 century, 2016 NYU Global Institute for Advanced Study

24. The Universal Declaration of Human Rights 1948

25. Dolinger J (2016) The failure of the universal declaration of human rights. University of Miami Inter-Am Law Rev 47:164–199

26. Zhong Y (2018) Customary international law and the rule against taking cultural property as spoils of war. Chinese J Int Law 17(4):943–986

27. International Convention on the Elimination of All Forms of Racial Discrimination 1963

28. Knox JH (2015) Human rights, environmental protection and the sustainable development goals. Washington Int Law J 24(3):517–536

29. von Bernstorff J (2004) International legal scholarship as a cooling medium in international law and politics. Eur J Int Law 25(4):977–990

30. Dogaru L (2014) Preserving the right to a healthy environment: European jurisprudence. Proc—Soc Behav Sci 141:1346–1352

31. Toreno V, Hanke G (2016) Chemical components entering the marine environment from sea-based sources: a review with a focus on economic seas. Marine Pollut Bull 112:17–38

32. Frank V (2007) The european community and marine environmental protection in the international law of the sea: implementing global obligations at the regional level. Leiden, Martinus Mijhoff, p 482

33. Hoag RW (2011) 'International covenant on economic, social, and cultural rights'. In: Dean Chatterjee K (ed), *Encyclopedia of global justice*, Springer, p 15

34. Howie E (2018) Protecting the human right to freedom of expression in international law. Int J Speech-Lang Pathol 20(1):12

35. The International Covenant on Economic, Social and political rights 1966

36. Howarth AJ et al (2015) 'Metal–organic frameworks for applications in remediation of oxyanion/cation-contaminated water'. Cryst Eng Comm 4

37. Ole NC, Philip HF (2017) 'Assessing the impact of the brent spar incident on the decommissioning regime in the north east Atlantic'. Hasanuddin Law Rev 3(2):141–147 at 145

38. See United Nations Human Rights, 'Committee on economic, social and cultural rights' <https://www.ohchr.org/en/hrbodies/cescr/pages/cescrindex.aspx>. Accessed 11 Apr 2020

39. UN Committee on Economic, Social and Cultural Rights (CESCR), *General Comment No. 14: The Right to the Highest Attainable Standard of Health (Art. 12 of the Covenant)*, the 11th of August 2000, E/C.12/2000/4 <https://www.refworld.org/docid/4538838d0.html>. Accessed 24 May 2020

40. The Vienna Law of Treaties 1969, Article 31 (2) (b)

41. Assensio H (2016) Article 31 of the vienna conventions on the law of treaties and international investment law. ICSID Rev-Foreign Invest Law J 31(2):366
42. Jeffrey MA (1992) Transboundary pollution and cross-border remedies. Canada-United States J 18(19):173–194
43. [1971] 401 U S 493, 495
44. Barret J, Barnes R (2016) *Law of the sea: unclos as a living entity*, British Institute of International and Comparative Law
45. Roach JA (2014) Today's customary international law of the sea. Ocean Dev Int Law 45(3):239–259
46. The United Nations Convention on the Law of the Sea 1982, Art 1, Art 21 (1) (f), Art 60, Art 80, Art 192, Art 194 (1) (4), Art 207, Art 210
47. Boczek BA (1984) 'Global and regional approaches to the protection and preservation of the marine environment'. Case W Res J Int' l L 16:39
48. Weidner E, Ciesielczyk F (2019) Removal of hazardous oxyanions from the environment using metal-oxide-based materials. Materials (Basel) 12(6):927
49. Ole NC (2017) 'The financial securities for the decommissioning of offshore installations in Nigeria: a review of the legal and contractual framework' 15 OGEL 1
50. Pappa M (2018) 'Private oil companies operating in contested waters and international law of the sea: a peculiar relationship' 16 OGEL 1
51. Jumbo I, Ole NC (2019) A critical analysis of the nigerian offshore oil risk governance regime (Post Macondo). African J Int Energy Environ Law 3(3):21
52. Omukoro DE (2017) Environmental degradation in Nigeria: regulatory agencies, conflict of interest and the use of unfettered discretion. OGEL 15(1):25
53. GCINT (2020) 'Everything you need to know about UN watercourses convention' <https://www.gcint.org/wp-content/uploads/2015/09/UNWC.pdf>. Accessed 29 May 2020
54. Daibes-Murad F (2005) *A new legal framework for managing the world shared groundwater*, IWA Publishing, p 84
55. The United Nations Convention on the Law of the Non-Navigational Uses of International Watercourses (UNWC) 1997, Article 2(a), Article 20, Article 21
56. Water Encyclopedia (2020) 'International cooperation' < http://www.waterencyclopedia.com/Hy-La/International-Cooperation.html>. Accessed 29 May 2020
57. The United Nations Convention on the Law of the Non-Navigational Uses of the International Water Courses 1997, UN Doc. A/RES/51/ 869
58. The United Nations (2019) 'Status' < https://treaties.un.org/Pages/ViewDetails.aspx?src=TREATY&mtdsg_no=XXVII-12&chapter=27&lang=en>. Accessed 29 May 2020
59. McIntyre O (2007) The environmental protection of international watercourses under international law, Ashgate Publishing, p 1
60. McCaffery SC, Sinjela M (1998) The 1997 United Nations convention on international watercourses. Am J Int Law 92(1):97–107
61. Querol M (2016) *Freshwater boundaries revisted: recent developments in international river and lake delimitation*, Brill, p 14
62. Stoa RB (2014) 'The UN watercourses convention on the dawn of entry into force'. Vand J Transnat'IL 47:1321,1333–1337
63. Chelkeba A (2018) The influence of the un watercourses convention on the development of the Nile River Basin cooperative framework agreement (CFA). Mizan Law Rev 12(1):166
64. Qureshi WA (2018) The IWT and the UNWC: commonalities and differences. Ocean Coastal Law J 23(1):87
65. McIntyre O (2015) Benefit-sharing and upstream/downstream cooperation for ecological protection of transboundary waters: opportunities for china as an upstream state. Water Int 40(1):48–70
66. Saul B (2011) 'China, Resources and International Law'. Sydney Law School, Legal Studies Research Paper No. 11/82 <http://ssrn.com/abstract=1954180>. Accessed 31 May 2020
67. McCaffrey SC (1989) 'The law of international watercourses: some recent developments and unanswered prayers'. 17 Denv J Int' l L. & Pol' y 505

68. Kraemer RA, Choudhury K, Kampa E (2001) 'Protecting water resources: pollution prevention.' Thematic background paper at the international conference on freshwater in Bonn, 3–7 Dec. <https://sswm.info/sites/default/files/reference_attachments/KRAEMER%20et%20al%202001%20Protecting%20Water%20Resources%20Pollution%20Prevention.pdf > . Accessed 1 May 2020

69. Bernstein JD 'Economic Instruments. In: Helmer R, Hespanhol I (eds), Water pollution control: a guide to the use of water quality management principles, E. & F. Spon 1997, p 526

70. Coria J, Köhlin G, Xu J (2019) On the use of market-based instruments to reduce air pollution in Asia. Sustainability 11(18):1–23

71. Blackman A, Harrington W (2000) The use of economic incentives in developing countries: lessons from international experience with industrial air pollution. J Environ Dev 9(1):5–44

72. Andersen MS (2001) 'Economic instruments and clean water: why institutions and policy design matter'. <https://www.oecd.org/gov/regulatory-policy/1910825.pdf>. Accessed 14 May 2020

73. Denne T (2006) 'Economic instruments for the environment waikato technical report'. <https://www.waikatoregion.govt.nz/assets/WRC/WRC-2019/tr06–23.pdf>. Accessed 14 May 2020

74. Olmstead S, Zheng J (2019) 'Policy instruments for water pollution control in developing countries'. < http://documents.worldbank.org/curated/en/547111576067554446/pdf/Policy-Instruments-for-Water-Pollution-Control-in-Developing-Countries.pdf>. Accessed 15 May 2020

75. London School of Economics (2018) What is the polluter pays principle? <http://www.lse.ac.uk/GranthamInstitute/faqs/what-is-the-polluter-pays-principle/>. Accessed 14 May 2020

76. Fishman Y, Becker N, Shechter M (2012) The polluter pays principle as a policy tool in an externality model for nitrogen fertilizer pollution. Water Policy 14(3):470–489

77. Baumol WJ (1972) On taxation and the control of externalities. Am Econ Rev 62(3):307–322

78. Caffera M (2011) The use of economic instruments for pollution control in Latin America: lessons for future policy design. Environ Dev Econ 16(3):247–273

79. Ashiqur Rahman M, Ancev T (2014) Economic analysis of alternative pollution abatement policies: the case of Buriganga River, Bangladesh. Interdisc Environ Rev 15(1):66–87

80. Das S, Mukhopadhyay D, Pohit S (2007) Role of economic instruments in mitigating carbon emissions: an indian perspective. Econ Political Weekly 42(24):2284–2291

81. Larsen H, Ipsen NH, 'Framework for water pollution control'. In: Helmer R, Hespanhol I (eds), *Water pollution control: a guide to the use of water quality management principles*, E. & F. Spon 1997, p 526

82. Sjödin J, Zaeske A, Joyce J (2016) 'Pricing instruments for sustainable water management'. Working paper Nr. SIWI, Stockholm. <https://www.siwi.org/wp-content/uploads/2016/07/Pricing-instruments-for-sustainable-water-management-DIGITAL-Final-1.pdf >. Accessed 30 May 2020

83. United Nations (2019) 'The human rights to water and sanitation in practice' <https://www.unece.org/fileadmin/DAM/env/water/publications/WH_17_Human_Rights/ECE_MP.WH_17_ENG.pdf>. Accessed 30 May 2020

84. United Nations (2010) 'Resolution adopted by the General Assembly' < https://documents-dds-ny.un.org/doc/UNDOC/GEN/N09/479/35/PDF/N0947935.pdf?OpenElement>. Accessed 30 May 2020

85. Adkins L, Lehtonen TK (2018) 'Price: an introduction'. Distinktion: J Soc Theory 19(2):109–116, https://doi.org/10.1080/1600910x.2018.1501588

86. Beckert J (2011) Where do prices come from? Sociological approaches to price formation. Socio-Econ Rev 9(4):757–786. https://doi.org/10.1093/ser/mwr012

87. Muldianto H, Andawayanti U, Suhartanto E, Soetopo W (2020) 'Determining water price for public water supply from Jatigede reservoir'. IOP Conf Series: Earth Environ Sci 437:1–8. https://doi.org/10.1088/1755-1315/437/1/012027

88. O'Connor D (1999) Applying economic instruments in developing countries: from theory to implementation. Environ Dev Econ 4(1):91–110

89. Coria J, Sterner T (2010) Tradable permits in developing countries: evidence from air pollution in Chile. J Environ Dev 19(2):145–170. https://doi.org/10.1177/1070496509355775

90. Global Water Partnership (2017) 'Tradable pollution permits (C7.03)'<
    https://www.gwp.org/en/learn/iwrm-toolbox/Management-Instruments/Eco
    nomic-Instruments/Tradable_pollution_permits/#:   ~   :text   =   Trad-
    able%20pollution%20permits%20are%20so,to%20firms%20that%20pollute%20more.&text
    = Credits%20are%20traded%20within%20defined%20trading%20areas >. Accessed 2 Jun
    2020
91. Di Falco S (2012) Economic incentives for pollution control in developing countries: what
    can we learn from the empirical literature? Politica Agricola Internazionale—Int Agric Policy
    2:1–17
92. Lallas PL (2001) The stockholm convention on persistent organic pollutant. Am J Int Law
    95(3):692–708
93. Nanda V, Ping GR (2012) International environmental law and policy for the 21st century,
    Martinus Nijhoff Publishers, p 62
94. Bellefontaine N, Johansson T (2015) 'The role of the international maritime organisation in
    preventing illegal oil pollution from the ships'. In: Carpenter A (ed) Oil pollution in the north
    sea, Springer, p 52
95. Welder RJ (1998) Listening to the sea: the politics of improving environmental protection,
    University of Pittursburgh Pre, p 143

Printed in the United States
by Baker & Taylor Publisher Services